Pure Honey: Assurance & Authentication

Muhammad Zafar • Mushtaq Ahmad
Meral Kekeçoğlu • Trobjon Makhkamov
Mozaniel Santana de Oliveira • Salman Majeed
Editors

Pure Honey: Assurance & Authentication

Springer

Editors
Muhammad Zafar
Department of Plant Sciences
Quaid-i-Azam University
Islamabad, Pakistan

Mushtaq Ahmad
Department of Plant Sciences
Quaid-i-Azam University
Islamabad, Pakistan

Meral Kekeçoğlu
Department of Biology
Düzce University
Düzce, Türkiye

Trobjon Makhkamov
Forestry and Landscape Design
Tashkent State Agrarian University
Tashkent, Uzbekistan

Mozaniel Santana de Oliveira
Postgraduate Program in Pharmaceutical Sciences (PPGCF)
Institute of Health Sciences, Federal University of Pará
Belém, Pará, Brazil

Salman Majeed
Department of Botany
University of Mianwali
Mianwali, Pakistan

ISBN 978-3-031-98912-4 ISBN 978-3-031-98913-1 (eBook)
https://doi.org/10.1007/978-3-031-98913-1

This Springer imprint is published by the registered company Springer Nature Switzerland AG
The registered company address is: Gewerbestrasse 11, 6330 Cham, Switzerland

We wish to express our gratefulness to all the authors who have contributed in this edited book. We thank them for their cooperation and knowledge. We also thank several colleagues for their help in many ways and for their suggestions from time to time during the finalization of this book. We are greatly thankful to the valuable assistance of supervisor Prof. Dr. Mir Ajab Khan, Ex Dean, Faculty of Biological Sciences, Quaid-i-Azam University, Islamabad, for his kindness, love, immense help, valuable guidance, keen interest, and encouragement throughout field collect work in beekeeping zones. I would also like to thanks Dr. Salman Majeed (IPFP Fellow, Assistant Professor at Department of Botany University of Mianwali) for their valuable input regarding formatting and suggestion during the editing of chapters files.

Foreword

This book offers an insightful and timely exploration of the significance of honey purity and the scientific and regulatory frameworks that support its authentication. The global demand for honey rises, so do concerns over adulteration, mislabeling, and the erosion of consumer trust. This volume brings together current methodologies, technologies, and practices used to ensure the integrity and authenticity of honey products. From traceability protocols and chemical profiling to legal standards and market implications, this work provides a comprehensive resource for producers, researchers, regulators, and consumers alike. The authentication into the multifaceted world of honey pollen grains, the book emphasizes the critical role of science, ethics, and transparency in sustaining the quality and credibility of one of nature's most valued products.

Muhammad Zafar
Palynology Lab, Department of Plant Sciences
Quaid-i-Azam University,
Islamabad, Pakistan

Mushtaq Ahmad
Plant Systematics and Biodiversity Lab, Department of Plant Sciences
Quaid-i-Azam University,
Islamabad, Pakistan

Meral Kekeçoğlu
Department of Biology
Düzce University,
Düzce, Turkey

Trobjon Makhkamov
Department of Forestry and Landscape Design
Tashkent State Agrarian University
Tashkent, Uzbekistan

Mozaniel Santana de Oliveira
Postgraduate Program in Pharmaceutical Sciences (PPGCF),
Institute of Health Sciences
Federal University of Pará,
Belém, Pará, Brazil

Salman Majeed
Department of Botany
University of Mianwali,
Mianwali, Pakistan

Introduction

The growing global demand for natural, nutritious, and authentic food products has intensified the need for quality assurance and authentication across the honey industry. *Pure Honey: Assurance & Authentication* is dedicated to exploring the multifaceted world of honey purity, the challenges associated with adulteration, and the scientific, regulatory, and technological advances aimed at maintaining the integrity of honey products. This book provides a comprehensive overview of methods used to identify and authenticate pure honey, ranging from traditional knowledge systems to cutting-edge laboratory analyses such as chromatography, spectroscopy, and DNA barcoding.

The primary objective of this volume is to bridge gaps in knowledge across disciplines—combining insights from apiculture, food science, analytical chemistry, policy, and trade. It addresses key issues such as the detection of contaminants and adulterants, certification standards, geographic and botanical traceability, and consumer awareness. The focus also extends to the socioeconomic impact of honey fraud on beekeepers and rural livelihoods, emphasizing the importance of trust and transparency in the honey value chain.

This book serves as a valuable resource for researchers, quality control professionals, regulatory agencies, and policymakers involved in ensuring food safety and authenticity. Additionally, it is a practical guide for beekeepers and honey producers seeking to uphold product integrity and meet international standards. By highlighting the significance of both indigenous beekeeping practices and modern analytical tools, *Pure Honey: Assurance & Authentication* advocates for an integrated approach to sustain honey quality in an increasingly globalized and complex food system. Through a rich combination of scientific insights, industry case studies, and visual documentation, this book aims to reinforce the credibility of honey as a pure, health-promoting substance and to support the development of reliable, ethical practices across the honey supply chain.

Key Features

- **Comprehensive Overview of Honey Authentication Techniques**

  This book presents in-depth coverage of modern analytical methods used to assess honey purity and authenticity, including spectroscopic techniques, chromatography, isotopic analysis, and DNA barcoding. It bridges scientific rigor with practical application for researchers and industry professionals.

- **Integrated Perspectives from Science and Industry**

  The combination from apiculture, food science, regulatory policy, and global trade, the book provides a multidisciplinary understanding of the challenges and solutions in ensuring honey integrity. Real-world case studies and examples enhance its relevance for practitioners and policymakers.

- **Focus on Quality Assurance, Adulteration Detection, and Consumer Trust**

  The book addresses the pressing issue of honey adulteration, outlining strategies for traceability, certification, and quality control. It highlights how ensuring the authenticity of honey supports public health, strengthens market confidence, and protects beekeepers and producers from economic loss.

Preface

Honey as a super natural food has long been revered not only for its nutritional and medicinal properties but also as a symbol of purity and natural goodness. However, the increasing incidence of honey adulteration poses significant challenges to consumer confidence, public health, and the integrity of global trade. This book aims to address these challenges by offering an in-depth examination of the processes, tools, and standards used to verify the authenticity and quality of honey. The book draws on interdisciplinary research, case studies, and current regulatory frameworks to provide a practical guide for stakeholders across the honey supply chain. It is intended for scientists, beekeepers, policymakers, and educators who are committed to ensuring the purity and traceability of honey. By fostering informed practices and evidence-based approaches, this work contributes to the sustainability and trustworthiness of the honey industry in an increasingly complex food system.

Islamabad, Pakistan	Muhammad Zafar
Islamabad, Pakistan	Mushtaq Ahmad
Düzce, Turkey	Meral Kekeçoğlu
Tashkent, Uzbekistan	Trobjon Makhkamov
Belém, Pará, Brazil	Mozaniel Santana de Oliveira
Mianwali, Pakistan	Salman Majeed

Acknowledgment

We are thankful to Prof. Dr. Mir Ajab Khan (Palynology Lab, Department of Plant Sciences) for providing the accurate identification of honey bee plants species and providing lab facilities during this research project.

Contents

Editors and Contributors

About the Editors

Muhammad Zafar did Ph.D. in Plant Sciences at Quaid-i-Azam University Islamabad-Pakistan in 2011. Currently, he is working as a Professor (Tenured), and Incharge Herbarium (ISL) in the Department of Plant Sciences, Quaid-i-Azam University, Islamabad, Pakistan. He is a renowned world scientist and botanist in the field of medicinal plants, plant biodiversity, palynology, and plant systematics and biodiversity. He is the founder of the first Melissopalynology at Quaid-i-Azam University Islamabad Pakistan. He has over 400 impact factor publications and ISI Web Sciences (citations ±7900, H-index 46) and Google Scholar Citation (citations ±16221, H-index 63, i10-index 327) in diverse fields of Plant Sciences. In addition to this he is the author of 13 international books and 23 book chapters published largely by Elsevier, Springer, Taylor & Francis, Wiley, etc. across the world. He is the editorial board member of various journals of repute. He has successfully supervised/produced 18 Ph.D. and 72 M.Phil. research scholars in advanced areas of Plant Systematics & Biodiversity.

Mushtaq Ahmad is a renowned world top Plant Scientist and Director of Sino-Pak Biodiversity Research Centre for Agro-Biological Resources, Quaid-i-Azam University, Islamabad, Pakistan. He is an Honorary Distinguished Professor in China and Uzbekistan. He is the author of 777 articles (citations ±22537, H-index 74, i10-index 440) and 60 international books and book chapters. He supervised 47 Ph.D.s and Post Doctorates, 180 M.Phil./M.Sc./BS research scholars. He is the recipient of the highest Civil Award (Tamgha-e-Imtiaz). He is a fellow member of IUCN, PAS, TWAS, AFOB, NAS, PIFI- CAS, Flora of Pan Himalaya-China, TUBITAK, MEVLANA (Turkey), HEC, PBS, and BSP.

Meral Kekeçoğlu's major research area focuses on honey bee, beekeeping, and bee products. She discovered the Yığılca honey bee ecotype, which is naturally found in the Yığılca district of Düzce, in the scope of her Ph.D. thesis on Turkey's honey bee biodiversity in 2007. After completing her doctorate at Agrıcultural University of Athens, she was appointed as Assistant Professor in Düzce University, Faculty of Arts and Sciences, Department of Biology in 2008. After beginning her academic career at Düzce University, she first went to work to establish a Beekeeping Development Center in the town. She has added value to the district with her social enterprise projects and scientific publications. The process involving her transformation of knowledge and scientific study results into a value for the development of rural area brought Turkey's first European Entrepreneurship Award by the project titled "Together We Can Do More" with the theme of beekeeping honey production forests and sustainable development in 2011, thereby bringing Turkey's prestige. She was selected as the "Changemaker in Turkey" by the Sabancı Foundation for the year 2012 due to her evaluation of science for the benefit of society and her activities focused on regional development. She is a member of the ISO Bee Products Mirror Committee and works in the commission on bee products standards.

Trobjon Makhkamov is a Professor in the Department of Forestry and Landscape Design, at Tashkent State Agrarian University, Tashkent, Uzbekistan. Previously, Trobjon Makhkamov was head of Department of the National Herbarium of Uzbekistan at Institute of Botany, and head of Department of Botany and Plant Physiology of the National University of Uzbekistan, head of laboratory of Introduction of Medicinal Plants at Tashkent Botanical Garden. Trobjon Makhkamov focuses on ecological and biological research of invasive plant species in Uzbekistan.

Mozaniel Santana de Oliveira holds a bachelor's degree in Chemistry from the Federal University of Pará (UFPA) and is a Licensed teacher in Chemistry from the Venda Nova do Imigrante University Center, UNIFAVENI. He has specializations in methods and management in food preservation and in teaching chemistry, both from the Brazilian Union of Colleges UNIBF. He holds a master's degree and a doctorate in Food Science and Technology (UFPA). His thesis focused on the extraction of essential oils from different plant matrices using supercritical fluid (CO_2) and conventional methods. He completed a Postdoctoral internship in the Postgraduate Program in Biological Sciences, with an emphasis on Tropical Botany at UFRA/MPEG. He has 247 academic (scientific) works, including 120 articles in international journals. He has published 12 books, 11 of which are internationally published by renowned publishers such as Bentham Science, IntechOpen, De Gruyter, CRC Press, Taylor & Francis, and Springer Nature, in addition to having contributed 59 chapters in international books and 62 abstracts in conferences. His work includes areas such as extraction of secondary metabolites from medicinal and aromatic plants using supercritical fluid, hydrodistillation, steam distillation, and solvent extraction. To date, he has edited over 300 scientific articles and has an H-index of 25 in Web of Science, 25 in Scopus, and 32 in Google Scholar.

Salman Majeed is a young scientist in the field of plant sciences and working as Assistant Professor at Department of Botany, University of Mianwali, Punjab, Pakistan, particularly known for his expertise on plant biodiversity and taxonomic studies of deserts. He has published 50 research articles in various journals of good repute in impact factor journals, Google Scholar Citation (citations ±704, H-index 19, i10-index 27), in diverse fields of plant sciences. He has authored two international books and five international book chapters published by Elsevier, Springer, and Taylor & Francis/CRC Press. He has participated in a number of conferences held at the national and international levels. His fields of interest are wild plant biodiversity of desert areas, plant taxonomy, food chemistry, nutraceuticals, palynology anatomy, melissopalynology, and medicinal plants ethnobotany. His research focuses on exploring the melliferous bee flora pollen and morpho-palynological features for the classification and taxonomy of several genera and species as well as the guiding principles for raising beekeeping products in developing countries.

Contributors

M. Adaouri National Institute of Agronomic Research of Algeria (INRAA), El-Harrach, Algiers, Algeria

Dipan Adhikari Post Graduate Department of Botany, Hooghly Mohsin College, Chinsurah, Hooghly, West Bengal, India

Muhammad Nauman Aftab Ikram-ul –Haq Institute of Industrial Biotechnology, Government College University, Lahore, Pakistan

Asif Ahmad Institute of Food and Nutritional Sciences, PMAS-Arid Agriculture University, Rawalpindi, Pakistan

Mushtaq Ahmad Department of Plant Sciences, Quaid-i-Azam University, Islamabad, Pakistan

Sidra Nisar Ahmed Department of Plant Sciences, Quaid-i-Azam University, Islamabad, Pakistan

Sara Aimen Department of Botany, The Women University Multan, Multan, Pakistan

Ceyda Aldag Faculty of Science, Department of Biology, Ege University, Izmir, Turkey

Nazish Mazhar Ali Microbiology Lab, Department of Zoology, Government College University, Lahore, Pakistan

Sikander Ali Ikram-ul –Haq Institute of Industrial Biotechnology, Government College University, Lahore, Pakistan

Maria Ameen Department of Botany, University of Chakwal, Chakwal, Punjab, Pakistan

Department of Plant Sciences, Quaid-i-Azam University, Islamabad, Pakistan

Saiqa Andleeb Department of Zoology, University of Azad Jammu and Kashmir, Muzzafarabad, Pakistan

Elsevar Sabir Asadov Nakhchivan State University, Nakhchivan, Azerbaijan

Shiza Atif Institute of Food and Nutritional Sciences, PMAS-Arid Agriculture University, Rawalpindi, Pakistan

Muhammad Farooq Azhar Department of Forestry and Range Management, Bahauddin Zakariya University, Multan, Pakistan

Abida Aziz Department of Botany, The Women University Multan, Multan, Pakistan

M. L. Berkani National Institute of Agronomic Research of Algeria (INRAA), El-Harrach, Algiers, Algeria

Duygu Nur Çobanoğlu Department of Crop and Animal Production, Vocational School of Food, Agriculture and Livestock, Bingol University, Bingol, Turkey

Sumreen Dawood Department of Botany, University of Mianwali, Mianwali, Punjab, Pakistan

İkranur Felek Department of Nutrition and Dietetics, Faculty of Health Sciences, Bingol University, Bingol, Türkiye

Adem Guner Faculty of Science, Department of Biology, Ege University, Izmir, Turkey

Hami Halima Department of Animal Production, National Higher School of Agronomy, Algiers, Algeria

Qurat Ul Ain Hyder Department of Botany, The Women University Multan, Multan, Pakistan

Sayyara Jamshid Ibadullayeva Institute of Botany, Ministry of Science and Education of the Republic of Azerbaijan, Baku, Azerbaijan

Zain Ul Iman University Institute of Information Technology, PMAS-Arid Agriculture University, Rawalpindi, Pakistan

Muhammad Iqbal Department of Botany, University of Chakwal, Chakwal, Punjab, Pakistan

Shaista Jabeen Department of Botany, University of Chakwal, Chakwal, Punjab, Pakistan

Sofia Khalid Environmental Sciences Department, Allama Iqbal Open University, Islamabad, Pakistan

Muhammad Rizwan Khan Department of Plant Sciences, Quaid-i-Azam University, Islamabad, Pakistan

Tunay Kilic Faculty of Science, Department of Biology, Ege University, Izmir, Turkey

Iram Liaqat Microbiology Lab, Department of Zoology, Government College University, Lahore, Pakistan

Nabeela Mahmood Department of Botany, University of Chakwal, Chakwal, Punjab, Pakistan

Salman Majeed Department of Botany, University of Mianwali, Mianwali, Punjab, Pakistan

Subrata Mitra Post Graduate Department of Botany, Hooghly Mohsin College, Chinsurah, Hooghly, West Bengal, India

Ayesha Mushtaq Environmental Sciences Department, Fatima Jinnah Women University, Rawalpindi, Pakistan

Sajida Naseem Department of Zoology, University of Education, Division of Science and Technology, Lahore, Pakistan

Mehwish Jamil Noor Environmental Sciences Department, Fatima Jinnah Women University, Rawalpindi, Pakistan

Pulakesh Parai Post Graduate Department of Botany, Hooghly Mohsin College, Chinsurah, Chinsurah, West Bengal, India

Aneta A. Ptaszyńska Institute of Biological Sciences, Maria Curie-Skłodowska University, Lublin, Poland

Huma Qureshi Department of Botany, University of Chakwal, Chakwal, Punjab, Pakistan

Sampriti Sarkar Alma Mater, School of Biosciences & Technology, Vellore Institute of Technology, Vellore, Tamil Nadu, India

Sukumar Sarkar Post Graduate Department of Botany, Hooghly Mohsin College, Chinsurah, Hooghly, West Bengal, India

Moutushi Sen Post Graduate Department of Botany, Hooghly Mohsin College, Chinsurah, West Bengal, India

Mursel Seyidov Nakhchivan State University, Nakhchivan, Azerbaijan

Rauf Lutfali Sultanov Department of Biology, Azerbaijan State Pedagogical University, Baku, Azerbaijan

İlginç Kızılpınar Temizer Department of Crop and Animal Production, Vocational School of Food, Agriculture and Livestock, Bingol University, Bingol, Turkey

Department of Medical Services and Techniques, High School of Health Services, Giresun University, Giresun, Turkiye

Mehtap Usta Tonya Vocational School, Trabzon University, Trabzon, Türkiye

Nefise Ulku Karabay Yavasoglu Faculty of Science, Department of Biology, Ege University, Izmir, Turkey

Muhammad Zafar Department of Plant Sciences, Quaid-i-Azam University, Islamabad, Pakistan

Chapter 1
Health Benefits of Honeybee Pollen Flora and Its Importance for Food Security

Salman Majeed, Muhammad Zafar, Mushtaq Ahmad, Muhammad Rizwan Khan, Sumreen Dawood, and Maria Ameen

1 Melliferous Flora

Honey plants were studied extensively in the twentieth century, and interest in this field is still alive today. The relevance of studying the melliferous group is linked to customers' desire to learn about the botanical composition of honey and its origin in order to detect falsifications. The territory's melliferous capacity is largely determined by the area where honey plants grow, rather than the diversity of species present. As a result, it is vital to determine the flora diversity that is significant for beekeeping (Kopytina et al., 2019). There is information on melliferous flora with common species useful for beekeeping in the previously cited regional studies; however, there is a lack of research on medicinally used melliferous flora at the level of municipalities, communities, or localities related to information on annual seasons or the beekeeping cycle (Martinez-Castillo et al., 2020). Honey plants are primarily nectar and pollen producers because bees and other insects rely on them for nourishment. These plants have been researched since ancient times, primarily by beekeepers and honey growers (Montoya-Bonilla et al., 2017). Honey species research is linked to the beginnings of beekeeping since beekeepers need to know which plants provide nectar and pollen for bees in order to find the best location for

S. Majeed (✉) · S. Dawood
Department of Botany, University of Mianwali, Mianwali, Punjab, Pakistan
e-mail: dmajeed@umw.edu.pk

M. Zafar · M. Ahmad · M. R. Khan
Department of Plant Sciences, Quaid-i-Azam University, Islamabad, Pakistan

M. Ameen
Department of Botany, University of Chakwal, Chakwal, Punjab, Pakistan

M. Zafar et al. (eds.), *Pure Honey: Assurance & Authentication*,
https://doi.org/10.1007/978-3-031-98913-1_1

their hives. The capacity of an area to provide food for the bees' families is known as its melliferous potential. Bees may be lured to flowers from a distant place and select the best nectar sources to produce a large amount of honey (Nabila et al., 2022).

2 Melissopalynology

Melissopalynology is one method for determining which flora are frequented by bees and so determining an area's apicultural potential. The presence of pollen grains from plant species visited by bees in search of supplies for the hive is regarded as a reference method for determining the botanical and geographical origin of honey. Furthermore, palynological study, along with other microscopic analyses, can be used to detect honey contamination, create quality standards, and add commercial value to bee-produced items (Bandeira & De Novais, 2020). Melissopalynological research aids in the discovery of plant species crucial to bee nutrition, as well as botanical knowledge on the products obtained from those insects' labors, and has the potential to provide commercial value (de Souza et al., 2019). Because of its affordable start-up costs, beekeeping helps to preserve natural resources and is used by communities as a source of revenue and survival. Understanding how bees interact with the environment is critical for understanding their ecology and building sustainable apiculture (Bosco & da Luz, 2018).

3 Honey

Honey is an unprocessed natural product that is characterized by its complex composition, which varies depending on the bee species, geographical region, and available floral supply. Honey is often split into two types: monofloral (honey produced by a single plant species) and polyfloral (honey made from the contributions of multiple plants). Honey is a natural source of antioxidants that may be used to improve human health and avoid harmful oxidative reactions in the body. Chemical composition, which is regulated by the floral source as well as environmental conditions, is directly linked to antioxidant qualities (Hailu & Belay, 2020). Honey is a sugary commodity that is utilized in food and beverage items as a natural sweetener. Honey is simply a water-based supersaturated monosaccharide solution. Among the elements are carbohydrates, amino acids, vitamins, organic acids, esters, volatile compounds, hydroxymethylfurfural, and enzymes. Honey is commonly utilized as a health food, but it is routinely mislabeled for profit due to its great economic value. Color, taste, flavor, and the quantity of physiologically active components can all be used to determine the botanical origin of honey (Ghorab et al., 2021). Plant communities, climate, soil, and apicultural methods all influence the quality of honey, making it one of the apian products that is more closely related to the place where it was produced. Honeybees rely on natural raw materials that can be recognized as honey. The main ingredients are pollen and nectar from various plant blooms, and pollen and nectar are the principal sources of nourishment for bees (Dhawan et al., 2018). The variety of local flora, floral phenology, honeybee forager species selection, and the timing of human harvesting operations are all factors that influence honey quality. Although honey is a high-quality product, it is susceptible to adulteration, mislabeling, and unethical mixing with low-quality honey, sugars, and other substances. Furthermore, there is a growing demand for natural food products due to their limited availability, proven therapeutic and healing powers, and increased population concerns about their health. The increased economic importance of honey makes it a possible target for adulteration and falsification (Fakhlaei et al., 2020).

Honey is a reliable source of national and state revenue for small marginal farmers, women, and other vulnerable community members who live near natural forests, woodlands, and riverine areas, as well as a source of nutritious food, medicines, and raw materials for the cosmetic and pharmaceutical industries (Adal et al., 2015). Pakistani honey is well-known for its unique flavor and great quality throughout the Middle East. Every year, Pakistan sells over 4000 tons of honey to Arab countries, valued at approximately $23 million. According to the agriculture department, arid regions have the capacity to produce more than 30,000 metric tons of standard honey every year. According to local traders, some 300 containers, each containing 20 tons of honey, are shipped to the Gulf States, Saudi Arabia, and the United Arab Emirates. The arid zone was the best province for apiculture due to the significant rise of bee floral plants as a result of the Billion Trees Afforestation Project (Gebretsadik, 2016).

4 Beekeeping

Beekeeping benefits the planet by protecting the environment and conserving natural resources. Bees, like other livestock species, require nutrition to produce and reproduce. Changes in nectar and pollen supply, as well as the causes of this trend, can be identified via honey analysis. Because honeybees do not visit all plants for nectar or pollen, it is critical to identify the plants that provide these resources, as well as plant communities and the phenological relationship between honeybee plants and honeybees, in order to practise beekeeping and assess an area's beekeeping potential (Ahmad et al., 2019). Beekeeping is a popular activity that varies depending on the flowering seasons in every area. Pollen morphology is a branch of botany that studies the botanical and geographical origins of plants. This approach is very helpful in determining the purity of honey (Ullah et al., 2021).

Plant species and flowering seasons vary from one region to another due to differences in geography, climate, and other cultural and farming practices. For successful beekeeping, a thorough understanding of the types, density, quality, and ethnomedicinal uses of bee flora is required. Such knowledge helps beekeepers to maximize their resources in order to harvest a plentiful quantity of honey and other bee floral products, as well as provide effective pollination that boosts agricultural yields. Such medical information is useful in compiling a list of extant bee flora and creating a floral calendar for arid regions.

5 Honeybee Flora

Botanical research and field observations of bee floral species were conducted in the community surrounding apiaries in order to identify melliferous plants and document information provided by local beekeepers. Botanical collections, on the other hand, were only used in this study to confirm the presence of melliferous species; these specimens were deposited in the Pakistan Herbarium (ISL).

Various plants bloomed at different times of the year, and honeybees rushed to them for nectar and pollen. Based on source reliability and diversity, 50 plant species were identified as important bee flora in arid regions. They were divided into three groups based on the frequency, number of bee visits, and abundance. For honeybees, 12 plant species were identified as significant sources, 32 species as medium sources, and 6 species as minor sources (Table 1.1).

Acacia modesta, Acacia nilotica, Brassica campestris, Cannabis sativa, Dalbergia sissoo, Eucalyptus camaldulensis, Helianthus annuus, Prosopis juliflora, Trifolium alexandrinum, Ziziphus oxyphylla, and *Ziziphus oxyphylla* were among the prominent plant species identified as potentially valuable bee flora. *Callistemon citrinus, Carthamus oxyacantha, Centaurium pulchellum, Chrozophora tinctoria, Cleome viscosa, Digera muricata, Eruca sativa, Euphorbia helioscopia, Indigofera heterantha, Malus domestica, Punica granatum, Sonchus asper*, and *Viola tricolor* were some of the medium and minor source plant species (Figs. 1.1 and 1.2).

The current study discovered the presence of important phytochemical substances in honeybee floral species, as well as their use as traditional remedies to treat a variety of ailments. The flora of bees was studied in 20 different families, which included 7 trees, 6 shrubs, and 37 herbs (Fig. 1.3). The dominant traditional medicinal plant family is *Asteraceae*, which contains 16 species, followed by 8 species of *Fabaceae*, 4 species of *Amaranthaceae* and *Brassicaceae*, and 2 species of *Euphorbiaceae* and *Myrtaceae*. (Fig. 1.4). Flowers (26%) were the most commonly used plant parts of bee flora for the preparation of herbal recipes, followed by leaves (21%), whole plants (20%), seeds (11%), roots (9%), and fruits (8%), as shown in Fig. 1.5. The least commonly used plant parts were the stem and bark (4%). Bees foraged frequently on trees and shrubs during autumn and on herbs in the spring.

Traditional remedies have been made from honey-producing bee vegetation. Honey has been utilized for therapeutic and healing purposes in a variety of ailments, and it even possesses antimicrobial properties (Jimenez-Arellanes et al., 2003). Arid regions' flora is critical for bees foraging and honey production. In arid landscapes, bees feed on vegetation, and honey is used as folk medicine to treat a variety of ailments. The majority of people in Pakistan are unaware of the flora of bees, their therapeutic benefits, and the source of honey production. This chapter contributes to a better understanding of the flowering seasons of bee floral plants visited by *Apis mellifera*. Scientists studying in the domains of plant ecology, plant taxonomy, genetics, pharmacology, molecular biology, and agriculture can benefit from the knowledge gained from local people (Qureshi, 2012).

Table 1.1 Bee floral plants' sampling, chemical compounds, and medicinal uses

Sr. No.	Taxa	Family	Vernacular names	Status	Habit	Part used	Phytochemical compounds	Medicinal aspects	Citation
1.	*Acacia modesta* wall	Fabaceae	Phulai	Wild	Tree	Leaves, fruit, root	Alkaloids, cyclitols, tannins	Cough, bacterial infection, leprosy	Sarwar (2016)
2.	*Acacia nilotica* (L.) Delile	Fabaceae	Prickly acacia	Wild	Tree	Flower	Phenolics, epicatechin, ellagic acids	Gastrointestinal problems, tuberculosis, antimutagenic, bronchitis	AbdErahman et al. (2018)
3.	Amaranthus spinosus L.	Amaranthaceae	Spiny amaranth	Wild	Herb	Leaves, seed	Betacyanins, carotenoids, phenolics,	Antidiabetic, diuretic, eczema, mouth ulcer, stomach disorder	Sarker and Oba (2019)
4.	*Bidens pilosa* L.	Asteraceae	Black-jack	Wild	Herb	Whole plant, seed, root	Aliphatics, flavonoids, terpenoids	Digestive complaints, wounds	Bartolome et al. (2013)
5.	*Brassica campestris* L	Brassicaceae	Field mustard	Cultivated	Herb	Flower	Flavonoids, phenolics, vitamins	Headache, muscular pain	Haider et al. (2021)
6.	*Calendula arvensis* M.Bieb	Asteraceae	Field marigold	Wild	Herb	Leaves, flower	Anthocyanins, flavonoids, tannins, triterpenoids	Antidiabetic, wound healing	Abudunia et al. (2020)
7.	*Callistemon citrinus* (Curtis) Skeels	Myrtaceae	Bottle brush	Wild	Tree	Root, fruit, leaves	Eucalyptol, monoterpenes, caryophyllene	Gastrointestinal disorders, infectious diseases, anti-inflammatory	Larayetan et al. (2017)

(continued)

Table 1.1 (continued)

Sr. No.	Taxa	Family	Vernacular names	Status	Habit	Part used	Phytochemical compounds	Medicinal aspects	Citation
8.	*Cannabis sativa* L	Cannabaceae	Hemp	Wild	Herb	Flower, stem, root	Phytocannabinoids, terpenoids, sterols	Bladder inflammation, antidiabetic, metabolic disorders	Shakil et al. (2021)
9.	*Carthamus oxyacantha* M.Bieb	Asteraceae	Wild safflower	Wild	Herb	Leaves, seed, flower	Cardiac glycosides, steroids, alkaloids	Antidiabetic, typhoid fever, swelling, cardiovascular disorders	Khalil and Al Ahmed (2017)
10.	*Celosia cristata* L	Amaranthaceae	Silver cock's comb	Wild	Herb	Whole plant, fruit, leaves, root	Asparagine, glycoproteins, kaempferol	Skin eruption, headache, ulcer, menstruation complaints	Islam et al. (2016)
11.	Centaurea iberica Trevir. Ex Spreng.	Asteraceae	Iberian star-thistle	Wild	Herb	Whole plants, stem	Steroid, fatty acids, lactones, flavones	Anti-inflammatory, wound healing, abdominal pain	Joujeh et al. (2019)
12.	*Centaurium pulchellum* (Sw.) Druce	Gentianaceae	Pink centaury	Ornamental	Herb	Leaves	Secoiridoid glycosides	Antidiabetic, cardiovascular disorders	Šiler and Mišić (2016)
13.	*Chenopodium album* L	Amaranthaceae	White goosefoot	Wild	Herb	Seed, leaves, flower	Alkaloids, phytosterols, phenols	Urinary complaints, rheumatism, treat burns	Saini and Saini (2020)
14.	*Chrozophora tinctoria* (L.) A.Juss	Euphorbiaceae	Dyers croton	Wild	Herb	Whole plant, flower	Alkaloids, chromones, diterpenoids,	Relieve chest pain, treat fever, warts	Al-Snafi (2015)

15.	*Cichorium intybus* L	Asteraceae	Chicory	Cultivated	Herb	Whole plant, leaves, root	Lactones, hydroxycoumarins, steroids	Pulmonary cough, liver complaints, flatulence	Al-Snafi (2016)
16.	*Cirsium arvense* (L.) Scop	Asteraceae	Creeping thistle	Wild	Herb	Whole plant, flower	Phenolics, tannins	Liver infection, sore throat	Hossain et al. (2016)
17.	*Cleome viscosa* L	Cleomaceae	Tick weed	Wild	Herb	Whole plant root, leaves	Polyphenols, flavonoids, flavone glycosides	Anti-diabetic, arthritis, malaria	Suresh et al. (2020)
18.	*Coreopsis tinctoria* Nutt	Asteraceae	Zaire gully	Ornamental	Herb	Whole plant, fruit	Terpenoids, flavonoids, polyacetylenes	Decrease blood glucose, sexual infection, and dysentery	Guo et al. (2015)
19.	*Dalbergia sissoo* DC.	Fabaceae	Indian rosewood	Wild	Tree	Whole plant, seed	Palmitoleic acid, triterpenoids, capric acid	Gonorrhea, dermal complaints, nausea	Ali et al. (2019)
20.	*Digera muricata* (L.) Mart	Amaranthaceae	False amaranth	Wild	Herb	Whole plant, flower	Tetracosanoic acid, rutin, hyperoside	Digestive disorders, nausea, antidiabetic	Manimekalai et al. (2020)
21.	*Erigeron canadensis* L	Asteraceae	Horseweed	Wild	Herb	Whole plant, seed	Flavonoids, tannins	Diuretic, prevent bleeding, inhibiting blood coagulation	Pawlaczyk et al. (2011)
22.	*Eruca sativa* mill	Brassicaceae	Rocket plant	Wild	Herb	Whole plant, leaves, flower	Fatty acids, terpenoids, saponins	Stomachache, liver complaints	Jaafar and Jaafar (2019)

(continued)

Table 1.1 (continued)

Sr. No.	Taxa	Family	Vernacular names	Status	Habit	Part used	Phytochemical compounds	Medicinal aspects	Citation
23.	*Eucalyptus camaldulensis* Dehnh	Myrtaceae	Lachae	Wild	Tree	Whole plant, bark, leaves	Tannins, saponins	Ease muscle, mouth sores, joint pain	Sani et al. (2014)
24.	*Euphorbia helioscopia* L	Euphorbiaceae	Sun spurge	Wild	Herb	Flower, leaves	Terpenoids, steroids, lipids, volatile oil	Dysentery, tuberculosis, cancer, cough	Yang et al. (2021)
25.	*Helianthus annuus* L	Asteraceae	*Sunflower*	Cultivated	Herb	Whole plant, flower	Tocopherols, flavonoids, phenolic acid	Heart disorder, pulmonary infections, whooping cough	Guo et al. (2017)
26.	*Indigofera heterantha* Brandis	Fabaceae	Himalayan indigo	Wild	Shrub	Flower, leaves	Aspirin, indigoferin, rutin	Hepatitis, whooping cough, antidiabetic	Khan et al. (2021)
27.	*Lantana camara* L	Verbenaceae	West Indian lantana	Wild	Shrub	Flower, leaves	Glycosides, steroids, saponins, quinine	Skin rashes, leprosy, asthma, measles	Kumar et al. (2020)
28.	*Launaea procumbens* (Roxb.) Ramayya & Rajagopal	Asteraceae	Creeping *launaea*	Wild	Herb	Whole plant, flower	Gallic acid, vanillic acid,	Gonorrhea, painful urination, sexual diseases	Khan et al. (2012)
29.	*Lepidium didymum* L	Brassicaceae	Lesser swinecress	Wild	Herb	Whole plant, flower	Cardiac glycosides, tannins, coumarins	Wound healing, relieving fever	Akhtar et al. (2017)
30.	*Ligustrum lucidum* W.T.Aiton	Oleaceae	White wax tree	Cultivated	Shrub	Leaves, fruit, seed	Phenylethanoids, triterpenoids	Promoting bone health	Che and Zong (2015)

31.	*Malus domestica* Borkh	Rosaceae	Saeb	Cultivated	Tree	Fruit, seed, whole plant	Polyphenols, phytosterols	Asthma, inflammation, antidiabetic	Patocka et al. (2020)
32.	*Medicago minima* (L.) L	Fabaceae	Small medik	Wild	Herb	Bud, root Leaves	Carotenoids, flavonoids, saponins, syringic acid	Constipation, fever, sore throat	Kabtni et al. (2020)
33.	*Melia azadirachta* L	Meliaceae	Tora shandy/ bikyna	Wild	Tree	Flower, leaves	Alkaloids, thiols	Dermatological infection, antidiabetic	Majid et al. (2021)
34.	*Melilotus indicus* (L.) all	Fabaceae	Uzmai	Wild	Herb	Whole plant, leaves	Coumarins, flavonoids	Bowel complaints, diarrhea	Marzouk (2021)
35.	*Oenothera rosea* L'Hér. Ex Aiton	Onagraceae	Sun cups	Wild	Herb	Flowers, leaves	Flavonoids, phenols	Anti-inflammatory	Márquez-Flores et al. (2018)
36.	*Parthenium hysterophorus* L	Asteraceae	Santa-Maria	Wild	Herb	Leaves, roots	Alkaloids, lactones, sterols, proteins	Skin infection, dysentery, muscular pain, constipation, insomnia	Kaur et al. (2021)
37.	*Prosopis juliflora* (Sw.) DC	Fabaceae	Ironwood	Wild	Shrub	Flower, seed	Flavonol glycoside, alkaloids, cinnamic acid	Asthma, diarrhea, skin problems, digestive complaints	de Brito Damasceno et al. (2017)
38.	*Punica granatum* L	Lythraceae	Anaar	Cultivated	Shrub	Fruit, flower	Gallic acid, quercetin	Antidiabetic, healing activity	Arun and Singh (2012)

(continued)

Table 1.1 (continued)

Sr. No.	Taxa	Family	Vernacular names	Status	Habit	Part used	Phytochemical compounds	Medicinal aspects	Citation
39.	*Silybum marianum* (L.) Gaertn	Asteraceae	Cardus marianus	Wild	Herb	Whole plant, flower	Flavonoids, polyphenolics, silybin,	Gallbladder disorders, kidney infection	Abenavoli et al. (2018)
40.	*Sisymbrium irio* L	Brassicaceae	Desert mustard	Wild	Herb	Whole plant seed, leaves	Flavonoids, glucosinolates	Fever, chest infection, asthma	Al-Qudah and Abu Zarga (2010)
41.	*Solanum surattense* Burm. f	Solanaceae	Wild eggplant	Wild	Herb	Fruit, leaves, flower	Glycoalkaloid, steroids, steroidal alkaloids	Antidiabetic, leprosy, throat infection, teeth pain	Tekuri et al. (2019)
42.	*Sonchus asper* (L.) hill	Asteraceae	Prickly sow-thistle	Wild	Herb	Whole plant, root, flower	Carotenoids, flavonoids, glycosides	Cardiac dysfunction, burns, gastrointestinal complaints	Upadhyay et al. (2013)
43.	*Taraxacum campylodes* G.E.Haglund	Asteraceae	Dandelion	Wild	Herb	Root, flower, leaves	Carotenoids, carotenoids, sesquiterpenoids, triterpenoids	Skin inflammation, clean blood, hypertension	Martinez et al. (2015)
44.	*Torilis leptophylla* (L.) Rchb.f	Apiaceae	Bristle fruit hedge parsley	Wild	Herb	Flower	Anthraquinones, glycosides, terpenoids	Joint inflammation, gastrointestinal complaints	Fatima et al. (2020)
45.	*Trifolium alexandrinum* L	Fabaceae	Spenigully shawtal	Wild	Herb	Whole plant, flower	Iso-flavonoids, steroids	Skin problems, pneumonia, stomach ache	Abdou and Abd Elkader (2021)
46.	*Verbascum thapsus* L	Scrophulariacae	Khardhag	Wild	Herb	Flower, seed	Aromatic acids, arabinose, saponins	Wound healing, sore throat	Prakash et al. (2016)

47.	*Verbesina encelioides* (Cav.) Benth. & Hook.f. ex A.Gray	Asteraceae	Gold weed	Wild	Herb	Whole plant, seed, flower	Flavonol, glycosides, triterpenoids	Antidiabetic, viral infections, antitumor	Bhati-Kushwaha and Malik (2013)
48.	*Viola tricolor* L	Violaceae	Gull	Ornamental	Herb	Whole plant seed	Flavonoids, auroxanthin	Bronchitis, anti-inflammatory	Muhammad et al. (2012)
49.	*Youngia japonica* (L.) DC	Asteraceae	Oriental false hawksbeard	Wild	Herb	Flower, leaves	β-Sitosterol, apigenin, stigmasterol	Anti-allergic, treat boils and snakebites	Rawat and Rao (2020)
50.	*Ziziphus oxyphylla* Edgew	Rhamnaceae	Marhkhaney	Wild	Shrub	Fruit, leaves, stem	Phenolic, alkaloids	Jaundice, gas troubles	Ahmad et al. (2017)

Keywords: +++ = major source, ++ = medium source, + = minor source

Fig. 1.1 Field pictorial photographs of bee floral species. (**a**) *Acacia modesta,* (**b**) *Acacia nilotica,* (**c**) *Amaranthus spinosus,* (**d**) *Bidens pilosa,* (**e**) *Brassica campestris,* (**f**) *Calendula arvensis,* (**g**) *Callistemon citrinus,* (**h**) *Cannabis sativa,* (**i**) *Carthamus oxyacantha,* (**j**) *Centaurea iberica,* (**k**) *Centaurium pulchellum,* (**l**) *Cichorium intybus,* (**m**) *Cirsium arvense,* (**n**) *Cleome viscosa,* (**o**) *Dalbergia sissoo, and* (**p**) *Digera muricata*

For many years, many rural communities in arid regions have employed the medicinal implications of honeybee flora and traditional plant knowledge (Chauhan et al., 2017). Traditional knowledge of medicinal plants is a part of Pakistani culture, and it is widely used throughout the country. People in ancient civilizations

Fig. 1.2 Field pictorial photographs of bee floral species. (**a**) *Erigeron canadensis,* (**b**) *Eruca sativa,* (**c**) *Eucalyptus camaldulensis,* (**d**) *Euphorbia helioscopia,* (**e**) *Helianthus annuus,* (**f**) *Indigofera heterantha,* (**g**) *Lantana camara,* (**h**) *Lepidium didymium,* (**i**) *Malus domestica,* (**j**) *Medicago minima,* (**k**) *Oenothera rosea,* (**l**) *Parthenium hysterophorus,* (**m**) *Prosopis juliflora,* (**n**) *Punica granatum,* (**o**) *Silybum marianum, and* (**p**) *Ziziphus oxyphylla*

employed bee foraging plants and honey as treatments for a variety of ailments such as cough, fever, stomach pain, dysentery, kidney issues, and respiratory disorders. When compared to other industrialized countries, knowledge of therapeutic applications of honey and plants has recently been utilized in Pakistan (Khan et al., 2025).

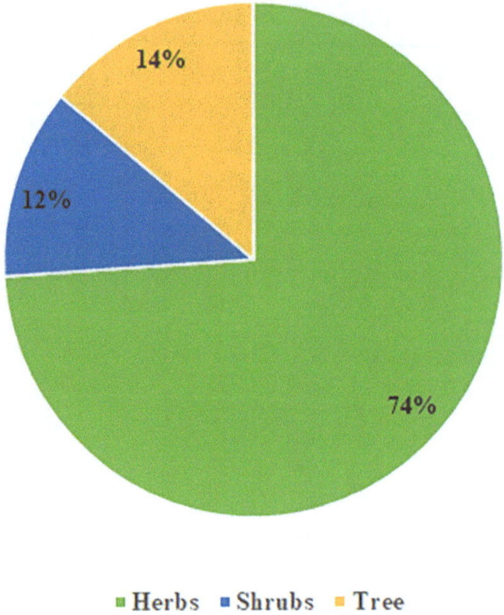

Fig. 1.3 Life form of a bee floral species

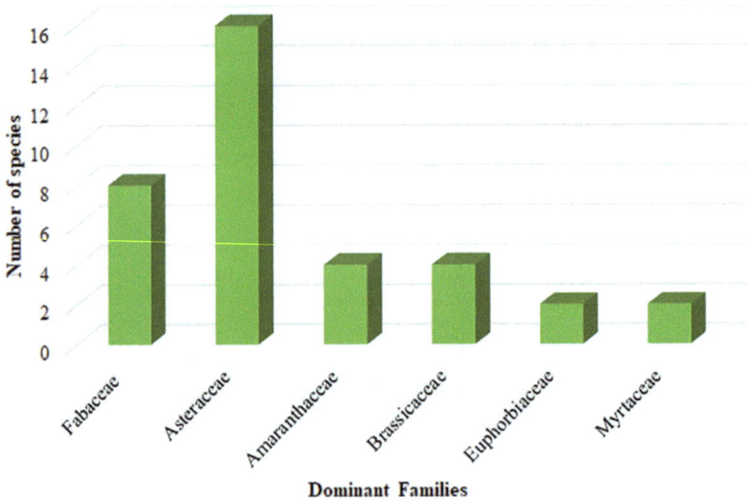

Fig. 1.4 Dominants reported plant families of the bee flora

The ethnomedicinal bee flora of the arid region is diverse and rich in botanical diversity. However, such research is limited to a few communities, resulting in gaps in data gathering and field surveys of existing flora. Traditional therapeutic benefits of bee flora were investigated in order to expand indigenous knowledge all over the world. The benefits of naturally prepared foods and their beneficial effects on health are being researched extensively around the world (Ranneh et al., 2021). The

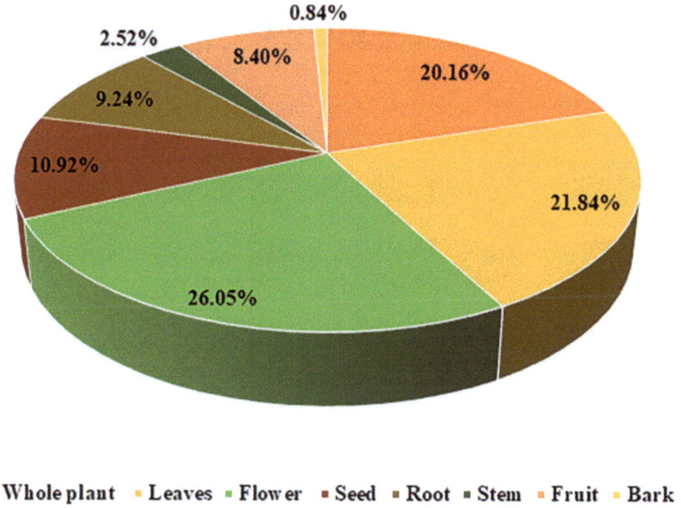

Fig. 1.5 Pie graph illustrates the plant parts used by bee floral species

utilization of ethnomedicinal knowledge in modern times has prompted scientists to pay close attention to its therapeutic value. Poor individuals in many rural villages cannot afford the costs of modern medications. As a result, people were recommended to consume honey and bee flora as cures for a variety of ailments (Kumar et al., 2024). However, in recent years, there has been a significant decline in traditional knowledge and its use in rural communities. For confirmation, data received from interviewers was matched to available literature.

This chapter also details the diversity of important phytochemical active chemicals found in bee floral species, as shown in Fig. 1.6. Aliphatics, alkaloids, anthocyanins, cyclitols, carotenoids, coumarins, eucalyptol, fatty acids, glycosides, lactones, phytocannabinoids, phenolics, flavonoids, steroids, terpenoids, tannins, thiols, and other significant compounds are included in Table 1.1. Alkaloids have been described in the literature to have an essential role in the treatment of several types of cancer, as well as to modulate human immunological processes (Mahani et al., 2022). Anthocyanins have been found to be a promising supplement therapy option for oxidative stress, inflammation, and dyslipidemia-related disorders (Habryka et al., 2021). According to a study by Ayoub et al. (2023), phenolic compounds are important for preventing or treating cardiovascular illnesses, atherosclerosis, and neurological disorders. Flavonoids have been examined for their potential use in the treatment of fatty liver disease and epilepsy (Masad et al., 2021). Steroids derived from plant chemical components have been shown to be anti-inflammatory agents in the treatment of rheumatoid arthritis and respiratory illnesses (Kumarasamy et al., 2021).

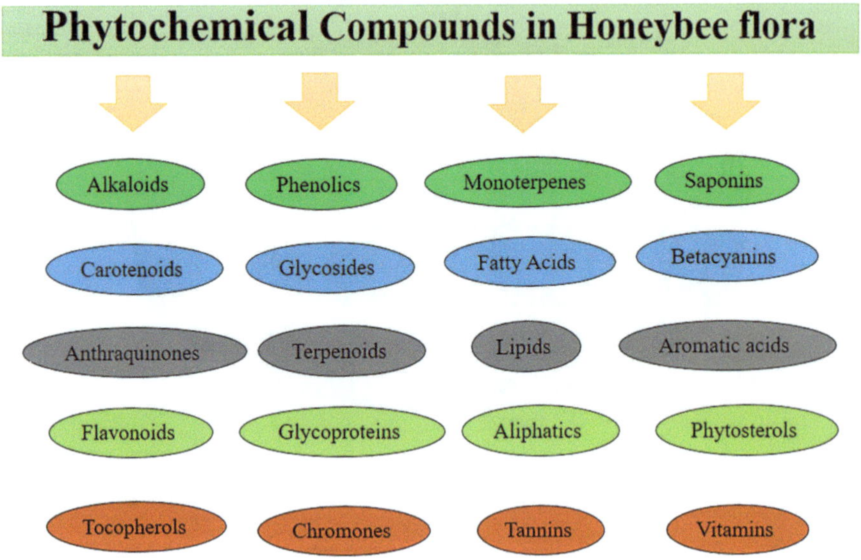

Fig. 1.6 Sketch of reported phytochemical compounds in bee floral species

6 Conclusion

The ethnomedicinal knowledge of bee flora and their active phytochemical components was detailed in this chapter to treat various diseases among indigenous people. According to the study, the arid regions province has a diverse bee flora and a large production of honey, which can be used to treat a variety of chronic ailments. Ethnomedicinal species are often used to treat gastrointestinal ailments, diarrhea, kidney problems, and respiratory issues. Accordingly, bee flora with active phytochemical components is diverse, valuable, and widely used as traditional medicine in many rural areas. Honeybees mostly foraged the Asteraceae, Fabaceae, and Amaranthaceae families, which were used to treat a variety of ailments. The ethnomedicinal perspective discussed in this chapter provides a brief summary of bee flora and their medicinal use, revealing secondary metabolites that will be further validated through pharmacological screening, leading to the development of useful beekeeping products.

Conflict of Interest The authors declare no potential conflict of interest regarding the publication of this chapter.

References

AbdErahman, A., Abayomi, O. O., Ahmed, A. E., Nour, A. H., Bin Mohd Yunus, R., Ibrahim, G. M., & Kabbashi, N. A. (2018). Comparative analysis of polyphenolic and antioxidant constituents in dried seedlings and seedless Acacia nilotica fruits. *Journal of Analysis and Testing, 2*(4), 352–355.

Abdou, H. M., & Abd Elkader, H. T. A. E. (2021). The potential therapeutic effects of Trifolium alexandrinum extract, hesperetin and quercetin against diabetic nephropathy via attenuation of oxidative stress, inflammation, GSK-3β and apoptosis in male rats. *Chemico-Biological Interactions, 109781*.

Abenavoli, L., Izzo, A. A., Milić, N., Cicala, C., Santini, A., & Capasso, R. (2018). Milk thistle (Silybum marianum): A concise overview on its chemistry, pharmacological, and nutraceutical uses in liver diseases. *Phytotherapy Research, 32*(11), 2202–2213.

Abudunia, A. M., Marmouzi, I., Kharbach, M., Jemli, M. E., Sayah, K., Bouyahya, A., & Ibrahimi, A. (2020). Hypoglycemic effect of Calendula arvensis flowers is mediated by digestive enzyme inhibition. *Current Bioactive Compounds, 16*(5), 588–592.

Adal, H., Asfaw, Z., Woldu, Z., Demissew, S., & Van Damme, P. (2015). An iconic traditional apiculture of park fringe communities of Borena Sayint National Park, north eastern Ethiopia. *Journal of Ethnobiology and Ethnomedicine, 11*, 1–17.

Ahmad, R., Ahmad, N., & Naqvi, A. A. (2017). Ziziphus oxyphylla: Ethnobotanical, ethnopharmacological and phytochemical review. *Biomedicine & Pharmacotherapy, 91*, 970–998.

Ahmad, M., Ozdimir, F. A., Zafar, M., Ahmad, S., & Sultana, S. (2019). Bee flora of Himalayan region (Pakistan) for sustainable rural development using SEM techniques for pollen identification. In *Uluslararasi Aricilik Arastirmalari ve Surdurulebilir Kirsal Kalkinma Stratijileri Kongresi Ekimbingul* (pp. 11–13).

Akhtar, N., Ali, S. S., Ahmed, S., Samin, J., Khan, M. A., & Khan, M. S. (2017). Phytochemical analysis of Lepidium didymum. *Pakistan Journal of Weed Science Research, 23*(2), 155–164.

Ali, I., Rizwani, G. H., Rasheed, M., Ali, M., Hassan, A., Hassan, S., & Rehman, S. (2019). Chemical analysis of Dalbergia sissoo (Roxb.) pod oil by (GC-MS)/GC-FID and evaluation of antioxidant potential. *Pakistan Journal of Pharmaceutical Sciences, 32*(5), 2175–2181.

Al-Qudah, M. A., & Abu Zarga, M. H. (2010). Chemical constituents of Sisymbrium irio L. from Jordan. *Natural Product Research, 24*(5), 448–456.

Al-Snafi, A. E. (2015). The chemical constituents and pharmacological importance of Chrozophora tinctoria. *International Journal of Pharmacology Review and Research, 5*(4), 391–396.

Al-Snafi, A. E. (2016). Medical importance of Cichorium intybus—A review. *IOSR Journal of Pharmacy, 6*(3), 41–56.

Arun, N., & Singh, D. P. (2012). Punica granatum: A review on pharmacological and therapeutic properties. *International Journal of Pharmaceutical Sciences and Research, 3*(5), 1240.

Ayoub, W. S., Zahoor, I., Dar, A. H., Farooq, S., Mir, T. A., Ganaie, T. A., & Altaf, A. (2023). Exploiting the polyphenolic potential of honey in the prevention of chronic diseases. *Food Chemistry Advances, 3*, 100373.

Bandeira, M. D. S. F., & De Novais, J. S. (2020). Melissopalynological characterization of honeys from the discovery coast. *Brazil. Palynology, 44*(3), 539–550.

Bartolome, A. P., Villaseñor, I. M., & Yang, W. C. (2013). Bidens pilosa L. (Asteraceae): Botanical properties, traditional uses, phytochemistry, and pharmacology. *Evidence-Based Complementary and Alternative Medicine*, 34021.

Bhati-Kushwaha, H., & Malik, C. P. (2013). Biopotential of Verbesina encelioides (stem and leaf powders) in silver nanoparticle fabrication. *Turkish Journal of Biology, 37*(6), 645–654.

Bosco, L. B., & da Luz, C. F. P. (2018). Pollen analysis of Atlantic forest honey from the Vale do Ribeira region, state of São Paulo. *Brazil. Grana, 57*(1–2), 144–157.

Chauhan, M. S., Farooqui, A., & Trivedi, A. N. J. A. L. I. (2017). Plants foraged by bees for honey production in northern India: The diverse flora of India and its implications for apiculture. *Acta Palaeobotanica, 57*(1), 119–132.

Che, C. T., & Zong, M. S. (2015). Ligustrum lucidum and its constituents: A mini-review on the anti-osteoporosis potential. *Natural Product Communications, 10*(12), 1934578X1501001242.

de Brito Damasceno, G. A., Ferrari, M., & Giordani, R. B. (2017). Prosopis juliflora (SW) DC, an invasive species at the Brazilian Caatinga: Phytochemical, pharmacological, toxicological and technological overview. *Phytochemistry Reviews, 16*(2), 309–331.

de Souza, R. R., de Abreu, V. H. R., & de Novais, J. S. (2019). Melissopalynology in Brazil: A map of pollen types and published productions between 2005 and 2017. *Palynology, 43*(4), 690–700.

Dhawan, A. S., Pawar, K. W., & Lokhande, A. A. (2018). Analysis of pollen grains in different honey samples from the region of Newasa tehsil in Maharashtra. *Journal of Pharmacognosy and Phytochemistry, 7*(3), 3438–3442.

Fakhlaei, R., Selamat, J., Khatib, A., Razis, A. F. A., Sukor, R., Ahmad, S., & Babadi, A. A. (2020). The toxic impact of honey adulteration: A review. *Food, 9*(11), 1538.

Fatima, J., Shaheen, B., Batool, S., Malik, T., Qammar, S. M., & Naureen, S. (2020). Evaluation of anti-arthritic effect of Torilis leptophylla and its comparison with indomethacin. *Journal of Fatima Jinnah Medical University, 14*(1), 25–29.

Gebretsadik, T. (2016). Assessment of major honey bee flora resources on selected districts of Sidama and Gedeo zones of south nations nationalities and peoples regional state, Ethiopia. *Journal of agricultural economics, extension and rural development, 4*(4), 49–63.

Ghorab, A., Rodríguez-Flores, M. S., Nakib, R., Escuredo, O., Haderbache, L., Bekdouche, F., & Seijo, M. C. (2021). Sensorial, melissopalynological and physico-chemical characteristics of honey from Babors Kabylia's region (Algeria). *Food, 10*(2), 225.

Guo, L., Zhang, W., Li, S., & Ho, C. T. (2015). Chemical and nutraceutical properties of Coreopsis tinctoria. *Journal of Functional Foods, 13*, 11–20.

Guo, S., Ge, Y., & Jom, K. N. (2017). A review of phytochemistry, metabolite changes, and medicinal uses of the common sunflower seed and sprouts (Helianthus annuus L.). *Chemistry Central Journal, 11*(1), 1–10.

Habryka, C., Socha, R., & Juszczak, L. (2021). Effect of bee pollen addition on the polyphenol content, antioxidant activity, and quality parameters of honey. *Antioxidants, 10*(5), 810.

Haider, A., Abdeldayem, S., & El-Esawi, M. (2021). Induction of hairy roots and phenolics biosynthesis of genetically-transformed Brassica campestris L. *The Egyptian Journal of Experimental Biology (Botany), 17*(1), 55–58.

Hailu, D., & Belay, A. (2020). Melissopalynology and antioxidant properties used to differentiate Schefflera abyssinica and polyfloral honey. *PLoS One, 15*(10), e0240868.

Hossain, M. L., Monjur-Al-Hossain, A. S. M., & Sadhu, S. K. (2016). HPLC profiling and evaluation of in-vitro antioxidant activity of Cirsium arvense L. *(Family: Asteraceae). Journal of Pharmacognosy and Phytochemistry, 5*(1), 272.

Islam, S., Shajib, M. S., & Ahmed, T. (2016). Antinociceptive effect of methanol extract of Celosia cristata Linn. In mice. *BMC Complementary and Alternative Medicine, 16*(1), 1–14.

Jaafar, N. S., & Jaafar, I. S. (2019). Eruca sativa Linn.: Pharmacognostical and pharmacological properties and pharmaceutical preparations. *Asian Journal of Pharmaceutical and Clinical Research, 12*(3), 39–45.

Jimenez-Arellanes, A., Meckes, M., Ramirez, R., Torres, J., & Luna-Herrera, J. (2003). Activity against multidrug-resistant mycobacterium tuberculosis in Mexican plants used to treat respiratory diseases. *Phytotherapy Research: An International Journal Devoted to Pharmacological and Toxicological Evaluation of Natural Product Derivatives, 17*(8), 903–908.

Joujeh, R., Zaid, S., & Mona, S. (2019). Pollen morphology of some selected species of the genus Centaurea L. (Asteraceae) from Syria. *South African Journal of Botany, 125*, 196–201.

Kabtni, S., Sdouga, D., Rebey, I. B., Save, M., Trifi-Farah, N., Fauconnier, M. L., & Marghali, S. (2020). Influence of climate variation on phenolic composition and antioxidant capacity of Medicago minima populations. *Scientific Reports, 10*(1), 1–15.

Kaur, L., Malhi, D. S., Cooper, R., Kaur, M., Sohal, H. S., Mutreja, V., & Sharma, A. (2021). Comprehensive review on ethnobotanical uses, phytochemistry, biological potential and toxicology of Parthenium hysterophorus L.: A journey from noxious weed to a therapeutic medicinal plant. *Journal of Ethnopharmacology, 281*, 114525.

Khalil, H. E., & Al Ahmed, A. (2017). Phytochemical analysis and free radical scavenging activity of Carthamus oxyacantha growing in Saudi Arabia: A comparative study. *International Journal of Pharmaceutical Science Review and Research, 45*(1), 51–55.

Khan, R. A., Khan, M. R., Sahreen, S., & Ahmed, M. (2012). Assessment of flavonoids contents and in vitro antioxidant activity of Launaea procumbens. *Chemistry Central Journal, 6*(1), 1–11.

Khan, A., Khan, A., Ahmad, M., Ali, M., Farooq, U., Khan, F. A., & Bukhari, S. M. (2021). Antiglycation potential of Indigoferin A, Indigoferin B and Indigoferin C natural products from Indigofera heterantha Brandis. *Clinical Phytoscience, 7*(1), 1–6.

Khan, B., Faiz, S., Khalid, S., Ullah, S., & Nafees, M. (2025). Comparative Pharmacognostic and pharmacological profiling of honey obtained from different plant sources: Honey from plants. *Biological Sciences-PJSIR, 68*(1), 84–98.

Kopytina, T., Nenasheva, G., & Ivanova, M. (2019). On the flora of honey plants of the northern and northwestern Altai regions within the Altai territory. In *BIO web of conferences* (Vol. 16, p. 00013). EDP Sciences.

Kumar, M., Verma, A. K., & Garkoti, S. C. (2020). Lantana camara and Ageratina adenophora invasion alter the understory species composition and diversity of chir pine forest in central Himalaya. *India. Acta Oecologica, 109*, 103642.

Kumar, S., Verma, M., Hajam, Y. A., & Kumar, R. (2024). Honey infused with herbs: A boon to cure pathological diseases. *Heliyon, 10*(1), e23302.

Kumarasamy, G., Ramli, R. R., Singh, H., & Abdullah, B. (2021). Tualang honey versus steroid impregnated nasal dressing following endoscopic sinus surgery: A randomized controlled trial. *Journal of Complementary and Integrative Medicine, 18*(2), 433–438.

Larayetan, R. A., Okoh, O. O., Sadimenko, A., & Okoh, A. I. (2017). Terpene constituents of the aerial parts, phenolic content, antibacterial potential, free radical scavenging and antioxidant activity of Callistemon citrinus (Curtis) Skeels (Myrtaceae) from eastern Cape Province of South Africa. *BMC Complementary and Alternative Medicine, 17*(1), 1–9.

Mahani, M., Wulandari, E., Lembong, E., & Adela, L. (2022). Correlation of alkaloid content and taste of honey from various provinces in Indonesia. *International Journal of Pharmacy and Pharmaceutical Sciences, 14*(12), 17–21.

Majid, A. P. W. J., Talab, T. A., & Al Omer, A. P. D. K. (2021). Constituents, pharmacological and toxicological effects of Melia azadirachta—A review. *Sapporo Medical Journal, 55*(1), 1–15.

Manimekalai, P., Kaveena, R., Naveena, S., Nivetha, S., & Nivetha, N. (2020). A review on pharmacognostical and phytochemical study of Digera muricata L. *Journal of Medicinal Plants, 8*(5), 52–56.

Márquez-Flores, Y. K., Meléndez-Camargo, M. E., García-Mateos, N. J., Huerta-Anaya, M. C., Pablo-Pérez, S. S., & Silva-Torres, R. (2018). Phytochemical composition and pharmacological evaluation of different extracts of Oenothera rosea L'Hér. Ex Ait (Onagraceae) aerial part. *South African Journal of Botany, 116*, 245–250.

Martinez, M., Poirrier, P., Chamy, R., Prüfer, D., Schulze-Gronover, C., Jorquera, L., & Ruiz, G. (2015). Taraxacum officinale and related species—An ethnopharmacological review and its potential as a commercial medicinal plant. *Journal of Ethnopharmacology, 169*, 244–262.

Martinez-Castillo, C., Astray, G., Mejuto, J. C., & Simal-Gandara, J. (2020). Random forest, artificial neural network, and support vector machine models for honey classification. *eFood, 1*(1), 69–76.

Marzouk, M. M. (2021). A comparative review on phytochemical constituents and biological effects of Melilotus indicus (L.) all. And Melilotus messanensis (L.) all. (Fabaceae): Evidence for chemosystematic analysis. *Egyptian Journal of Chemistry, 65*(2), 191–202.

Masad, R. J., Haneefa, S. M., Mohamed, Y. A., Al-Sbiei, A., Bashir, G., Fernandez-Cabezudo, M. J., & Al-Ramadi, B. K. (2021). The immunomodulatory effects of honey and associated flavonoids in cancer. *Nutrients, 13*(4), 1269.

Montoya-Bonilla, B. P., Baca-Gamboa, A. E., & Bonilla, B. L. (2017). The honey plants and its resources offer in five paths of the municipality of Piendamo, Cauca. *Biotecnología en el Sector Agropecuario y Agroindustrial, 15*(spe), 20–28.

Muhammad, N., Saeed, M., Aleem, A., & Khan, H. (2012). Ethnomedicinal, phytochemical and pharmacological profile of genus Viola. *Phytopharmacology, 3*(1), 214–226.

Nabila, A., Ali, M., Zafar, M., Sultana, S., Majeed, S., & Ahmad, S. (2022). Palynological diversity of Melliferous flora around apiaries from district Mardan Khyber Pakhtunkhwa-Pakistan. *The Botanical Review, 88*(3), 299–332.

Patocka, J., Bhardwaj, K., Klimova, B., Nepovimova, E., Wu, Q., Landi, M., & Wu, W. (2020). Malus domestica: A review on nutritional features, chemical composition, traditional and medicinal value. *Plants, 9*(11), 1408.

Pawlaczyk, I., Czerchawski, L., Kuliczkowski, W., Karolko, B., Pilecki, W., Witkiewicz, W., & Gancarz, R. (2011). Anticoagulant and anti-platelet activity of polyphenolic-polysaccharide preparation isolated from the medicinal plant Erigeron canadensis L. *Thrombosis Research, 127*(4), 328–340.

Prakash, V., Rana, S., & Sagar, A. (2016). Studies on antibacterial activity of Verbascum thapsus. *Journal of Medicinal Plants Studies, 4*(3), 101–103.

Qureshi, R. (2012). Medicinal flora of hingol national park, Baluchistan. *Pakistan. Pakistan Journal of Botany, 44*(2), 725–732.

Ranneh, Y., Akim, A. M., Hamid, H. A., Khazaai, H., Fadel, A., Zakaria, Z. A., & Bakar, M. F. A. (2021). Honey and its nutritional and anti-inflammatory value. *BMC Complementary Medicine and Therapies, 21*, 1–17.

Rawat, D., & Rao, P. B. (2020). Comparative evaluation of antioxidant profiling and quantitative phytochemicals of Ixeris polycephala Cass. And Youngia japonica (L.) DC. *(Asteraceae). The Pharma Innovation Journal, 9*(1), 5–9.

Saini, S., & Saini, K. K. (2020). Chenopodium album Linn: An outlook on weed cum nutritional vegetable along with medicinal properties. *Emergent Life Science Research, 6*, 28–33.

Sani, I., Abdulhamid, A., & Bello, F. (2014). Eucalyptus camaldulensis: Phytochemical composition of ethanolic and aqueous extracts of the leaves, stem-bark, root, fruits and seeds. *Journal of Scientific and Innovative Research, 3*(5), 523–526.

Sarker, U., & Oba, S. (2019). Nutraceuticals, antioxidant pigments, and phytochemicals in the leaves of Amaranthus spinosus and Amaranthus viridis weedy species. *Scientific Reports, 9*(1), 1–10.

Sarwar, W. (2016). Pharmacological and phytochemical studies on Acacia modesta wall; A review. *Journal of Phytopharmacology, 5*, 160–166.

Shakil, S. S., Gowan, M., Hughes, K., Azam, M. N. K., & Ahmed, M. N. (2021). A narrative review of the ethnomedicinal usage of Cannabis sativa Linnaeus as traditional phytomedicine by folk medicine practitioners of Bangladesh. *Journal of Cannabis Research, 3*(1), 1–12.

Šiler, B., & Mišić, D. (2016). Biologically active compounds from the genus Centaurium s.l. (Gentianaceae): Current knowledge and future prospects in medicine. *Studies in Natural Products Chemistry, 49*, 363–397.

Suresh, Y., Rajasekar, G., Lavanya, T., Lakshminarsimhulu, B., Reddy, K. S., & Reddy, S. R. (2020). Antioxidant and antidiabetic properties of isolated fractions from methanolic extract derived from the whole plant of Cleome viscosa L. *Future Journal of Pharmaceutical Sciences, 6*(1), 1–18.

Tekuri, S. K., Pasupuleti, S. K., Konidala, K. K., Amuru, S. R., Bassaiahgari, P., & Pabbaraju, N. (2019). Phytochemical and pharmacological activities of Solanum surattense Burm. f.—A review. *Journal of Applied Pharmaceutical Science, 9*(3), 126–136.

Ullah, S. A., Zafar, M., Ahmad, M., Ghufran, M. A., Bursal, E., Kilic, O., & Majeed, S. (2021). Microscopic implication and evaluation of herbaceous melliferous plants of southern Khyber Pakhtunkhwa-Pakistan using light and scanning electron microscope. *Microscopy Research and Technique, 84*(8), 1750–1764.

Upadhyay, H., Kumar, A., Gupta, M. K., Sharma, A., & Rahal, A. (2013). Validation of medicinal values of traditionally used Sonchus asper (prickly sow thistle) leaves for the treatment of skin ailments. *Advancement in Medicinal Plant Research, 1*(3), 29–35.

Yang, Y., Chen, X., Luan, F., Wang, M., Wang, Z., Wang, J., & He, X. (2021). Euphorbia helioscopia L.: A phytochemical and pharmacological overview. *Phytochemistry, 184*, 112649.

Chapter 2
Combination of Traditional and Modern Approaches in Honey Authentication

Maria Ameen, Mushtaq Ahmad, Muhammad Zafar, Sumreen Dawood, Shaista Jabeen, Salman Majeed, Muhammad Iqbal, Huma Qureshi, and Nabeela Mahmood

1 Introduction

Honey's authenticity has long been verified using traditional methods, which include sensory evaluation, physicochemical analysis, and pollen analysis. Sensory evaluation relies on the expertise of trained individuals to assess the aroma, taste, and texture of honey. Physicochemical analysis involves measuring properties such as pH, moisture content, and electrical conductivity. Pollen analysis, or melissopalynology, determines the botanical and geographical origins of honey by examining the pollen grains it contains. Although these methods are established and provide valuable information, they often lack the sensitivity and specificity needed to detect certain sophisticated adulterants, such as syrups or sugars, introduced through modern fraudulent practices (Trifković et al., 2017).

Modern analytical techniques such as chromatography, spectroscopy, and molecular analysis have significantly enhanced the sensitivity and specificity of honey authentication. Chromatography techniques, including gas chromatography (GC) and liquid chromatography (LC), allow for the separation and identification of various compounds within honey. Spectroscopy methods, such as nuclear magnetic

M. Ameen (✉)
Department of Botany, University of Chakwal, Chakwal, Punjab, Pakistan

Department of Plant Sciences, Quaid-i-Azam University, Islamabad, Pakistan
e-mail: maria.ameen@uoc.edu.pk

M. Ahmad · M. Zafar
Department of Plant Sciences, Quaid-i-Azam University, Islamabad, Pakistan

S. Dawood · S. Majeed
Department of Botany, University of Mianwali, Mianwali, Punjab, Pakistan, 42200

S. Jabeen · M. Iqbal · H. Qureshi · N. Mahmood
Department of Botany, University of Chakwal, Chakwal, Punjab, Pakistan

M. Zafar et al. (eds.), *Pure Honey: Assurance & Authentication*,
https://doi.org/10.1007/978-3-031-98913-1_2

resonance (NMR) and infrared (IR) spectroscopy, provide detailed molecular information, enabling the detection of adulterants at trace levels. Molecular analysis, particularly DNA barcoding, helps verify the botanical origin of honey by identifying specific plant DNA. These advanced techniques can detect minute discrepancies and anomalies indicative of adulteration, thus providing more robust and accurate assessments compared to traditional methods (Zhang & Abdulla, 2022).

The future of honey authentication lies in integrating traditional and modern methods to create a comprehensive approach that addresses the limitations of each. Combining sensory evaluation and physicochemical analysis with advanced chromatographic and spectroscopic techniques can improve the accuracy and reliability of authenticity assessments. This integrative approach also leverages the cultural significance and historical data provided by traditional methods while incorporating the precision and efficiency of modern technology (Tsagkaris et al., 2021). Moreover, there is a pressing need to develop harmonized global standards for honey authentication to tackle emerging challenges, such as detecting sophisticated adulterants and ensuring product integrity in a globalized market. This chapter presents case studies demonstrating successful integrations of these methods, highlighting their potential to enhance consumer trust and safeguard the reputation of honey producers worldwide.

2 Traditional Methods in Honey Authentication

2.1 Sensory Evaluation and its Limitations

Sensory evaluation involves the use of human senses to assess the quality and authenticity of honey. This method relies on the expertise of trained personnel to identify specific characteristics of honey, such as flavor, aroma, texture, and color. Despite its long-standing use, sensory evaluation has significant limitations. The subjective nature of this method can lead to inconsistencies and inaccuracies. Sensory evaluation cannot reliably detect subtle adulterants or differentiate between honeys of similar sensory profiles but different origins. Therefore, while valuable for initial assessments, sensory evaluation must be complemented by more objective and precise methods (Antônio, 2021).

2.2 Physicochemical Analysis: Techniques and Effectiveness

Physicochemical analysis encompasses a range of laboratory techniques to measure the physical and chemical properties of honey. Common tests include the determination of moisture content, electrical conductivity, pH, and the presence of specific sugars and enzymes. These analyses provide important information about the quality and authenticity of honey. However, physicochemical properties can be manipulated through sophisticated adulteration techniques. As such, while these methods are useful for routine quality control, they may not always detect adulterants introduced at trace levels or identify all forms of adulteration. The effectiveness of physicochemical analysis lies in its ability to provide a baseline for honey quality, but it is not foolproof against modern adulteration methods (Oroian & Ropciuc, 2017).

2.3 Pollen Analysis: Role and Constraints in Modern Authentication

Pollen analysis, also known as melissopalynology, involves examining the pollen grains present in honey to determine its botanical and geographical origin. This method can provide detailed insights into the floral sources and regions where the honey was produced. However, pollen analysis has its constraints. It requires specialized knowledge and expertise to accurately identify pollen types and their origins (Danieli & Lazzari, 2022). Additionally, the presence of pollen can be influenced by various factors, such as beekeeping practices and environmental conditions, which may lead to misinterpretation. Furthermore, modern adulteration techniques can remove or alter the pollen content, complicating the authentication process. While pollen analysis remains a valuable tool, it must be integrated with other advanced methods to ensure comprehensive authentication (Şengül, 2023; Soares et al., 2017).

3 Modern Techniques in Honey Authentication

3.1 Chromatography: Types and Applications in Detecting Adulterants

Chromatography is a powerful analytical technique used to separate and identify the components of a mixture. In honey authentication, various forms of chromatography, such as gas chromatography (GC) and high-performance liquid chromatography (HPLC), are employed to detect adulterants. These methods can identify and quantify specific sugars, organic acids, and other compounds in honey (Chin & Sowndhararajan, 2020). Chromatography offers high sensitivity and specificity, making it possible to detect even trace amounts of adulterants. For example, GC can distinguish between natural honey sugars and those added through adulteration. The application of chromatography in honey authentication enhances the ability to detect and prevent fraudulent practices, ensuring the purity and quality of honey (Taleuzzaman et al., 2020).

3.2 *Spectroscopy: Enhancing Sensitivity and Specificity*

Spectroscopy involves the interaction of light with matter to identify and quantify substances. Techniques such as nuclear magnetic resonance (NMR) spectroscopy, infrared (IR) spectroscopy, and ultraviolet-visible (UV-Vis) spectroscopy are commonly used in honey authentication. These methods provide detailed molecular information, enabling the detection of adulterants that might not be identified through traditional methods (Vogt et al., 2023). Spectroscopy enhances the sensitivity and specificity of honey analysis by offering a non-destructive means to analyze the composition of honey. For instance, NMR spectroscopy can detect minor adulteration by comparing the molecular fingerprint of the honey sample with authentic samples. Spectroscopic techniques significantly improve the accuracy of honey authentication, making them indispensable in modern quality control (Tripathi, n.d.).

3.3 *Molecular Analysis: Advantages in Identifying Trace Adulterants*

Molecular analysis involves examining the genetic material and molecular composition of honey to ensure its authenticity. Techniques such as DNA barcoding and polymerase chain reaction (PCR) are used to identify the botanical and geographical origins of honey by analyzing the DNA of pollen and other biological components. Molecular analysis offers unparalleled precision in detecting trace adulterants that might escape traditional and physicochemical methods. For example, DNA barcoding can confirm the presence of specific plant species in honey, verifying its claimed origin. The advantages of molecular analysis lie in its ability to provide conclusive evidence of honey's authenticity, thereby safeguarding against sophisticated adulteration techniques and ensuring consumer trust (Kowalczyk et al., 2021).

4 Integrating Traditional and Modern Approaches

Integrating traditional and modern approaches in honey authentication is essential to create a robust system capable of addressing the evolving challenges of honey adulteration. Traditional methods like sensory evaluation, physicochemical analysis, and pollen analysis provide valuable baseline data but often lack the specificity and sensitivity required to detect modern, sophisticated adulteration techniques. Modern analytical methods, such as chromatography, spectroscopy, and molecular analysis, offer higher sensitivity and specificity, allowing for the detection of adulterants at trace levels (Awal, 2020). Conventional and modern analytical methods used for honey authentication are sketched in Fig. 2.1.

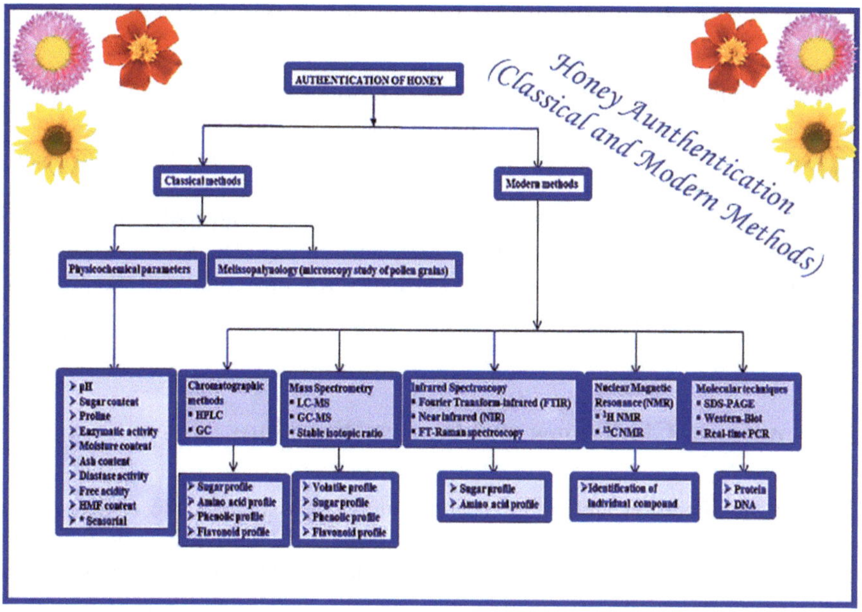

Fig. 2.1 Conventional and modern analytical methods used for honey authentication

4.1 Synergistic Potential of Combining Traditional and Modern Techniques

- Traditional sensory evaluations rely on human senses to assess honey's color, flavor, and aroma. While effective for detecting gross anomalies, they cannot identify specific adulterants.
- Physicochemical methods measure properties such as moisture content, pH, electrical conductivity, and sugar profile. For instance, pure honey typically has a moisture content of less than 20%, a pH of around 3.9, and an electrical conductivity of less than 0.8 mS/cm.
- Pollen analysis, or melissopalynology, determines the botanical and geographical origins of honey by examining the pollen grains present. However, this method can be labor-intensive and requires expert interpretation.
- Chromatographic techniques, such as gas chromatography (GC) and high-performance liquid chromatography (HPLC), separate and identify compounds in honey. GC, coupled with mass spectrometry (GC-MS), can detect adulteration with foreign sugars at levels as low as 0.1% (Se et al., 2019).
- Spectroscopic methods, including nuclear magnetic resonance (NMR) and infrared spectroscopy (IR), provide detailed molecular fingerprints of honey samples, offering high accuracy in detecting adulterants.
- Molecular techniques like DNA barcoding can identify plant species from honey's DNA, verifying its botanical origin with precision (Wu et al., 2017b).

4.2 *Case Studies: Successful Integration Examples*

- A study integrating traditional and modern techniques found that combining pollen analysis with HPLC improved the accuracy of identifying honey origins. The combined method could detect adulteration in samples where traditional methods alone failed.
- In another case, the integration of sensory evaluation with NMR spectroscopy allowed for the rapid screening of honey authenticity. The sensory evaluation provided initial assessments, while NMR spectroscopy confirmed the presence of specific adulterants (Ruiz-Matute et al., 2010).

5 Emerging Challenges and Future Directions

As honey adulteration techniques become more sophisticated, the challenges in ensuring honey authenticity also evolve. Emerging challenges include detecting adulterants that mimic honey's natural components, addressing the variability in honey composition due to environmental factors, and establishing harmonized global standards.

5.1 *Sophisticated Adulterants and Their Detection*

- Adulterants such as high-fructose corn syrup and rice syrup can closely mimic honey's natural sugar profile, making detection challenging. Advanced techniques like isotope ratio mass spectrometry (IRMS) can distinguish between plant-derived sugars and those from industrial sources, even when present at low levels.
- An analysis of 223 honey samples from around the world revealed that 27% contained added sugars not derived from nectar, highlighting the prevalence of sophisticated adulteration methods (Tarapoulouzi et al., 2023).

5.2 *Importance of Harmonized Global Standards*

- Currently, honey standards vary significantly between countries, complicating international trade and enforcement. For example, the Codex Alimentarius, the international food standards body, has set criteria for honey composition, but these are not universally adopted.
- A harmonized global standard would facilitate consistent testing and certification, improving consumer trust and trade transparency. The International Honey

Commission (IHC) is working toward developing such standards, aiming to establish common criteria for honey quality and authenticity.

5.3 *Future Trends in Honey Authentication Research and Technology*

- Research is focusing on developing portable and rapid testing methods to enable on-site honey authentication. Techniques like portable NMR and handheld IR spectrometers are being explored for their potential to provide quick and accurate results (Biswas & Chaudhari, 2024).
- Advances in machine learning and artificial intelligence are being applied to honey authentication, where algorithms can analyze complex data from various analytical techniques to identify patterns indicative of adulteration (Tsagkaris et al., 2021).

By integrating traditional methods with modern analytical techniques, the honey industry can more effectively combat adulteration, ensuring that consumers receive genuine, high-quality honey. As research and technology continue to advance, the future of honey authentication looks promising, with the potential for more accurate, rapid, and accessible methods to safeguard this valuable natural product.

6 Case Studies and Practical Examples

In the realm of honey authentication, practical examples and case studies provide valuable insights into the effectiveness of combining traditional and modern methods. These real-world instances illustrate how innovative approaches can enhance the reliability of honey authentication, thereby boosting consumer trust and safeguarding the reputation of honey producers.

One notable case study involves the use of nuclear magnetic resonance (NMR) spectroscopy in conjunction with traditional pollen analysis to authenticate Manuka honey from New Zealand. Manuka honey is highly prized for its unique medicinal properties, but its high market value also makes it a target for adulteration. By integrating NMR spectroscopy, which can detect the unique chemical fingerprint of Manuka honey, with traditional pollen analysis, researchers were able to achieve a 95% accuracy rate in identifying genuine Manuka honey (Zhang et al., 2023). This dual approach not only increased the reliability of the authentication process but also provided a robust defense against sophisticated adulteration techniques.

Another practical example is the application of high-performance liquid chromatography (HPLC) combined with sensory evaluation to authenticate acacia honey in Europe. Sensory evaluation, while useful for initial screening, often falls short in detecting subtle adulterations. HPLC, however, can identify specific compounds and their concentrations, which are unique to acacia honey. By comparing these chemical profiles with sensory characteristics, producers were able to detect adulterations with a sensitivity of up to 98%. This method proved particularly effective in differentiating genuine Acacia honey from those adulterated with sugar syrups, ensuring that only pure honey reached consumers (Bose & Padmavati, 2024).

In China, a comprehensive approach was taken to authenticate jujube honey, integrating gas chromatography-mass spectrometry (GC-MS) with traditional physicochemical tests. Jujube honey, known for its distinct flavor and therapeutic benefits, is often subjected to adulteration. GC-MS allowed for the precise identification of volatile compounds unique to jujube honey, while physicochemical tests measured parameters such as moisture content and acidity. This integrated method achieved a detection accuracy of 97%, significantly reducing the incidence of adulterated jujube honey in the market (Nhlapho, 2023).

A practical example from the United States involves the use of Fourier Transform Infrared (FTIR) spectroscopy in tandem with pollen analysis for clover honey authentication. Clover honey is one of the most commonly consumed honeys in the United States, and ensuring its purity is crucial. FTIR spectroscopy can quickly identify the molecular composition of honey samples, while pollen analysis verifies the floral source. This combined approach resulted in a detection accuracy of 96%, effectively differentiating pure clover honey from adulterated samples (Se et al., 2019; Wu et al., 2017a).

These case studies highlight the importance of integrating traditional and modern methods to enhance the accuracy and reliability of honey authentication. By leveraging the strengths of both approaches, producers can better detect adulteration, protect consumer interests, and maintain their market reputation.

7 Conclusion

The amalgamation of traditional and modern techniques in honey authentication represents a significant advancement in ensuring the purity and quality of honey. Traditional methods, such as sensory evaluation, physicochemical analysis, and pollen analysis, have been the cornerstone of honey authentication for decades. However, they often lack the sensitivity required to detect modern adulteration techniques. On the other hand, modern methods like chromatography, spectroscopy, and molecular analysis offer heightened sensitivity and specificity, enabling the detection of adulterants at trace levels and providing more accurate assessments.

The integration of these approaches not only enhances the reliability of authentication outcomes but also addresses the multifaceted challenges posed by increasingly sophisticated adulteration methods. For instance, the combination of NMR spectroscopy with pollen analysis for Manuka honey and HPLC with sensory evaluation for Acacia honey has proven to be highly effective, achieving detection accuracies of up to 98%. These integrated methods not only protect consumers but also help maintain the integrity of honey products in the market.

Furthermore, the need for harmonized global standards in honey authentication cannot be overstated. As the market for honey becomes increasingly interconnected, establishing consistent and universally accepted authentication methods is crucial. This would ensure that honey products meet the same quality standards worldwide, further protecting consumers and producers alike.

Looking ahead, the future of honey authentication lies in the continuous innovation and integration of traditional and modern techniques. Emerging technologies, such as advanced molecular analysis and real-time monitoring systems, hold great promise for further improving the accuracy and efficiency of honey authentication processes. Additionally, fostering collaboration between researchers, industry stakeholders, and regulatory bodies will be key to developing and implementing these advanced methods on a global scale.

In conclusion, the synergy between traditional and modern approaches in honey authentication represents a robust and effective strategy for combating adulteration, ensuring consumer trust, and safeguarding the reputation of honey producers. By embracing this comprehensive approach, the honey industry can continue to provide high-quality, authentic products to consumers worldwide.

Acknowledgments The authors acknowledge AI for the language editing.

References

Antônio, D.C., 2021. Use of multiway chemometric tools for honey and coffee analysis.

Awal, M. (2020). Integrating traditional and modern adaptation practices towards climate-smart agriculture in Bangladesh. *Discovery Agriculture, 6*(16), 113–126.

Biswas, A., & Chaudhari, S. R. (2024). Exploring the role of NIR spectroscopy in quantifying and verifying honey authenticity: A review. *Food Chemistry, 138712.*

Bose, D., & Padmavati, M. (2024). Honey authentication: A review of the issues and challenges associated with honey adulteration. Food. *Bioscience, 105004.*

Chin, N. L., & Sowndhararajan, K. (2020). A review on analytical methods for honey classification, identification and authentication. In *Honey Analysis—New Advances and Challenges* (pp. 1–33).

Danieli, P. P., & Lazzari, F. (2022). Honey traceability and authenticity. Review of current methods Most used to face this problem. *Journal of Apicultural Science, 66*(2), 101–119.

Kowalczyk, M., Staniszewski, A., Kamiñska, K., Domaradzki, P., & Horecka, B. (2021). Advantages, possibilities, and limitations of mitochondrial DNA analysis in molecular identification. *Folia Biologica (Kraków), 69*(3), 101–111.

Nhlapho, S. (2023). *Investigation of sugars in honey for authentication and detection of adulterants using selected analytical techniques*. University of Johannesburg.

Oroian, M., & Ropciuc, S. (2017). Honey authentication based on physicochemical parameters and phenolic compounds. *Computers and Electronics in Agriculture, 138*, 148–156.

Ruiz-Matute, A. I., Rodríguez-Sánchez, S., Sanz, M. L., & Martínez-Castro, I. (2010). Detection of adulterations of honey with high fructose syrups from inulin by GC analysis. *Journal of Food Composition and Analysis, 23*(3), 273–276.

Se, K. W., Wahab, R. A., Yaacob, S. N. S., & Ghoshal, S. K. (2019). Detection techniques for adulterants in honey: Challenges and recent trends. *Journal of Food Composition and Analysis, 80*, 16–32.

Şengül, M. (2023). Importance of pollen analysis in honey: Melissopalynology. In *Current research in agriculture, forestry and aquaculture*.
Soares, S., Amaral, J. S., Oliveira, M. B. P., & Mafra, I. (2017). A comprehensive review on the main honey authentication issues: Production and origin. *Comprehensive Reviews in Food Science and Food Safety, 16*(5), 1072–1100.
Taleuzzaman, M., Alam, M. J., Kala, C., & Rahat, I. (2020). Honey of authenticity: An analytical approach. *Therapeutic Applications of Honey and its Phytochemicals, 1*, 101–120.
Tarapoulouzi, M., Mironescu, M., Drouza, C., Mironescu, I. D., & Agriopoulou, S. (2023). Insight into the recent application of Chemometrics in quality analysis and characterization of bee honey during processing and storage. *Food, 12*(3), 473.
Trifković, J., Andrić, F., Ristivojević, P., Guzelmeric, E., & Yesilada, E. (2017). Analytical methods in tracing honey authenticity. *Journal of AOAC International, 100*(4), 827–839.
Tripathi, M. K. (n.d.). Introduction of spectroscopy. *Fundamental Concepts of Spectroscopy and Spectrometry for Chemical Analysis, 2*, 1.
Tsagkaris, A. S., Koulis, G. A., Danezis, G. P., Martakos, I., Dasenaki, M., Georgiou, C. A., & Thomaidis, N. S. (2021). Honey authenticity: Analytical techniques, state of the art and challenges. *RSC Advances, 11*(19), 11273–11294.
Vogt, C., Wondergem, C. S., & Weckhuysen, B. M. (2023). *Ultraviolet-visible (UV-Vis) spectroscopy, springer handbook of advanced catalyst characterization* (pp. 237–264). Springer.
Wu, L., Du, B., Vander Heyden, Y., Chen, L., Zhao, L., Wang, M., & Xue, X. (2017a). Recent advancements in detecting sugar-based adulterants in honey–a challenge. *TrAC Trends in Analytical Chemistry, 86*, 25–38.
Wu, Y., Yang, Y., Liu, M., Wang, B., Li, M., & Chen, Y. (2017b). Molecular tracing of the origin of six different plant species in bee honey using real-time PCR. *Journal of AOAC International, 100*(3), 744–752.
Zhang, G., & Abdulla, W. (2022). On honey authentication and adulterant detection techniques. *Food Control, 138*, 108992.
Zhang, X.-H., Gu, H.-W., Liu, R.-J., Qing, X.-D., & Nie, J.-F. (2023). A comprehensive review of the current trends and recent advancements on the authenticity of honey. *Food Chemistry, X*, 100850.

Chapter 3
Use of Phenolic Compounds in Honey Identification

Duygu Nur Çobanoğlu and İkranur Felek

Abbreviations

CZE	Capillary zone electrophoresis
DAD	Diodide array detector
NADPH	Nicotinamide adenine dinucleotide phosphate hydrogen
NMR	Nuclear magnetic resonance
HPLC	High-performance liquid chromatography
MS	Mass spectrometry
LC-ESI-MS	Liquid chromatography-electrospray ionization mass spectrometry
RP-UHPLC-ESI-MS	Reversed-phase ultra-high-performance liquid chromatography with electrospray ionization mass spectrometry
UHPLC-LTQ-Orbitrap-MS	Ultra-high-performance liquid chromatography (UHPLC) coupled with hybrid mass spectrometer which combines the linear trap quadrupole (LTQ) and OrbiTrap mass analyzer

D. N. Çobanoğlu (✉)
Department of Crop and Animal Production, Vocational School of Food, Agriculture and Livestock, Bingol University, Bingol, Türkiye
e-mail: dncobanoglu@bingol.edu.tr

İ. Felek
Department of Nutrition and Dietetics, Faculty of Health Sciences, Bingol University, Bingol, Türkiye

M. Zafar et al. (eds.), *Pure Honey: Assurance & Authentication*,
https://doi.org/10.1007/978-3-031-98913-1_3

1 Introduction to Honey Identification

The development of analytical methods for verifying food origin is a rapidly growing area. Honey, widely consumed around the world, is particularly valued for its numerous health benefits (Koulis et al., 2021). It is an essential natural sweetener that has been utilized for almost 6000 years; for a very long period, it was the only source of sweeteners. Bees gather honey from nectar of blossoms, exudates all living parts of plants other than flowers, or excretions of plant sucking insects (Jiang et al., 2020). The nectar is combined with honeybees' own unique substances, and then stored, dehyrated, and matured in honeycombs.

Honey can be broadly categorized into two types according to the botanical origin: floral (or blossom) honey and honeydew honey (Siddiqui et al., 2017). Honeys can be classified into two groups according to the nectar composition as monofloral (unifloral) and multifloral (polyfloral) honeys (Jiang et al., 2020). The term "monofloral honey" describes honey in which the major part of the nectar is obtained from a single plant species (Molan, 1998). Honeybees collect nectar from various plants for multifloral honey, and none of the nectar of these plants is dominant (Becerril-sánchez et al., 2021). Honey can be classified into different types based on botanical source from which nectar is derived, such as chestnut, linden, rape, heather, eucalyptus, thyme, buckwheat, and sunflower (Lawag et al., 2022). There are more than 320 distinct varieties of honey (Vîjan et al., 2023). Each type of honey has unique characteristics, including flavor, color, aroma, and texture, all of which are influenced by the specific flowers visited by the bees. In addition to carbohydrates like fructose and glucose, honey is rich in various other bioactive compounds and nutrients such as amino acids, enzymes, minerals, phenolics, and proteins (Da Silva et al., 2016). The unique composition of honey mainly originates from phenolic compounds and amino acids, contributing to a healthy, balanced diet (Zhu et al., 2024). Phenolic compounds, known as secondary metabolites of plants, are carried into honey by bees when they visit plants. Phenolic compounds act as powerful antioxidants to prevent oxidative stress, which is linked to a range of adverse outcomes and chronic diseases, including cancer, cardiovascular disease, metabolic disorders, obesity, and type 2 diabetes (Battino et al., 2021; Zammit Young & Blundell, 2023).

Honey's sensory qualities are significantly influenced by its source, which includes the geographical location of the apiary, botanical origin, and environmental conditions, all of which determine its chemical composition. As a result, the plant origin of honey is a key factor for its medicinal and nutritional value. Consumers' preferences for honey often differ based on its sensory characteristics and medicinal benefits (Pauliuc et al., 2020). Increasing consumer demand for high-quality honey varieties with health benefits, such as Manuka honey, has led to increased incidents of adulteration and mislabeling of honey by producers.

Therefore, accurately determining the botanical origin of honey is crucial. Beekeepers often determine the honey variety according to the time of nectar occurrence and the presence of nectar flows (Dzugan et al., 2018). With the globalization of the honey market, determining its origin and verifying authenticity has become a significant issue (Pauliuc et al., 2020). In recent years, interest in identifying the botanical origin of honey has grown due to consumers' preference for specific honey varieties. There are important reasons for knowing the plant origin of a honey, for example, for quality control in marketing, due to fair trade legislation requiring accurate disclosures on labels, and where there are regulatory concerns about the country of origin of honey (Molan, 1998). Adulteration can involve the addition of cheaper sweeteners like corn syrup or the misrepresentation of the honey's floral or geographic origin. Such practices not only deceive consumers but also undermine the credibility of honey producers and can potentially pose health risks (Siddiqui et al., 2017).

Authenticating honey is therefore crucial to ensure that consumers receive the product they expect, with all its associated health benefits and unique properties. Numerous studies have been published aiming to develop reliable methods to determine the floral origin of honey (Çobanoğlu et al., 2023; Rodopoulou et al., 2018; Sahin et al., 2022; Tsigouri et al., 2004). The melissopalynological method is the only laboratory method that provides certainty about the honey variety. Pollen grains from different plant species have a distinctive shape that allows identification of the family, genus, or often individual species of the plant by microscopic examination of the pollen (Molan, 1998). This technique is based on the microscopic quantitative identification of plant pollen present in the honey (Dzugan et al., 2018). Melissopalynological analysis involves identifying the pollen grains belonging to different taxa in the honey at the family, genus, and sometimes species level, counting these pollen grains, and classifying the honey according to the percentage of major pollen grains. For certain honeys like sunflower, raspberry, and rape, the predominant pollen should constitute at least 45% of the total pollen grains (Siddiqui et al., 2017). For some honey types, such as thyme, lavender, myosotis, castanea, and citrus honey, this percentage may vary (Karabagias et al., 2014a, b). Some of the difficulties with this method are related to the fact that the person performing the analysis must have extensive experience and knowledge of pollen morphology (Dzugan et al., 2018). Honey contains pollen from not only nectariferous plants but also wind-pollinated plant species and some insect-pollinated plants that do not produce nectar (Moar, 1985). There are also some honeys whose botanical origin can not be determined with melissopalynological analysis (e.g., honeydew honeys) (Molan, 1998). Due to the limitations of determining the origin of honey by melissopalynology and the misleading results that can be obtained, other methods are being sought to determine the origin of honey. Additionally, the lack of standardized methods for honey authentication can lead to inconsistencies in results, making it difficult to establish a reliable and universally accepted system for honey identification.

Researchers are increasingly looking for a robust and reliable method to overcome these challenges in honey authentication. The qualitative and quantitative profile of phenolic compounds in unifloral honeys, which is closely linked to various health-promoting properties, serves as a valuable tool for verifying their origin. In recent years, phenolic compounds in unifloral honeys have been studied to identify components that can be used for this purpose, either individually or more effectively in combination. Additionally, early efforts to trace the geographical origin of honey using phenolic profile differences have been reported in the literature.

Therefore, phenolic compounds can serve as botanical markers to identify the floral source and geographical origin of honey samples. The content of different phenolic compounds is considered to reflect the quality of honey well and to characterize its different brands appropriately because phenolic acids and flavonoids are natural chemical markers of floral origin (Sergiel et al., 2014). Accurately identifying honey, ensuring its purity, increasing consumer confidence, and supporting the integrity of the honey market are possible by determining the phenolic profiles associated with botanical species.

2 Phenolic Compounds: An Overview

The association of diet and the existence of bioactive compounds in foods with health has attracted great interest in recent years. Based on this, consumers tend to prefer food products that strengthen physical properties, have an effect on diseases, and meet nutritional needs (de Araújo et al., 2019). Free radicals are formed during the energy cycle in the human body. In addition, radiation, pesticides, viruses, ultraviolet rays, heavy metals, contaminated foods, industrial wastes, and chemical residues are other sources of free radicals. These radicals are highly active and carry unpaired electrons in their outer orbits (Langseth, 1995; Sies, 1991). As a result of metabolic activities, reactive oxygen species, which are an important group of free radicals, are formed in the body. Superoxide, hydrogen peroxide, and hydroxyl groups are the most common radicals. There is a system that can neutralize free radicals formed in the body. However, oxidative stress occurs due to exposure to free radicals in excess of the rate that this system can neutralize. As a result of this excess oxidative stress, many diseases such as cancer, cell damage, DNA damage, cardiovascular problems, diabetes, aging, and gastrointestinal problems occur (Halliwell, 2002; Storz & Imlayt, 1999; Tsao & Deng, 2004). Polyphenols are naturally synthesized by plants and are of great interest in the scientific world due to their therapeutic effects on health (de Paulo Farias et al., 2020). It is stated that they can retain and inhibit reactive oxygen species, prevent oxidative stress and inflammation, and show important activities in the prevention of many diseases such as diabetes, wound healing, cancer, obesity, cardiovascular problems, and neurodegenerative diseases (de Araújo et al., 2021).

Polyphenols are a diverse group of bioactive compounds that are secondary metabolites of plants and represent phytochemicals that exert beneficial health effects. They have important physiological roles in plants such as pigmentation, reproduction, growth, and resistance to pathogens. In addition, they form compounds beneficial for health in human nutrition (Rana et al., 2022). With their specific chemical properties, phenolic compounds show important biological activities such as antioxidant activity. Structure-dependent activities are affected by the variability of the number of hydroxyl groups and the position of the carboxyl group. The effects are associated with radical scavenging and metal chelating activities (Vuolo et al., 2019). Polyphenols protect against oxidative stress through these activities. They also inhibit enzymes such as xanthine oxidase, protein kinase C, mitochondrial succinoxidase, nicotinamide adenine dinucleotide phosphate hydrogen (NADPH) oxidase, and lipoxygenase, which are free radical sources (Lobiuc et al., 2023). Phenolic compounds are mostly in esterified or glycolysed form and some of them are found in free form. The activation or inhibition of enzymes is a factor affecting the amount of production of phenolic compounds. In addition, environmental factors such as temperature, light, soil properties, and irrigation can be effective in polyphenol synthesis. With such a wide spectrum in plants, they are present in high concentrations in human nutrition through beverages and plant-based foods (de la Rosa et al., 2019). Polyphenols have been studied by isolating over 8000 various polyphenolic compounds with chemical structures ranging from simple to complex high molecular weight polymers (Neto et al., 2018). Phenolic compounds are extracted by different extraction techniques such as soxhlet, ultrasonic, high pressure, and microwave extraction and maceration (El-Saadony et al., 2024). Phenolic compounds are benzene ring structures containing one or more hydroxyl groups, synthesized through the pentose phosphate, shikimate, and phenylpropanoid pathways. These compounds can be synthesized from basic phenolic molecules to more complex, highly polymerized structures (El-Saadony et al., 2024; Mandal et al., 2010).

The health benefits of phenolic compounds are largely attributed to their structural diversity, which includes simple phenolic acids, flavonoids, and complex polymers (Pérez-Jiménez & Torres, 2011; Rocchetti et al., 2022). This structural variety allows them to interact with various biological targets, enhancing their bioactivity and therapeutic potential (Prabhu et al., 2021). Moreover, numerous studies demonstrate the beneficial effects of foods and beverages that are rich in phenolic compounds. The bioavailability and bioaccessibility of phenolic compounds are critical factors influencing their health benefits. During digestion, phenolics can interact with dietary fibers and other macronutrients, affecting their release and absorption in the gastrointestinal tract (Quirós-Sauceda et al., 2014; Vaz et al., 2022). For instance, the gastrointestinal digestion process can facilitate the release of phenolics from the food matrix, which is essential for their bioactivity (Zhu et al., 2024). However, these interactions can also lead to the masking of antioxidant properties, as phenolic compounds may bind to proteins and carbohydrates, thereby reducing their availability for absorption (Sęczyk et al., 2021; Nemli et al., 2024). This suggests that it is important to apply food processing techniques that can enhance the release and absorption of phenolic compounds (Altuner & Tokuşoğlu, 2013; Ribas-Agustí et al., 2019). Phenolic compounds can exhibit different biological activities such as inhibiting tumor cell division and cholesterol, modulating cyclooxygenase and lipooxygenase enzymes, affecting cell regulation and platelet interactions (Magiera et al., 2022; Zhang et al., 2021). At the same time, they show effects against degenerative and chronic diseases with their chemical composition (Rudrapal et al., 2022).

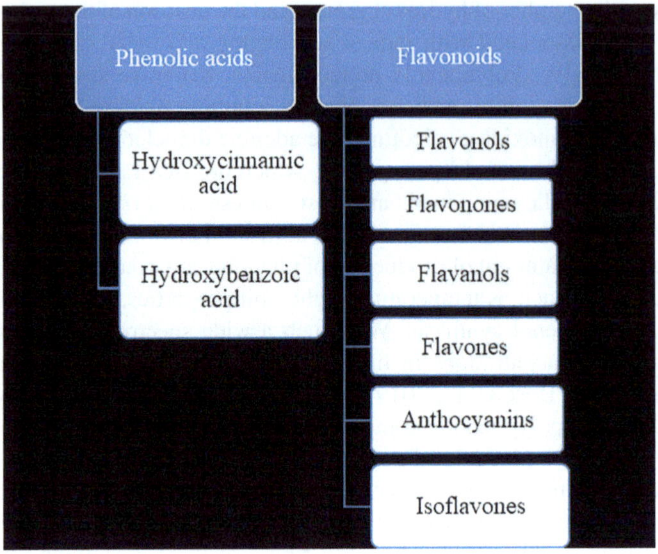

Fig. 3.1 Polyphenol classification

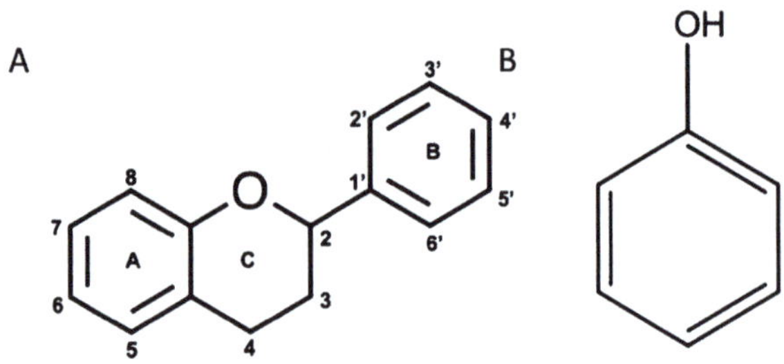

Fig. 3.2 Basic chemical structure of (**a**) flavonoid and (**b**) phenol groups (Al-Khalid & El-Naas, 2012; Uluad, 2017)

Phenolic compounds have various chemical properties such as water-soluble or insoluble in free form or conjugated with sugars and acids (Skrovankova et al., 2015). Based on their aglycone moieties and different chemical structures, phenolics are classified as phenolic acids, flavonoids, stilbenes, lignans, and xanthones (Figs. 3.1 and 3.2) (Vuolo et al., 2019). Lignans and stilbenes are a small class of phenolics that are usually found in edible foods in lower amounts. They are present in peanuts, berries, and grapes as trans-resveratrol (Hassan et al., 2022).

Flavonoids are polyphenolic compounds with two aromatic rings of 15 carbon atoms. Flavonoid biosynthesis is mainly initiated by the phenylpropanoid pathway (Kumar et al., 2023). Flavonoids are primarily based on an aglycone structure but are commonly found in glycoside forms. The biological activities of flavonoid glycosides, such as antioxidant activity, vary depending on their structural diversity and the pattern of glycosylation (El-Saadony et al., 2024). The chemical structure of flavonoids consists of a benzopyran skeleton, which includes an A and a B aromatic ring attached to a central heterocyclic pyran ring. There are six main categories of flavonoids: anthocyanins, isoflavones, flavonols, flavanones, and flavonols. (de la Rosa et al., 2019; Gómez-Caravaca et al., 2014; Song et al., 2022). Flavonoids play a crucial role in plants, contributing to color, taste, vitamin production, enzyme synthesis, and the inhibition of lipid peroxidation (Vuolo et al., 2019). They are produced in very high amounts in the leaf epidermis and fruit peel (Kumar et al., 2023). However, its amount varies depending on plant part, growing conditions, and degree of ripening (de la Rosa et al., 2019). In recent years, numerous studies have linked flavonoid consumption to various health benefits. It has been stated that consuming foods rich in flavonoids offers health benefits for individuals at risk of premature death (Bondonno et al., 2020). Additionally, it has been associated with reducing the risk of cancer and cardiovascular disease (Galatro et al., 2024). In a study, it was found that inflammation decreased with anthocyanin consumption (Cremonini et al., 2022), and in another study, green tea flavonoids were found to reduce diabetic nephropathy (Galatro et al., 2024).

Phenolic acids are natural phenolic compounds frequently found in fruits and vegetables (Karasawa & Mohan, 2018). Phenolic acids, which form part of phenolic compounds, are classified into hydroxycinnamic and hydroxybenzoic acid groups depending on the carbon skeleton form and the number and position of hydroxyl groups. Hydroxybenzoic acid (C6-C1) and its derivatives are characterized by one carboxyl group (COOH). Hydroxybenzoic acids can be conjugated with sugars, organic acids, or in cell-bound form. The most common hydroxybenzoic acid derivatives are gallic, procatechic, *p*-hydroxybenzoic, vanillic, salicylic, and elagic acid (Gómez-Caravaca et al., 2014; Rashmi & Negi, 2020). Hydroxycinnamic acids (C6-C3) are compounds characterized by a two-carbon skeletal structure that contains at least one hydrogen molecule, which can be replaced by a hydroxyl group (de Araújo et al., 2021; Heleno et al., 2015). Hydroxycinnamic acids are usually in conjugated form and are more commonly found in nature than hydroxybenzoic acids. *p*-Coumaric, cinnamic, caffeic, ferulic, sinapic, isoferulic, and *p*-hydroxycinnamic acid are hydroxycinnamic acid derivatives commonly found in nature (Rashmi & Negi, 2020). Phenolic acids have been associated with many biological functions such as antioxidant, antidepressive, antiulcer, antidiabetic, anti-inflammatory, anticancer, cardioprotective, neuroprotective, and antidiarrhoeal properties (de Araújo et al., 2021; Saibabu et al., 2015). In studies, it has been reported that phenolic acids have various health benefits such as antimicrobial, anti-inflammatory, antioxidant, antimperglycaemic, antiproliferative, and antiapoptotic effects (Rashmi & Negi, 2020; Semaming et al., 2015).

3 Phenolic Compounds in Honey

Honey contains small amounts of phenolics, which give it its color, sensory qualities, and many biological qualities like antibacterial, antioxidant, and radical-scavenging effects (Wang et al., 2022). Phenolics, including flavonoids and phenolic acids, represent the most important group of honey composition. The primary composition of honey polyphenols is flavonoids and their derivatives. When bees produce honey, they combine their bodily fluids with plant secretions or nectars, which are made up of water, carbohydrates, proteins, and phenolic compounds. This is why honey contains phenolic compounds (Jibril et al., 2019). The bioactive compounds have a major impact on the appearance of honey as well as its functional qualities (Alvarez-Suarez et al., 2010; Cianciosi et al., 2018; da Silva et al., 2016). In particular, the concentration of these antioxidants varies depending on the type of honey, the location and floral origins, honey age, the environment at the harvesting location, processing, and storage conditions (Becerril-Sánchez et al., 2021). According to studies, dark-colored honey has greater phenolic acid derivatives than light-colored honeys. However, dark-colored honey also contains fewer flavonoids (Hassan et al., 2022). Color variations are commonly observed in multifloral honeys, reflecting differences in the percentages of nectar used in their production. Occasionally, this variability in composition results in higher phenolic compound levels; however, some monofloral honeys contain even higher concentrations of phenolic compounds.

It was discovered that approximately 161 phenolics have been detected in honey (Lawag et al., 2022). The majority of these are flavonols, flavones, and hydroxybenzoic, and hydroxycinnamic acid derivatives (Lawag et al., 2023). The most common phenolic compounds in different types of honey are flavonoids such as catechin, chrysin, galangin, genistein, isorhamnetin, kaempferol, luteolin, myricetin, pinobanksin, pinocembrin, quercetin, rutin, and phenolic acids such as 2-cis, 4-trans abscisic, 2-hydroxycinnamic, caffeic, chlorogenic, cinnamic, ellagic, ferulic, gallic, *p*-coumaric, *p*-hydroxybenzoic, protocatechuic, sinapic, syringic, and vanillic acid (Cianciosi et al., 2018). Chrysin gives honeys additional value because it demonstrates many biological functions, such as antioxidant activity and anti-inflammatory activity in cell lines from lung, thyroid, breast, liver, and kidney cancer (Mani & Natesan, 2018).

4 Phenolic Compounds in Honey Authentication

The composition of phenolic compounds in honey can be affected by various factors, such as the bee variety, processing processes, seasonal and environmental factors, geographical origin, and the plant source from which the nectar was collected. The variations in the phenolic profiles of honeys from different botanical sources are directly linked to the natural diversity of these compounds in the plants from which they are derived. This variability forms the scientific foundation for two main research themes related to the study of the phenolic fraction in honey. The first approach focuses on evaluating the overall bioactive properties of honeys from different botanical and geographical origins. The second approach aims to determine the botanical and/or geographical origin of honey by identifying specific phenolic compounds as markers of origin based on their presence and abundance (Ciulu et al., 2016). These studies are important in both fields: honeys from different sources have shown various health benefits, and certain phenolic compounds have been suggested as potential markers for unifloral honeys. Studies have shown that specific phenolic compounds, particularly flavonoids and phenolic acids, are characteristic of certain floral types, making them valuable for distinguishing different honey varieties (Alvarez-Suarez et al., 2010; Hempattarasuwan et al., 2019). Compared to other honey floral source markers, phenolic compounds as floral markers are more common (J. Zhao et al., 2016).

In a study on the relationship between botanical origin and phenolic compound content of honey, it was reported that hesperedin, kaempferol, and quercetin could be markers of citrus, rosemary, and sunflower honey, respectively, in terms of botanical origin. In another study, it was reported that *p*-coumaric and *p*-hydroxybenzoic acids could be markers for buckwheat honey (Zhu et al., 2024). Numerous phenolic components have been found to serve as markers of the botanical origin of honey: hesperetin in citrus honey, coumaric and caffeic acids in chestnut honey, abscisic acid in strawberry tree honey, ellagic acid in acacia honey, and kaempferol in chestnut honey (Di Marco et al., 2018; Preti & Tarola, 2022).

There were several compounds found in Italian monofloral honeys that were possibly related to the flowers that produced them, such as caffeic acid, quercetin, and *p*-coumaric acid in *Castanea sativa* honey, myricetin in T*araxacum* sp. honey, apigenin, caffeic acid and chlorogenic acid in E*rica* sp. honey, and chlorogenic acid, quercetin, and *p*-coumaric acid in honeydew honey (Di Marco et al., 2018).

Many unifloral honeys from New Zealand (Stephens et al., 2010), Burkina Faso (Meda et al., 2005), Morocco (Petretto et al., 2017), Italy (Cavazza et al., 2013; Di Marco et al., 2018; Petretto et al., 2015; Preti & Tarola, 2022), India (Devi et al., 2018; Nayik & Nanda, 2016), Brazil (do Nascimento et al., 2018; Sousa et al., 2016), Bangladesh (Moniruzzaman et al., 2014), Australia (Goslinski et al., 2021; Moore et al., 2024), Poland (Jasicka-Misiak et al., 2012; Kedzierska-Matysek et al., 2021), Türkiye (Küçükaydın et al., 2023; Silici et al., 2014), Portugal (Escuredo et al., 2012), Romania (Paulic et al., 2020), Greece (Karabagias et al., 2016), and Serbia (Gasic et al., 2014) have been evaluated for their phenolic compounds' composition. Phenolic aicds and flavonoids are markers for the plant origin of some types of honeys. Table 3.1 reports a selection of phenolic compounds as chemical markers of the unifloral honey.

5 Methods for Analyzing Phenolic Compounds in Honey

When classifying honey according to its botanical origin, it is important to be supported by physicochemical and organoleptic properties together with melissopalynological analyses. Analytical methodologies are also utilized to determine botanical origin. The techniques include chromatographic, spectroscopic, and molecular biological methods. Physicochemical analyses can vary greatly between honey species and, in this case, a general interpretation of all results is needed to correlate botanical origin. Chromatographic methods allow the identification of some specific

Table 3.1 Dominant phenolics as a chemical marker found in several honey types

Phenolics as a chemical marker	Honey source	Geographical origin	Methodology	References
Ellagic acid and myricetin-3-*O*-methylether	Heather	Portugal	HPLC	Ferreres et al. (1996)
Rosmarinic acid	Thyme	Spain	CZE	Andrade et al. (1997)
Naringenin	Lavender			
Hesperetin	*Citrus*			
Homogentisic acid	(*Arbutus unedo*) Strawbery tree	Sardanian (Italy)	HPLC MS NMR	Cabras et al. (1999)
Myricetin, tricetin, and luteolin	*Eucalyptus*	Italy, Portugal, and Spain	HPLC-DAD	Martos et al. (2000)
Hesperetin	*Citrus*	Italy, France, Germany, Spain, Portugal, the Netherlands, and France	HPLC	Tomaanďs-Barberaanďn et al. (2001)
Kaempferol	Rosemary			
Quercetin	Sunflower			
Ellagic acid	Heather			
Caffeic, *p*-coumaric, ferulic acids	Chestnut, sunflower, lavender, and acacia			
Methyl syringate	*Asphodelus microcarpus*	Italy	HPLC-DAD HPLC-MS/MS	Tuberoso et al. (2009)
Hesperetin	*Citrus*		HPLC-DAD-UV	Cavazza et al. (2013)
Naringenin	*M. caesalpiniifolia*	Brazil	LC-ESI-MS	Silva et al. (2013)
Eriodictyol and Quercetin	Sunflower	Serbia	UHPLC-LTQ-Orbitrap-MS	Kečkeš et al. (2013)
Ellagic acid and myricetin-3-*O*-methylether	Heather	Portugal	HPLC	Karabagias, Vavoura, et al. (2014)
Kaempferol, chrysin, pinocembrin, caffeic acid, and naringenin	Rosemary	Spain	HPLC-DAD	Escriche et al. (2014)
Hesperetin	Citrus			
Myricetin, quercetin, galangin, and particularly *p*-coumaric acid	Honeydew			
Phenylacetic acid and dehydrovomifoliol	Heather	Poland	NMR spectroscopy	Zielinski et al. (2014)
Kaempferol	Sage (*Salvia officinalis*)	Caroatia	UHPLC-MS/MS Orbitrap	Gašić et al. (2015)
Apigenin, caffeic acid, chrysin, galangin, gallic acid, isorhamnetin, kaempherol, luteolin, myricetin, *p*-coumaric acid, pinocembrin, and quercetin	Acacia, honeydew, polyfloral, sunflower, and tilia	Romania	HPLC	Oroian and Ropciuc (2017)

(continued)

Table 3.1 (continued)

Phenolics as a chemical marker	Honey source	Geographical origin	Methodology	References
3-Phenyllactic acid, 2-methoxybenzoic acid, DL-*p*-hydroxyphenyllactic acid, and 2′-methoxyacetophenone	Manuka (*Leptospermum scopairum*)	New Zealand		McDonald et al. (2018)
Caffeic and coumaric acids, pinocembrin, chrysin, and luteolin	*Azara* spp.	Chile	HPLC-DAD UHPLC-MS/MS	Giordano et al. (2018)
Kaempferol, quercetin, isorhamnetin, rutin, apigenin	Rape	Romania	UHPLC-DAD-ESI/MS	Ciucure and Geană (2019)
Ferulic and *p*-hydroxybenzoic acids	Acacia			
p-Coumaric and chlorogenic acids	Polyfloral			
3,4-Dihydroxybenzoic, syringic and *trans*-cinnamic acids	Honeydew			
Rosmarinic acid	Prunella	China	HPLC-DAD/Q-TOF-MS	Wang et al. (2019)
Gallic, protocatechuic, 4-hydroxybenzoic, vanillic, chlorogenic, caffeic, *p*- coumaric, and rosmarinic acid, miricetin, quercetin, kaempferol	Mint, polyfloral, raspberry, rape, sunflower, and thyme	Romania	HPLC	Pauliuc et al. (2020)
Caffeic acid, datiscetin, and rhamnetin	Manuka honey	New Zealand	RP-UHPLC-ESI-MS	Gośliński et al. (2021)
Quinic acid, gentisic acid, 3-(2,5-dimethoxyphenyl) propanoicacid, 3-[[3-[(3S,4R,5S,6R)-3,4,5 trihydroxy-6 (hydroxymethyl)oxan-2 yl]phenoxy]methyl] benzoicacid	Greek honey	Greece	UPLC-QToF-MSCh	Koulis et al. (2021)
Chrysin, acacetin, sebabic acid, isorhamnetin	Polish	Poland		
Dehydrovomifoliol	Heather			
Gallic acid, *β*-resorcylic acid, and protocatechuic acid	Honeydew	Spain	LC-MS/MS	Vazquez et al. (2021)
4-anisaldehyde, 3-hydroxyphenylacetic acid, and 4-hydroxybenzaldehyde	Heather			
Methyl syringate	*Asphodelus microcarpus* (Zantaz)	Morocco	RP-HPLC LC/HRMS	Elamine et al. (2021)

(continued)

Table 3.1 (continued)

Phenolics as a chemical marker	Honey source	Geographical origin	Methodology	References
Safflomin A	*Carthamus tinctorius*	China	UHPLC/Q--TOF-MS	Zhao et al. (2022)
Caffeic acid and naringenin	Chestnut	Italy	HPLC-DAD	Preti and Tarola (2022)
Chrysin and coumaric acid	Honeydew			
Rutin, luteolin, and syringic and abscisic acids	Strawberry			
Hesperetin	Orange			
Coumaric, chlorogenic, and caffeic acid	Linden			
Taxifolin	Milkweed *Asclepias syriaca*	Hungary	HPLC-DAD-MS	Farkas et al. (2023)
Salicylic acid, syringic acid, naringenin, and galangin	Oak honeydew	Spain	HPLC-MS	Hernanz et al. (2023)
Syringic acid, vanillic acid, and coffee acid	Linden	Poland	HPLC	Puścion-Jakubik et al. (2022)
p-Coumaric acid and 4-hydroxybenzoic acid	Buckwheat			
Vanillic acid	Honeydew			
Protocatechuic acid, *p*-hydroxybenzoic acid, and benzoic acid	Pine	Türkiye	HPLC	Şahin et al. (2022)
Taxifolin	*Pinus* spp.	Türkiye	UHPLC-MS/MS	Çobanoğlu et al. (2023)
p-Hydroxybenzoic acid and *p*-coumaric acid	Buckwheat	America	LC-MS/MS	Zhu et al. (2024)
Hesperetin	Citrus			
Caffeic acid and isoferulic acid	Alfalfa			

minor components, making them more effective for honey classification (Pauliuc et al., 2020).

Analyzing phenolic compounds in honey is essential for understanding its quality, nutritional value, and potential health benefits. Phenolic compounds, including flavonoids and phenolic acids, contribute significantly to the antioxidant properties of honey, which are linked to various health-promoting effects (Cianciosi et al., 2018). Given the complex matrix of honey and the low concentrations of these compounds, several analytical methods have been developed to effectively extract, separate, and quantify phenolic compounds. The composition of phenolic compounds in honey varies depending on geographical and botanical origin. Phenolic compounds have a strong connection with the botanical origin of honey. In this respect, the determination of phenolic compounds is recommended as a marker for evaluating the geographical and botanical origin of honey (Becerril-Sánchez et al., 2021).

Previous studies have typically utilized one-dimensional liquid chromatography (1D-LC) combined with ultraviolet-visible spectroscopy (UV-Vis), fluorescence, and mass spectrometry (MS) detectors to analyze polyphenols in honey. However, the vast number of polyphenols with significant structural variability makes 1D-LC less advantageous. For this reason, there has been an interest in high-dimensional separation technologies in recent years. Better solubility technologies with higher resolving power attract attention. Especially, LC-LC and later UV-Vis detection have been successfully used for determining of polyphenols in a wide range of complex food matrices. This has led to better separation and resolution of major and minor polyphenols. However, it is noted that there is a need to develop methods for analyzing the polyphenolic composition of honeys, especially of unifloral origin (Vergara-Barberán et al., 2023). The analysis of individual honey samples for determining the phenolic content involves dilution, followed by an extraction clean-up step by solid phase extraction with ion exchange sorbents, and then a separation process including UV, diodide array detector (DAD), electrochemical, mass spectrometry, tandem MS (MS/MS) HPLC/ultra HPLC (UHPLC) analysis in reverse phase mode. Different trends such as the use of MS/MS detectors (ion trap, orbitrap, triple quadrupole, quadrupole time-of-flight, triple quadrupole ion trap mass spectrometer, and time-of-flight) that reduce the length of chromatographic runs such as UHPLC or increase method sensitivity and selectivity have been favored in HPLC in recent years. In addition, there is a study in which phenolic acids in honey powders were determined using GC-MS/MS (Valverde et al., 2022).

6 Conclusion

Phenolic compounds are among the most widely distributed secondary metabolites in nature. In recent years, the study of foods containing phenolic compounds has gained significant importance, primarily due to their beneficial effects on human health. Their presence in honey not only imparts biological and functional activity but also provides biomarkers that help determine the floral origin and some sensory characteristics. While the most effective method for identifying the phenol and flavonoid profiles in honey is still under evaluation, the phenolic composition is considered a potential marker of the botanical and geographical origin of honey. However, to confirm the identity of honey, not only the phenolic profile but also the phenolic profile analysis must be complemented by physicochemical, melissopalynological, and sensory analyzes.

Establishing a comprehensive global database of honey sample containing phenolic compounds is crucial for enhancing honey authentication, particularly for monofloral honey varieties. Such a database would allow the identification and comparison of characteristic markers, which are key indicators of honey origin and quality. Future research should not only focus on identifying these markers but also on quantifying their presence to ensure a more robust verification process. By incorporating both qualitative and quantitative analyses of these markers, we can develop a standardized system for routine monitoring, ensuring that honey products meet authenticity standards. This would protect both consumer interests and the integrity of the bee industry, safeguarding against adulteration and mislabeling while promoting transparency in the global honey market.

References

Al-Khalid, T., & El-Naas, M. H. (2012). Aerobic biodegradation of phenols: a comprehensive review. *Critical Reviews in Environmental Science and Technology, 42*(16), 1631–1690.

Altuner, E. M., & Tokuşoğlu, Ö. (2013). The effect of high hydrostatic pressure processing on the extraction, retention and stability of anthocyanins and flavonols contents of berry fruits and berry juices. *International Journal of Food Science & Technology, 48*(10), 1991–1997.

Alvarez-Suarez, J. M., González-Paramás, A. M., Santos-Buelga, C., & Battino, M. (2010). Antioxidant Characterization of Native Monofloral Cuban Honeys. *Journal of Agricultural and Food Chemistry, 58*(17), 9817–9824.

Andrade, P., Ferreres, F., Gil, M. I., & Tomás-Barberán, F. A. (1997). Determination of phenolic compounds in honeys with different floral origin by capillary zone electrophoresis. *Food Chemistry, 60*(1), 79–84.

Battino, M., Giampieri, F., Cianciosi, D., Ansary, J., Chen, X., Zhang, D., Gil, E., & Forbes-Hernández, T. (2021). The roles of strawberry and honey phytochemicals on human health: A possible clue on the molecular mechanisms involved in the prevention of oxidative stress and inflammation. *Phytomedicine : International Journal of Phytotherapy and Phytopharmacology, 86*.

Becerril-sánchez, A. L., Quintero-salazar, B., Dublán-garcía, O., & Escalona-buendía, H. B. (2021). Phenolic compounds in honey and their relationship with antioxidant activity, botanical origin, and color. *Antioxidants, 10*(11).

Bondonno, N. P., Lewis, J. R., Blekkenhorst, L. C., Bondonno, C. P., Shin, J. H., Croft, K. D., et al. (2020). Association of flavonoids and flavonoid-rich foods with all-cause mortality: The Blue Mountains Eye Study. *Clinical Nutrition, 39*(1), 141–150.

Cabras, P., Angioni, A., Tuberoso, C., Floris, I., Reniero, F., Guillou, C., & Ghelli, S. (1999). Homogentisic acid: A phenolic acid as a marker of strawberry-tree (Arbutus unedo) honey. *Journal of Agricultural and Food Chemistry, 47*(10), 4064–4067.

Cavazza, A., Corradini, C., Musci, M., & Salvadeo, P. (2013). High-performance liquid chromatographic phenolic compound fingerprint for authenticity assessment of honey. *Journal of the Science of Food and Agriculture, 93*(5), 1169–1175.

Cianciosi, D., Forbes-Hernández, T. Y., Afrin, S., Gasparrini, M., Reboredo-Rodriguez, P., Manna, P. P., et al. (2018). Phenolic compounds in honey and their associated health benefits: A review. *Molecules, 23*(9), 2322.

Ciucure, C. T., & Geană, E. I. (2019). Phenolic compounds profile and biochemical properties of honeys in relationship to the honey floral sources. *Phytochemical Analysis, 30*(4), 481–492.

Ciulu, M., Spano, N., Pilo, M. I., & Sanna, G. (2016). Recent Advances in the Analysis of Phenolic Compounds in Unifloral Honeys. *Molecules (Basel, Switzerland), 21*(4).

Çobanoğlu, D. N., Akyıldız, İ. E., Kızılpınar Temizer, İ., Damarlı, E., & Çelik, Ş. (2023). Phenolic compound, organic acid, mineral, and carbohydrate profiles of pine and blossom honeys. *European Food Research and Technology, 249*(6), 1503–1515.

Cremonini, E., Daveri, E., Iglesias, D. E., Kang, J., Wang, Z., Gray, R., et al. (2022). A randomized placebo-controlled cross-over study on the effects of anthocyanins on inflammatory and metabolic responses to a high-fat meal in healthy subjects. *Redox Biology, 51*, 102273.

da Silva, P. M., Gauche, C., Gonzaga, L. V., Costa, A. C. O., & Fett, R. (2016). Honey: Chemical composition, stability and authenticity. *Food Chemistry, 196*, 309–323.

de Araújo, F. F., Neri-Numa, I. A., de Paulo Farias, D., da Cunha, G. R. M. C., & Pastore, G. M. (2019). Wild Brazilian species of Eugenia genera (Myrtaceae) as an innovation hotspot for food and pharmacological purposes. *Food Research International, 121*, 57–72.

de Araújo, F. F., de Paulo Farias, D., Neri-Numa, I. A., & Pastore, G. M. (2021). Polyphenols and their applications: An approach in food chemistry and innovation potential. *Food Chemistry, 338*, 127535.

de la Rosa, L. A., Moreno-Escamilla, J. O., Rodrigo-García, J., & Alvarez-Parrilla, E. (2019). Phenolic compounds. In *Postharvest physiology and biochemistry of fruits and vegetables* (pp. 253–271). Woodhead Publishing.

de Paulo Farias, D., Neri-Numa, I. A., de Araujo, F. F., & Pastore, G. M. (2020). A critical review of some fruit trees from the Myrtaceae family as promising sources for food applications with functional claims. *Food Chemistry, 306*, 125630.

Di Marco, G., Gismondi, A., Panzanella, L., Canuti, L., Impei, S., Leonardi, D., & Canini, A. (2018). Botanical influence on phenolic profile and antioxidant level of Italian honeys. *Journal of Food Science and Technology, 55*, 4042–4050.

do Nascimento, K. S., Sattler, J. A. G., Macedo, L. F. L., González, C. V. S., de Melo, I. L. P., da Silva Araújo, E., et al. (2018). Phenolic compounds, antioxidant capacity and physicochemical properties of Brazilian Apis mellifera honeys. *Lwt, 91*, 85–94.

Dzugan, M., Tomczyk, M., Sowa, P., & Grabek-Lejko, D. (2018). Antioxidant Activity as Biomarker of Honey Variety. *Molecules (Basel, Switzerland), 23*(8).

Devi, A., Jangir, J., & Anu-Appaiah, K. A. (2018). Chemical characterization complemented with chemometrics for the botanical origin identification of unifloral and multifloral honeys from India. *Food Research International, 107*, 216–226.

Elamine, Y., Lyoussi, B., Miguel, M. G., Anjos, O., Estevinho, L., Alaiz, M., Girón-Calle, J., Martín, J., & Vioque, J. (2021). Physicochemical characteristics and antiproliferative and antioxidant activities of Moroccan Zantaz honey rich in methyl syringate. *Food Chemistry, 339, 128098.*

El-Saadony, M. T., Yang, T., Saad, A. M., Alkafaas, S. S., Elkafas, S. S., Eldeeb, G. S., et al. (2024). Polyphenols: Chemistry, bioavailability, bioactivity, nutritional aspects and human health benefits: A review. *International Journal of Biological Macromolecules, 277*(Pt. 3), 134223.

Escriche, I., Kadar, M., Juan-Borrás, M., & Domenech, E. (2014). Suitability of antioxidant capacity, flavonoids and phenolic acids for floral authentication of honey. Impact of industrial thermal treatment. *Food Chemistry, 142*, 135–143.

Escuredo, O., Silva, L. R., Valentão, P., Seijo, M. C., & Andrade, P. B. (2012). Assessing Rubus honey value: Pollen and phenolic compounds content and antibacterial capacity. *Food Chemistry, 130*(3), 671–678.

Farkas, Á., Horváth, G., Kuzma, M., Mayer, M., & Kocsis, M. (2023). Phenolic compounds in Hungarian acacia, linden, milkweed and goldenrod honeys. *Current Research in Food Science, 6*, 100526.

Ferreres, F., Andrade, P., & Tomás-Barberán, F. A. (1996). Natural Occurrence of Abscisic Acid in Heather Honey and Floral Nectar. *Journal of Agricultural and Food Chemistry, 44*(8), 2053–2056.

Galatro, A., Mas, A. L., Luquet, M., Fraga, C. G., & Galleano, M. (2024). Plants as a source of dietary bioactives: flavonoids and basis for their health benefits. *Aspects of Molecular Medicine*, 100048.

Gašić, U., Kečkeš, S., Dabić, D., Trifković, J., Milojković-Opsenica, D., Natić, M., & Tešić, Ž. (2014). Phenolic profile and antioxidant activity of Serbian polyfloral honeys. *Food Chemistry, 145*, 599–607.

Gašić, U. M., Natić, M. M., Mišić, D. M., Lušić, D. V., Milojković-Opsenica, D. M., Tešić, Ž. L., & Lušić, D. (2015). Chemical markers for the authentication of unifloral Salvia officinalis L. honey. *Journal of Food Composition and Analysis, 44*, 128–138.

Giordano, A., Retamal, M., Leyton, F., Martínez, P., Bridi, R., Velásquez, P., & Montenegro, G. (2018). Bioactive polyphenols and antioxidant capacity of Azara petiolaris and Azara integrifolia Honeys. *CyTA – Journal of Food, 16*(1), 484–489.

Gómez-Caravaca, A. M., Verardo, V., Segura-Carretero, A., Fernández-Gutiérrez, A., & Caboni, M. F. (2014). Phenolic compounds and saponins in plants grown under different irrigation regimes. In *Polyphenols in plants* (pp. 37–52). Academic Press.

Goślński, M., Nowak, D., & Szwengiel, A. (2021). Multidimensional comparative analysis of bioactive phenolic compounds of honeys of various origin. *Antioxidants, 10*(4), 530.

Halliwell, B. (2002). Effect of Diet on Cancer Development: Is Oxidative DNA Damage A Biomarker? *Free Radical Biology and Medicine, 32*(10), 968–974.

Hassan, N. H., Cacciola, F., Chong, N. S., Arena, K., Marriott, P. J., & Wong, Y. F. (2022). An updated review of extraction and liquid chromatography techniques for analysis of phenolic compounds in honey. *Journal of Food Composition and Analysis, 114*, 104751.

Heleno, S. A., Martins, A., Queiroz, M. J. R., & Ferreira, I. C. (2015). Bioactivity of phenolic acids: Metabolites versus parent compounds: A review. *Food Chemistry, 173*, 501–513.

Hempattarasuwan, P., Settachaimongkon, S., & Duangmal, K. (2019). Impact of botanical source and processing conditions on physicochemical properties and antioxidant activity of honey in the northern part of Thailand. *International Journal of Food Science & Technology, 54*(12), 3185–3195.

Hernanz, D., Jara-Palacios, M. J., Santos, J. L., Gómez Pajuelo, A., Heredia, F. J., & Terrab, A. (2023). The profile of phenolic compounds by HPLC-MS in Spanish oak (Quercus) honeydew honey and their relationships with color and antioxidant activity. *LWT, 180*, 114724.

Jasicka-Misiak, I., Poliwoda, A., Dereń, M., & Kafarski, P. (2012). Phenolic compounds and abscisic acid as potential markers for the floral origin of two Polish unifloral honeys. *Food Chemistry, 131*(4), 1149–1156.

Jiang, L., Xie, M., Chen, G., Qiao, J., Zhang, H., & Zeng, X. (2020). Phenolics and Carbohydrates in Buckwheat Honey Regulate the Human Intestinal Microbiota. *Evidence-based Complementary and Alternative Medicine, 2020*(1), 6432942.

Jibril, F. I., Hilmi, A. B. M., & Manivannan, L. (2019). Isolation and characterization of polyphenols in natural honey for the treatment of human diseases. *Bulletin of the National Research Centre, 43*(1), 4.

Karabagias, I. K., Badeka, A., Kontakos, S., Karabournioti, S., & Kontominas, M. G. (2014a). Characterization and classification of Thymus capitatus (L.) honey according to geographical origin based on volatile compounds, physicochemical parameters and chemometrics. *Food Research International, 55*, 363–372.

Karabagias, I. K., Vavoura, M. V., Nikolaou, C., Badeka, A. V., Kontakos, S., & Kontominas, M. G. (2014b). Floral authentication of Greek unifloral honeys based on the combination of phenolic compounds, physicochemical parameters and chemometrics. *Food Research International, 62*, 753–760.

Karabagias, I. K., Dimitriou, E., Kontakos, S., & Kontominas, M. G. (2016). Phenolic profile, colour intensity, and radical scavenging activity of Greek unifloral honeys. *European Food Research and Technology*, 242(8), 1201–1210.

Karasawa, M. M. G., & Mohan, C. (2018). Fruits as prospective reserves of bioactive compounds: a review. *Natural Products and Bioprospecting, 8*, 335–346.

Kečkeš, S., Gašić, U., Veličković, T. Ć., Milojković-Opsenica, D., Natić, M., & Tešić, Ž. (2013). The determination of phenolic profiles of Serbian unifloral honeys using ultra-high-performance liquid chromatography/high resolution accurate mass spectrometry. *Food Chemistry, 138*(1), 32–40.

Kędzierska-Matysek, M., Stryjecka, M., Teter, A., Skałecki, P., Domaradzki, P., & Florek, M. (2021). Relationships between the content of phenolic compounds and the antioxidant activity of polish honey varieties as a tool for botanical discrimination. *Molecules*, 26(6), 1810.

Koulis, G. A., Tsagkaris, A. S., Aalizadeh, R., Dasenaki, M. E., Panagopoulou, E. I., Drivelos, S., Halagarda, M., Georgiou, C. A., Proestos, C., & Thomaidis, N. S. (2021). Honey Phenolic Compound Profiling and Authenticity Assessment Using HRMS Targeted and Untargeted Metabolomics. *Molecules (Basel, Switzerland), 26*(9).

Küçükaydın, S., Tel-Çayan, G., Çayan, F., Taş-Küçükaydın, M., Eroğlu, B., Duru, M. E., & Öztürk, M. (2023). Characterization of Turkish Astragalus honeys according to their phenolic profiles and biological activities with a chemometric approach. *Food Bioscience, 53*, 102507.

Kumar, K., Debnath, P., Singh, S., & Kumar, N. (2023). An overview of plant phenolics and their involvement in abiotic stress tolerance. *Stress, 3*(3), 570–585.

Langseth, L. (1995). *Oxidants, antioxidants and disease prevention*. ILSI Press.

Lawag, I. L., Lim, L. Y., Joshi, R., Hammer, K. A., & Locher, C. (2022). A Comprehensive Survey of Phenolic Constituents Reported in Monofloral Honeys around the Globe. *Food, 11*(8), 1152.

Lawag, I. L., Islam, M. K., Sostaric, T., Lim, L. Y., Hammer, K., & Locher, C. (2023). Antioxidant activity and phenolic compound identification and quantification in Western Australian honeys. *Antioxidants, 12*(1), 189.

Lobiuc, A., Pavăl, N. E., Mangalagiu, I. I., Gheorghiţă, R., Teliban, G. C., Amăriucăi-Mantu, D., & Stoleru, V. (2023). Future antimicrobials: Natural and functionalized phenolics. *Molecules, 28*(3), 1114.

Magiera, A., Czerwińska, M. E., Owczarek, A., Marchelak, A., Granica, S., & Olszewska, M. A. (2022). Polyphenol-enriched extracts of Prunus spinosa fruits: Anti-inflammatory and antioxidant effects in human immune cells ex vivo in relation to phytochemical profile. *Molecules, 27*(5), 1691.

Mandal, S. M., Chakraborty, D., & Dey, S. (2010). Phenolic acids act as signaling molecules in plant-microbe symbioses. *Plant Signaling & Behavior, 5*(4), 359–368.

Mani, R., & Natesan, V. (2018). Chrysin: Sources, beneficial pharmacological activities, and molecular mechanism of action. *Phytochemistry, 145*, 187–196.

Martos, I., Ferreres, F., & Tomás-Barberán, F. A. (2000). Identification of flavonoid markers for the botanical origin of Eucalyptus honey. *Journal of Agricultural and Food Chemistry, 48*(5), 1498–1502.

McDonald, C. M., Keeling, S. E., Brewer, M. J., & Hathaway, S. C. (2018). Using chemical and DNA marker analysis to authenticate a high-value food, manuka honey. *npj Science of Food, 2*(1), 9.

Meda, A., Lamien, C. E., Romito, M., Millogo, J., & Nacoulma, O. G. (2005). Determination of the total phenolic, flavonoid and proline contents in Burkina Fasan honey, as well as their radical scavenging activity. *Food Chemistry, 91*(3), 571–577.

Moar, N. T. (1985). Pollen analysis of New Zealand honey. *New Zealand Journal of Agricultural Research, 28*(1), 39–70.

Molan, P. C. (1998). The limitations of the methods of identifying the floral source of honeys. *Bee World, 79*(2), 59–68.

Moore, G., Brooks, P., Pappalardo, L., & Boufridi, A. (2024). Phenolic profiles of Australian monofloral Eucalyptus, Corymbia, Macadamia and Lophostemon honeys via HPLC-DAD analysis. *Food Chemistry, 140900.*

Moniruzzaman, M., Yung An, C., Rao, P. V., Hawlader, M. N. I., Azlan, S. A. B. M., Sulaiman, S. A., & Gan, S. H. (2014). Identification of phenolic acids and flavonoids in monofloral honey from Bangladesh by high performance liquid chromatography: determination of antioxidant capacity. *BioMed research international, 2014*(1), 737490.

Nayik, G. A., & Nanda, V. (2016). A chemometric approach to evaluate the phenolic compounds, antioxidant activity and mineral content of different unifloral honey types from Kashmir, India. *LWT, 74*, 504–513.

Nemli, E., Ozkan, G., Gultekin Subasi, B., Cavdar, H., Lorenzo, J. M., Zhao, C., & Capanoglu, E. (2024). Interactions between proteins and phenolics: effects of food processing on the content and digestibility of phenolic compounds. *Journal of the Science of Food and Agriculture, 104*(5), 2535–2550.

Neto, M. L., Agra, K. L., Suassuna Filho, J., & Jorge, F. E. (2018). TDDFT calculations and photoacoustic spectroscopy experiments used to identify phenolic acid functional biomolecules in Brazilian tropical fruits in natura. *Spectrochimica Acta Part A: Molecular and Biomolecular Spectroscopy, 193*, 249–257.

Oroian, M., & Ropciuc, S. (2017). Honey authentication based on physicochemical parameters and phenolic compounds. *Computers and Electronics in Agriculture, 138*, 148–156.
Pauliuc, D., Dranca, F., & Oroian, M. (2020). Antioxidant Activity, Total Phenolic Content, Individual Phenolics and Physicochemical Parameters Suitability for Romanian Honey Authentication. *Foods (Basel, Switzerland), 9*(3).
Pérez-Jiménez, J., & Torres, J. L. (2011). Analysis of nonextractable phenolic compounds in foods: The current state of the art. *Journal of Agricultural and Food Chemistry, 59*(24), 12713–12724.
Petretto, G. L., Cossu, M., & Alamanni, M. C. (2015). Phenolic content, antioxidant and physicochemical properties of Sardinian monofloral honeys. *International Journal of Food Science & Technology, 50*(2), 482–491.
Petretto, G. L., Tuberoso, C. I. G., Fenu, M. A., Rourke, J. P., Belhaj, O., & Pintore, G. (2017). Antioxidant activity, color chromaticity coordinates, and chemical characterization of monofloral honeys from Morocco. *International Journal of Food Properties, 20*(9), 2016–2027.
Prabhu, S., Molath, A., Choksi, H., Kumar, S., & Mehra, R. (2021). Classifications of polyphenols and their potential application in human health and diseases. *International Journal of Physiology, Nutrition and Physical Education, 6*(1), 293–301.
Preti, R., & Tarola, A. M. (2022). Chemometric evaluation of the antioxidant properties and phenolic compounds in Italian honeys as markers of floral origin. *European Food Research and Technology, 248*(4), 991–1002.
Puścion-Jakubik, A., Karpińska, E., Moskwa, J., & Socha, K. (2022). Content of Phenolic acids as a marker of polish honey varieties and relationship with selected honey-quality-influencing variables. *Antioxidants (Basel, Switzerland), 11*(7).
Quirós-Sauceda, A. E., Palafox-Carlos, H., Sáyago-Ayerdi, S. G., Ayala-Zavala, J. F., Bello-Perez, L. A., Alvarez-Parrilla, E., et al. (2014). Dietary fiber and phenolic compounds as functional ingredients: interaction and possible effect after ingestion. *Food & Function, 5*(6), 1063–1072.
Rana, A., Samtiya, M., Dhewa, T., Mishra, V., & Aluko, R. E. (2022). Health benefits of polyphenols: A concise review. *Journal of Food Biochemistry, 46*(10), e14264.
Rashmi, H. B., & Negi, P. S. (2020). Phenolic acids from vegetables: A review on processing stability and health benefits. *Food Research International, 136*, 109298.
Ribas-Agustí, A., Martín-Belloso, O., Soliva-Fortuny, R., & Elez-Martínez, P. (2019). Influence of pulsed electric fields processing on the bioaccessible and non-bioaccessible fractions of apple phenolic compounds. *Journal of Functional Foods, 59*, 206–214.
Rocchetti, G., Gregorio, R. P., Lorenzo, J. M., Barba, F. J., Oliveira, P. G., Prieto, M. A., et al. (2022). Functional implications of bound phenolic compounds and phenolics–food interaction: A review. *Comprehensive Reviews in Food Science and Food Safety, 21*(2), 811–842.
Rodopoulou, M. A., Tananaki, C., Dimou, M., Liolios, V., Kanelis, D., Goras, G., & Thrasyvoulou, A. (2018). The determination of the botanical origin in honeys with over-represented pollen: combination of melissopalynological, sensory and physicochemical analysis. *Journal of the Science of Food and Agriculture, 98*(7), 2705–2712.
Rudrapal, M., Khairnar, S. J., Khan, J., Dukhyil, A. B., Ansari, M. A., Alomary, M. N., et al. (2022). Dietary polyphenols and their role in oxidative stress-induced human diseases: Insights into protective effects, antioxidant potentials and mechanism (s) of action. *Frontiers in Pharmacology, 13*, 806470.
Sahin, H., Ozkok, A., Tanugur Samanci, A. E., Onder, E. Y., & Kolayli, S. (2022). Identification of the Main Phenolic Markers in Turkish Pine Honeys and Their Biological Functions. *Chemistry & Biodiversity, 19*(12), e202200835.
Saibabu, V., Fatima, Z., Khan, L. A., & Hameed, S. (2015). Therapeutic potential of dietary phenolic acids. *Advances in Pharmacological and Pharmaceutical Sciences, 2015*(1), 823539.
Sęczyk, Ł., Sugier, D., Świeca, M., & Gawlik-Dziki, U. (2021). The effect of in vitro digestion, food matrix, and hydrothermal treatment on the potential bioaccessibility of selected phenolic compounds. *Food Chemistry, 344*, 128581.

Semaming, Y., Pannengpetch, P., Chattipakorn, S. C., & Chattipakorn, N. (2015). Pharmacological properties of protocatechuic acid and its potential roles as complementary medicine. *Evidence-based Complementary and Alternative Medicine, 2015*(1), 593902.

Sergiel, I., Pohl, P., Biesaga, M., & Mironczyk, A. (2014). Suitability of three-dimensional synchronous fluorescence spectroscopy for fingerprint analysis of honey samples with reference to their phenolic profiles. *Food Chemistry, 145*, 319–326.

Siddiqui, A. J., Musharraf, S. G., Choudhary, M. I., Rahman, A., & ur. (2017). Application of analytical methods in authentication and adulteration of honey. *Food Chemistry, 217*, 687–698.

Sies, H. (1991). Oxidative stress: From basic research to clinical application. *The American Journal of Medicine, 91*(3), S31–S38.

Silici, S., Sarioglu, K., Dogan, M., & Karaman, K. (2014). HPLC-DAD analysis to identify the phenolic profile of rhododendron honeys collected from different regions in Turkey. *International Journal of Food Properties, 17*(5), 1126–1135.

Silva, T. M. S., dos Santos, F. P., Evangelista-Rodrigues, A., da Silva, E. M. S., da Silva, G. S., de Novais, J. S., dos Santos, F., & de A. R., & Camara, C. A. (2013). Phenolic compounds, melissopalynological, physicochemical analysis and antioxidant activity of jandaíra (Melipona subnitida) honey. *Journal of Food Composition and Analysis, 29*(1), 10–18.

Skrovankova, S., Sumczynski, D., Mlcek, J., Jurikova, T., & Sochor, J. (2015). Bioactive compounds and antioxidant activity in different types of berries. *International Journal of Molecular Sciences, 16*(10), 24673–24706.

Song, H., Shen, X., Chu, Q., & Zheng, X. (2022). Pomegranate fruit pulp polyphenols reduce diet-induced obesity with modulation of gut microbiota in mice. *Journal of the Science of Food and Agriculture, 102*(5), 1968–1977.

Sousa, J. M., De Souza, E. L., Marques, G., Meireles, B., de Magalhães Cordeiro, Â. T., Gullón, B., et al. (2016). Polyphenolic profile and antioxidant and antibacterial activities of monofloral honeys produced by Meliponini in the Brazilian semiarid region. *Food Research International, 84*, 61–68.

Stephens, J. M., Schlothauer, R. C., Morris, B. D., Yang, D., Fearnley, L., Greenwood, D. R., & Loomes, K. M. (2010). Phenolic compounds and methylglyoxal in some New Zealand manuka and kanuka honeys. *Food Chemistry, 120*(1), 78–86.

Storz, G., & Imlayt, J. A. (1999). Oxidative stress. *Current Opinion in Microbiology, 2*(2), 188–194.

Tomaanđs-Barberaanđn, F. A., Martos, I., Ferreres, F., Radovic, B. S., & Anklam, E. (2001). HPLC flavonoid profiles as markers for the botanical origin of European unifloral honeys. *Journal of the Science of Food and Agriculture, 81*(5), 485–496.

Tsao, R., & Deng, Z. (2004). Separation Procedures for Naturally Occurring Antioxidant Phytochemicals. *Journal of Chromatography B, 812*, 85–99.

Tsigouri, A., Passaloglou-Katrali, M., & Sabatakou, O. (2004). Palynological characteristics of different unifloral honeys from Greece. *Grana, 43*(2), 122–128.

Tuberoso, C. I., Bifulco, E., JerkoviĆ , I., Caboni, P., Cabras, P., & Floris, I. (2009). Methyl syringate: a chemical marker of asphodel (Asphodelus microcarpus Salzm. et Viv.) monofloral honey. *Journal of Agricultural and Food Chemistry*, 57(9), 3895–3900.

Uluad, A. (2017). Antioxidant capacity assay and analysis of phenolics in artichoke (Cynara scolymus L.) leaf. In *Institute of Science, Chemistry Department* (pp. 1–54). Uludağ University.

Valverde, S., Ares, A. M., Elmore, J. S., & Bernal, J. (2022). Recent trends in the analysis of honey constituents. *Food Chemistry, 387*, 132920.

Vaz, A., Odriozola-Serrano, I., Oms-Oliu, G., & Martín-Belloso, O. (2022). Physicochemical properties and bioaccessibility of phenolic compounds of dietary fibre concentrates from vegetable by-products. *Food, 11*(17), 2578.

Vazquez, L., Armada, D., Celeiro, M., Dagnac, T., & Llompart, M. (2021). Evaluating the presence and contents of phytochemicals in honey samples: Phenolic compounds as indicators to identify their botanical origin. *Food, 10*(11).

Vergara-Barberán, M., Lerma-García, M. J., Simó-Alfonso, E. F., & García-Alvarez-Coque, M. C. (2023). Use of polyphenolic fingerprints established by comprehensive two-dimensional

liquid chromatography for the classification of honeys according to their floral origin. *Journal of Chromatography A, 1705*, 464138.
Vîjan, L. E., Mazilu, I. C., Enache, C., Enache, S., & Topală, C. M. (2023). Botanical Origin Influence on Some Honey Physicochemical Characteristics and Antioxidant Properties. *Foods (Basel, Switzerland), 12*(11).
Vuolo, M. M., Lima, V. S., & Junior, M. R. M. (2019). Phenolic compounds: Structure, classification, and antioxidant power. In *Bioactive compounds* (pp. 33–50). Woodhead Publishing.
Wang, K., Wan, Z., Ou, A., Liang, X., Guo, X., Zhang, Z., et al. (2019). Monofloral honey from a medical plant, Prunella Vulgaris, protected against dextran sulfate sodium-induced ulcerative colitis via modulating gut microbial populations in rats. *Food & Function, 10*(7), 3828–3838.
Wang, X., Chen, Y., Hu, Y., Zhou, J., Chen, L., & Lu, X. (2022). Systematic review of the characteristic markers in honey of various botanical, geographic, and entomological origins. *ACS Food Science & Technology, 2*(2), 206–220.
Zammit Young, G. W., & Blundell, R. (2023). A review on the phytochemical composition and health applications of honey. *Heliyon, 9*(2), e12507.
Zhang, S., Li, L., Chen, W., Xu, S., Feng, X., & Zhang, L. (2021). Natural products: The role and mechanism in low-density lipoprotein oxidation and atherosclerosis. *Phytotherapy Research, 35*(6), 2945–2967.
Zhao, J., Du, X., Cheng, N., Chen, L., Xue, X., Zhao, J., Wu, L., & Cao, W. (2016). Identification of monofloral honeys using HPLC–ECD and chemometrics. *Food Chemistry, 194*, 167–174.
Zhao, L., Ren, C., Xue, X., Lu, H., Wang, K., & Wu, L. (2022). Safflomin A: A novel chemical marker for Carthamus tinctorius L. (Safflower) monofloral honey. *Food Chemistry, 366*, 130584.
Zhu, Y., Liu, R., Mats, L., Zhu, H., Roasa, J., Khan, T., Ahmed, A., Brummer, Y., Cui, S., Sievenpiper, J., Dan Ramdath, D., & Tsao, R. (2024). A comprehensive characterization of phenolics, amino acids and other minor bioactives of selected honeys and identification of botanical origin markers. *Journal of Food Bioactives, 25*, 25–41.
Zieliński, Ł., Deja, S., Jasicka-Misiak, I., & Kafarski, P. (2014). Chemometrics as a tool of origin determination of Polish monofloral and multifloral honeys. *Journal of Agricultural and Food Chemistry, 62*(13), 2973–2981.

Chapter 4
Microbial Analysis in Honey Quality Assurance

Mehtap Usta

Abbreviations

NIR	Near-infrared spectroscopy
PCR	Polymerase chain reaction
PCA	Principal component analysis
CA	Cluster analysis
NGS	Next-generation sequencing
ELISA	Enzyme-linked immunosorbent assay
FAO	Food and Agriculture Organization of the United Nations
WHO	World Health Organization
USDA	United States Department of Agriculture
EU	European Union
FDA	Food and Drug Administration
FSSAI	The Food Safety and Standards Authority of India
HACCP	Hazard Analysis and Critical Control Points
AI	Artificial intelligence
NMR	Nuclear magnetic resonance

1 Introduction to Honey Quality Assurance

Honey, as one of nature's purest and oldest flavorings, has played an important role in human consumption for thousands of years. Known for both its nutritional value and health benefits, this valuable product also plays an important role in various

M. Usta (✉)
Tonya Vocational School, Trabzon University, Trabzon, Türkiye
e-mail: mehtapyakupoglu@trabzon.edu.tr

M. Zafar et al. (eds.), *Pure Honey: Assurance & Authentication*,
https://doi.org/10.1007/978-3-031-98913-1_4

cultural and religious practices. However, the increasing demands of modern consumers and the expansion of international trade require stricter control of the quality of honey (Mădaş et al., 2020; Bogdanov et al., 1999). Honey quality assurance involves a series of procedures aimed at maintaining the purity, naturalness, and nutritive value of honey from production to the final consumer. This process involves rigorous monitoring and evaluation of the chemical, physical, and microbial properties of honey (El Sohaimy et al., 2015). Factors affecting the quality of honey include geographical origin, plant source, processing and storage conditions, and, most importantly, microbial contamination. Microbial analysis has a critical role in the evaluation of honey quality. Although the inherently low water activity and acidic pH value of honey prevent the growth of some microorganisms, inappropriate conditions, especially in processing and storage processes, can lead to the growth of harmful microorganisms (Majtan et al., 2021; Mădaş et al., 2020). Therefore, microbial analysis methods are an integral part of the honey production chain to ensure that honey is safe and of high quality. Global standards for honey quality have been set by various international and national organizations to protect consumer health and enhance the commercial value of honey (De Beer et al., 2021). The standards require the limitation of microbial contaminants that may be present in honey and the fulfillment of certain quality criteria. Honey producers are continuously improving their quality assurance processes to comply with these standards and to provide a reliable product to the consumer. Consequently, honey quality assurance is vital not only for consumer satisfaction but also for product safety and marketing. Maintaining and improving the quality of honey provides a competitive advantage in both local and international markets and supports the sustainability of the honey industry (Bogdanov et al., 1999; De Beer et al., 2021).

2 Microbial Contamination in Honey

Although honey is considered an inherently low-microbial-risk food, the risk of microbial contamination from various sources cannot be completely ignored. Microbial contamination can degrade the quality of honey, lead to health risks, and negatively affect the commercial value of the product (Al-Waili et al., 2011, 2012). Therefore, understanding and controlling microbial contamination in honey production is critical to deliver a quality and safe product (Snowdon & Cliver, 1996).

2.1 Types of Microorganisms Found in Honey

Microorganisms that may be present in honey are usually classified as bacteria, yeasts, and molds. However, the natural structure of honey (low water activity and high sugar content) inhibits the growth of most pathogenic microorganisms.

Nevertheless, some resistant bacteria and spores can be present in honey and can multiply under certain conditions (Silva et al., 2017).

2.1.1 Bacteria

Bacillus and *Clostridium* species are among the most common bacteria found in honey. *Clostridium botulinum* spores in particular pose a serious health risk for infants, and therefore, honey should not be given to infants under one year of age (Devi et al., 2021; Martín-González et al., 2023).

2.1.2 Yeasts

Osmophilous yeasts can survive in honey despite its low water activity and can ferment and spoil the quality of the honey. Such yeasts can cause honey to become alcoholic and thus spoil it (Grabowski & Klein, 2017; Silva et al., 2017; Fratianni, 2024).

2.1.3 Molds

Mold spores can also be present in honey, but usually these spores do not become active when proper storage conditions are provided. However, mold contamination can grow on the surface of honey, especially in humid environments (Kačániová et al., 2009; Klutse & Adotey, 2022).

2.2 Sources of Microbial Contamination

Microbial contamination can occur at various stages of the honey production process (Adadi & Obeng, 2017; Guran et al., 2021):

2.2.1 Environmental Factors

Pollen from flowers, soil, and water sources can cause microbial contamination of honey. In addition, the hygiene conditions of the environment where the bees collect the honey can also increase this risk (Savarino et al., 2020; Dümen et al., 2013; Klutse & Adotey, 2022).

2.2.2 Processing and Storage

Unhygienic equipment used during honey processing increases the risk of microbial contamination. In addition, storage of honey in unsuitable conditions (e.g., in high humidity environments) may favor the proliferation of microorganisms (Scepankova et al., 2022; Khalil et al., 2012).

2.2.3 Human Factor

Failure of workers to follow hygiene rules during the processing and packaging of honey can lead to contamination of the honey. Therefore, compliance with food safety protocols is of great importance (Adadi & Obeng, 2017; Guran et al., 2021):

2.3 Impact of Microbial Contamination on Honey Quality

Microbial contamination can affect both the physical and chemical properties of honey. Fermentation can impair the taste and flavor of honey, which may lead to rejection of the product by the consumer (Mustafina et al., 2015). Furthermore, some microorganisms can alter the chemical composition of honey, reducing its nutritional value. Most importantly, the presence of pathogenic microorganisms can lead to serious health risks with the consumption of honey (Sinacori et al., 2014). Microbial contamination of honey must be meticulously monitored and regulated throughout the honey manufacturing process. Adherence to essential hygiene practices, routine microbiological testing, and appropriate storage conditions are crucial in mitigating these dangers (Scepankova et al., 2021).

3 Methods for Microbial Analysis in Honey

Since microbial contamination can directly affect honey quality and safety, accurate and reliable microbial analysis of honey is of great importance. Microbial analysis involves a combination of methods used to determine the type, quantity, and potential effects of microorganisms present in honey. These analyses can range from traditional methods to modern molecular techniques (Sinacori et al., 2014).

3.1 Traditional Microbial Testing Techniques.

Traditional microbial analysis methods are usually based on culturing microorganisms and include classical laboratory techniques such as microscopy. These methods have been in use for many years and allow direct observation and counting of microorganisms (Pimentel et al., 2013).

3.1.1 Culture-Based Methods

These methods involve the production of microorganisms by incubating honey samples on appropriate media. It is commonly used for the identification of bacterial and fungal species. This method results in the counting of colonies to determine the number of microorganisms. However, it can only detect culturable microorganisms and may miss some obligate anaerobic or difficult culturable species (Pimentel et al., 2013; Pomastowski et al., 2019; Gomes et al., 2010).

3.1.2 Microscopy

Microscopy allows direct observation of the microorganisms present in honey. For example, the morphology of bacteria can be examined with the Gram staining technique. This method provides a quick and visual analysis but does not give precise information about the type of microorganism (Wen et al., 2017).

3.2 *Modern Analytical Techniques.*

Due to the limitations of traditional methods, modern molecular and biochemical techniques are increasingly preferred for microbial analysis in honey. These methods provide faster, more sensitive, and culture-independent results (Aykas, 2023).

3.2.1 Polymerase Chain Reaction (PCR)

PCR enables the identification of microorganisms present in honey by amplifying their DNA. This method is particularly effective in the detection of pathogens and allows even low amounts of microorganisms to be detected. With PCR, dangerous pathogens such as *Clostridium botulinum* can be quickly detected in honey (Soares et al., 2018; Jain et al., 2013; Chin & Sowndhararajan, 2020).

3.2.2 Next-Generation Sequencing (NGS)

NGS is used to comprehensively analyze the microbial profile of honey. This method can determine the DNA sequences of all microorganisms present in a sample. NGS is a powerful tool to explore the diversity and abundance of honey's microorganism communities. It also allows the detection of rare microorganisms and difficult culturable species (Puścion-Jakubik et al., 2020; Chin & Sowndhararajan, 2020; Tsagkaris et al., 2021).

3.2.3 Enzyme-Linked Immunosorbent Assay (ELISA)

ELISA is a biochemical method used for the detection of specific pathogens or toxins. This technique is based on antibody-antigen interactions and can quickly determine the presence of certain pathogens in honey. It is particularly used for the detection of hazardous substances such as *Clostridium botulinum* toxins in honey (Joe et al., 2018; Valachová et al., 2016).

3.3 Advantages and Limitations of Different Methods

Each microbial analysis method has its advantages and limitations:

3.3.1 Culture-Based Methods

These low-cost and direct observation methods are effective for culturable microorganisms. However, they are time-consuming and may fail to detect difficult culturable microorganisms (Pimentel et al., 2013; Pomastowski et al., 2019; Gomes et al., 2010).

3.3.2 PCR

Fast and sensitive, PCR is particularly effective in the detection of certain microorganisms. However, its technical complexity and cost may limit its widespread use (Soares et al., 2018; Jain et al., 2013; Chin & Sowndhararajan, 2020).

3.3.3 NGS

Providing very comprehensive and detailed analyses, NGS is ideal for discovering microorganism diversity. However, its high cost and data analysis difficulties may be barriers to routine use (Puścion-Jakubik et al., 2020; Chin & Sowndhararajan, 2020; Tsagkaris et al., 2021).

3.3.4 ELISA

ELISA is fast and specific; it is useful for the detection of specific pathogens. However, it can only be applied to previously known pathogens, which prevents the detection of unknown microorganisms (Joe et al., 2018; Valachová et al., 2016).

3.4 Selection of Appropriate Method for Honey Analysis

Which method to use for microbial analysis of honey depends on the purposes of the analysis and the resources available. In general, traditional methods may be sufficient for routine screening and quantification, while modern techniques such as PCR and NGS are preferred for more detailed and specific analyses. It is also important to select appropriate analytical methods according to the target markets and regulatory requirements for honey (Chin & Sowndhararajan, 2020; Orfanakis et al., 2021).

4 Regulatory Standards and Microbial Limits

Honey appeals to a wide range of consumers as a food with both high nutritional and medicinal value. However, in order to protect the health of consumers and ensure the reliability of trade, strict regulations and standards are applied to the microbial quality of honey. These regulations include specific microbial limits to ensure the safe and quality production and distribution of honey (Malavi et al. 2018).

4.1 International Guidelines for Microbial Content in Honey

Regulations on the microbial content of honey worldwide are generally based on standards set by international and national organizations. These standards specify the maximum permissible levels of contamination of honey for various types of microorganisms and state that if these limits are exceeded, the product is commercially unsuitable (Singh & Singh, 2018; Ng et al., 2023).

4.1.1 Codex Alimentarius

Created by the Food and Agriculture Organization of the United Nations (FAO) and the World Health Organization (WHO), Codex Alimentarius sets international standards for many food products, including honey. According to Codex standards, certain pathogens and toxins are not allowed in honey. For example, *Clostridium botulinum* spores are considered unacceptable in honey (Ng et al., 2023; Sommano et al., 2020; Pauliuc et al., 2020).

4.1.2 European Union (EU) Standards

The European Union has established its own regulations regarding the microbial quality of honey. EU regulations contain strict rules limiting the presence and quantity of certain microorganisms in honey. In particular, due to the risk of infant botulism, honey must be free of *Clostridium botulinum* spores (Raweh et al., 2023; Kędzierska-Matysek, 2023).

4.1.3 United States Standards

In the United States, the microbial quality of honey is regulated by the United States Department of Agriculture (USDA) and the Food and Drug Administration (FDA). These organizations set the microbial limits required for the safe consumption of honey. In the classification system established by the USDA, microbial contamination is considered among the factors affecting the quality of honey (Musa et al., 2014; Wen et al., 2017; Gomes et al., 2010).

4.2 Microbial Limits Set by Different Countries

Microbial limits set for honey in different parts of the world may vary depending on local food safety concerns, climatic conditions, and production practices. Some countries may set stricter limits for certain microorganisms in honey, while others may adopt more flexible approaches (Hegazı et al. 2022).

4.2.1 Japan

Japan is very particular about the quality of honey and has set strict limits for microbial contamination in honey. In particular, honey must be tested for bacterial and yeast contamination (Wen et al., 2017).

4.2.2 Australia and New Zealand

Australia and New Zealand are known as countries with high standards in honey production. Manuka honey produced in these countries is recognized for its antimicrobial properties, and therefore special tests and standards are applied beyond microbial limits (Cokcetin et al., 2016).

4.2.3 India

India, as a major honey-producing country, has developed its own standards for the microbial quality of honey. The Food Safety and Standards Authority of India (FSSAI) has set limits for microbial contamination that may be present in honey and requires honey producers to comply with these standards (Saxena et al., 2010).

4.3 Compliance and Certification Processes

To ensure that honey conforms to national and international standards, honey producers and importers should regularly carry out microbial analyses and submit the results of these analyses to the competent authorities. This process is critical to improve the marketability of honey and consumer confidence (Testa et al., 2019; Palmieri et al., 2022).

4.3.1 Certification Schemes

Honey producers can participate in certain certification schemes to certify their products. For example, organic honey producers can obtain certificates showing that their products meet certain microbial and chemical limits (Raweh et al., 2023; Odekunle, 2023).

4.3.2 Quality Assurance Programs.

In some countries, honey producers must comply with food safety management systems (e.g., HACCP—Hazard Analysis and Critical Control Points) to prevent microbial contamination. These programs identify potential points of contamination in the production process and ensure that necessary precautions are taken (Odekunle, 2023).

4.3.3 Market Access

Enhance international marketing value and supply chain management of honey in a broader context. Therefore, honey exporters must understand the microbial standards of the target market and produce in accordance with these standards. Otherwise, the import of the product may be refused or recalled (Pippinato et al., 2020).

4.4 *Challenges in Meeting Regulatory Standards*

Compliance with microbial standards can be challenging, especially for small-scale producers. Consistently maintaining hygiene standards in production and processing, maintaining appropriate storage conditions, and conducting regular microbial analyses can be both time-consuming and costly. However, if these standards are not complied with, serious risks may arise for both the producer and the consumer (Soares et al., 2018; Odekunle, 2023).

5 Microbial Risks in Honey Production

Honey production, while aiming to obtain a natural and pure product, faces the risk of microbial contamination at various stages of the production process. Microbial risks can directly affect the quality and safety of honey, which can have serious consequences for both the producer and the consumer. Understanding and managing microbial risks in honey production is critical to deliver a safe and quality product (Silva et al., 2017; Grabowski & Klein, 2017; Wang et al., 2023).

5.1 *Common Microbial Hazards*

The most common microbial hazards encountered during honey production include bacteria, yeasts, and molds. These microorganisms can reduce the quality of honey and, in some cases, cause health risks (Mandal & Mandal, 2011).

5.1.1 Bacteria

One of the most common bacterial hazards in honey is *Clostridium botulinum* spores, which can cause infant botulism. *Bacillus* species and other spore-forming bacteria may also be present in honey and can multiply under unsuitable conditions (Nzeh et al., 2022).

5.1.2 Yeasts

Osmophilous yeasts can survive despite the low water activity of honey and ferment, leading to honey spoilage. Fermentation changes the flavor and odor of honey and shortens its shelf life (Naila et al., 2022).

5.1.3 Molds

Mold spores can be present in honey and can grow, especially in high humidity conditions. The growth of molds can form a moldy layer on the surface of honey, making it inedible (Scepankova, 2024).

5.2 Honey Harvesting and Processing Best Practices

Care must be taken at every stage of honey production to minimize microbial risks. This process starts from the beehives and continues until the honey is processed and packaged.

5.2.1 Hive Management

Keeping beehives in hygienic conditions reduces the risk of microbial contamination. Regular cleaning of hives and control of diseases are important to maintain the purity of the honey. In addition, the cleanliness of water sources around the hives should be carefully monitored (Negash et al., 2020; Warui et al., 2019).

5.2.2 Honey Extraction

Hygienic equipment used during honey harvesting plays a critical role in preventing microbial contamination. Equipment made of stainless steel provides a less favorable environment for the growth of microorganisms. In addition, honey must be protected from external factors such as excessive moisture and dust during harvesting (Grace et al., 2020).

5.2.3 Processing and Filtration

Filtration methods used during honey processing help to remove impurities and potential microorganisms from honey. However, it is also important to avoid excessive filtration to preserve the natural structure of honey. Also, not exposing the honey to excessive heat during processing helps to preserve the natural enzymes and microbial balance of the honey (Akbari et al., 2020; Preti & Tarola, 2021).

5.3 *Role of Temperature and Storage in Microbial Growth*

The microbial quality of honey also depends on storage conditions after processing. Suitable storage conditions can inhibit microbial growth and extend the shelf life of honey.

5.3.1 Temperature Control.

Since honey has naturally low water activity, the growth of microorganisms is usually limited. However, when exposed to high temperatures, fermentation of honey can begin, and the proliferation of yeasts can accelerate. Therefore, it is important to store honey in a cool and dry place (Dallagnol et al., 2022; Palmer-Young et al., 2023).

5.3.2 Humidity Management

The moisture content of honey directly affects microbial activity. High humidity can encourage the growth of molds and yeasts. Therefore, low humidity levels should be maintained in areas where honey is stored. In addition, a tightly closed lid prevents moisture in the air from reaching the honey (Baroyi et al., 2019).

5.3.3 Shelf Life Considerations

The shelf life of honey can be quite long when proper storage conditions are provided. However, over time, there may be changes in the microbial balance of honey, which may affect the quality of honey. To prolong the shelf life, it is recommended to store honey in a cool and dark place (Baroyi et al., 2019).

5.4 *Monitoring and Controlling Microbial Risks*

Continuous monitoring and control procedures must be in place to effectively manage microbial risks in honey production. Food safety management systems provide guidance for monitoring microorganisms and preventing contamination as part of this process.

5.4.1 Routine Testing

Honey producers should check the safety and quality of honey by performing microbial analyses on a regular basis. These analyses are necessary to confirm that honey is safe from microbial contamination (Savarino et al., 2020; Lima et al., 2020).

5.4.2 Hygiene Protocols

Implementing strict hygiene protocols in production facilities helps prevent microbial contamination. Attention to personal hygiene of employees and regular disinfection of equipment improves the safety of honey (Xiong et al., 2023).

5.4.3 Corrective Actions

When microbial contamination is detected, prompt and effective corrective actions should be taken. Separating contaminated honey and removing potential sources of contamination reduces future risks (Lima et al., 2020).

6 Impact of Microbial Contamination on Honey Properties

Microbial contamination can adversely affect the physical, chemical, and sensory properties of honey. These changes can reduce the quality and safety of honey, decrease consumer satisfaction, and significantly reduce the commercial value of honey. Therefore, understanding and managing the adverse effects of microbial contamination on honey is of great importance (Scepankova et al., 2022).

6.1 Changes in Physical Properties

The physical properties of honey can change significantly as a result of microbial activity. These changes can affect the appearance, consistency, and shelf life of honey.

6.1.1 Texture and Consistency

Microbial contamination can affect the consistency of honey. For example, carbon dioxide produced by the fermentation of osmophilous yeasts can cause honey to become foamy. Also, the fermentation process can change the consistency of honey by making it more liquid or increasing the rate of crystallization (Galhardo et al., 2020; Xiong et al., 2023).

6.1.2 Color

Microbial activity can also lead to changes in the color of honey. Fermentation or mold growth can cause honey to become cloudy or darker in color. These color changes can give consumers a negative impression of the freshness and purity of honey (Xiong et al., 2023).

6.1.3 Viscosity

The viscosity of honey can vary depending on the amount of water it contains and microbial activity. Fermentation can cause honey to become more viscous, which can adversely affect its quality and shelf life. High viscosity can lead to sweetening of honey, while low viscosity can be a sign of fermentation (Musa et al., 2014; Xiong et al., 2023).

6.2 Chemical Composition Alterations

Microbial contamination can lead to changes in the chemical composition of honey. These changes may affect the nutritional value and biochemical properties of honey.

6.2.1 pH and Acidity

Microbial activity can lower the pH level and increase the acidity of honey. For example, organic acids produced by yeasts during the fermentation process can strengthen the acidic nature of honey. Increased acidity can adversely affect the flavor and shelf life of honey (García-Tenesaca et al., 2017; Ramos et al., 2018; Álvarez-Suárez et al., 2014).

6.2.2 Sugar Composition

The sugars contained in honey can be altered by the action of microbial enzymes, especially during the fermentation process. Fermentation can lead to the conversion of simple sugars such as glucose and fructose into alcohol and carbon dioxide. This process causes marked changes in the flavor profile of honey and may reduce its nutritional value (Kaškonienė & Venskutonis, 2010; Brčina et al., 2021).

6.2.3 Enzymatic Activity

Microbial contamination can affect the activity of natural enzymes present in honey. In particular, the enzyme glucose oxidase produces hydrogen peroxide, which supports the natural antibacterial properties of honey. However, microbial activity can reduce the activity of these enzymes and weaken the natural protective properties of honey (Meister et al., 2021; İnanç, 2020; Erwan et al., 2020).

6.3 Deterioration of Sensory Qualities

The sensory properties of honey can be impaired by microbial contamination. These changes can adversely affect consumer perception by affecting the odor, taste, and texture of honey.

6.3.1 Flavor and Taste.

Microbial fermentation can cause sharp and unpleasant changes in the flavor profile of honey. Fermented honey can take on a sour or alcoholic taste, which masks the natural sweet and aromatic properties of honey. Consumers generally do not welcome such flavor changes, and this reduces the palatability of the honey (Razali et al., 2019; Šedík et al., 2018).

6.3.2 Aroma

The natural flavor of honey is due to the volatile components it contains. However, microbial activity can disrupt the chemical structure of these components and adversely affect the flavor of honey. Undesirable aromas caused by mold growth or fermentation give a negative impression of the freshness and quality of honey (Araujo et al., 2020; González et al., 2010).

6.3.3 Texture and Mouthfeel.

The texture and mouthfeel of honey can be altered by microbial contamination. For example, fermentation can cause honey to have a foamy or fizzy texture, which detracts from the smooth honey texture consumers expect (Saxena et al., 2023).

6.4 Reduction in Shelf Life

Microbial contamination can significantly shorten the shelf life of honey. This reduces the commercial value of honey and can lead to economic losses.

6.4.1 Spoilage

Honey can spoil due to microbial activity. Fermentation and mold growth, in particular, can cause honey to become inedible within a short time. This can lead to product recalls and customer complaints (Lage et al., 2012; Anwar et al., 2018).

6.4.2 Quality Degradation

Microbial contamination causes honey to degrade in quality over time. This deterioration in the physical, chemical, and sensory properties of honey has a negative impact on its commercial value (Truzzi et al., 2012; Singh & Singh, 2018).

6.5 Health Risks

Microbial contamination can threaten not only the quality of honey but also the health of consumers. In particular, the presence of pathogenic microorganisms can lead to serious health risks.

6.5.1 Botulism

Clostridium botulinum spores can cause botulism, especially in infants. Although these dangerous bacteria can be found naturally in honey, improper production and storage conditions may allow these spores to develop (Kumar et al., 2010; Cirik & Aksoy, 2020).

6.5.2 Allergic Reactions

Certain components that develop as a result of microbial contamination can cause allergic reactions in some consumers. This poses a risk to the consumer safety of honey (Stepanyan et al., 2022).

7 Case Studies in Microbial Contamination of Honey

The risks and effects of microbial contamination on honey can be better understood through various case studies. Such cases illustrate the difficulties encountered in honey production and processing, the problems caused by microbial contamination, and the measures taken against these problems (Bovo et al., 2018; Paluoja et al., 2024; Galanis et al., 2022). Several case studies on microbial contamination of honey are discussed below.

7.1 Case Study 1: Clostridium botulinum Contamination in Infant Honey

In one case, an infant in the United States was hospitalized with symptoms of botulism. Investigations revealed that the infant had consumed honey, which was contaminated with *Clostridium botulinum* spores. This case emphasizes that honey should not be consumed in inappropriate age groups because of the risk of botulism in infants (Harris & Dabritz, 2024).

7.1.1 Incident Details

The infant was taken to the doctor with symptoms of digestive problems and muscle weakness. After investigations, botulism was diagnosed, and it was determined that honey consumption had caused the condition. The infant required intensive medical care, and this case is a reminder to the public of the dangers of honey consumption for infants.

7.1.2 Lessons Learned

This case highlights the importance of raising parental awareness of the risk of infant botulism and the need for honey producers to display appropriate age warnings on their products. It also reaffirmed that *Clostridium botulinum* spores can naturally occur in honey, and therefore, infants should avoid honey consumption.

7.2 *Case Study 2: Fermentation Issues in Commercial Honey*

Another case concerns a widespread fermentation problem in a honey production facility. Due to the storage of honey in excessively humid conditions during the production process, osmophilous yeasts were activated, leading to fermentation of the honey (Disayathanoowat et al., 2020).

7.2.1 Incident Details

The plant produced and packaged a large batch of honey. However, after a few weeks, the honey jars showed signs of fermentation, which caused the honey to become frothy and the lids to burst. The products reached the consumer unnoticed in the distribution chain, leading to a serious recall.

7.2.2 Lessons Learned

This incident demonstrated the need for careful monitoring of honey production and storage conditions. Moisture control and appropriate storage temperatures play a critical role in preventing fermentation. It also emphasized the need for producers to continuously monitor the microbial quality of honey.

7.3 *Case Study 3: Pesticide-Induced Microbial Imbalance in Beehives*

Another case study addresses how the use of pesticides can disturb the microbial balance in hives and affect the microbial quality of honey. This led to microbial contamination of honey, adversely affecting its natural properties (Kakumanu et al., 2016).

7.3.1 Incident Details

In an apiculture enterprise, it was noticed that pesticides used to control crop pests were also killing beneficial microorganisms in the hives. This caused the natural microbial balance in the hives to be disturbed, and pathogenic microorganisms became dominant. As a result, undesirable microbial growth was observed in the collected honey.

7.3.2 Lessons Learned

This case illustrates the need for careful management of pesticide use in hives and the importance of preserving beneficial microorganisms. It is also recommended to use environmentally friendly alternatives that will not harm the health of bees and will not affect the microbial quality of honey.

8 Preventive Measures and Best Practices

Prevention of microbial contamination in honey production is vital to ensure the quality and safety of the product. To this end, preventive measures and best practices that can be implemented at each stage of the production process help to minimize microbial risks (Gomes et al., 2010; Herdiman, 2023). The following are the main preventive measures and best practices that can be applied to prevent microbial contamination in honey production.

8.1 Hygiene Management in Beekeeping

Ensuring hygiene in beekeeping is the first step in preventing microbial contamination. Keeping beehives and their surroundings clean is critical to maintain the natural purity of honey (Kuznetsov, 2024; Aziz & Alam, 2024).

8.1.1 Hive Cleanliness

Regular cleaning of hives prevents diseases and harmful microorganisms from entering the hives. Also, dead bees and other waste materials should be removed from the hives in a timely manner (Pietropaoli et al., 2020a; Formato et al., 2022).

8.1.2 Sanitization of Equipment

All equipment used during honey milking should be kept in hygienic conditions and sterilized regularly. Equipment made of stainless steel provides a less favorable environment for the growth of microorganisms (Zenunović et al., 2020; Rašović, 2022; Aziz & Alam, 2024).

8.1.3 Water Source Management

It is important that the bees' water sources are clean and free from microorganisms. Dirty water sources can carry microbial contamination to the beehives (Naila et al., 2022).

8.2 Controlled Harvesting and Processing

Careful and controlled processes should be followed to prevent microbial contamination during the harvesting and processing of honey.

8.2.1 Timing of Harvesting

The timing of harvesting honey should be carefully selected to minimize microbial risks. Honey should be harvested when weather conditions are not humid and the bees have built up sufficient honey stocks (Wu et al., 2022).

8.2.2 Proper Filtration

The filtration process applied during honey processing should be effective in removing impurities and microorganisms. However, excessive filtration should be avoided to preserve the natural structure and nutritional value of honey (Alam et al., 2012).

8.2.3 Pasteurization and Heat Treatment

Pasteurization or low heat treatment of honey may be necessary to reduce microbial contamination. This process helps to kill pathogens while preserving the nutritional value of honey (Ramly et al., 2021).

8.3 Optimal Storage Conditions

Storage conditions are extremely important for maintaining the microbial quality of honey. Improper storage can lead to honey spoilage and promote microbial growth.

8.3.1 Temperature Control

Storing honey at low temperatures slows microbial activity and inhibits fermentation. The ideal storage temperature should be between 10 and 20 °C. Extreme temperature fluctuations should be avoided, and honey should be stored in a cool, dry place (Chou et al., 2020).

8.3.2 Humidity Management

The humidity level of the environment where honey is stored should be low. High humidity can encourage the growth of microorganisms such as mold and yeast. Tightly closing the honey jars will prevent moisture from reaching the honey (Chen et al., 2021).

8.3.3 Shelf Life Monitoring

Microbial quality controls should be carried out regularly throughout the shelf life of honey. Producers should monitor the microbial load of honey and take necessary measures when any signs of spoilage appear (Bideci & Karasalihoğlu, 2022).

8.4 Regular Microbial Testing

Honey producers should conduct regular microbial tests for early detection and management of microbial contamination. These tests play a critical role in guaranteeing the safety and quality of honey.

8.4.1 Routine Sampling

Regular sampling and testing of honey at different stages of the production process is necessary. These tests allow for the identification of potential sources of microbial contamination and early intervention (Nurlila et al., 2023).

8.4.2 Advanced Analytical Techniques

Advanced microbial analysis methods such as PCR, ELISA, and other molecular biology techniques can be used to detect low levels of pathogens in honey. These methods help to assess the microbial quality of honey more precisely (Gaudin et al., 2020).

8.4.3 Enhance the Quality Assurance of Honey as a Food Safety Management System

Food safety management systems provide guidance to honey producers to monitor and prevent microbial contamination. The implementation of systems such as Hazard Analysis and Critical Control Points (HACCP) is an effective tool in the management of microbial risks (Yanovych et al., 2023; Bucekova et al., 2023).

8.5 Employee Training and Awareness

Raising awareness of microbial risks and hygiene protocols among honey production workers is critical to prevent microbial contamination.

8.5.1 Hygiene Practices

It should be ensured that employees pay attention to personal hygiene and comply with hygiene rules during the production process. Practices such as hand washing, wearing sterile clothes, and maintaining a hygienic working environment reduce the risk of microbial contamination (Zenunović et al., 2020).

8.5.2 Continuous Education

Employees should participate in regular training programs on microbial risks and best practices. These trainings increase the knowledge level of employees and enable them to take proactive measures against microbial contamination (Hamad et al., 2020).

8.5.3 Incident Response Training

Employees should be trained on how to act when microbial contamination is detected. Rapid and effective response prevents the spread of contamination and ensures the safety of the production process (Naila et al., 2022).

8.6 Collaboration with Regulatory Bodies

Honey producers should cooperate with local and international regulatory bodies to minimize the risks of microbial contamination. This co-operation ensures that standards and guidelines are followed to improve the quality and safety of honey (Hermanns et al. 2020).

8.6.1 Compliance with Regulations

Producers should comply with local and international regulations regarding the microbial quality of honey. Compliance with these regulations ensures the safety and commercial viability of honey (Odekunle et al., 2023).

8.6.2 Certification Programs

Honey producers who produce organically or in accordance with hygiene standards can certify the microbial safety and quality of their products by participating in certification programs. Such certificates increase consumer confidence and increase the competitiveness of products in the market (Civit et al., 2023).

8.6.3 Regular Audits

Regular audits by independent auditing firms or regulatory bodies assess whether production processes are safe in terms of microbial risks. These audits help manufacturers to continuously improve their processes (Yanovych et al., 2023).

9 Emerging Trends in Microbial Analysis of Honey

Microbial analysis methods are constantly evolving to ensure the safety and quality of honey. New technologies and research aim to detect microbial contamination more sensitively, speed up analysis processes, and improve the overall quality of honey. Some of the new trends in microbial analysis of honey are discussed below.

9.1 Molecular Techniques and Genomic Approaches

Molecular biology techniques and genomic approaches are increasingly used to improve the accuracy of microbial analysis. These methods enable rapid and accurate identification of microorganisms in honey (Tsagkaris et al., 2021; Soares et al., 2023; Zhang et al., 2024).

9.1.1 Next-Generation Sequencing (NGS)

NGS is an advanced genomic technique used to determine the microbial profile of honey. This method sequences the genetic material of all microorganisms present in honey, providing a complete picture of the microbial community. NGS provides a significant advantage, especially in the detection of low-abundant pathogens (Tsagkaris et al., 2021).

9.1.2 Polymerase Chain Reaction (PCR) Techniques

PCR and its derivatives (such as qPCR and RT-PCR) are widely used to determine the presence of specific microorganisms in honey. These techniques make it possible to analyze quickly and sensitively by amplifying specific microbial DNA or RNA sequences (Gaudin et al., 2020).

9.1.3 Metagenomics

This approach is used to comprehensively analyze the microbial ecosystem of honey. Metagenomic analysis allows us to understand how different microorganisms together form an ecosystem and the effects of this ecosystem on the quality of honey (Bovo et al., 2018).

9.2 *Rapid Detection Methods*

Accelerating microbial analysis processes is critical for maintaining the quality of honey and uninterrupted commercial processes. Rapid detection methods developed in recent years provide faster intervention by detecting contamination in the early stages (Johari et al., 2019; Pita-Calvo et al., 2017).

9.2.1 Biosensors

Biosensors are portable devices used for the rapid and sensitive detection of pathogens in honey. These sensors detect the presence of target microorganisms through electrical, optical, or thermal signals and provide results instantly (de Oliveira Neto et al., 2017).

9.2.2 Lab-on-a-Chip Technology

This technology performs many of the analyses that a laboratory can perform on a miniature chip. Developed to quickly analyze the microbial content of honey, this technology stands out as a time-saving innovation in honey production (Bagheri et al., 2021).

9.2.3 Immunoassays

Immunological tests, such as ELISA, are methods for the rapid detection of specific pathogens in honey. These tests provide fast results by detecting the presence of protein structures or antigens of microorganisms (Guillen et al., 2011).

9.3 Big Data and Artificial Intelligence (AI) in Microbial Analysis

Big data analysis and artificial intelligence offer new ways to assess the microbial quality of honey. By processing large data sets from microbial analysis, these technologies help to gain more in-depth insights into the safety and quality of honey (Chenchouni & Laallam, 2024).

9.3.1 Predictive Modeling

Artificial intelligence and machine learning algorithms can be used to predict the likelihood of microbial contamination. These models can predetermine microbial risks based on the production and storage conditions of honey (Escuredo et al., 2021).

9.3.2 Data Integration and Analysis

Big data analytics can provide more comprehensive and accurate conclusions about the overall microbial profile of honey by integrating data from different microbial analysis methods. This approach optimizes microbial analysis processes and helps producers make better decisions (Maione et al., 2019).

9.3.3 AI-Powered Quality Control Systems

Artificial intelligence can be used to automate quality control processes in honey production. AI-based systems can detect signs of microbial contamination in the early stages, enabling immediate intervention in production processes (Nath et al., 2024).

9.4 Probiotic Honey and Microbial Manipulation

In recent years, the production of probiotic honey and microbial manipulation have also attracted interest. These approaches aim to enhance the health benefits of honey by consciously regulating its microbial content (Mustar & Ibrahim, 2022; Schell et al., 2022).

9.4.1 Probiotic-Infused Honey

Some producers are working on the production of honey enriched with probiotic microorganisms. This type of honey is used both as a natural sweetener and offers probiotic benefits (Misra et al., 2022).

9.4.2 Microbial Manipulation Techniques

Research to intentionally regulate the microbial content of honey is examining the selective addition or removal of microorganisms to extend the shelf life and improve the nutritional value of honey (Luca et al., 2024).

9.5 Sustainability and Eco-friendly Practices in Microbial Analysis

Sustainability in honey production is also reflected in microbial analysis processes. The use of environmentally friendly methods contributes to both the conservation of natural resources and the safety of honey.

9.5.1 Green Chemistry in Testing

Green chemistry principles are adopted to reduce the environmental impact of chemicals used in microbial analysis. This encourages the use of less toxic and biodegradable chemicals (Pallone et al., 2018).

9.5.2 Eco-friendly Sampling and Storage

Sampling and storage methods used for microbial analysis of honey are also becoming more sustainable. These trends include sample containers made with reusable or biodegradable materials and storage solutions that consume less energy (Youssef et al., 2019).

9.5.3 Sustainable Beekeeping Practices

Sustainable beekeeping practices are also gaining importance to improve the microbial quality of honey. This includes approaches such as reducing pesticide use, protecting biodiversity, and supporting natural habitats (Pietropaoli et al., 2020b).

References

Adadi, P., & Obeng, A. (2017). Assessment of bacterial quality of honey produced in tamale metropolis (Ghana). *Journal of Food and Drug Analysis, 25*(2), 369–373. https://doi.org/10.1016/j.jfda.2016.07.005

Akbari, E., Baigbabaei, A., & Shahidi, M. (2020). Determination of the floral origin of honey based on its phenolic profile and physicochemical properties coupled with chemometrics. *International Journal of Food Properties, 23*(1), 506–519. https://doi.org/10.1080/10942912.2020.1740249

Alam, S., Sharma, D., Sehgal, V., Arora, M., & Bhatia, S. (2012). Development and evaluation of low cost honey heating-cum-filtration system. *Journal of Food Science and Technology, 51*(11), 3476–3481. https://doi.org/10.1007/s13197-012-0863-0

Álvarez-Suárez, J., Gasparrini, M., Forbes-Hernández, T., Mazzoni, L., & Giampieri, F. (2014). The composition and biological activity of honey: A focus on Manuka honey. *Food, 3*(3), 420–432. https://doi.org/10.3390/foods3030420

Al-Waili, N. S., Salom, K., Butler, G., & Al Ghamdi, A. A. (2011). Honey and microbial infections: A review supporting the use of honey for microbial control. *Journal of Medicinal Food, 14*(10), 1079–1096.

Al-Waili, N., Salom, K., Al-Ghamdi, A., & Ansari, M. J. (2012). Antibiotic, pesticide, and microbial contaminants of honey: Human health hazards. *The Scientific World Journal, 2012*(1), 930849.

Anwar, A., et al. (2018). Physicochemical characteristics of Malaysian honeys influenced by storage time and temperature. *Journal of Fundamental and Applied Sciences, 9*(2S), 841. https://doi.org/10.4314/jfas.v9i2s.52

Araujo, D., Pérez-Cacho, P., Serrano, S., Dios-Palomares, R., & Soldevilla, H. (2020). Sensory profile and physico-chemical properties of artisanal honey from Zulia, Venezuela. *Food, 9*(3), 339. https://doi.org/10.3390/foods9030339

Aykas, D. (2023). Determination of possible adulteration and quality assessment in commercial honey. *Food, 12*(3), 523. https://doi.org/10.3390/foods12030523

Aziz, M. A., & Alam, S. (2024). Diseases of honeybee (*Apis mellifera*). In *Melittology-new advances*. IntechOpen.

Bagheri, N., Al Lawati, H. A., & Hassanzadeh, J. (2021). Simultaneous determination of total phenolic acids and total flavonoids in tea and honey samples using an integrated lab on a chip device. *Food Chemistry, 342*, 128338.

Baroyi, S., Yusof, Y., Ghazali, H., Chin, N., Othman, S., Chang, L., & Ghazali, N. (2019). A novel method based on passive diffusion that reduces the moisture content of stingless bee (*Heterotrigona itama*) honey. *Journal of Food Process Engineering, 42*(6). https://doi.org/10.1111/jfpe.13221

Bideci, G., & Karasalihoğlu, S. (2022). A retrospective study: Physicochemical properties of the flower honey from the black sea region of Turkey in different years. *Food Science and Technology, 42*, 1–6. https://doi.org/10.1590/fst.58120

Bogdanov, S., Lüllmann, C., Martin, P., Von Der Ohe, W., Russmann, H., Vorwohl, G., & Vit, P. (1999). Honey quality and international regulatory standards: Review by the international honey commission. *Bee World, 80*(2), 61–69.

Bovo, S., Ribani, A., Utzeri, V. J., Schiavo, G., Bertolini, F., & Fontanesi, L. (2018). Shotgun metagenomics of honey DNA: Evaluation of a methodological approach to describe a multi-kingdom honey bee derived environmental DNA signature. *PLoS One, 13*(10), e0205575. https://doi.org/10.1371/journal.pone.0205575

Brčina, T., Halilčević, L., & Cvrk, R. (2021). Assessment of honey quality from the middle Podrinje area. *International Journal for Research in Applied Sciences and Biotechnology, 8*(1), 132–137. https://doi.org/10.31033/ijrasb.8.1.15

Bucekova, M., Godocikova, J., Kohutova, L., Danchenko, M., Barath, P., & Majtan, J. (2023). Antibacterial activity and bee-derived protein content of honey as important and suitable complementary tools for the assessment of honey quality. *Journal of Food Composition and Analysis, 123*, 105610. https://doi.org/10.1016/j.jfca.2023.105610

Chen, Y., Chuah, W., & Chye, F. (2021). Effect of drying on physicochemical and functional properties of stingless bee honey. *Journal of Food Processing and Preservation, 45*(4), 1–15. https://doi.org/10.1111/jfpp.15328

Chenchouni, H., & Laallam, H. (2024). Revolutionizing food quality assessment: Unleashing the potential of artificial intelligence for enhancing honey physicochemical, biochemical, and melissopalynological insights. *Journal of the Saudi Society of Agricultural Sciences, 23*(4), 312–325.

Chin, N. L., & Sowndhararajan, K. (2020). A review on analytical methods for honey classification, identification and authentication. In *Honey analysis – New advances and challenges* (pp. 1–33). https://doi.org/10.5772/intechopen.90232

Chou, W., Liao, H., Yang, Y., & Peng, C. (2020). Evaluation of honey quality with stored time and temperatures. *Journal of Food and Nutrition Research, 8*(10), 591–599. https://doi.org/10.12691/jfnr-8-10-8

Cirik, V., & Aksoy, B. (2020). Pedİatrİ hemşİrelerİnİn apİterapİ hakkindakİ bİlgİ, tutum ve deneyİmlerİnİn belİrlenmesİ: kesİtsel çok merkezlİ bİr çalişma. *Uludağ Arıcılık Dergisi, 20*(2), 157–171. https://doi.org/10.31467/uluaricilik.787299

Civit, D., Libonatti, C., & Agüeria, D. (2023). Verification of prerequisite programs implemented in a honey extraction establishment: A case study. *Food and Humanity, 1*, 258–262. https://doi.org/10.1016/j.foohum.2023.06.001

Cokcetin, N., Pappalardo, M., Campbell, L., Brooks, P., Carter, D., Blair, S., & Harry, E. (2016). The antibacterial activity of Australian leptospermum honey correlates with methylglyoxal levels. *PLoS One, 11*(12), e0167780. https://doi.org/10.1371/journal.pone.0167780

Dallagnol, A., Dallagnol, V., Vignolo, G., Lopes, N., & Brunetti, A. (2022). Flavonoids and phenylethylamides are pivotal factors affecting the antimicrobial properties of stingless bee honey. *Journal of Agricultural and Food Chemistry, 70*(39), 12596–12603. https://doi.org/10.1021/acs.jafc.2c04120

De Beer, T., Otto, M., Pretoruis, B., & Schönfeldt, H. C. (2021). Monitoring the quality of honey: South African case study. *Food Chemistry, 343*, 128527.

de Oliveira Neto, J. R., Rezende, S. G., Lobón, G. S., Garcia, T. A., Macedo, I. Y. L., Garcia, L. F., & de Souza Gil, E. (2017). Electroanalysis and laccase-based biosensor on the determination of phenolic content and antioxidant power of honey samples. *Food Chemistry, 237*, 1118–1123.

Devi, S., Parihar, A., Thakur, M., Thakur, B., & Sharma, H. (2021). Antibacterial potential of hive bees honey from Himachal Pradesh, India. *Archives of Microbiology, 203*(8), 5029–5041. https://doi.org/10.1007/s00203-021-02489-y

Disayathanoowat, T., Li, H., Supapimon, N., Suwannarach, N., Lumyong, S., Chantawannakul, P., & Guo, J. (2020). Different dynamics of bacterial and fungal communities in hive-stored bee bread and their possible roles: A case study from two commercial honey bees in China. *Microorganisms, 8*(2), 264.

Dümen, E., Akkaya, H., Öz, G., & Sezgin, F. (2013). Microbiological and parasitological quality of honey produced in İstanbul. *Turkish Journal of Veterinary and Animal Sciences, 37*, 602–607. https://doi.org/10.3906/vet-1301-46

El Sohaimy, S. A., Masry, S. H. D., & Shehata, M. G. (2015). Physicochemical characteristics of honey from different origins. *Annals of Agricultural Sciences, 60*(2), 279–287.

Erwan, E., Harun, M., & Muhsinin, M. (2020). The honey quality of *Apis mellifera* with extrafloral nectar in lombok West Nusa Tenggara Indonesia. *Journal of Science and Science Education, 1*(1), 1–7. https://doi.org/10.29303/jossed.v1i1.482

Escuredo, O., Rodríguez-Flores, M. S., Meno, L., & Seijo, M. C. (2021). Prediction of physicochemical properties in honeys with portable near-infrared (Micronir) spectroscopy combined with multivariate data processing. *Food, 10*(2), 317.

Formato, G., Rivera-Gomis, J., Bubnič, J., Martín-Hernández, R., Milito, M., Croppi, S., & Higes, M. (2022). Nosemosis prevention and control. *Applied Sciences, 12*(2), 783. https://doi.org/10.3390/app12020783

Fratianni, F. (2024). Chemical characterization and biological properties of leguminous honey. *Antioxidants, 13*(4), 482. https://doi.org/10.3390/antiox13040482

Galanis, A., Vardakas, P., Reczko, M., Harokopos, V., Hatzis, P., Skoulakis, E., & Patalano, S. (2022). Bee foraging preferences, microbiota and pathogens revealed by direct shotgun metagenomics of honey. *Molecular Ecology Resources, 22*(7), 2506–2523. https://doi.org/10.1111/1755-0998.13626

Galhardo, D., Garcia, R., Schneider, C., Ströher, S., Cerny, B., & Chambó, É. (2020). Microbiological quality of *Apis mellifera* L. honey samples from Western Paraná, southern Brazil. *Journal of Apicultural Science, 64*(2), 209–218. https://doi.org/10.2478/jas-2020-0024

García-Tenesaca, M., Navarrete, E., Iturralde, G., Granda, I., Tejera, E., Beltrán-Ayala, P., & Álvarez-Suarez, J. (2017). Influence of botanical origin and chemical composition on the protective effect against oxidative damage and the capacity to reduce *in vitro* bacterial biofilms of monofloral honeys from the Andean region of Ecuador. *International Journal of Molecular Sciences, 19*(1), 45. https://doi.org/10.3390/ijms19010045

Gaudin, V., Bodin, C., Hédou, C., Soumet, C., & Verdon, E. (2020). Development and optimization of an Amperometric Immunosensor for the detection of banned antibiotic residues in honey. *Proceedings, 60*(1), 48. https://doi.org/10.3390/iecb2020-07019

Gomes, S., Dias, L., Moreira, L., Rodrigues, P., & Estevinho, L. (2010). Physicochemical, microbiological and antimicrobial properties of commercial honeys from Portugal. *Food and Chemical Toxicology, 48*(2), 544–548. https://doi.org/10.1016/j.fct.2009.11.029

González, M., Lorenzo, C., & Pérez, R. (2010). Development of a structured sensory honey analysis: Application to artisanal Madrid honeys. *Food Science and Technology International, 16*(1), 19–29. https://doi.org/10.1177/1082013209351869

Grabowski, N. T., & Klein, G. (2017). Microbiology and foodborne pathogens in honey. *Critical Reviews in Food Science and Nutrition, 57*(9), 1852–1862.

Grace, E., Mantilla, S., Sunarharum, W., Ong, C., Waanders, J., D'Arcy, B., & Smyth, H. (2020). Sensory properties of yellow pea and macadamia honeys from conventional and flow hive extraction methods. *Journal of the Science of Food and Agriculture, 100*(5), 2027–2034. https://doi.org/10.1002/jsfa.10221

Guillen, I., Gabaldón, J. A., Núnez-Delicado, E., Puchades, R., Maquieira, A., & Morais, S. (2011). Detection of sulphathiazole in honey samples using a lateral flow immunoassay. *Food Chemistry, 129*(2), 624–629.

Guran, H., Durukan, G., & Mann, D. (2021). Assessment of the microbiological profile, species diversity and antimicrobial susceptibility of recovered bacteria from retail honeys in Turkey. *Journal of the Hellenic Veterinary Medical Society, 72*(1), 2765. 10.12681/jhvms.26763.

Hamad, G., Hafez, E., Abdelmotilib, N., Abdel-Hmeed, K., Ali, S., & Darwish, A. (2020). Quality assessment, functional potentials, and safety evaluation of stored Egyptian honey as an environmental pollution bioindicator. *Environmental Toxicology and Chemistry, 39*(10), 1894–1907. https://doi.org/10.1002/etc.4811

Harris, R. A., & Dabritz, H. A. (2024). Infant botulism: In search of *Clostridium botulinum* spores. *Current Microbiology, 81*(10), 306.

Hegazi, A. G., Al Guthami, F. M., Ramadan, M. F., Al Gethami, A. F., Craig, A. M., & Serrano, S. (2022). Characterization of Sidr (Ziziphus spp.) honey from different geographical origins. *Applied Sciences, 12*(18), 9295.

Herdiman, L. (2023). Design of a high-voltage pulse generator in a pulsed electric field for the pasteurization process of randu honey. *E3s Web of Conferences, 465*, 02025. https://doi.org/10.1051/e3sconf/202346502025

Hermanns, R., Mateescu, C., Thrasyvoulou, A., Tananaki, C., Wagener, F. A., & Cremers, N. A. (2020). Defining the standards for medical grade honey. *Journal of Apicultural Research, 59*(2), 125–135.

İnanç, B. (2020). The quality specialities in Turkiye's honies for apitherapy. *Journal of Pharmacopuncture, 23*(4), 194–200. https://doi.org/10.3831/kpi.2020.23.4.194
Jain, S., Jesus, F., Marchioro, G., & Araújo, E. (2013). Extraction of dna from honey and its amplification by PCR for botanical identification. *Food Science and Technology, 33*(4), 753–756. https://doi.org/10.1590/s0101-20612013000400022
Joe, T., Jeong, E., Yoo, G., Ahn, S., Rhee, H., & Lee, D. (2018). The monoclonal antibody for discrimination of natural honey against artificial honey. *Food Science & Nutrition, 6*(8), 2350–2354. https://doi.org/10.1002/fsn3.821
Johari, N. A., Ashaari, N. S., Mamat, M. R., & Muhamad, A. (2019). Simple and rapid screening test to detect fake honey product. *Journal of Agricultural Science and Technology, 9*, 330–338.
Kačániová, M., Pavličová, S., Haščík, P., Kociubinski, G., Kňazovická, V., Sudzina, M., & Fikselová, M. (2009). Microbial communities in bees, pollen and honey from Slovakia. *Acta Microbiologica et Immunologica Hungarica, 56*(3), 285–295. https://doi.org/10.1556/amicr.56.2009.3.7
Kakumanu, M. L., Reeves, A. M., Anderson, T. D., Rodrigues, R. R., & Williams, M. A. (2016). Honey bee gut microbiome is altered by in-hive pesticide exposures. *Frontiers in Microbiology, 7*, 1255.
Kaškonienė, V., & Venskutonis, P. (2010). Floral markers in honey of various botanical and geographic origins: A review. *Comprehensive Reviews in Food Science and Food Safety, 9*(6), 620–634. https://doi.org/10.1111/j.1541-4337.2010.00130.x
Kędzierska-Matysek, M. (2023). Microbiological quality of polish artisanal varietal honeys. *Food, 12*(18), 3349. https://doi.org/10.3390/foods12183349
Khalil, M., Moniruzzaman, M., Boukraâ, L., Benhanifia, M., Islam, M., Islam, M., & Gan, S. (2012). Physicochemical and antioxidant properties of Algerian honey. *Molecules, 17*(9), 11199–11215. https://doi.org/10.3390/molecules170911199
Klutse, C., & Adotey, D. (2022). Gamma irradiation effect on the microbial load and physicochemical properties of honey from Ghana. *Cell Biology and Development, 6*(2), 10.13057/cellbioldev/v060205.
Kumar, R., Lorenc, A., Robinson, N., & Blair, M. (2010). Parents' and primary healthcare practitioners' perspectives on the safety of honey and other traditional paediatric healthcare approaches. *Child Care Health and Development, 37*(5), 734–743. https://doi.org/10.1111/j.1365-2214.2010.01186.x
Kuznetsov, A. (2024). Zoohygienic assessment of carbohydrate feeds – an important component of beekeeping technology. *Legal Regulation in Veterinary Medicine, 4*, 143–146. https://doi.org/10.52419/issn2782-6252.2023.4.143
Lage, L., Coelho, L., Resende, H., Tavares, M., Campos, L., & Fernandes-Salomão, T. (2012). Honey physicochemical properties of three species of the Brazilian Melipona. *Anais da Academia Brasileira de Ciências, 84*(3), 605–608. https://doi.org/10.1590/s0001-37652012005000051
Lima, C., Nora, F., Seraglio, S., Silva, J., Marzoque, H., Santana, R., & Scussel, V. (2020). Antibiotic residues in honey: A public health issue. *Research Society and Development, 9*(11), e1739119604. https://doi.org/10.33448/rsd-v9i11.9604
Luca, L., Pauliuc, D., & Oroian, M. (2024). Honey microbiota, methods for determining the microbiological composition and the antimicrobial effect of honey–a review. *Food Chemistry, X*, 101524.
Mădaş, M. N., Mărghitaş, L. A., Dezmirean, D. S., Bobiş, O., Abbas, O., Danthine, S., & Nguyen, B. K. (2020). Labeling regulations and quality control of honey origin: A review. *Food Reviews International, 36*(3), 215–240.
Maione, C., Barbosa, F., Jr., & Barbosa, R. M. (2019). Predicting the botanical and geographical origin of honey with multivariate data analysis and machine learning techniques: A review. *Computers and Electronics in Agriculture, 157*, 436–446.
Majtan, J., Bucekova, M., Kafantaris, I., Szweda, P., Hammer, K., & Mossialos, D. (2021). Honey antibacterial activity: A neglected aspect of honey quality assurance as functional food. *Trends in Food Science & Technology, 118*, 870–886.

Malavi, D. N., Muzhingi, T., & Abong', G. O. (2018). Good manufacturing practices and microbial contamination sources in orange fleshed sweet potato puree processing plant in Kenya. *International Journal of Food Science*, 2018(1), 4093161.

Mandal, M., & Mandal, S. (2011). Honey: Its medicinal property and antibacterial activity. *Asian Pacific Journal of Tropical Biomedicine, 1*(2), 154–160. https://doi.org/10.1016/s2221-1691(11)60016-6

Martín-González, D., Bordel, S., Solis, S., Gutierrez-Merino, J., & Santos-Beneit, F. (2023). Characterization of B*acillus* strains from natural honeybee products with high keratinolytic activity and antimicrobial potential. *Microorganisms, 11*(2), 456. https://doi.org/10.3390/microorganisms11020456

Meister, A., Gutiérrez-Ginés, M., Maxfield, A., Gaw, S., Dickinson, N., Horswell, J., & Robinson, B. (2021). Chemical elements and the quality of mānuka (*Leptospermum scoparium*) honey. *Food, 10*(7), 1670. https://doi.org/10.3390/foods10071670

Misra, S., Mohanty, D., & Mohapatra, S. (2022). Food-based probiotics: Functional dietary ingredients. In *Probiotics* (pp. 257–275). Academic Press.

Musa, M., Elfaki, A., Eldin, S., & Mohammed, A. (2014). Microbiological characterization and physicochemical properties of sudanese honeys. *British Microbiology Research Journal, 4*(6), 715–722. https://doi.org/10.9734/bmrj/2014/5330

Mustafina, R., Maikanov, B., Wisniewski, J., Tracz, M., Anusz, K., Grenda, T., & Kwiatek, K. (2015). Contamination of honey produced in the Republic of Kazakhstan with C*lostridium botulinum. Bulletin of the Veterinary Institute in Pulawy, 59*(2), 241–246. https://doi.org/10.1515/bvip-2015-0036

Mustar, S., & Ibrahim, N. (2022). A sweeter pill to swallow: A review of honey bees and honey as a source of probiotic and prebiotic products. *Food, 11*(14), 2102.

Naila, A., Julika, W., Hannan, N., Flint, S., Sulaiman, A., & Mohamed, A. (2022). Microbiological and physiochemical quality of honey imported into the Maldives. *ACS Food Science & Technology, 2*(5), 836–843. https://doi.org/10.1021/acsfoodscitech.1c00478

Nath, P. C., Mishra, A. K., Sharma, R., Bhunia, B., Mishra, B., Tiwari, A., & Sridhar, K. (2024). Recent advances in artificial intelligence towards the sustainable future of agri-food industry. *Food Chemistry, 138945*, 138945.

Negash, D., Debara, M., Bekel, B., & Aregaw, A. (2020). Comparative evaluation of Mekonen beehive technology with Zander and ktb beehive types on honey yield and cost benefit analysis under sidama condition, Ethiopia. *International Journal of Biomedical Materials Research, 8*(2), 29. 10.11648/j.ijbmr.20200802.14.

Ng, L., Razak, S., Tan, S., & Ahmad, F. (2023). Low heating effects on the total microbial activity and physico-chemical properties of stingless bee (*Heterotrigona itama*) honey. *Malaysian Applied Biology, 52*(1), 43–52. https://doi.org/10.55230/mabjournal.v52i1.2442

Nurlila, R. U., Jumarddin, L. F., & Susniatin. (2023). Bacteriological quality test of Southeast Sulawesi honey to support functional food. *International Journal For Multidisciplinary Research, 5*(6), 1–8. https://doi.org/10.36948/ijfmr.2023.v05i06.10308

Nzeh, J., Quansah, L., & Dufailu, O. A. (2022). Physicochemical properties of imported and locally produced honey did not translate into its microbial quality and antibacterial activity. *Discover Food, 2*(1), 24. 10.21203/rs.3.rs-1513445/v1.

Odekunle, M. (2023). The effectiveness and oversight of quality in the harvesting and processing of honey. *International Journal of Agriculture Extension and Social Development, 6*(2), 31–34. https://doi.org/10.33545/26180723.2023.v6.i2a.448

Odekunle, M. P., Daniel, A., & Olusola, A. (2023). The effectiveness and oversight of quality in the harvesting and processing of honey. *International Journal of Agriculture Extension And Social Development, 6*(2), 31–34. https://doi.org/10.33545/26180723.2023.v6.i2a.448

Orfanakis, E., Markoulidakis, M., Philippidis, A., Zoumi, A., & Velegrakis, M. (2021). Optical spectroscopy methods combined with multivariate statistical analysis for the classification of cretan thyme, multi-floral and honeydew honey. *Journal of the Science of Food and Agriculture, 101*(13), 5337–5347. https://doi.org/10.1002/jsfa.11182

Pallone, J. A. L., dos Santos Caramês, E. T., & Alamar, P. D. (2018). Green analytical chemistry applied in food analysis: Alternative techniques. *Current Opinion in Food Science, 22*, 115–121.

Palmer-Young, E. C., Markowitz, L. M., Huang, W. F., & Evans, J. D. (2023). High temperatures augment inhibition of parasites by a honey bee gut symbiont. *Applied and Environmental Microbiology, 89*(10), e01023–e01023. https://doi.org/10.1101/2023.03.20.533504

Palmieri, N., Stefanoni, W., Latterini, F., & Pari, L. (2022). Italian consumer preferences for eucalyptus honey: An exploratory study. *Sustainability, 14*(13), 7741. https://doi.org/10.3390/su14137741

Paluoja, P., Krjutškov, K., Salumets, A., & Raime, K. (2024). Honey bulk DNA metagenomic analysis to identify honey biological composition and monitor honeybee pathogens. *bioRxiv*, 2024–07.

Pauliuc, D., Dranca, F., & Oroian, M. (2020). Antioxidant activity, total phenolic content, individual phenolics and physicochemical parameters suitability for romanian honey authentication. *Food, 9*(3), 306. https://doi.org/10.3390/foods9030306

Pietropaoli, M., Ribarits, A., Moosbeckhofer, R., Köglberger, H., et al. (2020a). Biosecurity measures in european beekeeping. *Revue Scientifique Et Technique De L Oie, 39*(3), 725–735. https://doi.org/10.20506/rst.39.3.3174

Pietropaoli, M., Skerl, M. S., Cazier, J., Riviere, M. P., Tiozzo, B., Eggenhoeffner, R., & Formato, G. (2020b). BPRACTICES project: Towards a sustainable European beekeeping. *Bee World, 97*(3), 66–69.

Pimentel, R., Costa, C., Albuquerque, P., & Duvoisin, S. (2013). Antimicrobial activity and rutin identification of honey produced by the stingless bee *Melipona compressipes manaosensis* and commercial honey. *BMC Complementary and Alternative Medicine, 13*(1). https://doi.org/10.1186/1472-6882-13-151

Pippinato, L., Blanc, S., Mancuso, T., & Brun, F. (2020). A sustainable niche market: How does honey behave? *Sustainability, 12*(24), 10678. https://doi.org/10.3390/su122410678

Pita-Calvo, C., Guerra-Rodriguez, M. E., & Vazquez, M. (2017). Analytical methods used in the quality control of honey. *Journal of Agricultural and Food Chemistry, 65*(4), 690–703.

Pomastowski, P., Złoch, M., Rodzik, A., Ligor, M., Kostrzewa, M., & Buszewski, B. (2019). Analysis of bacteria associated with honeys of different geographical and botanical origin using two different identification approaches: Maldi-tof ms and 16S rDNA PCR technique. *PLoS One, 14*(5), e0217078. https://doi.org/10.1371/journal.pone.0217078

Preti, R., & Tarola, A. (2021). Polyfloral honey from urban beekeeping: Two-year case study of polyphenols profile and antioxidant activity. *British Food Journal, 123*(12), 4224–4239. https://doi.org/10.1108/bfj-09-2020-0839

Puścion-Jakubik, A., Borawska, M., & Socha, K. (2020). Modern methods for assessing the quality of bee honey and botanical origin identification. *Food, 9*(8), 1028. https://doi.org/10.3390/foods9081028

Ramly, N. S., Sujanto, I. S. R., Abd Ghani, A., Huat, J. T. Y., Alias, N., & Ngah, N. (2021). The impact of processing methods on the quality of honey: A review. *Malaysian Journal of Applied Sciences, 6*(1), 99–110. https://doi.org/10.37231/MYJAS.2021.6.1.280

Ramos, O., Salomón, V., Libonatti, C., Cepeda, R., Maldonado, L., & Basualdo, M. (2018). Effect of botanical and physicochemical composition of argentinean honeys on the inhibitory action against food pathogens. *LWT, 87*, 457–463. https://doi.org/10.1016/j.lwt.2017.09.014

Rašović, M. (2022). Importance of controlling the hygienic correctness of honey and other bee products in Montenegro. *The Journal Agriculture and Forestry, 68*(3), 10.17707/agricultforest.68.3.02.

Raweh, H., Badjah-Hadj-Ahmed, A., Iqbal, J., & Alqarni, A. (2023). Physicochemical composition of local and imported honeys associated with quality standards. *Food, 12*(11), 2181. https://doi.org/10.3390/foods12112181

Razali, N., Mazlan, A., Salebi, M., Mohamed, H., & Ambran, S. (2019). Optical fiber tip sensor for glucose-adulterated honey detection. *Telkomnika (Telecommunication Computing Electronics and Control), 17*(5), 2445. 10.12928/telkomnika.v17i5.12813.

Savarino, A., Terio, V., Barrasso, R., Ceci, E., Panseri, S., Chiesa, L., & Bonerba, E. (2020). Occurrence of antibiotic residues in apulian honey: Potential risk of environmental pollution by antibiotics. *Italian Journal of Food Safety, 9*(1). https://doi.org/10.4081/ijfs.2020.8678

Saxena, S., Gautam, S., & Sharma, A. (2010). Microbial decontamination of honey of Indian origin using gamma radiation and its biochemical and organoleptic properties. *Journal of Food Science, 75*(1). https://doi.org/10.1111/j.1750-3841.2009.01405.x

Saxena, S., Yadav, A., & Tomar, S. (2023). Study of the hydroxymethylfurfural and sugar profile of *Apis melifera* honey in Madhya Pradesh, India. *International Journal of Environment and Climate Change, 13*(9), 1285–1290. https://doi.org/10.9734/ijecc/2023/v13i92356

Scepankova, H. (2024). The high pressure preservation of honey: A comparative study on quality changes during storage. *Food, 13*(7), 989. https://doi.org/10.3390/foods13070989

Scepankova, H., Pinto, C., Paula, V., Estevinho, L., & Saraiva, J. (2021). Conventional and emergent technologies for honey processing: A perspective on microbiological safety, bioactivity, and quality. *Comprehensive Reviews in Food Science and Food Safety, 20*(6), 5393–5420. https://doi.org/10.1111/1541-4337.12848

Scepankova, H., Pinto, C., Estevinho, L., & Saraiva, J. (2022). High-pressure-based strategies for the inactivation of *Bacillus subtilis* endospores in honey. *Molecules, 27*(18), 5918. https://doi.org/10.3390/molecules27185918

Schell, K. R., Fernandes, K. E., Shanahan, E., Wilson, I., Blair, S. E., Carter, D. A., & Cokcetin, N. N. (2022). The potential of honey as a prebiotic food to re-engineer the gut microbiome toward a healthy state. *Frontiers in Nutrition, 9*, 957932.

Šedík, P., Kňazovická, V., Horská, E., & Kačániová, M. (2018). Consumer sensory evaluation of honey across age cohorts in Slovakia. *Potravinarstvo Slovak Journal of Food Sciences, 12*(1), 673–679. https://doi.org/10.5219/938

Silva, M. S., Rabadzhiev, Y., Eller, M. R., Iliev, I., Ivanova, I., & Santana, W. C. (2017). Microorganisms in honey. *Honey Analysis, 500*, 233–257.

Sinacori, M., Francesca, N., Alfonzo, A., Cruciata, M., Sannino, C., Settanni, L., & Moschetti, G. (2014). Cultivable microorganisms associated with honeys of different geographical and botanical origin. *Food Microbiology, 38*, 284–294. https://doi.org/10.1016/j.fm.2013.07.013

Singh, I., & Singh, S. (2018). Honey moisture reduction and its quality. *Journal of Food Science and Technology, 55*(10), 3861–3871. https://doi.org/10.1007/s13197-018-3341-5

Snowdon, J. A., & Cliver, D. O. (1996). Microorganisms in honey. *International Journal of Food Microbiology, 31*(1–3), 1–26.

Soares, S., Grazina, L., Mafra, I., Costa, J., Pinto, M., Duc, H., & Amaral, J. (2018). Novel diagnostic tools for Asian (*Apis cerana*) and European (*Apis mellifera*) honey authentication. *Food Research International, 105*, 686–693. https://doi.org/10.1016/j.foodres.2017.11.081

Soares, S., Rodrigues, F., & Delerue-Matos, C. (2023). Towards DNA-based methods analysis for honey: An update. *Molecules, 28*(5), 2106.

Sommano, S., Bhat, F., Wongkeaw, M., Sriwichai, T., Udomwong, P., Chuttong, B., & Burgett, M. (2020). Amino acid profiling and chemometric relations of black dwarf honey and bee pollen. *Frontiers in Nutrition, 7*. https://doi.org/10.3389/fnut.2020.558579

Stepanyan, S., Khachatryan, M., & Pipoyan, D. (2022). Assessing copper risk in honey sold in city of Yerevan. *Agriscience and Technology, 2*, 202–206.

Testa, R., Asciuto, A., Schifani, G., Schimmenti, E., & Migliore, G. (2019). Quality determinants and effect of therapeutic properties in honey consumption. An exploratory study on Italian consumers. *Agriculture, 9*(8), 174. https://doi.org/10.3390/agriculture9080174

Truzzi, C., Annibaldi, A., Illuminati, S., Finale, C., Rossetti, M., & Scarponi, G. (2012). Determination of very low levels of 5-(hydroxymethyl)-2-furaldehyde (hmf) in natural honey: Comparison between the HPLC technique and the spectrophotometric white method. *Journal of Food Science, 77*(7). https://doi.org/10.1111/j.1750-3841.2012.02782.x

Tsagkaris, A. S., Koulis, G. A., Danezis, G. P., Martakos, I., Dasenaki, M., Georgiou, C. A., & Thomaidis, N. S. (2021). Honey authenticity: Analytical techniques, state of the art and challenges. *RSC Advances, 11*(19), 11273–11294. https://doi.org/10.1039/d1ra00069a

Valachová, I., Bucekova, M., & Juraj, M. (2016). Quantification of bee-derived peptide defensin-1 in honey by competitive enzyme-linked immunosorbent assay, a new approach in honey quality control. *Czech Journal of Food Sciences, 34*(3), 233–243. https://doi.org/10.17221/422/2015-cjfs

Wang, H., Li, L., Lin, X., Bai, W., Xiao, G., & Liu, G. (2023). Composition, functional properties and safety of honey: A review. *Journal of the Science of Food and Agriculture, 103*(14), 6767–6779. https://doi.org/10.1002/jsfa.12720

Warui, M., Hansted, L., Gikungu, M., Mburu, J., Kironchi, G., & Bosselmann, A. (2019). Characterization of Kenyan honeys based on their physicochemical properties, botanical and geographical origin. *International Journal of Food Science, 2019*, 1–10. https://doi.org/10.1155/2019/2932509

Wen, Y., Wang, L., Yugan, J., Zhang, J., Su, L., Zhou, J., & Li, Y. (2017). The microbial community dynamics during the vitex honey ripening process in the honeycomb. *Frontiers in Microbiology, 8*. https://doi.org/10.3389/fmicb.2017.01649

Wu, M. C., Wu, C. Y., Klaithin, K., Tiong, K. K., & Peng, C. C. (2022). Effect of harvest time span on physicochemical properties, antioxidant, antimicrobial, and anti-inflammatory activities of Meliponinae honey. *Journal of the Science of Food and Agriculture, 102*(13), 5750–5758. https://doi.org/10.1002/jsfa.11924

Xiong, Z., Sogin, J., & Worobo, R. (2023). Microbiome analysis of raw honey reveals important factors influencing the bacterial and fungal communities. *Frontiers in Microbiology, 13*. https://doi.org/10.3389/fmicb.2022.1099522

Yanovych, D., Zasadna, Z., Rydchuk, M., Plotytsia, S., Kislova, S., & Pazderska, O. (2023). Implementation of the system of honey safety control on the example of analyzing of residues of tetracycline group antibiotics. *Scientific and Technical Bulletin of State Scientific Research Control Institute of Veterinary Medical Products and Fodder Additives and Institute of Animal Biology, 24*(1), 232–245. https://doi.org/10.36359/scivp.2023-24-1.30

Youssef, G. A., El-Boraey, A. M., & Abdel-Tawab, M. M. (2019). Eco-friendly green synthesis of silver nanoparticles from egyptian honey: Evaluating its antibacterial activities. *Egyptian Journal of Botany, 59*(3), 709–721.

Zenunović, A., Srabović, E., & Keran, H. (2020). Influence of some physicochemical properties on the quality of honey. *International Journal for Research in Applied Sciences and Biotechnology, 7*(5), 271–276. https://doi.org/10.31033/ijrasb.7.5.35

Zhang, G., Zhang, Y., Yuan, B., Tiang En, R., Li, S., Zheng, H., & Hu, F. (2024). An innovative molecular approach towards the cost-effective entomological authentication of honey. *npj Science of Food, 8*(1), 24.

Chapter 5
Emerging Trends and Future Directions in Honey Authenticity

Ayesha Mushtaq, Mehwish Jamil Noor, and Sofia Khalid

Abbreviations

ANOVA	Analysis of variance
CA	Correspondence analysis
CDA	Canonical discriminant analysis
EA/IRMS	Elemental analyzer/Isotope ratio mass spectrometry
eDNA	Environmental DNA
E-sensors	Electronic sensors
FA	Factor analysis
FTIR	Fourier transform infrared
GC-MS	Gas chromatography-mass spectrometry
HCA	Hierarchical cluster analysis
HPLC	High performance liquid chromatography
HPLC-IR	High performance liquid chromatography-infrared detector
ICP-MS	Inductively coupled plasma-mass spectrometry
ICP-OES	Inductively coupled plasma-optical emission spectroscopy
LC/IRMS	Liquid chromatography/Isotope ratio mass spectrometry
LC-MS	Liquid chromatography-mass spectrometry
LDA	Linear discriminant analysis
MANOVA	Multivariate analysis of variance
NIR	Near infrared
NMR	Nuclear magnetic resonance
OPLS-DA	Orthogonal partial least squares discriminant analysis

A. Mushtaq · M. J. Noor (✉)
Environmental Sciences Department, Fatima Jinnah Women University, Rawalpindi, Pakistan
e-mail: mehwishjamil@fjwu.edu.pk

S. Khalid
Environmental Sciences Department, Allama Iqbal Open University, Islamabad, Pakistan

M. Zafar et al. (eds.), *Pure Honey: Assurance & Authentication*,
https://doi.org/10.1007/978-3-031-98913-1_5

PCA	Principal component analysis
PLS-DA	Partial least squares discriminant analysis
SCIRA	Stable carbon isotope ratio analysis
SDA	Stepwise discriminant analysis
SEM	scanning electron microscopy
SIMCA	Soft independent modeling of class analogy

1 Authenticity and Adulteration

According to Codex Alimentarius, honey should be free from any undesirable contaminants that compromise or alter its quality and is defined as a "natural sweet product of *Apis mellifera* bees which produce honey from the nectar of plants (honey) or excretions of plants (honeydew honey)" (Codex Alimentarius, 2001). Given the global importance of honey beekeeping for the economy, the authenticity of honey should be ensured using robust and efficient analytical methods that are reliable in determining the quality of bee products along with the geographical and botanical origin. Thus, to maintain the quality of honey for international marketing, honey authenticity should be ascertained. There are two aspects of honey authenticity: origin and quality. Fraudulent activities in labeling lead to mislabeling, while mixing low-quality honey and sugars with honey results in adulteration. For the development of the honey industry, reliable detection techniques are essential to monitor the change in the composition affecting honey's nutritional properties and compromising the consumer's confidence and revenue generation by the industry (Zhang & Abdulla, 2022).

The purpose of analytical methods is to analyze honeybee products for mislabeling and adulteration. To secure the quality of honey, innovations in analytical tools are of utmost importance for precision and discrimination of origin. To date, different technological innovations have been made in spectroscopic, separation, bioanalytical, elemental, isotopic ratio, and metagenomics technologies to evaluate authenticity in origin and purity (Table 5.1).

2 Spectroscopy Techniques

Spectroscopy records spectral information without any prior sample preparation. The spectroscopic methods depend upon fingerprint profiling which further makes it feasible by reducing the time for analysis and data treatment. Spectroscopy, along with chemometric tools, helps in the discrimination and differentiation of different honey types (Table 5.1). Spectroscopic methods provide information about the adulteration as well as the origin of honey. Near infrared (NIR) and mid-infrared regions provide reliable information about the botanical and entomological source of honey (Gan et al., 2016).

Table 5.1 The analytical tools for the determination of authenticity parameters

Country	Authenticity parameters	Analytical tools	Markers	Chemometric tools	Citation
Brazil	Physicochemical properties, pollen spectrum, and multi-element composition of honey samples	Physicochemical analysis, ICP-OES, microscopy	Moisture, reducing sugars, free acidity, Na, K, Ca, Mg, Cu, Fe, Mn, Zn, pollen	PCA showed Mg, Ca, Na, and K as key discriminators among samples. HCA indicated no clear clustering	Santos et al. (2020)
Argentina, Portugal, Angola	Physicochemical parameters, elemental analysis	Physicochemical analysis, ICP-OES	Moisture, sugar content (°Brix), pH, free acidity, ash, and water activity, Na, K, Ca, Mg, P, Fe, Zn, Mn	PCA showed that minerals like K, P, and Ca had the highest discriminatory power, explaining 67% of the data variability. HCA grouped samples effectively based on their origin, with clear regional distinctions	Rios et al. (2024)

(continued)

Table 5.1 (continued)

Country	Authenticity parameters	Analytical tools	Markers	Chemometric tools	Citation
Greece	Mineral profile	ICP-OES	Ag, Al, As, B, Ba, Be, Ca, Cd, Co, Cr, Cu, Fe, Hg, Mg, Mn, Mo, Ni, Pb, Sb, Se, Si, Ti, Tl, V, Zn LDA and SDA achieved 100% correct classification of honey samples, highlighting variables like Ca, Mg, Mn, and Pb as key discriminators	FA reduced data into four components, explaining 94.1% of the total variance	Louppis et al. (2019)
Malaysia	18 multi-trace metals, classify honey samples geographically	ICP-OES	Ag, Al, As, Ba, Be, Cd, Co, Cr, Cu, Mn, Ni, Pb, Se, TI, Th, U, V, Zn	PCA revealed distinct groupings for honey samples from different regions during both rainy and dry seasons	Dzolin et al. (2023)
Canada, Europe, Tasmania, Mexico, Thailand, New Zealand, Uruguay, Germany	Lithium detection in honeybee products	ICP-OES	Lithium	–	Richter et al. (2024)

Turkey	Multi-element composition with botanical and geographical distribution	ICP-OES	Al, B, Ba, Ca, Cd, Co, Cr, Cu, Fe, K, Mg, Mn, Na, Ni, Pb, Sr, and Zn	Elements like K, Mg, and Mn were identified as the most critical variables for distinguishing honey samples based on their origins. CA grouped the honey samples into three main clusters, reflecting their botanical and geographical origins. Samples within each cluster shared similar elemental compositions	Yücel and Sultanoğlu (2013)
Ethiopia	Essential and toxic metals in honey	ICP-OES	Fe, Cu, Mn, Zn, Co, Pb, Cd, As, Hg	–	Goroya et al. (2021)
Japan, Spain, Italy, France New Zealand, China, Canada, Hungary EU, Argentina, Australia, Bulgaria, Kuba, Mexico, Romania, Taiwan, and United States	To detect adulteration in honey using SCIRA (stable carbon isotope ratio analysis)	EA/IRMS and LC/IRMS	δ13C, protein, C4 sugar	–	Kawashima et al. (2019)

(continued)

Table 5.1 (continued)

Country	Authenticity parameters	Analytical tools	Markers	Chemometric tools	Citation
Bangladesh	To detect adulteration and identify the botanical and geographical origin of honey	IRMS (isotope ratio mass spectrometry)	Carbon (δ13C), nitrogen (δ15N), and oxygen (δ18O)	The chemometric analysis of carbon and nitrogen isotopes revealed discrimination among floral types of Bangladeshi authentic honey. The geographical distinction between Bangladeshi and overseas-origin honey was observed using canonical discriminant analysis (CDA) of carbon and nitrogen isotopes	Khatun et al. (2024)
Australia and 18 other countries from Africa, Asia, Europe, North America, and Oceania	Detection of adulteration, geographical origin, and discrimination of honey	EA-IRMS and ICPMS	$\delta^{13}C$, 64 trace elements	Significant differences ($p < 0.05$) in trace element concentrations were found between honeys from different Australian regions and continents. PCA and CDA revealed distinct clusters according to geographic origin. The C5.0 model highlighted trace elements Sr, P, Mn, and K as effective markers for regional differentiation	Zhou et al. (2018)

Thailand	To identify the presence of undeclared added sugars	EA-IRMS, cavity ring-down spectrometry (CRDS)	$\delta^{13}C$	–	El Hawari et al. (2021)
Romania	To characterize Romanian unifloral and multifloral honeys by using isotopic fingerprinting	IRMS, CF-IRMS, SNIF-NMR	2H, 18O, and 13C	PCA and CDA highlighted (D/H)I and $\delta^{13}C$ as the most discriminatory variables	Dinca et al. (2015a, b)
Romania	Isotopic ratio, geographical origin of honey	IRMS	$\delta^{13}C$, $\delta^{15}N$	Multivariate analyses (CA, PCA, LDA) were conducted to explore the data further and potentially differentiate honey based on origin. Homogeneous group tests identified groups of honey types with statistically similar isotopic ratios. Multiple range tests (ANOVA) confirmed significant differences in $\delta^{13}C$ and $\delta^{15}N$ (protein fraction) for several honey-type comparisons	Dinca et al. (2015a, b)

(continued)

Table 5.1 (continued)

Country	Authenticity parameters	Analytical tools	Markers	Chemometric tools	Citation
Romania	Botanical source and harvesting year distribution	FT-Raman spectroscopy	Spectral waves for fructose, glucose, and sucrose	The best performance of the PLS-DA harvesting year classification models, before and after variable selection, based on the most suitable initial spectral range found.	David and Magdas (2024)
Romania	To detect botanical authenticity	i-Raman Spectrometry	Spectral waves	Linear discriminant analysis (LDA) correctly cross-validated 83.33% of the honey samples	Oroian and Ropciuc (2018)
Germany, USA, Turkey, Canada, Argentina, Peru, Mexico, Hungary, Brazil, Greece, Italy, Spain, Lebanon	To authenticate commercial honey using Raman spectroscopy and pattern recognition analysis	HPLC-RI, Raman spectrometer	Spectral waves for Sugars	SIMCA clusters for samples with unique sugar profiles. Raman algorithms correctly predicted non-authentic samples, confirmed by HPLC-RI	Aykas et al. (2020)

Romania	To detect honey adulterated with fructose (F), glucose (G), inverted sugar (IS), hydrolyzed inulin syrup (IN), and malt must (M)	i-Raman spectrometer	Fructose, glucose, proteins, amino acid	The classification of honey authenticity has been made using the partial least square linear discriminant analysis (PLS-LDA); a total accuracy of 96.54% (authentic honey vs. adulterated honey) was observed in authentic honey, while in the case of adulterated honey, a total accuracy of 90.00% was observed.	Oroian et al. (2018)

(continued)

Table 5.1 (continued)

Country	Authenticity parameters	Analytical tools	Markers	Chemometric tools	Citation
Romania and France	Potential of Raman spectroscopy for honey authenticity control, focusing on geographical and botanical origin	Raman spectrometer	Spectral waves for carbohydrates, amino acids	Applying the SIMCA classification method for botanical differentiation allowed a good simultaneous separation among the investigated honey groups. On comparison, the two classification models (SIMCA and ML) used for geographical discrimination provided comparable results (both models provided wrong attributions for some French samples)	Magdas et al. (2021)
Cyprus	To determine pesticide residues in honey	LC-MS/MS, GC-MS/MS	Pesticides: Coumaphos and N-(2,4-dimethylphenyl) formamide		Oymen et al. (2022)
Brazil	Determination of pyrolizidinealkaloids in honey	LC-MS/MS	Pyrrolizidine alkaloids	The statistical significance of the equations was examined with an ANOVA test for each response	Valese et al. (2021)

Indonesia	To identify added chemicals to detect frauds and adulteration	LC-MS/MS	Carbohydrate, glycosides, steroid sulphate, organic compound, phosphatidyle thanolamine, phosphatidylserine, lipid, amino alcohol, terpenoid, sterol lipid, fatty acid esters, phosphatidylserine, benzenes, flavonoids, triterpenoids, prodtanoid, antifungal compounds	–	Huyop et al. (2024)

(continued)

Table 5.1 (continued)

Country	Authenticity parameters	Analytical tools	Markers	Chemometric tools	Citation
Turkey	To identify honey for important and risky antibiotics	LC-MS/MS	Dihydrostreptomycin, streptomycin, erythromycin, sulfadimidine (sulfamethazine), and enrofloxacin	Statistical evaluation was made with one-way ANOVA using the Duncan method for brands with homogeneous distribution and with the Tamhane method for brands with nonhomogeneous distributions. The normality of the distribution of the sulfadimidine (sulfamethazine) data was determined by the Shapiro-Wilk test, and the Mann-Whitney *U* test was used for the statistical comparison of brands	Er Demirhan and Demirhan (2022)
China	To identify the authenticity of acacia honey	FTIR, SEM	Fluorescence spectra, pollens	ANOVA, PCA	Hao et al. (2021)
China	To identify adulteration of honey by sugars	IR	Fluorescence spectra	ANOVA	Yan et al. (2022)
Serbia	Detection of adulterated honey	IR	Fluorescence spectra	LDA	Dramićanin et al. (2018)

China	Detection of authentic and adulterated honey	NMR	NMR spectra for component analysis	PCA, LDA, non-targeted orthogonal partial least squares discriminant analysis (OPLS-DA)	He et al. (2020)
Brazil	Discrimination of geographical origin and fraud detection	UV-VIS, NIR, NMR	Carbohydrates, organic acids, amino acids, and alcohols	Scott & Knott test	Nunes et al. (2022)
India	To identify botanical origin of honey	DNA barcoding	DNA barcode (pollen)		Manivanan et al. (2018)
China	To identify geographical and botanical origin of honey	DNA metabarcoding	DNA barcode (Pollen)	PCA	Liu et al. (2022)

In a study in China, a significant difference in emission spectra was found between five different types of honey and sugar syrups with a considerable difference in frequency-doubled peak (FDP) intensity at a wavelength of 740 nm which declined in honey spiked with sugar syrup $\geq$10% (Yan et al., 2022). In another study, the Fourier transform infrared (FTIR) was used to identify the authenticity of acacia honey. The fluorescence intensities varied in different honey samples increasing from 2019 to 2020. The unprocessed honey had higher fluorescence intensities than the concentrated honey. The honey samples had fluorescence intensities significantly higher than sugar syrups. Sugar syrups showed extremely reduced intensities (Hao et al., 2021).

The nuclear magnetic resonance (NMR) is an efficient tool as it collects spectra in less than 5 min to identify adulteration by the quantification and structural details of different compounds in a mixture in a single run with outstanding reproducibility and repeatability. In ^{1}H, Nuclear magnetic resonance (NMR) is divided into aliphatic, carbons carbohydrate, and aromatic regions with chemical shifts in the range of 0.5–3.0 ppm, 3.0–5.5 ppm, and 5.5–10 ppm, respectively. A study discriminated adulterated honey from pure honey by identifying 65 components in honey combined with multivariate analysis. The obtained potential markers were xylobiose, melezitose, uridine, lysine, turanose, proline, β-glucose, and one unknown component (He et al., 2020). Raman spectrometry is also an efficient technique for detecting adulteration in honey. Studies show the detection of sugars, amino acids, and proteins from the spectral waves (David & Magdas, 2024; Oroian & Ropciuc, 2018; Magdas et al., 2021; Aykas et al., 2020; Oroian et al., 2018). The spectroscopic techniques are more reliable and preferred tools for nondestructive analysis, with no use of chemical reagents, generating harmful residues for pretreatment of samples. Spectroscopic techniques are effectively used with chemometrics (principal component analysis [PCA], linear discriminant analysis [LDA], orthogonal partial least squares discriminant analysis [OPLS-DA]) to discriminate and classify honey based on botanical and geographical origin (Valinger et al., 2021; Dramićanin et al., 2018; He et al., 2020; Nunes et al., 2022).

3 Separation Techniques

Separation techniques are the most widely used analytical tools in the food industry due to the ease of separating components or analytes from a complex matrix, sensitivity, accuracy, and precision. To determine the authenticity of honey, chromatographic techniques are widely used. Liquid chromatography (LC) is a preferred chromatographic technique over gas chromatography (GC). GC can only detect thermostable volatile compounds (Tsagkaris et al., 2021). Table 5.1 shows the use of separation techniques (liquid chromatography-mass spectrometry [LC-MS], high performance liquid chromatography [HPLC], gas chromatography-mass spectrometry [GC-MS]) in Cyprus, Brazil, Indonesia, and Turkey for the determination of insecticide, pyrrolizidine alkaloids, and antibiotics along with important compositional components to detect frauds and adulteration (Aykas et al., 2020; Oymen et al., 2022; Valese et al., 2021; Huyop et al., 2024; Er Demirhan & Demirhan, 2022). Honey formation is highly influenced by the quality of forage area as honey bees are exposed to many pollutants during pollen and nectar collection for beehives. One of the potential pollutant is insecticides that honeybees come in contact with during foraging and is dangerous at a very high concentration (Paradis et al., 2014).

To comply with national and international regulations regarding permissible limits of pesticide residues in honey, analytical techniques are continuously modified for rapid and better analysis, focusing mainly on simple sample preparation procedures. A recent development is QuEChERS. This is a quick, easy, cheap, effective, rugged, and safe multiclass multi-residue analytical approach to scan pesticide residues and other unknown or untargeted residues, thus providing information on all the pollutants. Thus, technological innovation that replace the traditional sample preparation method to enhance the sensitivity and quality of analytical methods are continuously improving (Prasanth et al., 2022). The use of GC and LC in combination is also effective in providing more accurate results for pesticides and other chemical pollutants. This combined technique was used in Northern Cyprus to extract and identify pesticides as well as acaricides, which are veterinary drugs, in honey. Results revealed coumaphos and N-(2,4-dimethylphenyl) formamide as the most frequent pesticides in the study area (Oymen et al., 2022).

4 Elemental Techniques

The elemental or mineral profiling of honey is highly influenced by the botanical and geographical origin of honey. The mineral content is highly influenced by the type of foraged flora and the geographical formation of a region. The elemental profiling of honey gives an insight into the macro minerals, microminerals, trace metals, and heavy metals. The elemental content varies depending on floral types, season, chemical composition, and location of apiaries. Apart from natural sources, there are also anthropogenic sources (waste disposal, industrial, and agricultural practices) affecting the elemental content (Wahid et al., 2022). Inductively coupled plasma-mass spectrometry (ICP-MS) and inductively coupled plasma-optical emission spectroscopy (ICP-OES) are the two most reliable and rapid techniques providing a multi-element profile within a single run following prior microwave digestion (Tsagkaris et al., 2021).

The elemental analysis of six types of Greek honey using ICP-OES showed a significant variation in total mineral content in all selected types. The results demonstrated that fir honey had the highest, while Asfaka honey had the lowest mineral content with respective concentrations of 182.13 mg/kg and 40.50 mg/kg. The highest Si content (16.08 mg/kg) was observed in flower honey. Some samples also showed Pb concentration higher than the permissible limit (0.1 mg/kg) as set by EU safety limits. Chemometric analysis [multivariate analysis of variance (MANOVA), factor analysis (FA), LDA, stepwise discriminant analysis (SDA)] identified significant mineral differentiation, reduced data dimensionality, and classified honey samples depending on mineral content (Louppis et al., 2019). The ICP-MS technique was employed to identify 64 trace elements in the honey samples from Australia and 18 other countries in Africa, Europe, Asia, North America, and Oceania (Zhou et al., 2018). The elemental analysis of honey quantifies essential minerals, assesses natural and anthropogenic sources of contamination, and develops a link between the botanical and geographical origin of honey studied globally (Santos et al., 2020; Rios et al., 2024; Dzolin et al., 2023; Richter et al., 2024; Yücel & Sultanoğlu, 2013; Goroya et al., 2021). The markers and chemometric techniques are presented in Table 5.1.

5 Isotopic Technique

The isotopic fingerprints are used to identify adulterated honey with sugar syrups and study the origin of honey. The adulteration can be identified based on the sources of sugars as the sugars in honey are biosynthesized through the Calvin-Benson photosynthetic cycle (C3 cycle) while the sugarcane and corn syrups follow the C4 cycle for biosynthesis. Both these types of sugar differ in the δ13C values with respective ranges of −32 to −21% and −19 to 12% for C3 and C4 sugars, respectively (Khatun et al., 2024).

The stable carbon isotope ratio analysis (SCIRA) was applied to detect protein, δ13C, and C4 sugar in honey by using elemental analyzer/isotope ratio mass spectrometry (EA/IRMS) and liquid chromatography/isotope ratio mass spectrometry (LC/IRMS). The results revealed that out of 116 samples, 39 were adulterated. A range of −33.6 to −26.5% was obtained for organic acids and δ13C values of glucose and organic acids showed a strong correlation. Hence, LC/IRMS was a useful technique for the determination of δ13C values of different components in honey (Kawashima et al., 2019). This method has its limitation as EA/IRMS cannot detect adulteration of honey with C3 sugars (rice and beet sugar). It also requires the extraction of proteins which is a time-consuming process. Other studies were also conducted to detect carbon (δ13C), nitrogen (δ15N), and oxygen (δ18O) along with chemometric analysis to identify origin (geographical and botanical) and detect adulteration and discrimination based on origin (Table 5.1).

6 DNA Barcoding Techniques

For the melissopalynological analysis, light microscopy and scanning electron microscopy (SEM) are widely used methods for the identification of pollens in honey based on the morphological features of pollens. This procedure is time-consuming, tedious, and requires expert botanical knowledge about plant species.

The slight differences in morphological features of pollen become problematic in the exact identification of flora (Galimberti et al., 2014). To avoid the problems with the identification of pollens, DNA barcoding technique is used which is an emerging technique for the identification of melliferous flora and entomological origin. Polymerase chain reaction is the technique used for the identification of specific flora. For the melliferous flora, the plastid genome is used as a biomarker for identification. This method has a limitation that one needs a DNA barcode database for the floristic diversity of the region to identify the pollens in the honey of that region. Studies have been conducted to identify pollen species using DNA barcoding for different types of honey with different botanical origins (Khansaritoreh et al., 2020; Bruni et al., 2015; Galimberti et al., 2014). This method is also helpful in identifying the flora available throughout the year during growth, dry, and dearth periods (Park & Nieh, 2017).

Another interesting and most recent bioanalytical approach is metagenomics which detects many species that come into contact with honeybees during forage activity. These organisms contain DNA that is termed "environmental DNA" (eDNA). The eDNA in honey could be from plants, microbiota from bee gut and hive organisms including honeybees, and pathogens (Galanis et al., 2022). Honey shotgun metagenomics identified pathogens as well as the botanical origin of honey with the help of eDNA (Bovo et al., 2018; Wirta et al., 2021).

7 Electronic Sensors

E-sensors (Iviz., e-tongue and e-nose) are increasingly used in the honey industry to detect adulteration. E-tongue is an electrochemical device that uses model recognition tools to process sensor signals. This can classify and quantify the physicochemical properties of honey by generating a prediction model. E-nose uses gas sensors made from polymers or metal oxides to interact with volatile compounds in honey which are responsible for the aroma. This sensor produces an electrical or optical signal that can be processed or amplified. E-sensors can classify honey based on geographical and botanical origin, detect adulteration, determine physicochemical properties, and perform pollen profiling (Veloso, et al., 2018). Besides these sensors, omics technology is another emerging tool used to analyze protein and other markers of the geographical and botanical origin of honey for authenticity. Omics is the combination of metabolomics, transcriptomics, genomics, and proteomics thus combining biological sciences, food chemistry, and bioinformatics techniques.

8 Conclusion

Along with analytical methods, technological innovations like the development of portable devices ensuring onsite testing of honey, blockchain technology, and machine learning are emerging tools for authenticity. With the emerging analytical methods and technological innovations, the enactment of a regulatory framework is mandatory to comply with the legislation and achieve an advanced and clearer understanding of authenticity with harmonized standards at a global level for better trade and quality assurance. Internationally accepted standards developed to ensure the quality and authenticity of honey are Codex Alimentarius, European regulation, and US FDA regulation. For a large dataset, chemometrics are applied by regulatory authorities and industries to monitor the authenticity of the final product, raw materials, and processes to find the desired discrimination. Honey authenticity markers and chemometrics are the most recent and advanced tools for securing the quality of honey in the global market. The authenticity of honey is an integrated approach that ensures integrity, promotes fairness, and confirms consumers' well-being by improving the quality of honey globally.

References

Aykas, D. P., Shotts, M. L., & Rodriguez-Saona, L. E. (2020). Authentication of commercial honeys based on Raman fingerprinting and pattern recognition analysis. *Food Control, 117*, 107346.

Bovo, S., Ribani, A., Utzeri, V. J., Schiavo, G., Bertolini, F., & Fontanesi, L. (2018). Shotgun metagenomics of honey DNA: Evaluation of a methodological approach to describe a multi-kingdom honey bee derived environmental DNA signature. *PLoS One, 13*(10), e0205575.

Bruni, I., Galimberti, A., Caridi, L., Scaccabarozzi, D., De Mattia, F., Casiraghi, M., & Labra, M. (2015). A DNA barcoding approach to identify plant species in multiflower honey. *Food Chemistry, 170*, 308–315.

Codex Alimentarius. (2001). Revised codex standard for honey. *Codex Alimentarius*.

David, M., & Magdas, D. A. (2024). Authentication of honey origin and harvesting year based on Raman spectroscopy and chemometrics. *Talanta Open, 10*, 100342.

Dinca, O. R., Ionete, R. E., Costinel, D., Popescu, R., Miricioiu, M. G., Stefanescu, I., & Radu, G. L. (2015a). Evaluating the origin of honey by stable isotope ratio (C and N) using multivariate statistical analysis. *Smart Energy and Sustainable Environment, 18*(1), 73.

Dinca, O. R., Ionete, R. E., Popescu, R., Costinel, D., & Radu, G. L. (2015b). Geographical and botanical origin discrimination of Romanian honey using complex stable isotope data and chemometrics. *Food Analytical Methods, 8*, 401–412.

Dramićanin, T., Lenhardt Acković, L., Zeković, I., & Dramićanin, M. D. (2018). Detection of adulterated honey by fluorescence excitation-emission matrices. *Journal of Spectroscopy, 2018*(1), 8395212.

Dzolin, S., Ibrahim, W. A. W., Mahat, N. A., Ismail, D., & Shadan, A. F. (2023). Multi-trace metals determination of peninsular Malaysia stingless bee honeys from different regions and seasons using inductively coupled plasma-optical emission spectrometry and chemometric techniques. *Malaysian Journal of Analytical Sciences, 27*(1), 216–230.

El Hawari, K., Al Iskandarani, M., Jaber, F., Ezzeddine, R., Ziller, L., Perini, M., & Camin, F. (2021). Evaluation of honey authenticity in Lebanon by analysis of carbon stable isotope

ratio using elemental analyzer and liquid chromatography coupled to isotope ratio mass spectrometry. *Journal of Mass Spectrometry, 56*(6), e4730.
Er Demirhan, B., & Demirhan, B. (2022). Detection of antibiotic residues in blossom honeys from different regions in Turkey by LC-MS/MS method. *Antibiotics, 11*(3), 357.
Galanis, A., Vardakas, P., Reczko, M., Harokopos, V., Hatzis, P., Skoulakis, E. M., et al. (2022). Bee foraging preferences, microbiota and pathogens revealed by direct shotgun metagenomics of honey. *Molecular Ecology Resources, 22*(7), 2506–2523.
Galimberti, A., De Mattia, F., Bruni, I., Scaccabarozzi, D., Sandionigi, A., Barbuto, M., et al. (2014). A DNA barcoding approach to characterize pollen collected by honeybees. *PLoS One, 9*(10), e109363.
Gan, Z., et al. (2016). Using sensor and spectral analysis to classify botanical origin and determine adulteration of raw honey. *Journal of Food Engineering, 178*, 151–158.
Goroya, K. G., Abadi, G., & Asresahegn, Y. A. (2021). Determination of heavy and trace metals in honey using inductively coupled plasma optical emission spectroscopy (ICP-OES) in South Eastern Zone of Tigray Region, Northern Ethiopia. *The Journal of Pure and Applied Chemistry Research, 10*(2), 113–122.
Hao, S., Li, J., Liu, X., Yuan, J., Yuan, W., Tian, Y., & Xuan, H. (2021). Authentication of acacia honey using fluorescence spectroscopy. *Food Control, 130*, 108327.
He, C., Liu, Y., Liu, H., Zheng, X., Shen, G., & Feng, J. (2020). Compositional identification and authentication of Chinese honeys by 1H NMR combined with multivariate analysis. *Food Research International, 130*, 108936.
Huyop, F., Hussin, H., Zulkifli, R. M., Huda, N., Antara, N. S., Wahab, R. A., et al. (2024). LC-MS/MS-QTOF dataset of chemical compounds detected in honey samples from Bali and Lombok, Indonesia. *Data in Brief, 57*, 110884.
Kawashima, H., Suto, M., & Suto, N. (2019). Stable carbon isotope ratios for organic acids in commercial honey samples. *Food Chemistry, 289*, 49–55.
Khansaritoreh, E., Salmaki, Y., Ramezani, E., Azirani, T. A., Keller, A., Neumann, K., et al. (2020). Employing DNA metabarcoding to determine the geographical origin of honey. *Heliyon, 6*(11).
Khatun, M. A., Yoshimura, J., Yoshida, M., Suzuki, Y., Huque, R., Kelly, S. D., & Munshi, M. K. (2024). Isotopic characteristics (δ13C, δ15N, and δ18O) of honey from Bangladesh retail markets: Investigating sugar manipulation, botanical and geographical authentication. *Food Chemistry, 435*, 137612.
Liu, S., Lang, D., Meng, G., Hu, J., Tang, M., & Zhou, X. (2022). Tracing the origin of honey products based on metagenomics and machine learning. *Food Chemistry, 371*, 131066.
Louppis, A. P., Karabagias, I. K., Papastephanou, C., & Badeka, A. (2019). Two-way characterization of beekeepers' honey according to botanical origin on the basis of mineral content analysis using ICP-OES implemented with multiple chemometric tools. *Food, 8*(6), 210.
Magdas, D. A., Guyon, F., Berghian-Grosan, C., & Molnar, C. M. (2021). Challenges and a step forward in honey classification based on Raman spectroscopy. *Food Control, 123*, 107769.
Manivanan, P., Rajagopalan, S. M., & Subbarayalu, M. (2018). Studies on authentication of true source of honey using pollen DNA barcoding. *Journal of Entomology and Zoology Studies, 6*(3), 255–261.
Nunes, A., Azevedo, G. Z., Dos Santos, B. R., Borges, C. V., Lima, G. P. P., Crocoli, L. C., et al. (2022). Characterization of Brazilian floral honey produced in the states of Santa Catarina and São Paulo through ultraviolet–visible (UV–vis), near-infrared (NIR), and nuclear magnetic resonance (NMR) spectroscopy. *Food Research International, 162*, 111913.
Oroian, M., & Ropciuc, S. (2018). Botanical authentication of honeys based on Raman spectra. *Journal of Food Measurement and Characterization, 12*, 545–554.
Oroian, M., Ropciuc, S., & Paduret, S. (2018). Honey adulteration detection using Raman spectroscopy. *Food Analytical Methods, 11*, 959–968.
Oymen, B., Aşır, S., Türkmen, D., & Denizli, A. (2022). Determination of multi-pesticide residues in honey with a modified QuEChERS procedure followed by LC-MS/MS and GC-MS/

MS. *Journal of Apicultural Research, 61*(4), 530–542. https://doi.org/10.1080/00218839.2021.2017540
Paradis, D., Bérail, G., Bonmatin, J. M., & Belzunces, L. P. (2014). Sensitive analytical methods for 22 relevant insecticides of 3 chemical families in honey by GC-MS/MS and LC-MS/MS. *Analytical and Bioanalytical Chemistry, 406*, 621–633.
Park, B., & Nieh, J. C. (2017). Seasonal trends in honey bee pollen foraging revealed through DNA barcoding of bee-collected pollen. *Insectes Sociaux, 64*, 425–437.
Prasanth, J., Vincy, M. V., & Brilliant, R. (2022). Screening and quantitation of pesticide residues in Indian honey samples by LC-MS/MS and GC-MS/MS. *Indian Journal of Science and Technology, 15*, 1112–1123.
Richter, A., Rüdiger, M., Schumacher, D. M., & Kürbis, C. (2024). ICP-OES analysis of Lithium in honey, royal jelly, bee bread, propolis, and bees following microwave-assisted sample preparation. *Journal of Consumer Protection and Food Safety*, 1–6.
Rios, F. T., Lobo, M. O., Castanheira, I., Delgado, I., Nascimiento, A., & Sammán, N. (2024). Chemometric and multielement techniques for the exploratory analysis of honey quality from different geographical origin. *Journal of Agricultural Food Science and Biotechnology, 2*(1), 54–63.
Santos Júnior, A. D. F., Brandão, G. C., Santos Júnior, M. C., Santos, F. A. D., Magalhaes, H. I., & Korn, M. G. A. (2020). Multi-element composition, physicochemical and pollen attributes of honeys from the Paraguaçu River (Bahia, Brazil) by inductively coupled plasma-optical emission spectrometry (ICP OES). *Anais da Academia Brasileira de Ciências, 92*, e20181196.
Tsagkaris, A. S., Koulis, G. A., Danezis, G. P., Martakos, I., Dasenaki, M., Georgiou, C. A., & Thomaidis, N. S. (2021). Honey authenticity: Analytical techniques, state of the art and challenges. *RSC Advances, 11*(19), 11273–11294.
Valese, A. C., Daguer, H., Muller, C. M. O., Molognoni, L., da Luz, C. F. P., de Barcellos Falkenberg, D., et al. (2021). Quantification of pyrrolizidine alkaloids in Senecio brasiliensis, beehive pollen, and honey by LC-MS/MS. *Journal of Environmental Science and Health Part B, 56*(7), 685–694.
Valinger, D., Longin, L., Grbeš, F., Benković, M., Jurina, T., Kljusurić, J. G., & Tušek, A. J. (2021). Detection of honey adulteration–The potential of UV-VIS and NIR spectroscopy coupled with multivariate analysis. *LWT, 145*, 111316.
Veloso, A. C., Sousa, M. E., Estevinho, L., Dias, L. G., & Peres, A. M. (2018). Honey evaluation using electronic tongues: An overview. *Chemosensors, 6*(3), 28.
Wahid, D. N. A., Jawan, R., Nazarie, W. F. W. M., Gansau, J. A., Syahir, A., & Sabullah, M. K. (2022, August). Comparison of macro, trace and ultra-trace minerals contents of kelulut honey (Heterotrigona itama sp) from the West Coast of Sabah. *Journal of Physics: Conference Series, 2314*(1), 012002.
Wirta, H., Abrego, N., Miller, K., Roslin, T., & Vesterinen, E. (2021). DNA traces the origin of honey by identifying plants, bacteria and fungi. *Scientific Reports, 11*(1), 4798.
Yan, S., Sun, M., Wang, X., Shan, J., & Xue, X. (2022). A novel, rapid screening technique for sugar syrup adulteration in honey using fluorescence spectroscopy. *Food, 11*(15), 2316.
Yücel, Y., & Sultanoğlu, P. (2013). Characterization of Hatay honeys according to their multi-element analysis using ICP-OES combined with chemometrics. *Food Chemistry, 140*(1–2), 231–237.
Zhang, G., & Abdulla, W. (2022). On honey authentication and adulterant detection techniques. *Food Control, 138*, 108992.
Zhou, X., Taylor, M. P., Salouros, H., & Prasad, S. (2018). Authenticity and geographic origin of global honeys determined using carbon isotope ratios and trace elements. *Scientific Reports, 8*(1), 14639.

Chapter 6
Honey: The "Marvel Molecule" of Mother Nature Augments Some Cutting-Edge Technologies for Analytical Validation

Sampriti Sarkar, Moutushi Sen, Pulakesh Parai, Subrata Mitra, Sukumar Sarkar, and Dipan Adhikari

1 Introduction

Evidence from Stone Age paintings demonstrates that the use of bee products, such as honey, for medicinal purposes dates back 8000 years. Honey has been used as a medication for a long time, as evidenced by ancient scrolls, tablets, and books such as the Bible, Veda (a literature written by the Hindus), Hippocrates (460–357 BC), Egyptian papyri (1900–1250 BC), and Sumerian clay tablets (6200 BC) (Rahman et al., 2014). The bees were given the command by the Lord to establish their colonies in hills, on trees, and inside human homes. Their bodies produce a drink that comes in a variety of colors and is said to be therapeutic for humans. This is undoubtedly a sign for people who are thoughtful. In this century, man is infected with different deadly diseases and chemotherapy-based current stream treatment techniques mimic multidrug resistance and a number of other negative effects (Anand et al., 2022). This compels one to look for alternatives. Natural goods are being considered as a workable substitute strategy to lessen the growing burden of illnesses and some of its inevitable negative effects (Dzobo, 2022). Researchers have recently become interested in honey as a supplementary and alternative medicine because of its high medicinal properties (Miguel et al., 2017). According to the European Union, honey is a viscous, sugary liquid that is naturally created by honeybees (*Apis mellifera*) from flower sap or emissions of any plant parts that they gather and this golden elixir created by nature's crafty pollinators; change after mixed up to specific components of themselves, deposit, desiccate, stock, and then place in honeycombs

S. Sarkar
Alma Mater, School of Biosciences & Technology, Vellore Institute of Technology, Vellore, Tamil Nadu, India

M. Sen · P. Parai · S. Mitra · S. Sarkar · D. Adhikari (✉)
Post Graduate Department of Botany, Hooghly Mohsin College, Chinsurah, Hooghly, West Bengal, India

M. Zafar et al. (eds.), *Pure Honey: Assurance & Authentication*,
https://doi.org/10.1007/978-3-031-98913-1_6

to grow and ripen gradually. According to Almeida-Muradian (2013), nectar is an aqueous solution of sugars that is produced by the hypopharyngeal glands of bees. The enzyme invertase hydrolyzes the disaccharide sucrose in nectar to monosaccharides of glucose and fructose, turning nectar into honey. After being placed in the cell, honey is concentrated by removing water from it, then sealed and kept in cerumen pots or beeswax combs (Vit et al., 1998, 2004, 2013). Since prehistoric times, honey has captivated human attention for almost 5500 years. The first written records of honey were discovered in 2100–2000 BC on a Sumerian tablet. The majority of ancient civilizations, including the Mayans, Babylonians, Greeks, Chinese, Egyptians, Romans, and Egyptians, used honey for both nutritional and therapeutic purposes (Samarghandian et al., 2017). Combining traditional knowledge with modern scientific research has led to the discovery that this "golden tincture" possesses a complex tapestry of pharmacological and therapeutic qualities. Honey is the only insect-derived natural creation, and it has nutritional, cosmetic, therapeutic, and industrial values (Samarghandian et al., 2017). Honey is reviewed as a balanced diet and equally popular among males and females of all ages (Palma-Morales et al., 2023). Honey does not need to be refrigerated, it never spoils, and it can also be stored unopened at room temperature in a dry place (Hassapidou et al., 2006). One of honey's most beloved applications is as a natural sweetener, as it contains fructose and has been used for centuries. These days, honey is touted as a cure-all for a variety of human ailments in general publications, journals, and natural product brochures, among other unidentified uses. Honey may offer protective effects on the neurological, gastrointestinal, circulatory, and respiratory systems in addition to its antioxidant, anti-inflammatory, antimicrobial, antiviral, and antibacterial properties (Miguel et al., 2017).

1.1 Classification of Honey

Honey can be classified based on where it comes from or how it was harvested and processed. It is divided into blossom, honeydew, monofloral, and multifloral honeys based on where it originated. The primary source of blossom honey is floral juice.

(a) Bees gather honeydew from plant fluids to produce honeydew, often known as forest honey.
(b) Monofloral honey, also known as plant-like citrus, manuka, and acacia honey, is mostly obtained from a single plant source with a total pollen concentration of more than 45% from the same plants (Ling & Sowndhararajan, 2020).
(c) According to de Toledo and Dechechi Chambó (2020), multifloral honey is defined as polyfloral honey with many plant origins, none of which are major, such as meadow blossom honey and woodland honey.
(d) Honey is divided into distinct categories and given distinct descriptions depending on how it is packaged, produced, or presented. These include raw,

ultra-filtered, granulated, chunk, pasteurized, dried, strained, ultrasonicated, comb, and whipped honey (Zielinski et al., 2014).

(e) Beehive contents can be extracted, settled, or strained to get raw honey, which can be produced with very little processing and no heat treatment (Suchart, 2023).

1.2 *Crystallization of Honey*

The texture and color of honey are affected by crystallization, but the flavor, quality, or nutritional status remains unchanged. Crystallization causes honey to become lighter in color because it forms glucose crystals. The amount of water also influences the pace of crystallization; the more glucose present, the faster the process is; the less water, the faster the process is (Olaitan et al., 2007). The rate of crystallization is also influenced by the honey's trace components, specific sugar blend, and storage circumstances. While goldenrod honey frequently crystallizes in the comb itself, tulepo and acacia-type honeys crystallize incredibly slowly. Heat affects the physical properties of honey, including the previously mentioned crystallization. Thus, honey can crystallize between 10 and 21 °C, but it can also be obtained more smoothly by indirectly heating honey to a temperature of 49 degrees Celsius. Honey is an industrial food sweetener that doesn't require prior processing, and its production is expanding globally (Ajibola, 2016).

1.3 *Strained Honey*

The procedure of straining honey removes dirt, waxes, plant exudates, and propolis; it does not remove minerals, pollen, or enzymes from the honey combination. Applying 65–77 °C will result in ultra-filtered honey, which easily melts and goes through a fine filter under high pressure to remove pollen grains and unnecessary particles (Bogdanov et al., 2008).

1.4 *Ultrasonicated Honey*

The non-heating process used to obtain ultrasonicated honey is said to stop fungus development and prevent crystallization. Large-mouthed bottles containing chunks of comb honey immersed in melted honey have been filled with chunk honey.

Comb honey is made from the wax comb of honeybees and is often acquired by cutting the comb into chunks using wooden frames before packaging it in honey supers (Ajibola, 2016). Aeration and anticaking chemicals are used to create completely compact, nonsticky granules from dried honey, which lacks moisture.

To create honey that is uniformly smooth and spreadable, whipped or creamed honey is made out of a large number of tiny crystals.

A directive from the European Union Council stipulates that baker's honey can also be used industrially or as a component in the food sector.

2 Honey: Composition and Properties

Due to its growing body of scientific knowledge and historical connection to human health, honey is an intriguing topic for further research. By delving into its pharmacological intricacies and shedding light on the mechanisms behind its therapeutic effects, we want to bridge the gap between conventional knowledge and modern scientific understanding. Further research is required to fully grasp the medicinal potential of honey. This study attempts to give a comprehensive summary of the current body of knowledge regarding the pharmacological characteristics and therapeutic applications of honey.

As of right now, around 300 types of honey are recognized (Samarghandian et al., 2017). The numerous varieties of nectar that bees collect are linked to varieties. Ninety-five to ninety-seven percent of the dry weight of honey is composed of carbohydrates, which are mostly made up of the monosaccharides fructose (38%) and glucose (31%) as the two primary sugars. In addition to fructose and glucose, honey also includes smaller amounts of disaccharides, galactose, alpha- and beta-trehalose, gentiobiose, laminaribiose, and trisaccharides (melezitose, maltotriose, 1-ketose, panose, isomaltose, glucose, erlose, isomaltotriose, theanderose, centose, isopanose, and maltopentaose). Important compounds found in honey include vitamins, minerals, proteins, amino acids, and organic acids (Fig. 6.1).

Many of these sugars are formed during the honey ripening and maturation times. Gluconic acid, a product of glucose oxidation, is the main organic acid that is present in honey. It has been shown to increase calcium absorption (Samarghandian et al., 2017); in addition, small amounts of acetic, formic, and citric acids have been found. These organic acids are responsible for the acidic (pH between 3.2 and 4.5) property of honey (Ratiu et al., 2020). These organic acids altogether constitute the typical tartness to the flavor and taste of honey samples. Other organic acids found in honey are malic, lactic, formic, fumaric, malonic, aspartic acid, methylmalonic, succinic, citric, oxalic, butyric, acetic, glutaric, acetoglutaric, gluconic, glutamic, galacturonic, shikimic, isocitric, quinic, etc. (Chen et al., 2000; Mato et al., 2006).

All of the water-soluble vitamins exist in honey, with vitamin C being the most frequent (Fig. 6.2). The enzymes included in honey, including glucose oxidase, saccharase, catalase, and diastase, are important for both the production of honey and its nutritional value (Cianciosi et al., 2018). Glucose oxidase transforms glucose to δ-gluconolactone, which hydrolyzes to gluconic acid, the main acid, and potentially antibacterial H_2O_2. Invertase breaks down sucrose into fructose and glucose. The enzyme amylase breaks down long starch chains to produce maltose and dextrin, which improves the quality of honey. Another enzyme that converts H_2O_2 into O_2

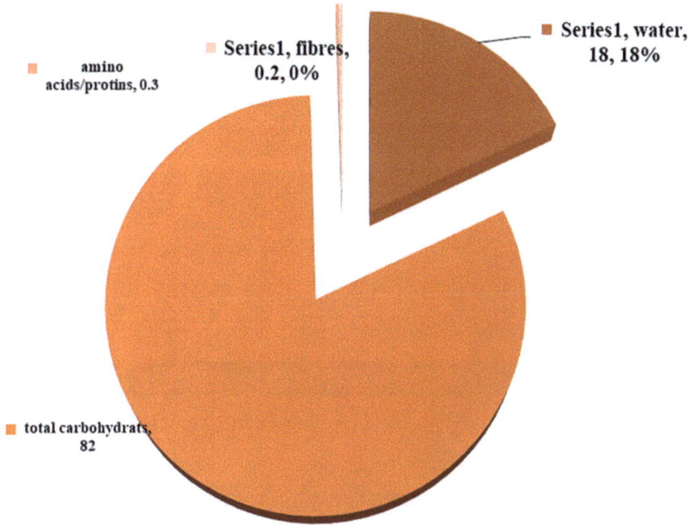

Fig. 6.1 Pie chart depicts the different percentages of dietary components (% wise)

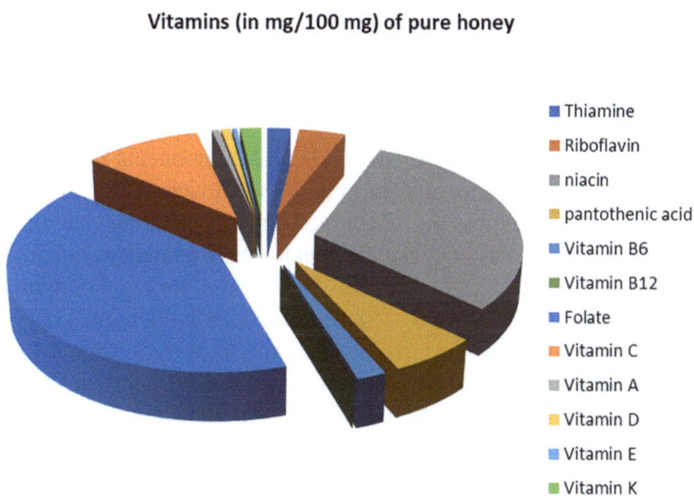

Fig. 6.2 Pie chart depicts the relative proportion of vitamins (in mg/100 g) of pure honey

and H_2O is called catalase. The organic acids in honey are what give it its unique flavor, stability, and acidity (Molan, 1999, 2001).

Additionally, the amount of critical minerals varies based on the region, processing methods, and source of the plant. The elements that are most prevalent are calcium, phosphorus, copper, magnesium, potassium, sodium, zinc, iron, manganese, and selenium (Fig. 6.3) (Khan et al., 2018).

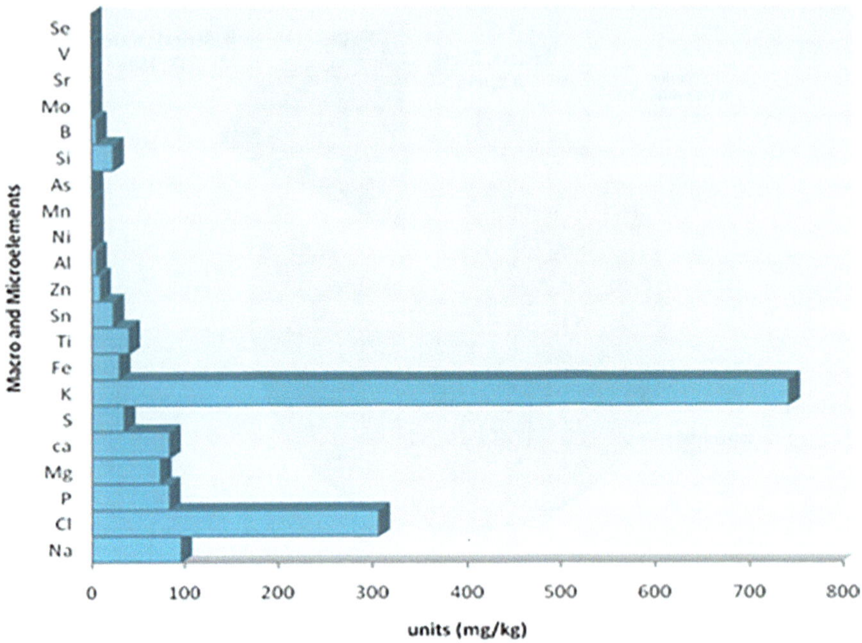

Fig. 6.3 Bar graph represents the relative units (mg/kg) of different macro and micro essential mineral compositions in pure organic honey

Honey has varying quantities of minerals and components depending on the source of the plant and its geographical region (Vincēviča-Gaile, 2010). Minerals are created spontaneously by the decomposition of plant and animal remains through geological processes that take place in the environment as inanimate, compact elements (Belitz et al., 2009). Minerals are essential for regulating the metabolism of living things (Gopalan et al., 1989). According to the body's requirements, they are divided into three groups: major, trace, and ultra-trace elements.

Major minerals are those that are required in amounts greater than 50 mg per day, while trace elements are those that are needed in amounts less than 50 mg per day. Ultra-trace elements are those that are required in values even lower than 1 μg/g and are often found in diets at less than 50 ng/g (Belitz et al., 2009). For good strength, those that start naturally in specific micro- or trace minerals are beneficial (Fig. 6.3). Alternatively, if the source is an inorganic or metallic component that constitutes heavy metals, a fivefold specific gravity of H_2O develops toxicity (Ajibola et al., 2012). Because they accumulate in the biological system, heavy metals (Fig. 6.4) are toxic or deadly at smaller doses (Zugravu et al., 2009).

Honey has varying quantities of minerals and components depending on the source of the plant and its geographical region (Vincēviča-Gaile, 2010). Many essential trace components are detected in honey, such as silicon (Si), rubidium (RB), vanadium (V), zirconium (Zr), lithium (Li), and strontium (Sr). However,

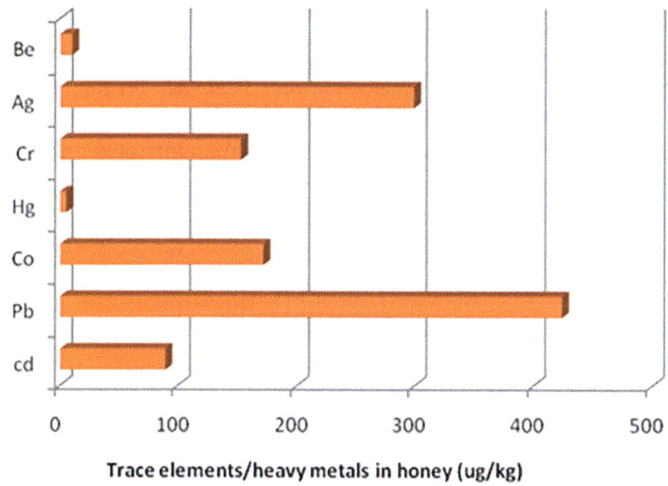

Fig. 6.4 Bar graph represents the relative units (μg/kg) of different trace and heavy metals in pure organic honey

some heavy metals such as lead (Pb), cadmium (Cd), and arsenic (As) are present as pollutants (Vorlova & Pridal, 2002).

Honey has about 600 volatile compounds, which may be involved in some of its possible medicinal benefits (Ajibola et al., 2012). Aldehydes, alcohols, hydrocarbons, ketones, acid esters, benzene and its derivatives, pyran, terpene and its derivatives, norisoprenoids, sulfur, furan, and cyclic compounds are among the volatile compounds in honey, albeit they are normally low in concentration (Manyi-Loh et al., 2011; Barra et al., 2010). Peña et al. (2004) and Castro-Vázquez et al. (2006) identified the following monoterpenes and monoterpenoids in multiple honey samples: dihydrocitronellol, β-terpineol, β-citronellol, citronellal, geranyl acetone, limonene, β-pinene, tetrahydrogeraniol, carvacrol, p-cymene, 1,8-cineol, camphor, isoborneol, and p-cymenol. These substances are well-known for their potency against a variety of microbes, including viruses, fungi, and Gram-negative and Gram-positive bacteria, according to Abd El-Moaty (2010).

2.1 Flavonoids and Polyphenols

According to Cianciosi et al. (2018), flavonoids—quercetin, pinocembrin, pinobanksin, chrysin, galangin, kaempferol, and luteolin—are the main phenolic components found in honey. Aromatic acids have an organic acid function (C6–C1 skeleton) and an aromatic ring. Because they have an organic carboxylic acid activity and a phenolic ring, phenolic compounds are an example of aromatic acids. Numerous plant species contain phenolic acids (Olas, 2020). According to Dixon and Pasinetti (2010), flavonoids are specialized metabolites found in plants that serve a variety of purposes, including plant coloring, UV filtration, and symbiotic nitrogen fixation. Plants contain a large variety of flavonoids, which have the basic molecular structure 2-phenyl-1,4-benzopyrone. Benzoic, ferulic, gallic, chlorogenic, caffeic, p-coumaric, ellagic, and syringic acids are examples of phenolic acids obtained from plants. Antibacterial, anti-inflammatory, and antioxidant properties are exhibited by phenolic substances. The following is a list of the constituents of two significant classes of phytochemicals, phenolic acids, and flavonoids, derived from several significant honey variations across the globe (Fig. 6.5).

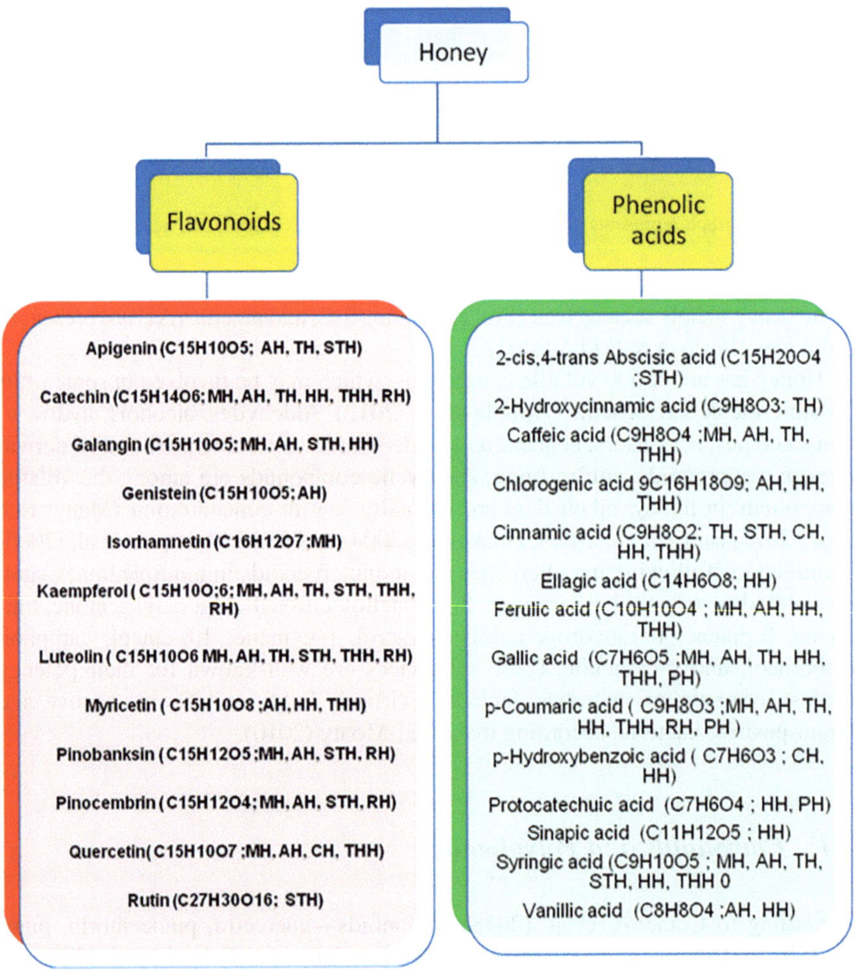

Fig. 6.5 Composition and source of different polyphenolics and flavonoids in honey, manuka honey (MH); acacia honey (AH); tualang honey (TH); strawberry tree honey (STH); clover honey (CH); heather honey (HH); thyme honey (THH); rosemary honey (RH); pine honey (Cianciosi et al., 2018)

2.2 *Volatile Compounds*

In honey, there are about 600 volatile organic compounds (VOCs) known to exist (Table 6.1). Organic compounds classified as volatiles have a high vapor pressure at room temperature. Honey has been shown to include seven main groups, which include aldehydes, ketones, acids, alcohols, esters, hydrocarbons, and cyclic compounds. Although the percentage of volatile organic compounds (VOCs) in honey is low, the taste, smell, color, and texture of the honey are all influenced by the kinds of plants and flowers that bees visit (Manyi-Loh et al., 2011). Some VOCs are produced during the manufacturing or storing of honey, while others are derived from plants or nectar sources (Machado et al., 2020). Heat treatment of honey causes the Maillard reaction, which is a nonenzymatic browning reaction between sugars and amino acids that produces or transforms volatile organic compounds (VOCs) (Charnock et al., 2022).

2.2.1 Hydroxymethylfurfuraldehyde (HMF)

In trace amounts, hydroxymethylfurfuraldehyde (HMF) is also present. Honey microbial contamination (HMF) has been linked to honey adulteration because it can result from the breakdown of fructose in the presence of acid. However, studies have shown that even fresh honey contains trace amounts of HMF (Rahman et al., 2017), which can easily rise if the honey is stored at moderate or high temperatures. As a result, honey should be stored in a refrigerator or other cool place to minimize HMF levels, as HMF is a major factor in honey quality and marketing (Fig. 6.6).

2.2.2 Methylglyoxal (MGO)

Honey contains a natural protein glycation agent called MGO (CH_3-CO-CH=O or $C_3H_4O_2$). According to Daniels et al. (2016), MGO is the main antimicrobial ingredient in manuka monofloral honey, which is made from the *Leptospermum scoparium* tree in New Zealand. Manuka honey is highly regarded for its potent antibacterial properties, also referred to as "non-peroxide" properties (Al-Kafaween et al., 2023). This honey's MGO content (Fig. 6.7), which Atrott and Henle (2009) identified as its active compound, is strongly correlated with its antibacterial properties. MGO is produced over time by converting DHA, a naturally occurring compound that is present in large amounts in the nectar of manuka flowers, during the ripening and storage of honey (Rabie et al., 2016). MGO is a chemical that destroys microorganisms, which induces the alteration of bacterial flagella, thus impeding adhesion to the surfaces and accordingly suppressing the formation of bacterial biofilms (Rabie et al., 2016).

Table 6.1 Different classes of volatile compounds adopted (after Rahman et al., 2017)

Chemical class	Aroma profile	Citation
1 hydrocarbons	Hexane (gasoline-like odor) Cyclohexane (mild sweet odor) n-decanal (citrus odor) Benzene (gasoline-like odor) Alpha-pinene (odor of pine) Hexanal (strong green grass odor) Heptanal (fatty, pungent odor) Octanal (strong, fruity odor)	Cajka et al. (2009), Silici, (2011) and Wolski et al. (2006)
Acids and esters	Nonanoic acid (coconut aroma) Propanoic acid (slightly pungent disagreeable rancid odor) Butyric acid (unpleasant, rancid odor) Benzoic acid (odorless or with a slight benzaldehyde odor) Methyl salicylate (wintergreen-like odor) Phosphoric acid (odorless) Fumaric acid (odorless) Benzeneacetic acid (floral odor) Butanoic acid (unpleasant, rancid odor) Formic acid (pungent, penetrating odor) Acetic acid sour (vinegar-like odor). 2-methyl propanoic acid (sharp, butter-fat-like odor) 2-ethyl-hexanoic acid (mild odor) 3-methyl butanoic acid (disagreeable, rancid-cheese odor)	Cajka et al. (2009), Silici (2011) and Wolski et al. (2006)
Ketones	2-Cyclopentanone (distinctive ethereal odor, somewhat like peppermint acetone fruity odor) 2,5-Furandione (irritating, choking odor) 1-Phenylethanone (orange blossom or jasmine-like) 2,3-Butanodione (odorless) 3-Hydroxy-2-butanone (bland, woody, yogurt odor)	Cajka et al. (2009), Silici (2011) and Wolski et al. (2006)
Benzene derivatives	Toluene (sweet, pungent benzene-like odor) Vinylbenzene (sweet, floral odor) 1,2-dimethyl benzene (sweet) 1,3-dimethyl benzene (sweet odor; like benzene: Characteristically aromatic) 1-Methyl-4-propan-2-yl-benzene (sweetish aromatic odor)	(Barra et al. (2010), Cajka et al. (2009) and Wolski et al. (2006)
Terpenes and its derivatives	Linalool (floral, spicy, wood odor) α pinene (odor of turpentine) B-pinene (piney, turpentine-like odor), limonene (odor of orange peels (orange oil sweet) Camphene (camphor-like) Myrtenol Trans-anethol (anise oil odor) p-cymene (sweetish aromatic odor) Nerol (sweet rose odor) Cumene (gasoline-like odor)	Montenegro et al. (2009)

(continued)

Table 6.1 (continued)

Chemical class	Aroma profile	Citation
Furan derivatives	Furfuryl alcohol (faint burning odor) 5-Hydroxymethylfurfural (odor of chamomile flower) Furfural (peculiar odor resembling the odor of benzaldehyde) 2-Furancarboxaldehyde (almond-like odor) 2-Furanmethanol (faint burning odor) Butyrolactone (pleasant odor)	Barra et al. (2010), Cajka et al. (2009), Odeh et al. (2013), Silici (2011) and Wolski et al. (2006)
Alcohols	Phenylethyl alcohol (characteristic rose-like odor), ethanol (mild, rather pleasant; like wine or whiskey), 1-butanol (harsh fusel odor with banana), methanol (characteristic pungent odor), 1,3-butanediol (odorless when pure), 2-methyl-1-propanol (sweet, musty odor), 2-furanmethanol (faint burning odor), benzyl alcohol (faint aromatic odor), linalool (floral, spicy, wood odor), isoborneol (pungent camphor-like), nonanol (floral odor), thymol/carvacrol (odor reminiscent of thyme), hexan-1-ol (characteristic; sweet alcohol; pleasant), heptan-1-ol (faint, aromatic, fatty), 2-phenylethanol (characteristic rose-like odor), decan-1-ol (resemble orange flowers), octan-1-ol (fresh, orange-rose odor), p-cresol (phenolic odor), (benzenol)/phenol (somewhat sickening sweet and acrid odor), acetophenone (sweet, pungent odor of acacia)	Barra et al. (2010), Cajka et al. (2009), Odeh et al. (2013), Silici (2011) and Wolski et al. (2006)
Pyran derivative	2,3-Dihydro-3,5-dihydroxyl-6-methyl-4H-pyran-4-one—Valerolactone	Barra et al. (2010) and Wolski et al. (2006)
Others	Thiophene (slight aromatic odor resembling benzene), propanenitrile (pleasant, sweetish, ethereal odor). Acetamide (odorless when pure, but frequently has a mousy odor), guanidine	Silici (2011) and Wolski et al. (2006)

2.2.3 Hydrogen Peroxide (H_2O_2)

An effective antibacterial is hydrogen peroxide, often known as oxygenated water (H_2O_2) (Zhu et al., 2017). In varied amounts, H_2O_2 can be found in all varieties of honey (Guttentag et al., 2021a, b). According to Kus et al. (2016), glucose oxidase generates H_2O_2 when glucose is oxidized enzymatically (Fig. 6.8).

As to Nolan et al. (2019), the primary inhibitor present in the majority of honeys is this antimicrobial component. The fact that this inhibin is exclusive to unripe honey should be noted (Brudzynski, 2023). However, mature honey has very little H_2O_2, which very marginally inhibits bacterial growth. Thus, oxygenated water that is synthesized has two sources: vegetal, as glucose is derived from foraged nectar, and animal, since worker bee hypopharyngeal glands secrete glucose oxidase during the nectar conversion process into honey (Desmouliere et al., 2013). Additionally, the gluconic acid that is produced (Fig. 6.9) raises the acidity of honey.

This inhibits the growth of harmful bacteria (Kwakman & Zaat, 2012). Free radicals, which are thought to be extremely potent cytotoxic oxidants capable of

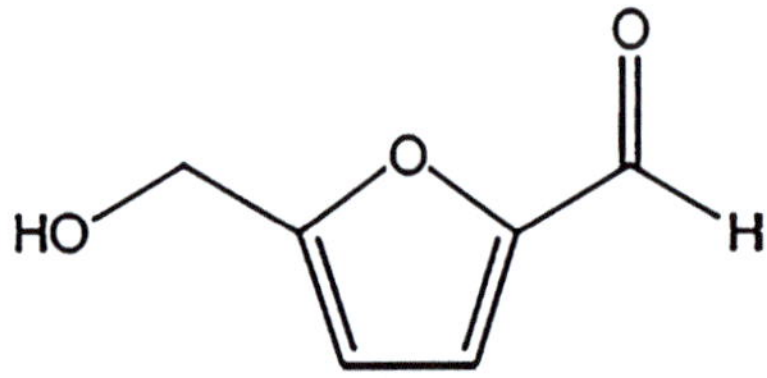

Fig. 6.6 Chemical structure of hydroxymethylfurfuraldehyde (HMF). (After Rahman et al., 2017)

Fig. 6.7 Chemical structure of methylglyoxal (MGO). (After Alvarez-Suarez et al., 2009)

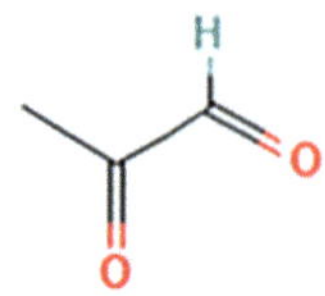

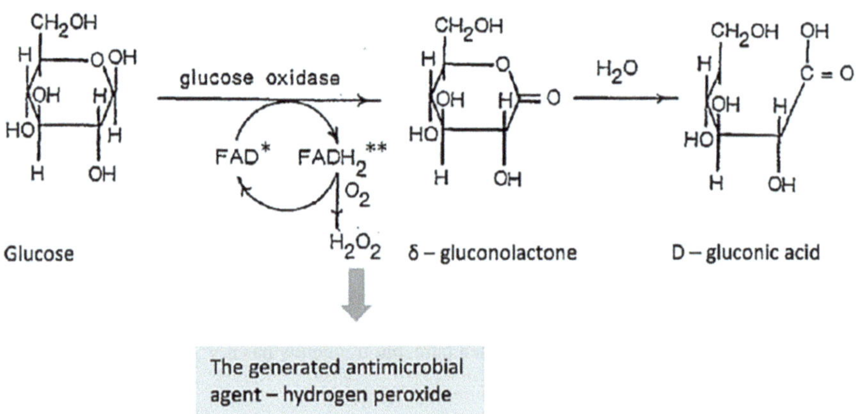

Fig. 6.8 Enzymatic conversion of glucose to D-gluconic acid by glucose oxidase. (After Kus et al., 2016)

Fig. 6.9 Structure of D-gluconic acid, Source Pubchem

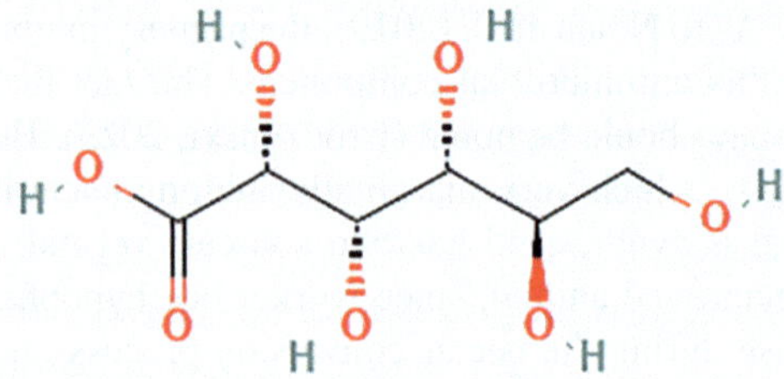

rupturing bacterial deoxyribonucleic acid (DNA) strands (Adhikari, 2024), are responsible for the antibacterial qualities (Brudzynski & Lannigan, 2012). According to a study, H_2O_2 is mostly responsible for honey's antibacterial effect (Brudzynski, 2023). Certain honeys possess an attribute known as "non-peroxidasic" activity, meaning they maintain their potent antibacterial properties even when their peroxidasic activity is neutralized by heat or by catalase.

2.2.4 Bee-Derived Antimicrobial Peptides and Short and Cyclic Peptides in Honey

Using the LC-MS and chemical structure databases, three Ziziphus honeys (*Ziziphus spina-christi*) and samples of Acacia honey (*Acacia tortilis* and *Acacia hamulosa*) were recently evaluated for their short and cyclic peptide contents. Three short peptides were found in the *A. tortilis honey*: HWCC, DSST, and ECH (Fig. 6.10). The *A. tortilis* honey contained three short peptides, which were:

1. Chemical formula $C_{23}H_{29}N_7O_5S_2$ is the formula for His-Trp-Cys-Cys (HWCC) Royal Society of Chemistry (2021), and it can be found in the ChemSpider database (www.chemspider.com). Cross connections between His-Cys and Trp-Cys may have an impact on host enzyme control (Fujieda, 2020). Tryptophan, cysteine, tyrosine, or methionine is found in dipeptides that have antioxidant capacity (Zheng et al., 2016).
2. The molecular formula for DSST (Asp-Ser-Ser-Thr), which was $C_{14}H_{24}N_4O_{10}$, was obtained from the Molbase database (MOLBASE, 2021). According to reports, cations and the substrates that go along with them are bound and transported by proteins and short peptides that contain aspartic acid, serine, and threonine (Simonin et al., 2015; Sun et al., 2020). Therefore, the *A. tortilis* honey's mineral content may be indicated by the presence of the DSST short peptide.
3. When looking up the molecular formula of $C_{14}H_{22}N_6O_5S$ on PubChem, two short peptide isomers with three amino acids and distinct sequences were discovered: Gln-Cys-His (ECH) and Cys-His-Gln (CHE) (NCBI CIDs 145454757 and 145455020). On the other hand, acetyl transfer processes in enzyme-mimic systems are known to be attributed to Cys-His (Takahashi et al., 2016).

In *Acacia hamulosa*, honey sample contained five short peptides and one cyclic peptide (Fig. 6.11). The short peptides of the *A. hamulosa* honey were:

1. According to a PubChem search, the chemical formulas of two short peptides, Ac-GMGHG-OH (Ac-Gly-Met-Gly-His-Gly-OH) and Ac-MGGHG-OH (Ac-Met-Gly-Gly-His-Gly-OH), are comparable to that of $C_{19}H_{29}N_7O_7S$. According to Gooding et al. (2001) and Karami and Akbari-Adergani (2019), peptides with the Gly-Gly-His sequence are known to function as copper-binding proteins. Furthermore, tyrosine, tryptophan, methionine, lysine, cysteine, and histidine have all been shown to enhance a short peptide's antioxidant activity.
2. A short peptide with the molecular formula $C_{38}H_{65}N_5O_{10}S$, which is comparable to the molecular formula found in

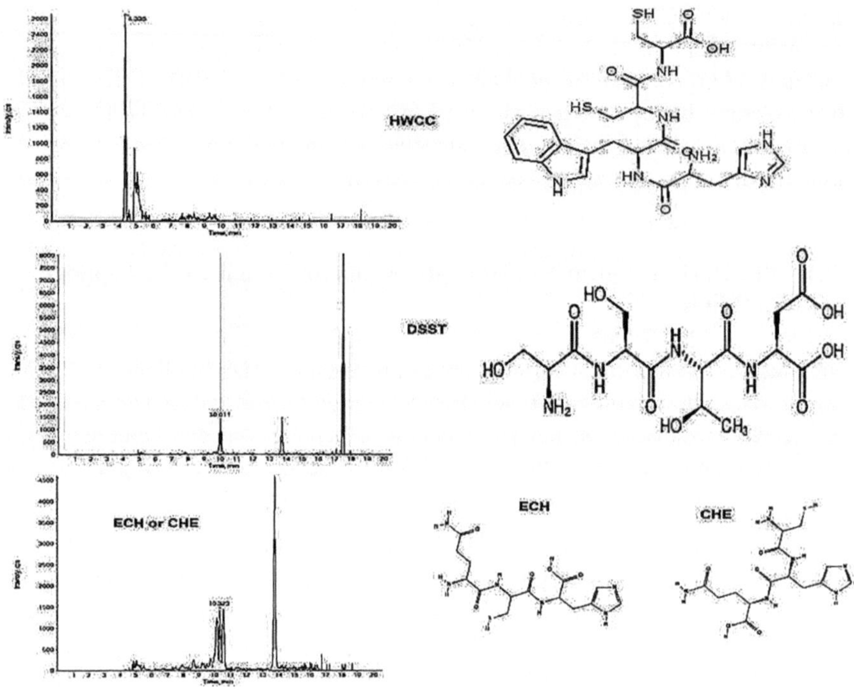

Fig. 6.10 The LC-MS spectra of the short peptides in the *Acacia tortilis* honey sample. The three short peptides were HWCC, DSST, and ECH (CHE). The molecular structures of the HWCC were retrieved from the ChemSpider, Molbase, and PubChem databases

A. hamulosa honey, was synthesized by Eisele (2000). The short peptide was Nα-tert-butyloxycarbonyl-(Nδ,ω-diallyloxycarbonyl)-L-arginyl-(S-palmitoyl) or Boc-R(Aloc)2-C (Pal)-OH (Boc-Arg(Aloc)2-Cys(Pal)-OH).-L-cysteine in order to determine how viral proteins and their altered versions work. Because of the direct interaction between SH and radicals, cysteine residues in short peptides are known to possess antioxidant properties (Yang et al., 2018). It has been demonstrated that certain short peptides in royal jelly samples containing arginine have antioxidant properties (Karami & Akbari-Adergani, 2019). This short peptide may allow the honey from *A. hamulosa* to function as an antioxidant diet because it contains both cysteine and arginine.

3. The third formula was $C_{14}H_{30}N_4O_2S_2$, which in the PubChem database (2021) corresponds to H–C (1)–NEt2•H–C (1)–NEt2 (L-cysteine diethylamide (1 → 1′)-disulfide compound with L-cysteine diethylamide). This dipeptide was created and investigated as a potential treatment for cystinuria, a means of preventing cystine renal stones, and an inhibitor of L-cystine crystallization. Cysteine renal stone development may be inhibited by the honey derived from *A. hamulosa.*
4. AAPP (Ala-Ala-Pro-Pro) in PubChem and APAP (Ala-Pro-Ala-Pro) in the ChemSpider database. The formula for the molecules was $C_{16}H_{26}N_4O_5$. It is well-

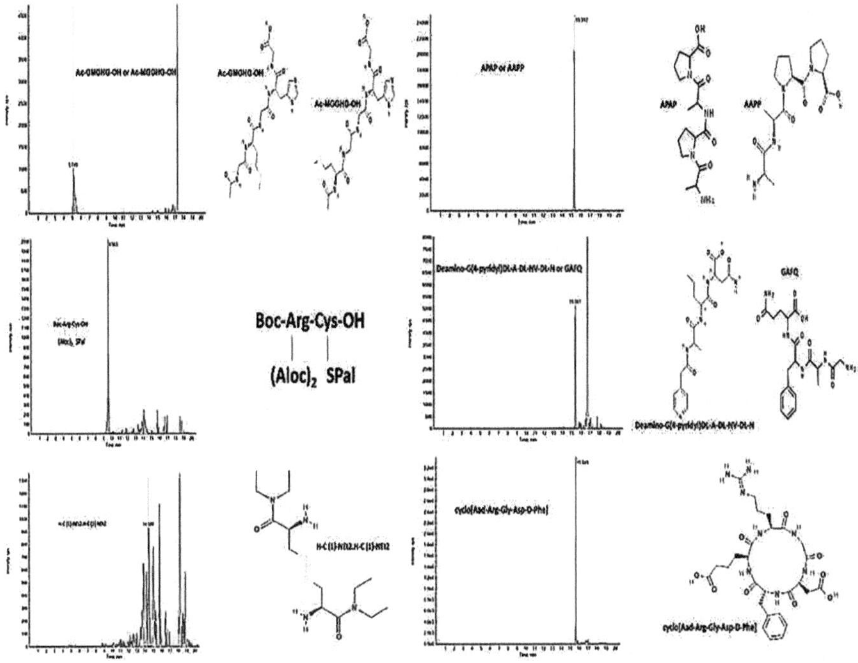

Fig. 6.11 The LC-MS spectra of the short and cyclic peptides in the *Acacia hamulosa* honey sample. (After ALaerjani et al., 2021)

known that proline-rich short peptides have antioxidant properties (Mendis et al., 2005). Strong hydroxyl radical scavenging ability was noted for the dipeptides containing alanine.

5. The chemical formula of the fifth short peptide found in *A. hamulosa* honey was $C_{19}H_{27}N_5O_6$, which matches the sequences found in the PubChem database and the ChemSpider database for deamino-G(4-pyridyl)DLA-DL-NV-DL-N (deamino-Gly(4-pyridyl)-DL-Ala-DL-Nva-DL-Asn-OH). Furthermore, the chemical formula of alfuzosin hydroxy acid is $C_{19}H_{27}N_5O_6$. But for benign prostatic hyperplasia (BPH), alfuzosin and its metabolites are standard medications (Sanki & Mandal, 2012). The sequence of short peptides is made easier to attach to fatty acids by the presence of hydrophobic amino acids like valine, glycine, and alanine (Karami & Akbari-Adergani, 2019). This suppression of lipid oxidation results from this process. A short peptide's potential to scavenge radicals is indicated by the presence of phenylalanine (Zaky et al., 2022), whereas short peptides containing asparagine or glutamine in the mid exhibited antihypertensive activity (Suetsuna & Nakano, 2000).

With a molecular weight of $C_{27}H_{38}N_8O_9$, the cyclic peptide cyclo[L-alpha-homoglutamyl-L-arginyl-glycyl-L-alphaaspartyl-D-phenylalanyl] (cyclo[Aad-RGD-D-F]) is present in *A. hamulosa* honey. The PubChem database was searched for the LC-MS molecular formula.

There were discovered to be five short peptides: Ac-GMGHG-OH or Ac-MGGHG-OH, Boc-R(Aloc)2-C(Pal)-OH, H–C (1)-NEt2.APAP or AAPP, deamino-G(4-pyridyl)DL-A-DL-NVDL-N or GAFQ, and H–C (1)–NEt2.

- The cyclic peptide (cyclo[Aad-Arg-Gly-Asp-DPhe]) is cyclo[L-alpha-homoglutamyl-L-arginyl-glycyl-L-alpha-aspartyl-D-phenylalanyl].

In the integrin receptor bioassay, the cyclic peptide is employed (NCBI, 2021). This peptide's cyclic structure and amino acid composition suggest that it has potential applications as an antioxidant, antibacterial, anticancer, and weight-loss-inducing peptide (Akbarian et al., 2022; Joo, 2012).

2.2.5 The Short Peptide of the *Ziziphus* Honey Samples Are

The honey sample from *Ziziphus spina-christi* had a short peptide with the chemical formula $C_{40}H_{64}N_6O_6$. The short peptide, Auristatin-6-AQ, is a dipeptide with two methylated valine residues that is registered in PubChem (NCBI, 2021). The molecular formula for Almiramide B, a short peptide comprising three methylation valine residues, one methyl alanine, and phenylalaninamide, was found using the ChemSpider and PubChem searches (Fig. 6.12; Chemspider, 2021; NCBI, 2021). Almiramide B, a short peptide that functions as an anti-*Leishmania donovani* and anti-*Trypanosoma brucei brucei* agent, was found in *Z. spina-christi* honey (Jones et al., 2013; Sanchez et al., 2010). Strongly active against human cancer cell lines is Auristatin-6-AQ (Newman & Cragg, 2017).

3 Honey as a Modulator of Biogenic Therapeutics: Molecular Aspects

3.1 Antioxidant Properties of Organic Honey

In addition to other materials, including organic acids, vitamins, and enzymes, honey has a range of phytochemicals that can be used as dietary sources of antioxidants (Zammit Young & Blundell, 2023). Antioxidants stop molecules from oxidizing, which is essential for a cell's proper functioning. Oxidation would gradually damage organs, tissues, and ultimately the body's ability to function physiologically. The body has an intricate system that keeps antioxidants in check. Antioxidant-rich foods have been shown to assist in regulating this system and hence improve health. In vivo investigations have demonstrated that honey reduces oxidative stress in a variety of organ systems, a finding that has not yet been validated by clinical trials (Meo et al., 2017). The kind and quantity of these antioxidant chemicals are primarily determined by the honey's floral variety or source. Utilizing an evaluation method called oxygen radical absorbance capacity, or ORAC, researchers at the

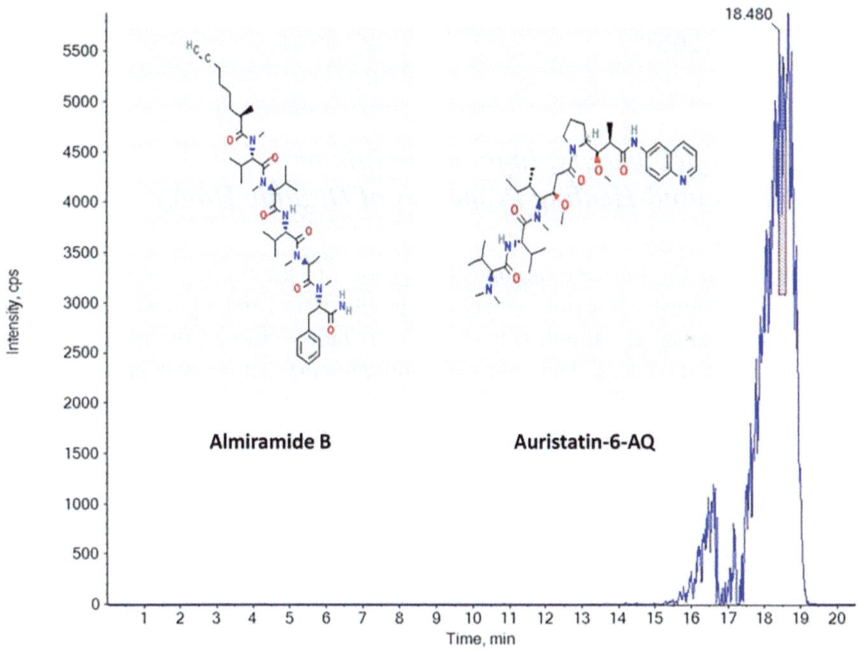

Fig. 6.12 The short peptide of the *Z. spina-christi* honey. The short peptide was defined as Auristatin-6-AQ or Almiramide B. (After ALaerjani et al., 2021)

University of Illinois Champaign/Urbana investigated the antioxidant content of 14 unifloral honeys in comparison to a sugar substitute. The acacia honey had an ORAC value of 3.0 μmol TE/g, while the Illinois buckwheat honey had an ORAC value of 17.0 μmol TE/g. Antioxidant activity was absent from the sugar analogue (Swellam et al., 2003). Reactive oxygen species (ROS) and free radicals have been linked to the aging and disease processes. The effects of 1.5g/kg body weight of buckwheat honey or corn syrup on the reducing and antioxidant capacities of plasma in adult, healthy humans are described in the report. There is speculation that these substances could strengthen oxidative stress defenses and even shield individuals from oxidative stress (Schramm et al., 2003). Since humans are thought to consume more than 70 kg of sweets annually on average, replacing certain traditional sweeteners in food with honey may strengthen an adult's antioxidant defenses (Hunter et al., 2021). Using spectrophotometric and photochemiluminescence techniques, researchers examined the antioxidant properties of volatile oil of propolis (VOP) from India. Based on IC50 values, they were able to conclude that VOP was more effective at scavenging ABTS radicals than other radicals (Orsolić et al., 2003). While classical examples of antioxidants such as vitamins C and E become pro-oxidants themselves when in large dosages, honey has been found not to exhibit this behavior when given in larger levels. This is believed to be caused by the fact that honey contains many chemicals with antioxidant activity, which aid in transforming any molecules that turn pro-oxidants back into their active, antioxidant forms (Erejuwa et al., 2010).

3.2 Anti-inflammatory, Immunomodulatory, and Wound-Healing Properties of Organic Honey

Inflammation is the intricate biological response of vascular tissues to damaging stimuli. This defensive reaction is displayed by the tissues and organism to drive out infections or damaging stimuli (Fig. 6.13). It is believed to be the main cause of several chronic disorders. Thus, the anti-inflammatory action is believed to help treat chronic diseases like cancer, liver, and kidney problems. A plethora of variables, including cytokines, mitogens, macrophages, cyclooxygenases (COXs), lipoxygenases (LOXs), TNF factors, and several other inflammatory pathway components, might contribute to a proinflammatory response (Ahmed et al., 2018; Adhikari & Karmakar, 2018, Adhikari et al., 2007). Two aspects of the inflammatory system that become active during disease are the nuclear factor kappa B (NF-κB) and mitogen-activated protein kinase (MAPK) pathways. The activation of MAPK and NF-κB eventually leads to the induction of numerous additional inflammatory mediators, enzymes, cytokines, proteins, and genes, such as TNF-α, lipoxygenase 2 (LOX-2), cyclooxygenase-2 (COX-2), C-reactive protein (CRP), interleukins (IL-1, IL-6, and IL-10), and CRP. Each of these proinflammatory action markers is known to have a major role in the etiology of disorders associated to inflammation and angiogenesis (Abdulkhaleq et al., 2018, Ahmed et al., 2018). Honey's anti-inflammatory properties have been demonstrated by recent in vivo research, which also shows that honey reduces plasma levels of proinflammatory cytokines such as COX-2, TNF-α, PGE2, NO, iNOS, and edema. According to Ranneh et al. (2021) honey (Gelam) negatively influences the translocation of transcription factor NF-κB into the nucleus and inhibits the degradation of IκBα (inhibitor of kappa B). Proinflammatory enzymes, including prostaglandins, cyclooxygenase-2 (COX-2), and inducible nitric oxide synthase (iNOS), have been shown to be inhibited by phenolic acids and flavonoids including chrysin, quercetin, and galangin (Silva et al., 2021). Studies have indicated that the flavonoid concentration of honey inhibits the development of matrix metallopeptidase 9, or MMP-9, an inflammatory mediator responsible for long-term inflammation. The expression of anti-inflammatory cytokines like IL-1 and IL-10, as well as growth factors like TGF-β and PDGF (platelet-derived growth factor), can be strongly inhibited by honey. Another theory suggests that neutrophils, monocytes, and macrophages all create reactive oxygen species, which in turn promote inflammation (Fig. 6.13). Such cells are not released by honey, which has an anti-inflammatory impact. In addition, it suppresses leukocyte and keratinocyte formation to lessen inflammation (Ahmed et al., 2018).

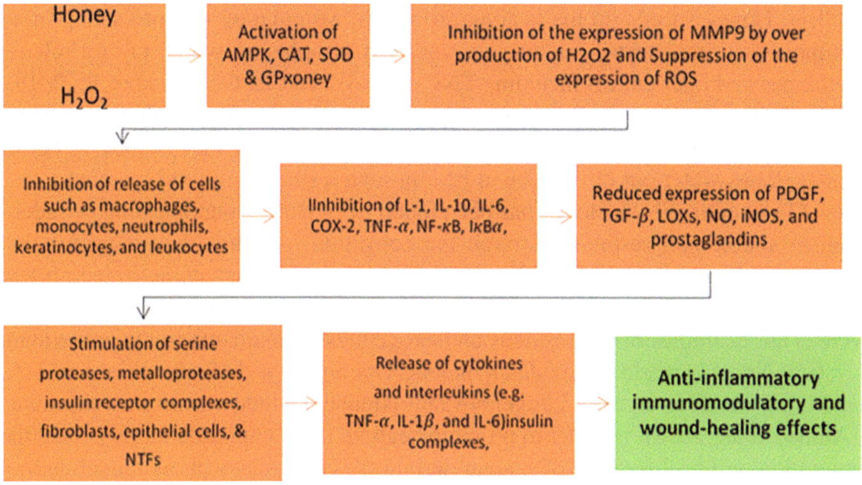

Fig. 6.13 Flow chart depicts the anti-inflammatory and wound healing properties of organic honey

It has been shown that honey's ability to produce H_2O_2 during an inflammatory reaction promotes the development of fibroblasts and epithelial cells to help heal the damage caused by the inflammation. Honey is a novel agent to modulate a disease because of its anti-inflammatory properties (Scepankova et al., 2021). Inflammation is related to abnormal metabolism of arachidonic acid. Arachidonic acid is metabolized by LOXs into leukotrienes (LTs). Lipoxygenase (LOX) isozymes come in three different varieties: 12-LOX, 15-LOX, and 5-LOX. Twelve-LOX causes allergic and inflammatory conditions; 15-LOX produces the anti-inflammatory 15-HETE (15-hydroxyeicosatetraenoic acid); and 5-LOX produces 5-HETE (5-hydroxyeicosatetraenoic acid) and LTs. Many polyphenols in honey have been reported to suppress LOXs (Scepankova et al., 2021). The anti-inflammatory effect of honey can be attributed to its phenolic compounds and flavonoids (Silva et al., 2021). During primary and secondary immunological reactions, honey stimulates the production of T and B lymphocytes, antibodies, eosinophils, neutrophils, monocytes, and natural killer cells. Honey contains a sugar called nigerooligosaccharides, which has been shown to have immunopotentiating properties (Brown et al., 2023). Immunomodulation is also caused by non-sugar components in honey (Masad et al., 2021).

According to Gonzalez et al. (2016), the process of normal wound healing consists of several steps that happen simultaneously, such as coagulation, inflammation, cell proliferation, tissue remodeling, and replacement of injured tissue. According to Taskandi (2021), honey is frequently used to heal a variety of chronic wounds, including burn, necrotic, diabetic foot, and split skin wounds following surgery. Honey promotes the remodeling phase, reduces bacterial development, and helps remove necrotic tissues during the inflammatory phase of wound healing (Scepankova et al., 2021). A recent study shows that throughout the healing process in IL-6-deficient mice, honey at the wound site increases the production of IL-6 and

TNF-α. Honey speeds up the healing process by stimulating and producing more lymphocytes, phagocytes, monocytes, and/or macrophages, which in turn release cytokines and interleukins including TNF-α, IL-1β, and IL-6 (Ahmed et al., 2018). Honey's high sugar content and osmolarity both aid in healing. If there is enough blood circulation at the wound site to support this action, the osmotic effect of honey draws water out of the wound bed through a straightforward lymphatic outflow (Tashkandi, 2021). Honey has been found to enhance wound healing by inducing an antioxidant response through the activation of antioxidant enzymes that reduce oxidative stress and AMPK (5′ adenosine monophosphate-activated protein kinase). Superoxide dismutase (SOD), catalase (CAT), and glutathione peroxidase (GPx) are examples of endogenous and exogenous antioxidants in the antioxidant system. Nonenzymatic antioxidants include vitamins E and C, glutathione (GSH), and a few small molecules. Exogenous antioxidants include certain micronutrients (Ahmed et al., 2018). In order to promote healing, these antioxidants also boost the migration and proliferation of human dermal fibroblasts as well as mitochondrial activity (Ahmed et al., 2018).

The presence of serine proteases and matrix metalloproteases, two different types of protein-digesting enzymes, at wound sites is explained by another method. Because of the presence of certain inhibitors, these protease enzymes are often inactive. When H_2O_2 renders the inhibitors ineffective, the proteases become active. H_2O_2 therefore functions as a physiological switching stimulus for the oxidative activation and inactivation of these enzymes. Honey is said to increase and accelerate the creation of H_2O_2. Active proteases break down germs and wound debris. In order to repair the damage caused by inflammation, H_2O_2 also promotes the development of epithelial and fibroblast cells. In a similar manner, H_2O_2 activates nuclear transcription factors (NTFs) to promote wound healing and cell division (Ahmed et al., 2018). In order to repair tissues, honey also promotes the release of cytokines from monocytes and lymphocyte proliferation. Reactive oxygen species are produced when mitogens or honey activate monocytes, which intensifies the inflammatory response. It results in edema in the surrounding tissue, which blocks capillary circulation. As a result, cells receive less oxygen and nutrition. In the end, it inhibits the formation of cells needed to replace tissues and heal wounds (Scepankova et al., 2021).

3.3 *Cardioprotective and Neuroprotective Properties of Organic Honey*

Honey's therapeutic properties have been valued by numerous traditional medical systems. Honey has roots in ancient human civilization and is a functional food that is commonly consumed. It contains a wide range of bioactive compounds with pharmacological significance, including sugars, flavonoids, organic acids, enzymes, and phenolic acids. Because of its low glycemic index and several health advantages, it is now preferred by many over sugar as a natural sweetener (Nikhat & Fazil, 2022). These phenolic acids are essential for cardioprotection because they prevent low-density lipoprotein (LDL) from oxidizing, which is a crucial step in the development of atherosclerosis. Honey lowers the incidence of plaque formation and coronary artery disease by decreasing LDL oxidation (Bt HjIdrus et al., 2020). Significant proof-of-concept data has been presented by a number of animal studies, indicating that quercetin may be useful in lowering blood pressure in patients with hypertension as well as in lowering the risk of stroke and coronary heart disease (Yamagata, 2019; Dabeek & Marra, 2019; Zhang et al., 2023). The compound kaempferol is essential in reducing the accumulation of low-density lipoprotein (LDL) cholesterol, which is a primary cause of heart problems. The main mechanisms by which polyphenols prevent heart attacks and strokes are through their suppression of LDL cholesterol oxidation, enhancement of coronary vasodilation, and inhibition of platelet aggregation in the blood (Chen et al., 2022). These mechanisms work together to effectively mitigate atherosclerosis, the disease that narrows and hardens the arteries. Some research indicates that chysin, which is found in different kinds of honey, has anti-hyperlipidemic qualities by lowering LDL and total cholesterol levels, improving lipid profiles overall, and protecting the heart (Nguyen et al., 2019). In addition to phenolic substances, honey's cardioprotective qualities are also greatly influenced by organic acids such as acetic, citric, and gluconic acids, which control lipid metabolism and boost the antioxidant capacity of the honey's bioactive components. These acids work in concert with flavonoids and phenolic acids to prevent lipid peroxidation and shield vascular endothelial cells. Moreover, the natural acidity of honey contributes to the preservation of these bioactive substances, guaranteeing their effectiveness in reducing risk factors for cardiovascular disease (CVD) (Pasupuleti et al., 2017). Further boosting honey's cardiovascular advantages, glucose oxidase produces hydrogen peroxide, which has antibacterial properties and reduces vascular inflammation. It also keeps honey's antioxidant balance intact (Pasupuleti et al., 2017).

Because of its anti-inflammatory, neurotrophic, and antioxidant qualities, honey has the potential to be a neuroprotective agent and is therefore a good choice for controlling and preventing neurodegenerative illnesses. Further studies will focus on identifying important bioactive chemicals and their potential for working in concert with other medicines in order to create therapeutic applications and clarify the specific molecular mechanisms underlying honey's effects.

The hallmarks of neurodegenerative illnesses, such as multiple sclerosis (MS), PD, and AD, include oxidative stress, inflammation, and poor cellular signaling, which lead to progressive neuronal death. The shortcomings of available pharmacological therapies have sparked a renewed interest in the quest for naturally occurring, bioactive substances with neuroprotective potential. Honey is known to include polyphenols, flavonoids, and other bioactive substances that may have neuroprotective properties. Honey's potential for managing and preventing neurodegenerative illnesses stems from its anti-inflammatory, antioxidant, and neurotrophic qualities. Flavonoids such as galangin have been demonstrated to effectively modulate neuroinflammatory responses by suppressing the synthesis of nitric oxide (NO), interleukin-6 (IL-6), tumor necrosis factor-alpha (TNF-α), and interleukin-10 (IL-10) in lipopolysaccharide (LPS)-stimulated BV2 microglia. Galangin has the ability to significantly reduce neuroinflammation by targeting these important inflammatory mediators. This makes it a viable candidate for development as a pharmacological drug or as a dietary supplement that protects against AD and PD. As per Fadzil et al. (2023), apigenin, which is found in honey, used to treat Parkinson's disease and AD, has been shown to mimic neurotrophic action and encourage the formation of new neurons (Namsi et al., 2018). By blocking acetyl cholinesterase (AChE) and butyryl cholinesterase (BChE), respectively, phenolic compounds and flavonoids in samples of Algerian and buckwheat honey can be prevented, which aids in neurological issues associated with AD. Anti-inflammatory, anti-AChE, and anti-α-glucosidase activities are also demonstrated by the samples (Baranowska-Wójcik et al., 2020). Additionally, goldenrod, thyme, heather, buckwheat, and acacia honey have been shown to have inhibitory activities for both BChE and AChE (Szwajgier et al., 2022). In addition to gallic acid, epicatechin, syringic acid, and catechin, methyl syringate, which makes up more than 50% of the honey's total polyphenols, protects against an unidentified neurological illness. Tualang honey heals hypoxia-induced memory deficits in stressed adult male Sprague Dawley rats by reducing neuronal damage in the hippocampus and cholinergic system (Qaid et al., 2020). In a rat model of Alzheimer's disease, Iranian thyme honey shows important behavioral, cellular, and molecular effects as a powerful preventive and therapeutic drug (Rosli et al., 2020). According to studies conducted by Asari et al. (2019), tualang honey and fish oil rich in DHA have the ability to decrease the generation of pro-inflammatory cytokines in the rat brain after prolonged stress exposure. This suggests that they may be useful in reducing neuroinflammation and stress-related brain damage.

3.4 Antimicrobial/Antifungal/Antiviral Properties of Organic Honey

Honey has a proven antibacterial effect against a broad spectrum of bacterial species including aerobes and anaerobes, Gram positives, and Gram negatives including *Escherichia coli*, *Pseudomonas aeruginosa*, *Klebsiella pneumoniae*, *Staphylococcus aureus*, *Bacillus subtilis*, and *Listeria monocytogenes*, and their multidrug-resistant counterparts (Laallam et al., 2015). Interestingly, it has been observed that no organism has gained resistance to honey (Table 6.2).

Table 6.2 List of different polyphenolic compounds reported to have strong antimicrobial, antifungal, and antiviral activities

Phenolic acids	Mechanism	Flavonoids	Mechanism
2-cis,A-trans abscisic acid	Not discovered	Apigenin	Inhibits DNA gyrase
2-Hydroxycinnamic acid	Not discovered	Catechin	Hydrogen peroxide generation
Caffeic acid	Oxidative stress	Naringenin	Unknown
Chlorogenic acid	Increase in membrane permeability resulting in cytoplasmic and nucleotide leakage	Galangin	Inhibition of peptidoglycan and ribosome synthesis
Cinnamic acid	Not discovered	Genistein	Disruption to topoisomerase-II DNA Cleavage complex
Ellagic acid	Not discovered	Isorhamnetin	Unknown
Ferulic acid	Cell membrane dysfunction and changes in cell morphology	Kaempferol	Inhibits DNA gyrase
Gallic acid	Cell membrane disruption resulting in pore formation and intracellular leakage	Luteolin	Inhibits FAS-I in mycobacteria and inhibits DNA helicase DnaB and RecBCD
p-Coumaric acid	Cell membrane disruption and binding to bacterial DNA	Myricetin	Inhibits DNA B helicase
p-Hydroxybenzoic acid	Not discovered	Pinobanksin	Not discovered
Protocatechuic acid	Not discovered	Pinobanksin	Not discovered
Syringic acid	Not discovered	Quercetin	Disrupts membranes, transport, and motility
Vannilic acid	Not discovered	Rutin	Induces topoisomerase IV-mediated DNA cleavage

After Nolan et al. (2019), all cross references are given in the paper for further readings

3.4.1 Mechanisms of the Antibacterial Effect of Honey

The capacity of certain honeys to break down bacterial DNA is credited for their antibacterial properties. Due to the high sugar concentration in honey, bacteria's ability to suck water out of their cells and become dehydrated is inhibited, preventing further growth. Additionally, because honey has a high sugar content, bacterial proteins become denaturated, which impairs their functionality and stops germs from growing. The viscous nature of honey can act as a physical barrier, keeping germs from growing and sticking to surfaces. The bactericidal effect of honey was based on the disruption of cell division. These effects were also linked to significant lysis, changes in the shape and surface area of pathogenic bacteria, and considerable cell death (loss of structural integrity) (Fig. 6.14). Honey has been shown to prevent bacteria from adhering to tissues, which is a critical first stage in the infection process, and to prevent biofilm development, which shields germs from antibiotics (Kot et al., 2020). An enzyme that bees generate, called lysozyme, has bacteriostatic properties (Bruneau, 2006). In the peptidoglycan of the bacterial cell wall, the lysozyme hydrolyzes the β-1.4 link that connects the acid residues of N-acetylmuramic acid and N-acetyl-D-glucosamine (Molan, 1992).

Furthermore, lactic acid bacteria are a very diverse group that can be found in flower nectar, pollen grains, vegetative surfaces, animal indigenous microflora, and pollen grains. According to Bhattacharya et al. (2022), lactic bacteria produce a variety of metabolite compounds with antibacterial qualities, including organic acids, H_2O_2, carbon dioxide, reuterin, diacetyl, and bacteriocins. Lactic bacteria are present in the microbiome of all bees (Iorizzo et al., 2022). Antimicrobial action of these bacteria is well-known, especially the genus *Lactobacillus*. A microbiota of lactic bacteria from both genera (*Lactobacillus* and *Bifidobacterium*) that coexist in symbiosis in the stomach of honey bees (*Apis mellifera*) has been isolated (Zhang et al., 2022). Several bioactive metabolites, including organic acids, lactic acids, formic acid, acetic acid, and H_2O_2, are produced by these bacteria. According to Olofsson et al. (2016), volatile substances such as benzene, toluene, n-octane,

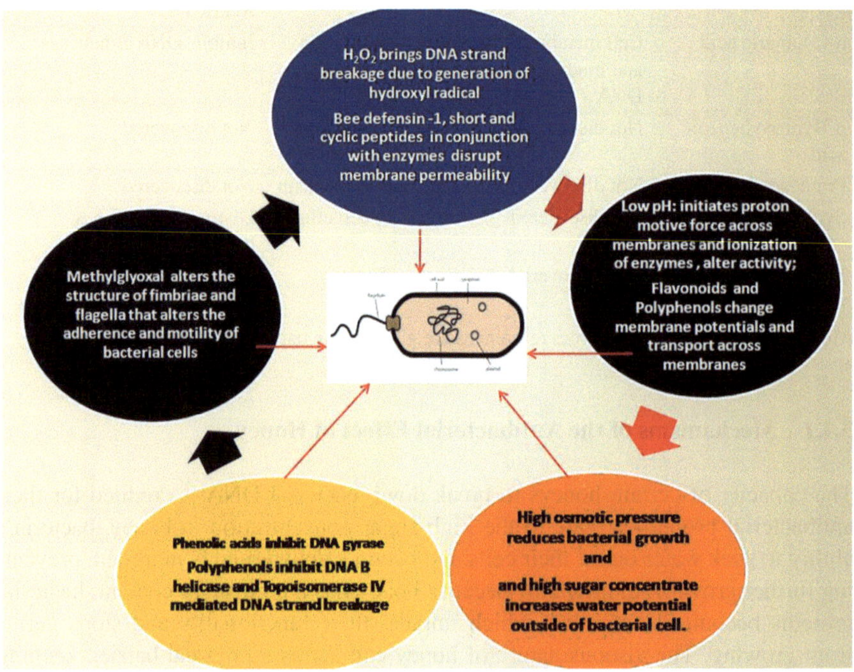

Fig. 6.14 Antibacterial activity of organic honey: mechanism of action

xylene, ethylbenzene, and n-nonane have some antibacterial and antifungal properties against bacteria that are resistant to antibiotics.

3.4.2 Mechanism of Antifungal Effect of Organic Honey

It has been discovered that some honeys exhibit antifungal qualities against *Penicillium* and *Aspergillus* species (Sampath Kumar et al., 2010). Furthermore, it has been observed that a number of yeasts of the genus *Candida*, including *Candida albicans*, *Candida glabrata*, *Candida dubliniensis*, *Candida tropicalis*, *Candida krusei*, and *Candida parapsilosis*, are resistant to these honeys (Bulgasem et al., 2016; Fernandes et al., 2020). According to Al-Waili and Haq (2004), honey inhibits the growth of fungi and can stop the formation of toxins when it is diluted. Mulu et al. (2013) reported that Ethiopian honey was used to treat HIV patients for resistant strains of *Candida albicans*. According to Guttentag et al.'s in vitro research from 2021a, Australian honey interferes with *Trichophyton rubrum* hyphae's structure and inhibits conidial germination. El-Bondkly et al. (2022) have reported that certain Egyptian honeys featuring solitary blooms have antifungal and antimycotic properties against specific species of *Epidermophyton*, *Trichosporon*, and *Microsporum*.

Honey has shown antifungal properties through a range of mechanisms of action. Fructose and glucose are the two main sugars found in honey. Fungal cells are effectively dehydrated and destroyed by an osmotic mechanism that draws water out of them due to the high concentration of sugar. Some types of honey contain an enzyme called glucose oxidase, which produces hydrogen peroxide when honey comes into contact with physiological fluids. Hydrogen peroxide has antifungal properties that can help stop the growth of fungus (Fig. 6.15). Many forms of fungus cannot survive in the acidic environment created by honey's low pH of 3–4.5. Polyphenols and phytochemicals, both of which have been shown to have antifungal properties, are abundant in honey. These compounds possess the capacity to cause harm to the fungal cell membrane, hinder the division of cells, and inhibit the development and synthesis of fungal enzymes. Honey's immunomodulatory qualities may be indirectly linked to its antifungal activity. Immune system fortification could lead to improved efficacy in the battle against fungal diseases. Honey contains some peptides that are antimicrobial. These peptides are generated when honey comes into contact with human fluids and have the power to prevent the growth of fungus.

3.4.3 Mechanisms of Antiviral Effects of Honey

Honey has been used as a medicine for numerous ailments since ancient times. It is a crucial treatment for respiratory infections, such as viruses that induce coughing. Numerous investigations have proven that honey has antiviral properties against a wide spectrum of infections. Depending on the kind of honey and the particular virus involved, its efficacy can change. According to a study by Behbahani (2014),

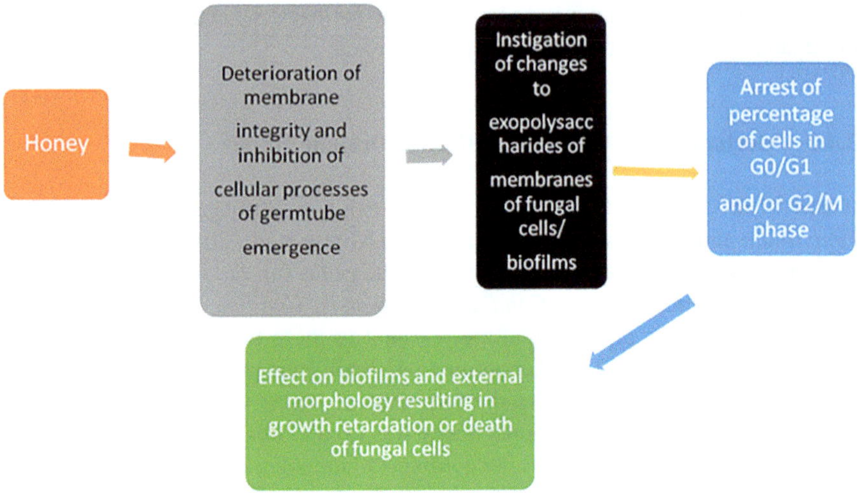

Fig. 6.15 Flow chart illustrates the typical antifungal activity of organic honey: mechanism of action

Iranian monofloral honey had strong anti-HIV-1 properties. Hossain et al. (2020) have reported on the antiviral efficacy of manuka honey against the severe acute respiratory syndrome coronavirus 2 (SARS-CoV-2). According to Kwon et al.'s (2023) study results from 2023, honey inhibits the herpes simplex virus type 1 infection.

According to recent research, honey may have antiviral properties. According to Hossain et al. (2020), the presence of ascorbic acid, flavonoids, and H_2O_2 produced by honey also inhibits the proliferation of viruses by stopping their transcription and translation. Research conducted in vitro has demonstrated the antiviral properties of honey against a variety of viruses, including varicella-zoster, herpes simplex, COVID-19 infection (Soares et al., 2022) and rubella (Ahmed et al., 2018). The secretions from the pharyngeal and salivary glands on the honeybee's head make up honey. Recently, the salivary gland's section has been found to include nitrite, nitrate, and nitric oxide (NO) metabolites (Rosier et al., 2020). The fact that NO is an energetic chemical that generates host defense against viruses—both DNA and RNA viruses—has long been known. NO works by stopping the replication of viruses and slowing down the growth of viral lesions. NO works by interfering with viral polymerase, nucleic acid, and/or capsid proteins to suppress replication. It has also been observed that the flavonoid content of honey inhibits the transcription and replication of viruses (Ahmed et al., 2018). The potential processes underlying honey's antiviral action are shown in Fig. 6.16. In order to comprehend the true impact of honey on viruses and processes, further research is planned to chart the course (Table 6.3).

4 Identification of Different Classes of Contaminants in Honey

One of the biggest issues facing society now is environmental degradation (Manisalidis et al., 2020). Around the hive, honey bees have strong interactions with the surrounding plants, soil, air, and water. Around the hive, in an area of roughly 7 km^2, they gather nectar and pollen (Papa et al., 2022). Bees are exposed to various plants and their toxins when they venture outside of their hives in pursuit of food (Leska et al., 2021; Martinello et al., 2021). They give us important information about pollution in their surroundings because, in addition to being exposed to pollutants during nectar gathering, their body feathers and wings are also exposed to air pollution while they are in flight (Martinello et al., 2021). They give us important information about pollution in their surroundings because, in addition to being exposed to pollutants during nectar gathering, their body feathers and wings are also exposed to air pollution while they are in flight (Martinello et al., 2021). Collectively, the bees in the hive exhibit a significant degree of resistance to these pollutants, notwithstanding their individual susceptibility (Cunningham et al., 2022). They thus contribute to the build-up of dangerous pollutants in bee products and offer continuous observation (Mair et al., 2023). According to Catalano et al. (2024), beehives situated in contaminated locations can offer a variety of data for ongoing pollution monitoring and source identification.

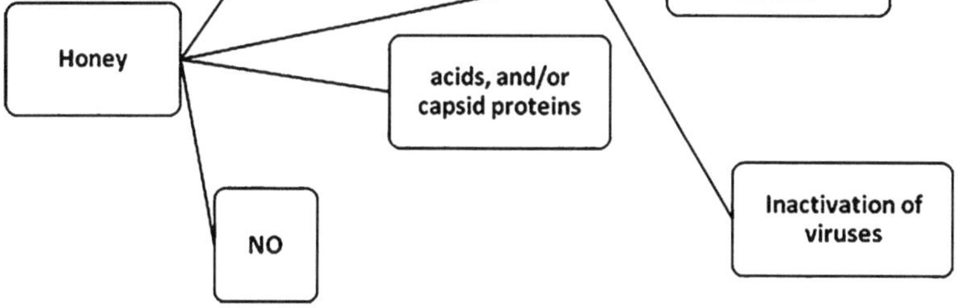

Fig. 6.16 Flow chart illustrates the antiviral activity of bioactives in honey: mechanism of action

Table 6.3 List of different classes of antiviral components present in honey and their mechanism of actions

Compound	Discretion	Mechanisms of antiviral action	References
Methylglyoxal	Dicarbonyl resulted from the conversion of DHA during the ripening of honey	Blocks formation of virion assembly and maturation	Behbahani (2014)
Hydrogen peroxide	Produced mainly during glucose oxidation	Viral inactivation	Dembinski et al. (2014)
Chrysin	Flavonoid	Inhibition of viral protease enzymes	Kim et al. (2021)
CAPE	Polyphenolic ester	Inhibition of viral protease enzymes	Taysi et al. (2023)
Galangin	Flavonoid	Inhibition of viral protease enzymes	Badshah et al. (2021)
Caffeic acid	Flavonoid	Inhibition of viral protease enzymes	Cruz-Arreola et al. (2022)
Hesperidin	Flavonoid	Inhibition of viral protease enzymes—binding to S-RBD and then blocking the interaction with ACE2	Zalpoor et al. (2022)
Rosmarinic acid	Polyphenolic hydroxycinnamic acid	Inhibition of viral protease enzymes	Krüger et al. (2023)
Isoquercetin	Flavonoid	Reduction of viral load	Mbikay and Chrétien (2022)
Rutin	Flavonoid	Reduction of viral load	Rahman et al. (2021)
Quercetin	Flavonoid	Reduction of viral load	Di Petrillo et al. (2022)
3-hydroxy-sebacic acid	Fatty acid	Unknown	Al-Hatamleh et al. (2020)
Ascorbic acid	Sugar acid	Activation of antiviral immune responses	Laubert et al. (2022)
p -Coumaric acid	Phenolic acid	Reduce the entrance of the viruses into the cells to protect the cells from the virus destruction and abate virus replication	Kwon et al. (2019)
Benzoic acid	Aromatic carboxylic acid	Could suppress NP and M1 protein expression levels during the late stages of viral biosynthesis and inhibit neuraminidase (NA) or the matrix protein 2 (M2) ion channel activity in influenza A virus, which may influence virus release	Guo et al. (2019)
Pinocembrin	Flavonoid	Pinocembrin inhibits viral RNA production and envelope protein synthesis in Zika virus (ZIKV), being a mosquito-borne virus	Lee et al. (2019)

4.1 Microbial Contamination

The health of the entire colony is adversely affected by microorganisms, which also play a role in colony collapse disorder (CCD). Protists, fungus, and bacteria are microorganisms associated to bees; some of these species are significant pathogens for bees.

- The round-ended, flagellated, spore-forming bacteria that make up *Paenibacillus* larvae are Gram-positive. These are certain diseases that infect bee larvae. The larva's midgut is where they grow.

- It is not possible for tube cells to divide and, with the aid of flagella, enter the body cavity and propagate in hemolymph. A bacterial illness that affects the entire body kills bee larvae (Poppinga & Genersch, 2015).
- *Melissococcus plutonius* is a Gram-positive anaerobic bacterium. When larvae consume bacterially contaminated food, infection results. *M. plutonius* colonizes the midgut of the larvae. Bee broods typically perish 4–5days after infection (Takamatsu, 2023).
- An animal pathogen that is nosocomial is *Serratia marcescens*. It is an anaerobic, rod-shaped, motile, Gram-negative bacterium. An accelerated course of CCD may result from infection with an opportunistic disease in bees that are already compromised by other causes. Sepsis and death are frequent results. According to Leska et al. (2021), *S. marcescens* does not cause the production of phenoxide or antimicrobial peptides, indicating that it has defense mechanisms against the immune system of bees.
- The filamentous, heterothallic fungus *Ascosphaera apis* infects bees and produces spores. *A. apis* attacks bee larvae with the chalkbrood disease. The hindgut is an anaerobic habitat where fungal spores develop. *Ascosphaera* sp. enters the abdomen after mycelia have developed, where it grows aerobically before penetrating the cuticle. The larva is spongy and white at first, but as the illness worsens, it hardens and takes on the appearance of chalk. This fungus produces sporocarps, which change color to gray, black, or brown as they develop on the body surface or integument of larvae (Mráz et al., 2021).
- *Aspergillus* species are saprophytic, filamentous fungus. *Aspergillus* sp. infections can cause symptoms strikingly similar to CCD and even lead to the collapse of the entire bee colony. After consuming contaminated food, the bee contracts *Aspergillus* sp. through its digestive tract. The skin might become infected as well. Aflatoxin production is likely the primary factor in larval mortality following infection in the majority of *Aspergillus* sp. (Bush et al., 2024).
- The severe bee disease nosemosis is caused by parasitic fungus belonging to the genus *Nosema*. *Nosema apis* and *Nosema ceranae* are the two species of microsporangia that cause it most frequently. Spores of both species are found in the outside world, but the vegetative form is found only within the bee's body. *Nosema ceranae* modifies metabolic and hormonal processes to impact gene expression. The body of the bee experiences more oxidative stress, and its

nutrient-sensitive components shrink. Because they are hungry, infected bees do not share food with the other members of the colony. *Nosema* sp. harms digestive tissue, weakens bees, and when combined with other elements, ultimately results in a bee's death (Marín-García et al., 2022).

- An ectoparasitic mite species known as *Varroa destructor* targets honey bee hives. The ectoparasitic mite *V. destructor* is frequently linked to occurrences of CCD and increases colony mortality. Adult and larval bees' fat body tissue is the food source for *V. destructor*. Furthermore, mites spread infections to the bodies of their hosts by weakening the bees' bodies through blood loss (Reams & Rangel, 2022).
- Even sometimes, honey is primarily contaminated by bacteria, yeasts, and molds (Rosiak et al., 2021). The types and levels of these microorganisms can vary depending on the source of the honey, its handling, and storage conditions. Bacterial contamination in honey is relatively rare due to its low water activity and high sugar content, which inhibits the growth of many bacteria. However, some bacteria can still be present in honey, particularly spore-forming bacteria like *Clostridium botulinum*, which can cause botulism, a severe and potentially fatal illness. Other bacterial contaminants may include *Bacillus* species, *Escherichia coli*, and *Staphylococcus aureus*. Yeasts are the most common microorganisms found in honey. They can cause fermentation, especially in honey with higher moisture content. Fermentation not only alters the taste and quality of honey but can also pose health risks if consumed in large quantities. Common yeast species found in honey include *Saccharomyces*, *Zygosaccharomyces*, and *Candida.* Molds are less common in honey due to its acidic pH and low water activity. However, they can still be present, particularly if the honey has been stored under conditions that promote mold growth. Some molds produce mycotoxins, which can be harmful to human health. Common molds found in honey include species of *Aspergillus*, *Penicillium*, and *Rhizopus* (Leska et al., 2021).

4.2 Environmental Pollutants

- Burning municipal waste, evaporating from contaminated lakes, and landfilling obsolete electrical equipment are some of the ways that polychlorinated biphenyls, or PCBs, a class of persistent pollutants, are released into the environment (Cunningham et al., 2022). PCBs accumulate in human tissues and impair immunity. They are also a global problem because they increase the risk of cardiovascular diseases and cancer (Montano et al., 2022). Studies show that PCBs accumulate in honey first, then in bees, and lastly, in pollen to a lesser extent (Cunningham et al., 2022).
- Polycyclic aromatic hydrocarbons (PAHs), which are air pollutants, are found naturally in gasoline, coal, and crude oil (Låg et al., 2020). The production of iron, steel, aluminum, cement, coal tar pitch, paint, rubber, asphalt, fungicides,

and pesticides is the main industrial sources of PAHs. Energy production and refinery exhaust emissions have an impact on PAH emissions as well (Patel et al., 2020). The presence of PAHs in bees or bee products is one of the best indicators of environmental pollution, according to Margaoan et al. (2024). In an Italian investigation, different quantities of PAHs were discovered in honey and honey bees collected from two different locations (Murcia-Morales et al., 2023). A French study revealed that PAHs were present in pollen, honey, and bees themselves. These results were connected to environmental pollution (Mair et al., 2023).

4.3 Pesticides

Pesticides and fertilizers are widely utilized to boost agricultural product productivity (Sharma & Singhvi, 2017). In the natural world, pesticides help fruits and vegetables flourish without causing harm. But excessive usage of them endangers the health of people and other living things (Baweja et al., 2020). The colony has a chance of surviving based on the pesticide dosage. Bees and the matrices they are linked with thus make it possible to monitor pesticides over both the short and long term (Teeters et al., 2012). In one study, the concentration of 64 different pesticide kinds in beeswax, freshly stored pollen, and in-hive bees was assessed. The highest concentration of agrochemicals was found in beeswax, whereas the biggest diversity of pesticide species was found in pollen. According to this study, intensive agricultural settings had higher pesticide content in pollen than did rural or grassland environments. This finding raises the possibility that pesticides in stored pollen may be a reflection of regional variations in pesticide contamination in the environment (Calatayud et al., 2018, Leska et al., 2021).

4.4 Heavy Metals

There is significant worry about the effects of heavy metal contamination in places with high human population densities, particularly in relation to industrial operations (Adhikari, 2021a, b; Ali et al., 2019). The burning of fossil fuels, the use of agricultural chemicals, and exhaust fumes are the main human causes of heavy metal pollution. The heavy metals that are produced are extremely harmful (Adhikari, 2019; Alloway, 2013). Bee products like honey, pollen, propolis, and beeswax, as well as the bees themselves, may contain heavy metals. While environmental heavy metals can have detrimental effects on bees' ability to produce offspring, navigate their surroundings, and survive, they frequently do not cause colony collapse (Cunningham et al., 2022). Bioavailability in the environment is made possible by this. For instance, pollen, propolis, and beeswax samples from hive products and honey bees collected in high-traffic areas have higher levels of heavy metals compared to nonurban areas (Matuszewska et al., 2021).

5 Technologies in Honey Analysis

Detection of Microbial Contaminants in Honey

Finding microbiological contaminants in honey is essential to guaranteeing its quality and safety. Many techniques, both conventional and contemporary, are employed to find these pollutants.

5.1 Conventional Techniques in Microbiology

Conventional techniques for finding microorganisms in honey entail cultivating the samples of honey on particular media and classifying the bacteria according to their morphological and biochemical traits. These methods include:

- **Total Plate Count (TPC):** This technique counts all of the live microorganisms in a sample of honey. It entails counting the colonies that emerge during incubation by plating diluted honey samples on nutrient agar.
- **Yeast and Mold Count:** This technique is comparable to TPC, but it makes use of medium that is specifically tailored to yeasts and molds, including Sabouraud dextrose agar. After incubation, the colonies are counted under the proper circumstances.
- **Detection of Specific Pathogens:** Bacteria including *Escherichia coli*, *Staphylococcus aureus*, and *Clostridium botulinum* can be identified using biochemical tests and selective media.

- While traditional methods are reliable, they are time-consuming and labor-intensive, often requiring several days to obtain results.

5.2 Molecular Methods Offer Faster and More Sensitive Detection of Microbial Contaminants in Honey

- **Polymerase Chain Reaction (PCR):** PCR is an extremely sensitive method that may identify even minute amounts of microbiological pollutants by amplifying particular DNA sequences. By focusing on its distinct genetic markers, PCR can be used to identify particular pathogens in honey, such as *Clostridium botulinum*.

- **Real-Time PCR (qPCR):** This particular PCR variant makes it possible to measure DNA in real time. When it comes to identifying and measuring particular microbial contaminants in honey, qPCR is quite helpful.
- **Next-Generation Sequencing (NGS):** NGS is a cutting-edge method that makes it possible to thoroughly examine the microbial community in honey. With the use of targeted RNA sequencing, targeted methyl sequencing, whole exome sequencing, amplicon sequencing, anchored multiplex PCR, sequencing by ligation (SBL; a technique for massively sequencing DNA immobilized on arrayed beads), Illumina sequencing (which uses clonal amplification and sequencing-by-synthesis to quickly read large stretches of DNA), and pyrosequencing (an early NGS method that detects pyrophosphate released when nucleotides are added to the DNA chain), it can identify a wide variety of microorganisms, even ones that are challenging to culture using conventional methods. Therefore, NGS is a cutting-edge technique for genetic material analysis that can quickly sequence big data sets of DNA or RNA. It permits the concurrent sequencing of millions of tiny DNA fragments.

5.3 Immunological Methods

Immunological techniques, including the enzyme-linked immunosorbent test (ELISA), are employed in honey analysis to identify particular microbial antigens or chemicals. These techniques are very helpful for finding poisons that bacteria like *Clostridium botulinum* create.

5.4 Markers for Detection of Microbial Contaminants in Honey

Markers are specific molecules or characteristics used to detect the presence of microbial contaminants in honey. These markers can be biochemical, genetic, or immunological in nature.

5.5 Biochemical Markers

Specific metabolites or enzymes produced by bacteria that can be identified to signify contamination are known as biochemical markers. For example:

- **Diastase Activity:** Honey bees manufacture the enzyme diastase, although microbial contamination can alter the enzyme's activity. According to Alaerjani et al. (2021), low diastase activity in honey may be a sign of microbial deterioration.

- **Hydroxymethylfurfural (HMF):** Honey can develop the chemical molecule HMF as it ages or is heated. According to Shapla et al. (2018), high levels of HMF may be a sign of microbial activity or unfavorable storage conditions.

5.6 *Genetic Markers*

- Genetic markers are specific DNA sequences that can be used to identify and locate microbiological pollutants. These markers are used in molecular techniques such as PCR and NGS. The 16S rRNA gene is a crucial genetic marker that offers details on the identification and classification of bacteria in the study of microbial communities within honey (Tran et al., 2017). This method has become increasingly important in the study of the safety, quality control, and microbiological diversity of honey. The following is a comprehensive analysis, divided into multiple important sections, of the fundamental ideas underlying the use of the 16S rRNA gene as a genetic marker in honey. The bacterial genome's 16S ribosomal RNA (rRNA) gene, which is roughly 1500 base pairs long and codes for a part of the 30S small subunit of prokaryotic ribosomes, is a highly conserved area. The conservation of the 16S rRNA gene and the existence of hypervariable areas make it a popular tool for phylogenetic and taxonomy research in microbiology. According to Bukin et al. (2019), it functions as a trustworthy genetic marker for genus and species-level identification and classification of bacteria.

5.6.1 Relevance of 16S rRNA in Honey Microbiome Studies

- Because honey is a natural product, it contains a wide variety of microorganisms, such as yeasts and bacteria. It is crucial to comprehend the microbiological makeup of honey for a number of reasons:
- Quality Control: Determining the presence of microbiological pollutants can aid in evaluating the safety and quality of honey.
- Authentication: The botanical and geographical origins of honey can be verified using its microbiological profile.
- Safety: Finding harmful microorganisms, including spores of *Clostridium botulinum*, is essential to guaranteeing the safety of honey, particularly for susceptible groups like young children.

5.6.2 Principles of 16S rRNA Gene Sequencing

- **Conserved and Variable Regions:** There are both variable and conserved regions in the 16S rRNA gene. The polymerase chain reaction (PCR) can be used to universally amplify the gene in many bacterial species due to the conserved areas. Sequence differences found in the variable regions (V1-V9) give them the selective capacity required for bacterial identification.
- **PCR Amplification and Sequencing:** Primers that target the conserved areas surrounding one or more hypervariable regions are commonly used in 16S rRNA gene PCR amplification. Following amplification, the products are sequenced using either conventional Sanger sequencing or, increasingly used these days, next-generation sequencing (NGS) technology. To determine whether species of bacteria are present in the sample, the sequences that are obtained are compared to reference databases.
- **Bioinformatics and Data Analysis:** Following sequencing, the raw data is processed using bioinformatics methods, such as sequence alignment, taxonomy designation, and quality filtering. To represent the bacterial taxa in the sample, amplicon sequence variants (ASVs) or operational taxonomic units (OTUs) are created. Sequence comparison and identification are frequently performed using databases such as SILVA, Greengenes, and RDP (Ribosomal Database Project) (Ptaszyńska et al., 2021).

5.7 Applications of 16S rRNA Gene Sequencing in Honey

- **Microbial Diversity Profiling:** The 16S rRNA gene sequencing offers a thorough picture of the bacterial community that exists in honey. According to studies, honey contains a range of bacteria, some of which are potentially harmful as well as beneficial species like *Lactobacillus*, *Bifidobacterium*, and *Bacillus*.
- **Quality and Authenticity Assessment:** By linking certain microbial communities to specific floral or geographic sources, the microbial profile derived from 16S rRNA sequencing can be utilized to confirm the authenticity of honey. This is especially helpful in the fight against fraud and adulterated honey.
- **Detection of Pathogens:** Pathogens that can cause botulism, including *Clostridium botulinum*, are a major worry when it comes to honey. The sensitive identification of these dangerous bacteria is made possible by the 16S rRNA gene sequencing method, guaranteeing the safety of honey for human consumption.
- **Impact of Processing and Storage:** The impact of processing and storage on the microbial makeup of honey can be evaluated by researchers by comparing the microbial communities in raw and processed honey. Pasteurization, e.g., may lower the microbial load in honey, but it may also have an impact on the good bacteria in the product (Bovo et al., 2020).

Stavropoulou et al. (2023) investigated the bacterial populations in different honey kinds using 16S rRNA gene sequencing. The study found significant differences in the microbial composition, with some bacterial taxa unique to specific types of honey, based on the honey's floral source. 16S rRNA sequencing was used by Pomastowski et al. (2019) to confirm the geographical provenance of honey samples. The microbiological profiles showed promise as a means of verifying the authenticity of honey, as they were sufficiently varied to differentiate honeys from various locations.

5.8 Detection of Heavy metals in Honey

The use of molecular indicators to identify heavy metal pollution in honey bees has also been studied. The maintenance of essential heavy metal homeostasis and the body's defense against dangerous heavy metals depend on metal-binding proteins known as metallothioneins (Yang et al., 2024). According to Purać et al. (2019), feeding forager bees sucrose solutions containing sub-lethal doses of Cd, Cu, and Pb increased the expression of AmMT, the gene encoding for metallothionein, in a dose-dependent manner. Through field-based tests, they looked at the expression profile of AmMT in forager bees from two locations in Serbia: the industrial zone of ZajaŇca and the city of Belgrade. Foragers in ZajaŇca expressed more AmMT than those in Belgrade, which was in line with the greater levels of heavy metals present in the industrial environment. Purtic et al. (2019) suggest that AmMT in *A. mellifera*, based on expression profiles observed in laboratory and field-based investigations, may be used as a bioindicator of environmental Cd, Cu, and Pb pollution. The studies in this subsection have demonstrated the relationship between heavy metal concentrations, Pb isotopes, and molecular biomarkers such as AmMT in honey bees and/or their hive matrices, and geographical variations in the sources of metal pollution. Monitoring the environment for heavy metals on a regular basis could help determine the level and source of pollution, which could have a big impact on protecting the environment and public health (Das et al., 2023). Measurements of metal concentrations, isotopic ratios, and/or relevant biomarkers in honey, beeswax, and honey bees can be used to do routine monitoring.

5.9 Technologies in Assessment of the Physicochemical Characteristics of Honey

Analysis of honey considering the physicochemical properties (color, pH, viscosity and rheology, moisture content, electric conductivity [EC], carbohydrate, amino acids, proline content, phenol content, organic acids, vitamin, mineral compositions, pollen content, enzymes, sensory analysis, ash content, hydroxy methyl furfur, total solids soluble [°Brix], and microbial content), biochemical functionality (diastase activity, vitamin C, and antioxidant activity), and presence of organic acids, honey proteins, in addition to honey bee gut microbiota assessment can provide myriad of significant information regarding honey composition and authenticity. The significant physicochemical attributes of honey and their significance in quality assessment are illustrated in Table 6.4.

The instruments commonly used are ET-AAS electro thermal atomic absorption spectroscopy, FAAS flame atomic absroption spectroscopy, *FT-MIR-ATR* Fourier transform mid-infrared spectrometry with attenuated total reflectance, *GF-AAS* graphite furnace atomic absorption, *HPAEC-MS* high-pressure anion-exchange chromatography coupled with mass spectroscopy, *ICP-OES* inductively coupled plasma optical emission, *ICP-MS* inductively coupled plasma mass spectrometry, *IR* infrared spectrometer, *LC-ECD* Liquid chromatography–electrochemical detection.

6 Chromatography Coupled Spectroscopic Analysis of Honey

6.1 Mass Spectrometry

Carbohydrates hold the major constituency in the composition of honey. The plethora of carbohydrates provides honey with its sweet taste and significant caloric value. Identification and quantitative analysis of existing carbohydrates, especially higher order oligosaccharides are crucial and can be done by chromatography coupled mass spectroscopy, namely HPAEC-MS, UHPLC/ESI Q-Orbitrap MS, Nano-ESI-MS, UHPLC-Q-ToF-HRMS and more frequently utilized GC-MS/MS, LC-MS/MS, and UHPLC-MS.

6.1.1 HPAEC-MS

In high-performance anion-exchange chromatography coupled with mass spectrometry (HPAEC-MS), carbohydrate profiling of honey can be done with significant specificity. Employing HPAEC-MS in analyzing monofloral, polyfloral, and honeydew honey, it has been identified that, the major carbohydrates in honey are glucose and fructose (73% to 89%). Depending on the source of the honey, variability in the carbohydrate composition and concentration has been observed. Higher concentration of melezitose (3%), lactose (3%), turanose (2%), and erlose (1%) was found in multifloral honey, whereas in monofloral honey considering Acacia, prevalent carbohydrates are lactose (2%), turanose (2%), and erlose (2%); in Dandelions, lactose (4%), turanose (3%), and palatinose (2%); in Rhododendron, erlose (6%), sucrose (3%), lactose (2%), and turanose (2%); and in Honeydew,

Table 6.4 Physicochemical properties and their significance in quality assessment

Physicochemical properties	Importance in quality evaluation	References
Color	Varies from medium light to dark amber, owing to plant pigments (anthocyanins, carotene, xanthophylls), polyphenols, flavonoids, amino acids and minerals, USDA considered optical density range from 0.0945 to 3.008.	Pascual-Mate et al. (2018) and Smetanska et al. (2021)
pH	Acidic (3.5–4.5), presence of organic acids (including gluconic, galacturonic, pyruvic, malic, citric and malonic acids, etc.); pH meter	El Sohaimy et al. (2015)
Viscosity and rheology	Function of moisture content (14–24%), range 14.73 Pa. s to 4.17 Pa. s (25 °C), varies due to handling and storing temperature also protein content, measured by rheometer	Garcia et al. (2005) and Bogdanov et al. (2008)
Moisture content	Usually less than 20%; honey collected in humid season has high moisture content, causes fermentation; measured by refractometric (Abbe & Electronic) method, also IR, FT-NIR, FT-MIR-ATR, and FTIR	Singh and Singh (2018), Ananias et al. (2013) and Pataca et al. (2007)
Electric conductivity (EC)	Mineral salts, organic acids, and amino acids are determining factor of EC, more than 0.8 in honeydew honey and less than 0.8 $ms.cm^{-1}$ in blossom honey; measured by dissolved honey in electrical conductivity cell	Kaškonienė et al. (2010)
Carbohydrate	80–85% of honey, mainly glucose and fructose, also signature carbohydrates including melezitose, turanose, and erlose; percentage depends on the honey type, HPAEC-MS, HPLC, and GC	De La Fuente et al. (2011) and Tedesco et al. (2020)
Amino acids	Major source of amino acid is pollen, amino acids in flower honey (0.1%–1.5%) and honeydew honey (3.0%) ^{1}H NMR, LC-ECD	Cotte et al. (2004) and Iglesias et al. (2006)
Proline content	Proline amount to almost 50–85% of the total amino acid, 873.00 mg/kg; LC–MS/MS	Iglesias et al. (2006) and Kowalski et al. (2017)
Phenol content	4-Dimethylaminobenzoic acid, p-coumaric acid, caffeic acid, gallic acid, vallinic acid, syringic acid, chlorogenic acid; evaluated by CE, GC, TLC, HPLC, and colorimetric assays	Alvarez-Suarez et al. (2009) and Trautvetter et al. (2009)
Organic acids	Gluconic, malic, and citric acids, approximately 0.57% of honey; measured by stable isotope labeling assisted LCMS	An et al. (2020)
Vitamin	Vitamin A, B1, B2, B3, B5, B6, B7, B9, C, E, and K; around 0.05–0.11 g/100 g, analyzed by HPLC-RP	Bogdanov (2012)
Mineral compositions	Potassium, calcium, magnesium, sodium, Sulphur, and phosphorus. Trace elements: Iron, copper, zinc, and manganese, almost 0.5–0.9 g/100 g; evaluated by acid digestion followed by FAAS, GF-AAS, ET-AAS, ICP-OES, and ICP-MS	Solayman et al. (2016) and Pohl et al. (2012)

(continued)

Table 6.4 (continued)

Physicochemical properties	Importance in quality evaluation	References
Pollen content	Taxonomical origin, biomonitoring, source of amino acid, microscopy, and DNA sequencing	De-Melo and de Almeida-Muradian (2017)
Enzymes	Diastase activity more than 8 DN Invertase activity range 0.8 to 25.9 IN Glucose oxidase range 25.58 to 402.47 μg H_2O_2/h·g	Sakač and Sak-Bosnar (2012), Serra Bonvehí J. et al. (2000) and Sahin et al. (2020)
Sensory analysis	Electrical nose and electrical tongue	Veloso et al. (2018) and Gonçalves et al. (2023)
Ash content	About 0.1–0.4 g/100 g, depend on the mineral content and determined by thermogravimetry	Felsner et al. (2004)
Hydroxy methyl furfur	Less than 40 mg/kg, although fresh honey has a minimal HMF (0–0.2 mg/kg). Heat treatment increases HMF content, due to dehydration of fructose; evaluated by reverse phase HPLC coupled with ultraviolet detector, ^{1}H NMR	Anese et al. (2013), Shapla et al. (2018), Chernetsova and Morlock (2012) and del Campo et al. (2016)
Total solids soluble (°brix)	Calculated from the table of correspondence between the refractive index (20 °C) and °brix	Albu et al. (2021)
Microbial content	Yeasts and bacteria; least colony-forming units per gram (cfug)	Adenekan et al. (2012)

melezitose (12%), erlose (3%), and raffinose (2%) have been detected (Tedesco et al., 2020).

6.1.2 LC-MS/MS

Organic acids in honey can be quantified utilizing biochemical experiments (enzymatic) and coupled chromatography and mass spectrometric analysis. Liquid chromatography-tandem mass spectrometry (LC-MS/MS) is instrumental in the identification and quantification of organic acids in honey. The presence of gluconic acid (2995.6 ± 1602.3mg/kg), malic acid (148.6 ± 346.4 mg/kg), citric acid (73.4 ± 84.2 mg/kg), in addition to succinic acid (28.6 ± 25.2 mg/kg) and tartaric acid (8.0 ± 4.6 mg/kg), was found in honey samples from different geographical regions, including China, Japan, Canada, Argentina, Romania, United States, Spain, and New Zealand (Suto et al., 2020).

6.1.3 UHPLC/ESI Q-Orbitrap MS

Metabolite profiling of honey provides valuable insights regarding honey composition, geographical source, entomological origin and species foraging behavior. Ultrahigh-performance liquid chromatography electrospray ionization quadrupole orbitrap high-resolution mass spectrometry and ultrahigh performance liquid chromatography coupled with triple quadrupole tandem mass spectrometry (UHPLC/ESI Q-Orbitrap MS) successfully detected the presence of 3-amino-2-naphthoic acid and methyl indole-3-acetate from *A. cerana* honey and kynurenic acid from *A. mellifera* honey. This highly sensitive metabolite profiling of honey is useful in identifying molecular authenticity markers for confirmative entomological origin (Wang et al., 2022).

6.1.4 Nano-ESI-MS

Nanoliter electrospray ionization mass spectrometry (Nano-ESI-MS) with UPLC-Q/TOF-MS multivariate analysis of honey provides the presence of a differential metabolic biomarker, which includes entomological biomarkers, geographical region-specific biomarkers, and taxonomical biomarker. Metabolite profiling was performed on *Castanopsis* honey (CH), *Triadica cochinchinensis* honey (TH), *Eurya* honey (EH), and *D. dentiger* honey (DH). Potential biomarkers of specific honey type have been identified, phenethylamine for CH, tricoumaroyl spermidine for EH, and (+/−)-abscisic acid for DH (Yu et al., 2023).

6.1.5 Nano-ESI-MS

Manuka honey is produced by bees foraging the manuka tree (*Leptospermum scoparium*), mainly found in Pacific regions (New Zealand and Australia). Possessing antibacterial and antioxidant properties, manuka honey holds huge nutritional and economic significance. Liquid chromatography coupled with coupled with quadrupole-time-of-flight and high-resolution mass spectrometry provides metabolic biomarkers of honey. This analytical detection of honey-specific biomarkers possesses enormous importance in the economic significance of honey. In manuka honey, the presence of manuka pollen DNA marker, methylglyoxal, dihydroxyacetone (precursor of methylglyoxal) along with 19 other manuka markers have been identified, these are methyl syringate, methyl syringate dimer and trimer, leptosperin, lepteridine, leptosperintriglycoside, acetosyringone, 3,4,5-trimethoxybenzoic acid, lumichrome, 3-hydroxy-1-(2-methoxyphenyl) penta-1,4-dione, α-hydroxy-2-methoxy-γ-oxobenzene butanoic acid, 2′-hydroxyacetophenone, paeonol, kojic acid, unedone, 3,4-dihydroxyphenylacetic acid (DOPAC), 3-(3,4-dihydroxyphenyl)-4-phenyldihydrofuran-2(3H)-one, and p-hydroxyphenylacetic acid (Díaz-Galiano et al., 2023).

6.1.6 GC-MS/MS in Volatile Compound Detection

Both qualitative and quantitative analysis of volatile and gaseous compounds can be performed by gas chromatography coupled with MS (GC-MS), while GC assistances in separating the constituents of a sample, MS facilitates the detection of compounds. Honey authentication and adulteration evaluation performed by GC-MS helps in the identification and quantification of the volatile compounds present in honey with recognition of unifloral signature aromatic profile and floral markers to ensure nutritional validation and appropriate value assessment (Tian et al., 2018, An et al., 2020). Comprehensive GC–MS analysis of cotton, Erica, fir, pinus, and thyme honey displayed a unique volatile profile chromatogram, and thyme honey possesses the highest number of volatile compounds (Tananaki et al., 2021). Analysis of 51 honey samples from 12 monofloral honey indicated characteristic VOCs in honey samples; these VOCs can be informative botanical biomarkers in indicating geographical variation of monofloral honey (Machado et al., 2020, 2021). VOCs function as a pivotal indicator of honey's altered aroma profile during prolonged storage, i.e., *cis*- and *trans*-linalool oxide along with hotrienol (da Silva et al., 2020). Signature monofloral volatile compounds are listed in Table 6.5.

6.1.7 GC-MS/MS, LC-MS/MS, and UHPLC-MS in Pesticide Detection

The harmful effects of pesticides include but are not limited to the destruction and elimination of honey insects and harmful pests. Colony collapse disorder (CCD) is an example of this, where a large amount of honey bee colonies get abolished due to pesticides. The limit of detection (LOD, the lowest detectable analyte concentration) and limit of quantification (LOQ, the lowest quantifiable analyte concentration) values indicate the sensitivity of the detection method. Regarding the occurrence of pesticide residues in honey, maximum residue levels (MRLs) have been employed by the European Union (EU) to ensure consumer health safety and honey quality assurance.

Liquid chromatography-tandem mass spectroscopy (LC-MS/MS) and gas chromatography-tandem mass spectroscopy (GC-MS/MS) analysis have been employed for the detection and quantification of pesticides in honey at ppb (parts per billion) level. The presence of various pesticide residues including neonicotinoids, organochlorines, organophosphates, triazoles, carbamates, dicarboximides, and dinitroaniline in honey has a negative impact on consumer health and economic valuations of honey. LC-MS/MS screening for pesticide residues in honey indicates the presence of amitraz, parathion, parathion methyl, deltamethrin, anthraquinone, and 2-phenylphenol above the LOQ levels in the honey collected from several districts of Kerala (Prasanth et al., 2022). LC-MS/MS and GC-MS/MS analysis of commercial honey samples from six Brazilian states indicated that among all the detected pesticides, thiabendazole, carbendazim, azoxystrobin, chlorpyrifos, imidacloprid, and thiamethoxam have been observed to be present in more honey samples (Almeida et al., 2020). Utilizing GC-MS/MS and UHPLC-MS, several

Table 6.5 Characteristic volatile compounds of monofloral honey

Honey category	Characteristic volatile compounds	References
Acacia honey	Cis-linalool oxide, 3-methyl-3-buten-1-ol, and heptanal	Machado et al. (2020)
Buckwheat honey	3-Methylbutanal, 2-methylbutanal, and 3-methylbutyric acid	Panseri et al. (2013) and Machado et al. (2020)
Chestnut	Benzaldehyde, 2-aminoacetophenone, acetophenone, 1-phenylethanol, phenylacetic acid, 3-hexen-1-ol, and 2-methyldihydrofuranone	Soria et al. (2009) and Machado et al. (2020)
Cotton	1-nonanal, phenylacetaldehyde, and phenylethyl alcohol	Machado et al. (2020)
Dandelion	3-Methylpentanenitrile and phenylacetonitrile	Jerković et al. (2015) and Machado et al. (2020)
Eucalyptus	Phenylacetaldehyde, nonanol and nonanoic acid, terpenes, linalool and linalool oxides, norisoprenoids	Castro-Vázquez et al. (2009) and Machado et al. (2020)
Pine	Nonanal, nonanol, decanal, and octanal	Tananaki et al. (2007) and Machado et al. (2020)
Strawberry	α-Isophorone, β-isophorone, 4-oxoisophorone, 2,5-dimethylfuran, and 2,3-butanedione	Seisonen et al. (2015)
Honeydew	Pinene, octane, and nonanal	Tananaki et al. (2021)
Erica	Isophorone and furfural	
Thyme	Benzeneacetaldehyde, benzaldehyde, and benzyl nitrile	
Orange	α-Pinene, cis-myrcenol, methyl anthranilate	Machado et al. (2021)
Strawberry	2,3,5-trimethylphenol, 3,4,5-trimethylphenol	

contaminants have been identified from pollen and honey samples. These include 2,4-D (herbicide), azoxystrobin (fungicide), boscalid (fungicide), coumaphos (insecticide-acaricide), cyprodinil (fungicide), fludioxonil (fungicide), MCPA (herbicide), propargite (insecticide-acaricide), quizalofop (herbicide), tau-fluvalinate (insecticide-acaricide), trifluralin (herbicide), DDAC (fungicide), and propargite (insecticide-acaricide) in the real honey and pollen samples collected from the Island of Ireland. Among these, propargite was found to be the most commonly occurring insecticide (Díaz et al., 2023). UHPLC-MS/MS with modified QuEChERS preparation of honey samples from Apiário Cambará, Rio Grande do Sul, Brazil, indicated the presence of insecticides in honey. Although insecticide residues, including cypermethrin, permethrin, imidacloprid, clothianidin, chlorpyrifos, and dimethoate, have been detected in honey, however, their concentrations were lower than LOQ. Regular monitoring of these insecticide residues is required for honey quality assurance and proper nutritional value maintenance (Souza et al., 2021).

6.2 *Fluorescence Spectroscopy*

Fluorescence emission spectra of unifloral, Acacia, and Sidr honey indicate the presence of fluorophores, such as phenolic compounds including phenolic acids, and flavonoids, along with aromatic amino acids and vitamins. Comparative analysis of adulterated honey made with different concentrations of sugar syrup with Sidr honey provides evidence of existing fluorophores in Sidr honey, and increasing right shift toward wavelength of 507 nm of fluorescence emission spectra with increasing concentration of Sidr honey (507 nm Sidr honey fluorescence emission maxima) (Ali et al., 2022). Comparative analysis of Sidr honey with unifloral honey from *Elettaria cardamomum, Citrus reticulata*, and *Grewia asiatica*, polyfloral honey, and commercial samples provides valuable insights regarding the presence of fluorophore markers as botanical identifiers (Ali et al., 2020). Analysis of unifloral honey samples derived from the four individual taxonomical sources (thyme, pine, fir, and citrus) and three-dimensional-emission excitation matrices (3D-EEM) spectra of phenolic compounds validates the presence of a plethora of phenolic compounds (Xagoraris et al., 2021). Fluorescence spectroscopy combined with multivariate curve resolution alternating least squares (MCR–ALS) was employed to elucidate the spectral components ratio generated from phenol and proteins (Ph/Pr) components of honey (Stanković et al., 2019). Fluorescence spectroscopy with soft independent modeling of class analogy (SIMCA) analysis distinguishes real honey, fake honey, and HFCS-55 (high fructose corn syrup). Fluorescence spectra of real honey (SBH) exhibited a main peak at the wavelength of 480 nm; fake honey and HFCS-55 demonstrate identical peaks at the wavelength of 378nm (Suhandy et al., 2024). The presence of sulfonamide and its derivatives can also be identified by utilizing combined liquid chromatography-fluorescence detection (LC-FLD) (Shi et al., 2024).

6.3 *Ultraviolet-Visible Spectroscopy*

UV–vis spectroscopy, combined with multivariate statistical analysis, is employed for the detection of colorants and artificial sweeteners, and identification and quantification of polyphenols such as polyphenols and flavonoids in honey (Uçar et al., 2023). To detect honey adulteration, UV–vis spectroscopy-based detection of the percentage presence of aromatic composition is crucial in honey analysis. According to the literature, the absorbance range of sugars, specifically glucose and fructose, and phenolic compounds is 200–260 nm, the absorbance range of amino acids, mainly tryptophan, proteins, and phenolic compounds is 260–300 nm, and the absorbance range of flavonoids is 300–340 nm (Suhandy & Yulia, 2021). A decrease in absorbance intensity has been indicated in the honey with high fructose corn syrup (HFCS), whereas spectral data of pure honey indicates higher absorption at a wavelength window between 250 and 300 nm; the distinguishable change is observed at 280 nm (Suhandy et al., 2022). Corn syrup, agave syrup, sugarcane molasses, and other sweeteners can be present in honey as adulteration; the detection of these fillers is crucial in honey authentication and quality assurance. UV–vis spectroscopic examination coupled with one-class pattern recognition methods, namely OC-PLS and DD-SIMCA, provides highly sensitive and accurate results in detecting sugar-based adulterants present in honey (de Souza et al., 2021).

6.4 Fourier Transform Infrared Spectroscopy (FTIR) Spectroscopy

Fourier transform infrared spectroscopy (FTIR) analysis of honey provides botanical information by representing FTIR spectral band attribution of specific groups and compounds. FTIR spectroscopy utilizes the frequencies of molecular vibrations of single bonds (C-H, O-H, N-H), double bonds (C-O, C-N), triple bonds (C-C, C-N), and fingerprint zones to reveal distinct peaks and characteristic spectral features. Group-specific frequency range (cm^{-1}) in FTIR spectra provides insights regarding honey components, e.g., carbohydrates, organic acids, amino acids, and honey proteins, along with the ratio of oligosaccharides from the spectral peak intensity (Prata & da Costa, 2024). FTIR coupled with principal component analysis (PCA) and cluster analysis (CA) is utilized for the detection of adulteration in honey, and spectral region analysis has been considered between 4000 and 650 cm^{-1} and cluster analysis between 1800 and 650 cm^{-1}. The spectral range from 1800 to 750 cm^{-1} is considered highly significant in distinguishing pure and adulterated honey consisting of simple sugar. The wave number (cm^{-1}) of pure honey and the shifted wave number of adulterated honey help to identify specific functional groups and their altered molecular bond vibration (Damto et al., 2023). FTIR absorption spectra are also useful in determining the origin of honey, i.e., unifloral and multifloral honey exhibited significant FTIR spectral changes, indicating variation in honey composition. In the study's comparative analysis, FTIR absorption spectra showed Honeydew honey having the highest absorbance among the sample group (Kędzierska-Matysek et al., 2018). FTIR spectral region of $1580–1500\,cm - {}^{1}$ had been set, and Amide II spectral absorption peak has been observed at 1547 cm^{-1}. An increase in absorbance indicates an increase in royal jelly content in honey; this observation exhibits a synonymous result with HPLC (R2 = 0.99) (Cebi et al., 2020).

6.5 Attenuated Total Reflectance (ATR)-FTIR Spectroscopy

ATR-FTIR spectroscopy is a type of FTIR utilized in the identification and quantification of the organic functional groups and inorganic ions in aerosol samples (Coury & Dillner, 2009). Analysis of honey samples is performed by the spectral deviations in the 4000–650 cm^{-1} spectral region. To identify monofloral honey from different taxonomical sources the FTIR spectra have been segregated into five regions (Regions 1–5, Region 1: 3000–2800 cm^{-1}, Region 2: 1700–1600 cm^{-1}, Region 3: 1540–1175 cm^{-1}, Region 4: 1175–940 cm^{-1}, Region 5: 940–700 cm^{-1}); each region lays in specific wave number and represent particular bond movements in macromolecules present in honey (Gok et al., 2015). ATR-FTIR combined with chemometric analysis of 80 four unifloral honey samples (strawberry, asphodel, thistle, and eucalyptus) from Sardinia (Italy) indicated characteristic phenolic fraction in the ATR-FTIR spectra at region 1540–1175 cm^{-1} (Ciulu et al., 2020).

6.6 Isotope Ratio Mass Spectrometry (IRMS)

Element analyzer (EA) and liquid chromatography (LC) coupled with isotope ratio mass spectrometry (IRMS) analysis can differentiate carbon isotope ratios 13C/12C, produced by different photosynthesis cycles (C4 and C3plants). Analysis of δ13C of 33 honey samples can determine the carbon stable isotope ratio of individual sugars (δ13Cglucose [δ13Cg], δ13Cfructose [δ13Cf], δ13Cdisaccharide [δ13Cds], and δ13Ctrisaccharide [δ13Cts]). The authenticity of honey has been determined by employing the maximum absolute differences between all δ13C values (Δδ13Cmax); more adulteration is indicated by a larger Δδ13Cmax value. LC IRMS results of 33 honey samples from several Lebanese regions indicated 13 samples of adulterated honey, and EA IRMS results showed a higher percentage of C4 sugar (high fructose corn syrup) in adulterated honey (El Hawari et al., 2021). The Δδ13Cmax (abs) less than or equal to ±2.1% would be considered pure honey. In the EA/LC-IRMS analysis, Δδ13Cmax (%) value outside the range of ±2.1% had been considered as adulterated. Analyzing Acacia honey (n = 29), Rape honey (n = 33), Vitex honey (n = 11), Cotton honey (n = 10), Linden honey (n = 4), Jujube honey (n = 3), and Sunflower honey (n = 4) indicated about 70% of the honey samples was unable to qualify the δ13C-LC-IRMS test (Xu et al., 2020a, b). IRMS analysis provides honey saccharide profiling and is mainly employed to determine the presence of C4 sugar adulteration in honey, in addition, δ13C and δ15N values for honey and proteins provide useful information in differentiate different floral origins of honey (e.g., Buckwheat, Clover, Heather, Linden, Rapeseed, Willow, and Polyfloral) (Aries et al., 2021).

6.7 Stable Isotope Ratio Mass Spectrometry (SIRMS)

Comprehensive EA/IRMS and LC/IRMS analysis have been performed on 116 commercial honeys from Japan (34), Spain (13), Italy (13), France (10), New Zealand (10), China (9), Canada (4), Hungary (4), EU (4), Argentina (1), Australia (1), Bulgaria (1), Kuba (1), Mexico (1), Romania (1), Taiwan (1), and United States

(1). δ13C values of protein, glucose, fructose, disaccharides, trisaccharides, and organic acids in pure honey as well as δ13C value of pure honey have been obtained and elucidated in Table 6.6 (Kawashima et al., 2019). SIRMS analysis can be highly accurate in detecting C4 sugars (cane and corn syrup) based on honey adulteration. ISCIRA method-based C4 sugar adulteration assessment has been performed on available honey samples from the Philippine market, and results showed the percentage of C4 adulteration in local online-bought and local store-bought honey to be 86.5% and 75%, respectively. This differential analysis has been performed in comparison with authentic Philippines honey samples from *Apis mellifera* (11), *Tetragonula biroi* (13), and *Apis breviligula* (14). Utilizing a continuous-flow EA-IRMS system, carbon-13 isotope abundance (δ13C) of all 38samples has been calculated to be from −29.57 to −23.79‰, the feature of C3 plants. However, in (64 of 74) Philippines online purchased honey, Bulk honey δ13C (‰) ranges from −10.71 to −18.73, and apparent C4 sugar content (%) found to be from 7 to 100% [from 7 to 10 (3), from 11 to 50 (1), from 51 to 70 (1), 71 to 80 (1), 81 to 90 (7), 91 to 100 (51)] (Lao et al., 2021). Two-dimensional (2D)-LC has been observed to be more efficient in detecting stable isotopes. Heart-cutting two-dimensional LC/IRMS method for the analysis of honey samples has been utilized in the detection of organic acid in honey with high precision. 2D-LC/IRMS analysis of 25 purchased honey samples from Japan (10), New Zealand (5), China (3), Spain (2), Canada (2), Romania (1), United States (1), and Argentina (1) indicated the presence of gluconic, malonic, and citric acids in chromatograms; however, solely gluconic acid have been detected, elucidating the presence of other organic acids in lesser concentration. 2D-LC/IRMS results procured for gluconic acid showed the δ13C value −31.7 to −28.5‰ (mean: −30.0 ± 0.7‰) and concentration 415 to 6043 mg kg^{-1} (mean: 2674 ± 1635 mg kg^{-1}) (Suto et al., 2019).

6.8 *Inductively Coupled Plasma Mass Spectrometry (ICP-MS)*

ICP-MS is a highly accurate analytical technique in food adulteration assessment and authenticity screening. ICP-MS is useful in the determination of multi-elemental composition, quantification, and isotopic abundances of various components (Mazarakioti et al., 2022). Inductively coupled plasma quadrupole mass

Table 6.6 The stable isotope ratio of individual sugars and their δ13C values

Stable isotope	δ13C values	References
Glucose	From −27.0 to−23.8‰	Kawashima et al. (2019)
Fructose	From −28.4 to−23.8‰	
Disaccharides	From −30.2 to−24.0‰	
Trisaccharides	From −29.2 to−22.8 ‰	
Organic acids	From −33.6 to −26.5 ‰	
Pure honey	From −29.1 to −24.3 ‰	

spectrometry (ICP-Q-MS) elemental profile of Romanian honey samples of linden, acacia, rape, and sunflower has been examined by the concentration of 18 elements (Al, As, Ba, Ca, Cd, Co, Cr, Cu, Mg, Mn, Na, Ni, K, Pb, Sr, Ti, V, and Zn). The most frequent elements are found to be K (248.70 mg kg^{-1}), Ca (59.97 mg kg^{-1}), Mg (20.54 mg kg^{-1}), and Na (11.92 mg kg^{-1}); the mineral fingerprint profile of honey can be a useful determinant of origin and botanical identification (Voica et al., 2020). In another study, screening of 18 elements of diluted honey samples of Poland has been analyzed; results indicated distinguishable concentrations of major elements (K, Mg, Mn, and Na) and trace element (Cu) (Gręda et al., 2024). Monofloral honey of buckwheat and rape honey and honeydew from Poland have been analyzed for 13 elements (Al, B, Ba, Ca, Cd, Cu, K, Mg, Mn, Na, Ni, Pb, and Zn), coupled with CA and PCA analysis to determine mineral concentration profiles as an indicator of botanical sources of honey samples. Principal components analysis (PCA) elucidated Al, Cd, K, Ni (PC 1), Ba, Na, Pb (PC 2), Cu, Mg (PC 3), and Zn (PC 4) to be major elements. K, Al, Ni, Cd, and Zn profiles are characteristic of honeydew honey, and Ba, Na, and Pb profiles are characteristic of rape honey (Chudzinska & Baralkiewicz, 2010). Mineral elements as a possible pollution indicator in honey samples were analyzed using the HPLC-ICP-MS technique. A total of 90 honey samples from Poland and Ukraine have been collected and analyzed to identify and quantify water-soluble arsenic and arsenic species, including As(III), As(V), AsB, DMA, and MMA present in honey. In Polish multiflowered honey, the mean arsenic concentration was determined to be 6.1 μg kg^{-1}; in Polish acacia honey, this concentration is 13 μg kg^{-1}; however, in Ukrainian multiflowered and acacia honey, the mean arsenic concentration is found to be 1.0 μg kg^{-1} and 2.5 μg kg^{-1}, respectively (Jakkielska et al., 2024). ICP-MS-based ionomics analysis has been performed to determine the geographical origin of honey samples. HPLC-DAD (diode array detector) technique combined with multivariate data analysis of mineral element composition of honey samples has been performed to differentiate *A. cerana* honey from different geographical regions. Honey samples from Liuba, Yangxian, and Longxian showed distinguishable characteristics in the heat map analysis (Wu et al., 2021).

6.9 NMR Spectroscopy

Detection and quantification of potentially threatening components existing in honey samples is a significant parameter for honey authentication and consumer health safety. Employing benchtop nuclear magnetic resonance (NMR) spectrometers has been utilized in the detection of contaminants in honey, such as pharmaceutical ingredients and drug residues. Analysis of 50 French honey samples revealed the presence of tadalafil (37), sildenafil (5), and a combination of flibanserin and tadalafil (1) (Pujol et al., 2024). ^{1}H NMR coupled with the multivariate data analysis provides insights regarding the geographical origin of honey, depending on the varying percentage presence of sugars, amino acids, and carboxylic acid (Spiteri et al., 2020). Moreover, the presence of C3 and C4 adulterants in stingless bee honey (SBH) can also be determined by ^{1}H NMR metabolomics analysis (Yong et al., 2022). Recognition and quantification of non-sugar components (NSC) in monofloral honey samples including jujube honey (15), chaste honey (16), and locust honey (19) have been accomplished by HPLC and ^{1}H-NMR combined with chemometrics analysis. Results showed NSC in jujube honey to be 780.77 ± 8.98 mg/100 g, in locust honey to be 707.94 ± 7.86 mg/100 g, and in chaste honey to be 292.43 ± 5.67 mg/100 g (Ren et al., 2019). Standard ^{1}D ^{1}H-NMR spectra of honey worker bee body parts can discriminate between seasonal bees. Among 28 metabolites, fructose, sucrose, and trehalose have been identified as important indicators of seasonal variation in honey bees. Winter bees have more adenosine monophosphate (AMP), oxidized nicotinamide adenine dinucleotide (NAD+), fructose, sucrose, trehalose, and O-phosphocholine, and summer bees have increased free amino acids. The result exemplifies seasonal change-based alterations in physiological requirements (Lee et al., 2022). The 1D and 2D ^{1}H and ^{13}C-NMR spectra analysis of North Macedonian (16) and Bulgarian (22) honey samples can be beneficial and convenient technique in detecting botanical and geographical origin. The 2D score PCA and hierarchical clustering along with the signature carbohydrate profiles of honey can be efficacious determinants in origin identification and authentication (Gerginova et al., 2020). The ^{1}H NMR-based metabolomic evaluation of honey samples (9) from Vietnam has been performed by considering the following metabolites: HMF, citric acid, acetic acid, lactic acid, ethanol, alanine, tyrosine, and phenylalanine. Since honey has been considered to possess both nutritional and medicinal properties, critical monitoring of the freshness of honey (absence of HMF) and quality assurance (no artificial additives) can be instrumental in authenticity screening (Luong et al., 2019).

6.10 Raman Spectroscopy

The phytochemical properties of honey including pH, moisture, total soluble solids (TSS), lactonic acidity, total acidity, electrical conductivity, redox potential, hydroxymethylfurfural (HMF), and ash content can be instrumental in honey authentication. One hundred eighty-nine honey samples from Campeche, Mexico, have been subjected to Raman spectroscopy-based low-cost analysis to determine the physicochemical characteristics. Raman spectra considered for analysis were in ranges of 201 cm^{-1} to 2000 cm^{-1}; however, significant spectral changes have been detected from 330 cm^{-1} to 1460 cm^{-1}, with 13 characteristic segments associated with specific bond vibrations (stretching and bending) (Anguebes-Franseschi et al., 2019). Analysis of honey sugar profile, detection of honey adulteration with high fructose corn syrup and maltose syrup, and differentiation of honey of various sources can also be done by Raman spectroscopy. Spectral analysis of honey can be done with different types of model establishment, principal component analysis (PCA), principal component regression (PCR), partial least square (PLS), and partial least square discrimination analysis (PLS-DA) (Xu et al., 2020). From 12 different countries, 97 pure honey samples have been sourced and analyzed with HPLC-RI with Soft Independent Modeling of Class Analogy (SIMCA) data analysis to determine possible adulteration. Raman spectra considered for analysis range from 200 to 2500 cm^{-1}, indicating the presence of adulteration in the spectra fingerprint region from 400 to 1400 cm^{-1} in 17% of the samples (Aykas et al., 2020). Detection of methylglyoxal (MGO) in manuka honey has been performed with an intrinsic Raman signal amplification technique for quick identification by utilizing silver-coated gold nanoparticles (Au@Ag NPs) in addition to a highly accurate SERS probe, 8-thioguanosine (8-TG). 8-TG probed Au@Ag NPs exhibited a characteristic Raman peak at 700 cm^{-1}, and the presence of MGO showed a peak at 631 cm^{-1} because 8-TG reacts with MGO to form N2-(1-carboxyethyl)-thioguanosine (CETG); thus, analysis has been analyzed by Raman intensity ratio (I631/I700), and MGO promotes increase in the Raman intensity ratio. The limit of detection of the assay in detecting MGO was indicated to be 0.392 $\mu g\ mL^{-1}$ (Xia et al., 2021). From Romania, 56 pure honey samples (acacia, honeydew, sunflower, tilia, and polyfloral) and 900 adulterated honey samples (with increasing concentrations of fructose, glucose, inverted sugar, malt must, and hydrolyzed inulin syrup) have been analyzed with Raman spectroscopy in the range of 250 to 2339 cm^{-1} to identify specific spectral changes depending on adulterant concentration. Among the spectral profiles of different adulterants, increasing concentration of malt must exhibit significant spectral alteration (Oroian et al., 2018). Comparative analysis of raw honey samples with heat-treated and water-diluted honey exemplifies reliable and convenient green Raman spectroscopy analysis. Discrete spectral changes have been depicted in acacia, honeydew, raspberry, thyme, sunflower, lime, coriander, and rape honey (Molnar et al., 2020).

7 Advanced Techniques in Honey Analysis

In addition to conventional palynology, advanced analytical techniques including genetic marker-based identification technologies, such as PCR, transcriptome sequencing, and high throughput sequencing of DNA fragments present in honey could provide increasingly accurate and highly sensitive comprehensive information encompassing identification of molecular markers for floristic genes, geographical region, entomological origin, bees foraging behavior, plant species diversity monitoring, influence of varying climatic changes including possible effects of global warming. DNA metabarcoding, enzyme-linked immunosorbent assay (ELISA), zymography of honey, and metagenomic analysis of honey bee gut microbiota studies elucidated in Table 6.7 are the significant techniques in honey analysis that ensure honey authenticity while monitoring environmental determinants and biodiversity.

Table 6.7 Advanced techniques in honey analysis

Techniques	Principle	Applications in honey analysis	References
DNA metabarcoding	Amplification and sequencing of environmental DNA (eDNA)	Identification of taxonomical source of honey	Wirta et al. (2021)
ELISA	Assay based on affinity of ab to quantify ab or ag by use of an enzyme-linked antibody by forming a colored reaction product	Detection of antibiotic contaminants and possible allergens in honey	Poungmalai et al. (2021)
Proteomics and zymography	Protein purification and SDS PAGE, enzymatic activity assay	Recognize signature protein and enzymatic profile of honey	Alaerjani et al. (2022)
Metagenomic analysis of honey bee gut microbiota	Extraction, amplification, and sequencing of 16S rRNA gene for detecting microorganism evolution and bioinformatics-based gene ontology study	Identification of entomological origin and microorganism biodiversity	Ellegaard et al. (2020)

7.1 DNA Metabarcoding

Following pollen DNA extraction from 196 honey samples, amplification has been performed by utilizing a marker of internal transcribed spacer (ITS2) and the large subunit of ribulose bisphosphate carboxylase, RuBisCO (*rbcL*), for identification of plant species and their diversity in a discrete geographical region (Khansaritoreh et al., 2020). The rbcL Illumina MiSeq was performed with primers rbcLaf and rbcLr506 for sequencing pollen DNA to identify taxonomical biomarkers. Distinguishable seasonal preferences and changes in the foraging behavior of honey bees have been observed. In April, bees prefer *Salix* spp., *Prunus* spp., *Ulex* spp., and *Brassica* spp.; in June, July, and August, *Rubus* spp. found to be the primary preferential selection, in addition to *Trifolium repens*, *Cirsium/Hypochaeris/Centaurea* spp., *Rosa* spp.; and in September, *Hedera helix*, alongside *I. glandulifera* and *Camellia* spp., was preferred by the honey bees. Season-specific alteration in honey has been concluded from this pollen DNA barcoding observation, indicating seasonal variation in the nutritional value of honey (Jones et al., 2022). Sequencing-based biomonitoring has been performed utilizing two metabarcoding markers (ITS2 and trnL) to analyze the foraging behavior of European honey bees (*Apis mellifera*) in the Jerrabomberra wetlands, an urban reserve in Canberra, Australia. Pollen DNA metabarcoding revealed *Plantago* (75% of plots), *Hypochaeris* (62%), and *Sonchus* (50%) to be the three most frequently detected genera, and *Eucalyptus* (26.7%) to be the most widespread genus (Milla et al., 2022). The trnL and ITS2 sequencing data of pollen DNA metabarcoding of honey from different regions of Australia revealed *Fabaceae, Asteraceae, Myrtaceae, Brassicaceae, Macadamia integrifolia*, and *Melaleuca nodosa* to be the most distinguished taxa (Milla et al., 2021). Comparative exploration of *Apis cerana* honey

(ACH) and *A. mellifera* honey (AMH) indicated species-specific flower selectivity in honey bees. The result depicted that out of 56 identified taxa, the occurrence of unique taxa (ACH 23 and AMH 15) and shared taxa (predominantly *Actinidia, Diospyros, Flueggea,* and *Anacardiaceae*) thus elucidated inter-species competition pattern (Mohamadzade et al., 2022). Metabarcoding of environmental DNA (eDNA) of different sources obtained from honey includes microorganism-specific (16 s rRNA), botanical origin-specific (chloroplast trnL), and honey bee-specific (mitochondrial cytochrome oxidase subunit I (COI) genes; amplification and sequencing of these genes illustrated the presence of pathogens, geographical, and floral source of honey (Pathiraja et al., 2023). Honey samples collected from the northern European countries (Estonia, Finland, Sweden) were analyzed for DNA traces of bacteria (16S rRNA gene), fungus (ITS), and plant (ITS2, rbcLa, and trnL); results indicated microbial and taxonomical diversity (Wirta et al., 2021). In Finland, from June to August 2021, honey and beebread samples were accumulated from 43 beehives and consequently analyzed for pollen DNA metabarcoding. The result elucidated the floral selectivity of honey bees, amongst 73 genera identified, only 32 and 30 were found in honey and beebread, respectively. In addition, seasonal variation in honey metabolites was also indicated (Leponiemi et al., 2023). Noteworthy molecular markers in honey analysis are represented in Table 6.8. Environmental DNA metabarcoding with integrated sequencing technologies, bioinformatics, and statistical tools provides myriads of information regarding environmental samples. Ecosystem health, monitoring biodiversity, geographical provenance of honey, pollinator species diversity, foraging habits, and seasonal alterations can be distinguished by the molecular marker-based technologies with high accuracy and specificity.

7.2 Elisa

The presence of antibiotics and antibiotic residues in edible products can pose severe consequences, and have the ability to interfere with existing medicinal interventions, causing serious health ailments. Antibiotics present in health products can be detected by utilizing monoclonal antibody (mAb)-based ELISA technologies. Possessing specificity and sensitivity, ELISA has proven to detect antibiotic residues with high accuracy and at a significantly lesser detection limit. Chlortetracycline (CTC) has been detected to be present in honey with indirect competitive ELISA (icELISA), with a detection limit of 0.1 ng/ml (Poungmalai et al., 2021). Screening for chloramphenicol (CAP) in 40 honey samples showed the presence of CAP in all of the samples (the reporting limit is 0.1 $\mu g\ kg^{-1}$), the highest concentration was observed to be 2.2 $\mu g\ kg^{-1}$ (Rimkus et al., 2020). The study of 80 honey samples with semi-quantitative ELISA assay revealed the presence of sulfonamide antibiotic residue in 23.75% (19) of the samples from varying concentrations of antibiotic range within 10–120 ng/g (Mehrabi et al., 2022). The ic-ELISA screening of commercially available honey samples for Sparfloxacin (SPFX) showed the IC50 to be

Table 6.8 List of common and advanced molecular markers in analysis of organic honey

Gene code	Gene name and source	Significance in honey analysis	References
RuBisCO (*rbcL*)	Large subunit of ribulose bisphosphate carboxylase; plant	Taxonomical & diversity biomarkers	Khansaritoreh et al. (2020)
trnL	Chloroplast; plant		Milla et al. (2021) and Milla et al. (2022)
ITS2	Internal transcribed spacer; plant		
COI	Mitochondrial cytochrome oxidase subunit I; honeybee	Entomological origin	Pathiraja et al. (2023)
ITS	Nuclear ribosomal internal transcribed spacer; fungi	Microbial diversity biomarkers	Wirta et al. (2021)
16 s rRNA	16S ribosomal RNA, component of the 30S subunit of prokaryotic ribosome; microorganism		Pathiraja et al. (2023)

0.12 ng/mL and also indicated existing cross-reaction (CR) with relevant antibiotic compounds, including flumequine, fleroxacin, enrofloxacin, ciprofloxacin, and ractopamine (Li et al., 2018). Development of ic-ELISA able to detect both salinomycin (SLM) and methyl salinomycin (MLN) in honey samples exhibited IC50 of 0.86 ng/mL, and detection limit of 0.28 ng/mL in addition to being able to cross-react (CR) with narasin, maduramicin, monensin, nigericin, and lasalocid (Xu et al., 2019). Detection of tulathromycin (TULA), a respiratory disease antibiotic for cattle, if consumed, can induce allergies with hypersensitivity and the emergence of antibiotic-resistant strains. The immunochromatographic test strip prepared for TULA detection has demonstrated the lower detection limits to be 4 ng/mL and the cut-off limits of 10 ng/mL (Liu et al., 2018). The novel honey antibiotic detection system (HADS) based on lateral-flow immunochromatographic assay (LFIA)-based strip can simultaneously detect sulfonamides (SAs) and tetracyclines (TCs), exhibiting detecting limits of 0.4 μg/kg (Li et al., 2021). Detection of thiacloprid (insecticide) residues in 40 samples of honey detection range has been found to be 0.003 to 0.06 mg/kg. For the *ic*-ELISA, the IC50 and IC20 were indicated to be 0.38 ng/mL and 0.097 ng/mL, respectively (Zhu et al., 2024). These screening techniques demonstrate several advantages in immunoassay-based methodologies in honey analysis, including simple experimentation, rapid procedure, cost effectiveness, highly sensitive and accurate detection, quick data analysis, and interpretation. The requirement of stringent quality testing and continuous surveillance for maintenance of the efficacy of nutritional benefits of honey can be easily achieved by these analytical techniques.

7.3 Proteomics and Zymography

Major enzymatic components of honey are diastase (honey quality factor), invertase (honey storage and processing factor), glucose oxidase (antibacterial), proteases (antimicrobial and antioxidant), and acid phosphatase (floral origin biomarker and fermentation indicator) (Alaerjani et al., 2022). Essential optimization of variables (temperature, time, and pH) and their cumulative effect on honey enzymatic activity is essential for honey processing and storage for prime quality (Kamboj et al., 2019). Analysis of 110 honey samples exhibited increasing temperature and heating duration negatively impact diastase activity (Huang et al., 2019). Invertase and glucose oxidase secreted by the hypopharyngeal glands of the honeybee function as a quality biomarker for honey nutritional value and quality assurance. Turkish honey samples (Anzer blossom, Pine, and Oak) showed decreasing Invertase activity and varying glucose oxidase activity as a function of temperature and time, which are pivotal to estimate ideal extraction temperature (Sahin et al., 2020). The 24 honey samples from Castilla y León, Spain, indicated glucose oxidase (GOx) and catalase (CAT) mediated antimicrobial activity of honey against *Staphylococcus aureus* (Osés et al., 2024). Protein profiling can also be a useful indicator of geographical origin of honey. SDS-PAGE analysis of the 45 honey samples from Belgium, France, Italy, Romania, and Spain showed signature protein profiles as palynological and geographical biomarkers (Mureşan et al., 2022). In monofloral honey, floral nectar chitinase can be employed as an authenticity biomarker; xylosidase and thaumatin, two types of chitinases identified in the analyzed honey (Song et al., 2019).

7.4 Metagenomic Analysis of Honey Bee Gut Microbiota

Honey bee gut-associated microorganism diversity screening and profiling have proven to be an excellent indicator of significant parameters in ensuring honey's nutritional and medicinal properties. Diversity between host-specific bacterial populations can be a useful determinant of the entomological origin, and health condition of honey bees, along with environmental changes including environmental pollution, climate changes, and global warming. Comparative metagenomic analysis of gut microbiota between *A. mellifera* and *A. cerana* showed the existence of species-specific and shared bacterial communities. In the study, a total of 40 bees (20 per species) hindguts were analyzed employing qPCR assay-based 16S rRNA gene analysis. *A. mellifera*-specific strains are found to be *Bartonella apis*, *Commensalibacter* sp., and *Frischella perrara*, whereas the *A. cerana*-specific strain is *Apibacter* sp. Bacterial strain diversity has proven to be more abundant in *A. mellifera*. However, these two species are closely related. *A. mellifera* exhibits a heavier hindgut, a diverse pollen diet, increased colony size, and widespread global distribution (Ellegaard et al., 2020). Common bacterial populations in *A. cerana* and *A. mellifera* guts were found to be *Lactobacillus, Bifidobacterium, Snodgrassella,*

Bartonella, Gilliamella, and *Frischella* (Wu et al., 2022). Diversity in honey bees' gut microbiota can function as a signature of environmental changes and the evolution of microorganisms. KEGG gene ontology study showed possible effect of diet characteristic carbohydrate transport and metabolism (Su et al., 2022). Considering four seasonal points (summer, early-, mid- and late-winter), gut metabolomics indicated a distinguishable decline in flavonoids (kaempferol, keracyanin, and quercitrin) and pollen specific metabolites, including 9,10-dihydroxystearic acid (sporopollenin), spermidine, and tricoumaroyl spermidine (exosporium) would indicate reduced levels of nutrients in honey bees' gut, which can further influence the consortium of gut microbiomes in honey bee population in certain areas. Comparative study between *Apis mellifera* sub-species, *A. m. carnica*, *A. m. ligustica*, and *A. m. mellifera*, indicated seasonal variation in honey bee gut microbiome. Furthermore, gene ontology and pathway analysis showed predominance of TCA cycle, glyoxylate, decarboxylate, porphyrin, and chlorophyll metabolism in winter (Li et al., 2022). Detection of variation in honey gut microbiota due to Jujube (*Ziziphus jujuba* Mill) flower nectar can also be determined by extraction and amplification of the bacterial 16S rRNA (V3–V4 region). Jujube flower disease-affected honey bees showed a relative increase of Firmicutes and Actinobacteria, and a relative decrease of Proteobacteria, compared to healthy bees (Ma et al., 2020). Amplification of the V4 region of 16S rRNA gene from Switzerland honey bees gut microbiome showed the abundance of *Lactobacillus* Firm-4 and *Bartonella* in winter bees, while the prevalence of *Gilliamella* and *Snodgrassella* in forager bees (Kešnerová et al., 2020). *Apis mellifera* from subtropical regions of Uruguay, Latin America, exhibited in existing microorganisms.

8 Conclusion

Since ancient times, honey has been valued for its therapeutic properties, purity, nutritional efficacy, ability to nourish life at any age, and use as a human elixir. It is considered an ancient treasure of the living world. Nowadays, honey is opening up a wide range of intricately related networks among diversified fields of research and development on Ayurveda science, Apiculture, Plant science, Floriculture, Wildflower gardening, Entomology, Forestry and Silviculture especially in tropical and subtropical areas with to some extent temperate and mangrove regions, Chemical ecology with respect to trophic relationship among the members (plants and honey bees and other associated invaders), Honey Biotechnology, Cottage and wax and polishing industry and the food and brewing industry. Honey is no longer just a symbolic or metaphoric term used overzealously rather emotionally biased sense. Establishing the goals and objectives of research on honey in relation to adulteration and authentication with the expectation of satisfactory results in progress appears to be fragmented and discrete until all of the aforementioned chapters are in equilibrium with one another and with the adaptation of cutting-edge research from nature to the laboratory and back again to nature. This would be extremely noteworthy. Therefore, to isolate the adulterants from such a honey sample and to define and authenticate pure honey, numerous contemporary chromatographic separation techniques have been used. Other than atomic absorption, these methods for identifying different functional groups include high-pressure liquid chromatography (HPLC), high-pressure thin layer chromatography (HPTLC), gas chromatography with mass spectroscopy (GC-MS), liquid chromatography with mass spectroscopy (LC-MS), infrared spectroscopy (IRS), near infrared spectroscopy (NIRS), and Fourier transform infrared spectroscopy (FTIR). The detection of heavy metals such as lead (Pb), mercury (Hg), arsenic (As), and cadmium (Cd) in honey has been shown to be harmful to biological specimens. However, it is also puzzling that these metals may have entered the human food chain through honey and are present as poison. Spectroscopy is necessary to detect these metals.

While many harmless plants that produce honey are extremely selective and cautious in where their nectar glands are located within the flower during the biology of flower development, they also rarely produce poisonous alkaloids, reserving the bitter alkaloids for the underground rhizomes and older leaves. Therefore, the only irrational adulteration by different sugars in aqueous solvent that results in really bad honey quality is caused by very selfish, unholy, and unethical manmade forces that are impossible to eradicate in this situation. Other species that frequently visit nectar-bearing flowers and contaminate pure honey, such as butterflies, beetles, spiders, snails, hummingbirds, and bats, may occasionally share a great deal of the responsibility. Another harmful intruder of honey combs and nectar is fungus spores and mycelia, which alter the chemical and physical characteristics of real honey by secreting different mycotoxins. But in cases where two nodal organisms—plants and honey bees—interact and are bound by an incredibly sacred relationship, the best honey production should be thoroughly studied for its evolutionary progression in vivo, in addition to aspects of molecular biology and in vitro mutation research. The flowers and nectars they select during different seasons and diurnal cycles in various phytogeographic regions determine the quality assurance of honey.

Due to their fascination with particular food habits and flower selection, honey bee diversity at the subspecies and variety levels may play a significant role in preserving the quality of honey. This is because honey produced from single-species flowers is known as monofloral honey, while honey produced from multiple-species flowers is known as polyfloral honey. In addition to these two primary variables, there are other challenging environmental aspects as well, such as temperature, humidity, rainfall, fresh air, vegetation thickness, pollution-free zone, tourism restrictions, afforestation, and forest fires. Pure honey is made from nectar by the bees' digestive enzymes and the microbiota found in their guts and crops. In this sense, floral nectar synthesis and pollen maturation are co-programmed to the mutual benefit of both plant and bee, so initiating the biology of pollination by honey bees. Their reproductive biology, in turn, began almost simultaneously on tree branches with beehive sculpturing (architectures of bee hives with mostly large numbers of hexagonal venters geometrically clustered in a big nest circumference are nevertheless important parameter for the identification of species and subspecies of honey bees). The various sovereign sugars—such as glucose, fructose, maltose, sucrose, erythrose, xylulose, and trehalose—and their monomeric, dimeric, and oligomeric forms—present as essential components of honey, serve as the foundation for the entire narrative. Strong osmoregulants, these sugars cause water stress when laid by honey bees and provide protection against bacterial and fungal infections during the metamorphosis from egg to larvae to adult. Both the glycolytic and pentose phosphate pathways, as well as other metabolic pathways, rely on sugars as a fundamental substrate. These pathways supply phosphorus, nucleic acids, vitamins A, B, and C, as well as a variety of antioxidants such as flavonoids, carotenes, phenolics, sterols, and minerals (in combination with enzymes or redox compounds) that nourish and revitalize bee larvae, pupae, juveniles, and adults.

According to the genetics of nectar-related research, the relevant genes are primarily switched on during the spring and summer flowering seasons, but they are hardly ever turned on during the winter or rainy months. One of the main justifications for using proteomics technologies in conjunction with data analysis based on the enzymomics and metabolomics of honeybees is to investigate the peak period of quality honey production. The study of pollen biology has greatly benefited from interest in honey studies, particularly in melisopalynology. In contrast to polypollenic honey, which is a mixture of different types of pollen scavenged from various flowers, monopollenic honey denotes the reservation of one type of bee for only one type of blossom. Since nectar is actually a homogeneous mixture of these sugars with some aliphatic hydrocarbon and an aromatic flower scent, it has a high specific gravity and viscosity (allowing it to hang from a honeycomb against another achievement for an extended period of time). However, when honey bees consume nectar, it is combined with digestive enzymes and gastric juices in the stomach and crop, which causes the bees to vomit and feed their larvae. Specific gravity and density are then determined by the fanning of their wings. Finally, the use of pure honey as an honest biomarker of an environment free from ambient pollution reveals the biodiversity of flora, which is reflected in the biodiversity of pollens. However, the absence of any one particular type of pollen in the honey sample should raise doubts because the extinction of flora could occur if it occurs repeatedly for few years. Even now, these events have had a disastrous impact on both honey bees and humans. The main obstacles to the best quality assurance of pure honey and the better genetic race of honey bees are the precariously alarming effect of noxious gases like carbon monoxide (CO), nitrogen oxides (Nox), sulfur dioxide (SO_2), hydrogen (H_2S), methane (CH_4), ethylene (C_2H_4), acid and alkali vapors, paper mills, tanneries, and industrial effluents, phosphate and chlorine emissions, and fumigation of insect repellents on crop fields from the fertilizer and pesticide industries. Therefore, the terms "honey adulteration" and "authentication" are mutually exclusive. Adulteration refers to the immoral and unscientific practice of mixing sugar or jaggery with honey in an aqueous solution to the permissible limit of commercial acceptance, while authentication is inversely related to this agreement and supports the level of purity that could reasonably satisfy people's needs. Not to mention, the pure honey of honeycomb and floral nectars both have exceptional biochemical efficiency and long-lasting self-preservation properties. They also enhance the metabolic and genomic research of both plants and honey bees, greatly expanding their potential openings and trajectories to a wide range of fields. Thus, it can be said "Pure honey is a 'subset of testament' for clean and pollution-free environment"—which is the clarion call of this hour for all environmentalists across the globe.

References

Abd El-Moaty, H. I. (2010). Essential oil and iridoide glycosides of *Nepeta septemcrenata* Erenb. *Journal of Natural Products, 3*, 103–111. https://doi.org/10.1080/22311866.2024.2358785

Abdulkhaleq, L. A., Assi, M. A., Abdullah, R., Zamri-Saad, M., Taufiq-Yap, Y. H., & Hezmee, M. N. M. (2018). The crucial roles of inflammatory mediators in inflammation: A review. *Veterinary World, 11*(5), 627–635. https://doi.org/10.14202/vetworld.2018.627-635

Adenekan, M. O., Amusa, N. A., Okpeze, V. E., & Owosibo, A. O. (2012). Nutritional and microbiological components of honey samples obtained from Ogun state, southwestern Nigeria. *European Journal of Sustainable Development, 1*(2), 271–271.

Adhikari, D. (2019). Augmentation of Mitodepressive and cytogenetic effects of Lead upon acute exposure on grass pea (*Lathryrus sativus* L.) root tip cells. American. *Journal of Biological Sciences, 1*(1), 14–22. ISBN: 2688-1055.

Adhikari, D. (2021a). A curtain raiser on natural supplements being targeted choice of surveillance as potential anti-viral drugs: Neutraceutical supplementations vis-à-vis immune boosters. In *Hand book of agriculture & plant sciences* (pp. 146–168) ISBN: 978-93-91002-25-1.

Adhikari, D. (2021b). Chromium induces genotoxicity in root tip cells of grass pea (*Lathyrus sativus* L., variety Nirmal): A ROS mediated acute toxicity study. *Journal of Stress Physiology and Biochemistry, 17*(2), 98–120. ISSN: 1997-0838.

Adhikari, D. (2024). Biotic and abiotic stresses and reactive oxygen species (ROS)-mediated Signalling in higher plants: Regulatory mediators and their mechanism of actions. *International Journal of Trend in Scientific Research and Development (IJTSRD)*. eISSN: 2456–6470.

Adhikari, D., & Karmakar, S. (2018). *In-vitro* pro-inflammatory activity and associated pharmacological activities of *Paracondylactis indicus* Dave. Nematocyst Venom. *Pharmacologia, 9*, 114–121. https://doi.org/10.5567/pharmacologia.2018.114.121

Adhikari, D., Samanta, S., Roy, A., Dutta, A., Vedasiromoni, J., & Sen, T. (2007). In vitro hemolysis and lipid peroxidation inducing activity of the tentacle extract of the sea anemone (*Paracondylactis indicus* Dave.) on rat erythrocytes. *Indian Jr of Pharmacology, 39*(3), 155–159. https://doi.org/10.4103/0253-7613.33436

Ahmed, S., Sulaiman, S. A., Baig, A. A., Ibrahim, M., Liaqat, S., Fatima, S., Jabeen, S., Shamim, N., & Othman, N. H. (2018). Honey as a potential natural antioxidant medicine: An insight into its molecular mechanisms of action. *Oxidative Medicine and Cellular Longevity, 2018*, 8367846. https://doi.org/10.1155/2018/8367846

Ajibola, A. (2016). Physico-chemical and physiological values of honey and its importance as a functional food. *International Journal of Food and Nutritional Sciences, 2*(6), 1–9. https://doi.org/10.15436/2377-0619.15.040

Ajibola, A., Chamunorwa, J. P., & Erlwanger, K. H. (2012). Nutraceutical values of natural honey and its contribution to human health and wealth. *Nutrition & Metabolism, 9*, 61. https://doi.org/10.1186/1743-7075-9-61

Akbarian, M., Khani, A., Eghbalpour, S., & Uversky, V. N. (2022). Bioactive peptides: Synthesis, sources, applications, and proposed mechanisms of action. *International Journal of Molecular Sciences, 23*(3), 1445. https://doi.org/10.3390/ijms23031445

ALaerjani, W., Abu-Melha, S., Khan, K., Ghramh, H., Alalmie, A., Alshareef, R., AL-Shehri, B., & Mohammed, M. (2021). Presence of short and cyclic peptides in acacia and *Ziziphus* honeys may potentiate their medicinal values. *Open Chemistry, 19*(1), 1162–1173. https://doi.org/10.1515/chem-2021-0106

Alaerjani, W. M. A., Abu-Melha, S., Alshareef, R. M. H., Al-Farhan, B. S., Ghramh, H. A., Al-Shehri, B. M. A., Bajaber, M. A., Khan, K. A., Alrooqi, M. M., Modawe, G. A., & Mohammed, M. E. A. (2022). Biochemical reactions and their biological contributions in honey. *Molecules, 27*(15), 4719. https://doi.org/10.3390/molecules27154719

Albu, A., Radu-Rusu, C. G., Pop, I. M., Frunza, G., & Nacu, G. (2021). Quality assessment of raw honey issued from eastern Romania. *Agriculture, 11*(3), 247. https://doi.org/10.3390/agriculture11030247

Al-Hatamleh, M. A. I., Hatmal, M. M., Sattar, K., Ahmad, S., Mustafa, M. Z., Bittencourt, M. D. C., & Mohamud, R. (2020). Antiviral and immunomodulatory effects of phytochemicals from honey against COVID-19: Potential mechanisms of action and future directions. *Molecules, 25*(21), 5017. https://doi.org/10.3390/molecules25215017

Ali, H., Khan, E., & Ilahi, I. (2019). Environmental chemistry and ecotoxicology of hazardous heavy metals: Environmental persistence, toxicity, and bioaccumulation. *Journal of Chemistry, 6730305*, 14. https://doi.org/10.1155/2019/6730305

Ali, H., Khan, S., Ullah, R., & Khan, B. (2020). Fluorescence fingerprints of Sidr honey in comparison with uni/polyfloral honey samples. *European Food Research and Technology, 246*, 1829–1837. https://doi.org/10.1007/s00217-020-03536-6

Ali, H., Rafique, K., Ullah, R., Saleem, M., & Ahmad, I. (2022). Classification of Sidr honey and detection of sugar adulteration using right angle fluorescence spectroscopy and chemometrics. *European Food Research and Technology, 248*(7), 1823–1829. https://doi.org/10.1007/s00217-022-04008-9

Al-Kafaween, M. A., Alwahsh, M., Mohd Hilmi, A. B., & Abulebdah, D. H. (2023). Physicochemical characteristics and bioactive compounds of different types of honey and their biological and therapeutic properties: A comprehensive review. *Antibiotics (Basel, Switzerland), 12*(2), 337. https://doi.org/10.3390/antibiotics12020337

Alloway, B. J. (2013). Sources of Heavy Metals and Metalloids in Soils. In: Alloway, B. (eds) Heavy Metals in Soils. Environmental Pollution, vol 22. Springer, Dordrecht. https://doi.org/10.1007/978-94-007-4470-7_2

Almeida, M. O., Oloris, S. C. S., Faria, V. H. F., Ribeiro, M. C. M., Cantini, D. M., & Soto-Blanco, B. (2020). Optimization of method for pesticide detection in honey by using liquid and gas chromatography coupled with mass spectrometric detection. *Food, 9*(10), 1368. https://doi.org/10.3390/foods9101368

Almeida-Muradian, L. B. (2013). Tetragoniscaangustula pot-honey compared to *Apis mellifera* honey from Brazil. https://api.semanticscholar.org/CorpusID:81306085

Alvarez-Suarez, J. M., Tulipani, S., Romandini, S., Vidal, A., & Battino, M. (2009). Methodological aspects about determination of phenolic compounds and *in vitro* evaluation of antioxidant capacity in the honey: A review. *Current Analytical Chemistry, 5*(4), 293–302. https://doi.org/10.2174/157341109789077768

Al-Waili, N. S., & Haq, A. (2004). Effect of honey on antibody production against thymus-dependent and thymus- -independent antigens in primary and secondary immune responses. *Journal of Medicinal Food, 7*, 491–494.

An, N., Cai, W. J., Zhu, Q. F., Wang, W., Hussain, D., & Feng, Y. Q. (2020). Metabolic profiling of organic acids in honey by stable isotope labeling assisted liquid chromatography-mass spectrometry. *Journal of Food Composition and Analysis, 87*, 103423. https://doi.org/10.1016/j.jfca.2020.103423

Anand, U., Dey, A., Chandel, A. K. S., Sanyal, R., Mishra, A., Pandey, D. K., De Falco, V., Upadhyay, A., Kandimalla, R., Chaudhary, A., Dhanjal, J. K., Dewanjee, S., Vallamkondu, J., & Pérez de la Lastra, J. M. (2022). Cancer chemotherapy and beyond: Current status, drug candidates, associated risks and progress in targeted therapeutics. *Genes & Diseases, 10*(4), 1367–1401. https://doi.org/10.1016/j.gendis.2022.02.007

Ananias, K. R., Melo, A. A. M. D., & Moura, C. J. D. (2013). Analysis of moisture content, acidity and contamination by yeast and molds in *Apis mellifera* L. honey from Central Brazil. *Brazilian Journal of Microbiology, 44*, 679–683.

Anese, M., Manzocco, L., Calligaris, S., & Nicoli, M. C. (2013). Industrially applicable strategies for mitigating acrylamide, furan, and 5-hydroxymethylfurfural in food. *Journal of Agricultural and Food Chemistry, 61*(43), 10209–10214. https://doi.org/10.1021/jf305085r

Anguebes-Franseschi, F., Abatal, M., Pat, L., Flores, A., Córdova Quiroz, A. V., Ramírez-Elias, M. A., Pedro, L. S., May Tzuc, O., & Bassam, A. (2019). Raman spectroscopy and chemometric modeling to predict physical-chemical honey properties from Campeche, Mexico. *Molecules (Basel, Switzerland), 24*(22), 4091. https://doi.org/10.3390/molecules24224091

Aries, E., De Rudder, O., Kaklamanos, G., Maquet, A., & Ulberth, F. (2021). Results of an interlaboratory comparison of a liquid chromatography–isotope ratio mass spectrometry method for the determination of 13C/12C ratios of saccharides in honey. *Journal of AOAC International, 104*(6), 1698–1702. https://doi.org/10.1093/jaoacint/qsab091

Asari, M. A., Zulkaflee, M. H., Sirajudeen, K. N. S., Mohd Yusof, N. A., & Mohd Sairazi, N. S. (2019). Tualang honey and DHA-rich fish oil reduce the production of pro-inflammatory cytokines in the rat brain following exposure to chronic stress. *Journal of Taibah University Medical Sciences, 14*(4), 317–323. https://doi.org/10.1016/j.jtumed.2019.06.004

Atrott, J., & Henle, T. (2009). Methylglyoxal in Manuka Honey - Correlation with Antibacterial Properties. *Czech Journal of Food Sciences, 27*(Special Issue 1), S163–165. https://doi.org/10.17221/911-CJFS

Aykas, D. P., Shotts, M. L., & Rodriguez-Saona, L. E. (2020). Authentication of commercial honeys based on Raman fingerprinting and pattern recognition analysis. *Food Control, 117*, 107346. https://doi.org/10.1016/j.foodcont.2020.107346

Badshah, S. L., Faisal, S., Muhammad, A., Poulson, B. G., Emwas, A. H., & Jaremko, M. (2021). Antiviral activities of flavonoids. *Biomedicine & Pharmacotherapy, 140*, 111596. https://doi.org/10.1016/j.biopha.2021.111596

Baranowska-Wójcik, E., Szwajgier, D., & Winiarska-Mieczan, A. (2020). Honey as the potential natural source of cholinesterase inhibitors in Alzheimer's disease. *Plant Foods for Human Nutrition (Dordrecht, Netherlands), 75*(1), 30–32. https://doi.org/10.1007/s11130-019-00791-1

Barra, M. P., Ponce-Diaz, M. C., & Venegas-Gallegos, C. (2010). Volatile compounds in honey produced in the central valley of Nuble province, Chile. *Chilean Journal of Agricultural Research, 70*, 75–84.

Baweja, P., Kumar, S., Kumar, G. (2020). Fertilizers and Pesticides: Their Impact on Soil Health and Environment. In: Giri, B., Varma, A. (eds) Soil Health. Soil Biology, vol 59. Springer, Cham. https://doi.org/10.1007/978-3-030-44364-1_15

Behbahani, M. (2014). Anti-HIV-1 activity of eight monofloral Iranian honey types. *PLoS One, 9*(10), e108195. https://doi.org/10.1371/journal.pone.0108195

Belitz, H., Grosch, W., & Schieberle, P. (2009). *Food chemistry*. 4th revised and extended edn. Springer, 1070 pages.

Bhattacharya, D., Nanda, P. K., Pateiro, M., Lorenzo, J. M., Dhar, P., & Das, A. K. (2022). Lactic acid bacteria and Bacteriocins: Novel biotechnological approach for biopreservation of meat and meat products. *Microorganisms, 10*(10), 2058. https://doi.org/10.3390/microorganisms10102058

Bogdanov, S. (2012). Honey as nutrient and functional food. *Proteins, 1100*, 1400–2700. www.bee-hexagon.net

Bogdanov, S., Jurendic, T., Sieber, R., & Gallmann, P. (2008). Honey for nutrition and health: A review. *Journal of the American College of Nutrition, 27*(6), 677–689. https://doi.org/10.1080/07315724.2008.10719745

Bovo, S., Utzeri, V. J., Ribani, A., et al. (2020). Shotgun sequencing of honey DNA can describe honey bee derived environmental signatures and the honey bee hologenome complexity. *Scientific Reports, 10*, 9279. https://doi.org/10.1038/s41598-020-66127-1

Brown, B., Ojha, V., Fricke, I., Al-Sheboul, S. A., Imarogbe, C., Gravier, T., Green, M., Peterson, L., Koutsaroff, I. P., Demir, A., et al. (2023). Innate and adaptive immunity during SARS-CoV-2 infection: Biomolecular cellular markers and mechanisms. *Vaccine, 11*(2), 408. https://doi.org/10.3390/vaccines11020408

Brudzynski, K. (2023). Unexpected value of honey color for prediction of a non-enzymatic H_2O_2 production and honey antibacterial activity: A perspective. *Metabolites, 13*(4), 526. https://doi.org/10.3390/metabo13040526

Brudzynski, K., & Lannigan, R. (2012). Mechanism of honey bacteriostatic action against MRSA and VRE involves hydroxyl radicals generated from honey's hydrogen peroxide. *Frontiers in Microbiology, 3*(36). https://doi.org/10.3389/fmicb.2012.00036. PMID: 22347223; PMCID: PMC3273858.

Bruneau E. (2006). Antibiotics in honey (Antibiotiques dans le miel). *Abeille and Cie, 110*, 26–28. (in French).

Bt HjIdrus, R., Sainik, N. Q. A. V., Nordin, A., Saim, A. B., & Sulaiman, N. (2020). Cardioprotective effects of honey and its constituent: An evidence-based review of laboratory studies and clinical trials. *International Journal of Environmental Research and Public Health, 17*(10), 3613. https://doi.org/10.3390/ijerph17103613

Bukin, Y., Galachyants, Y., Morozov, I., et al. (2019). The effect of 16S rRNA region choice on bacterial community metabarcoding results. *Scientific Data, 6*, 190007. https://doi.org/10.1038/sdata.2019.7

Bulgasem, Y., Bulgasem, M., Nizam, L., Zaiton, H., Wan, M., Wan, Y., & Sumaya, G. F. (2016). Antifungal activity of lactic acid bacteria strains isolated from natural honey against pathogenic *Candida* species. *Mycobiology, 44*, 302–309.

Bush, D. S., Calla, B., & Berenbaum, M. R. (2024). An *aspergillus flavus* strain from bee bread of the Western honey bee (*Apis mellifera*) displays adaptations to distinctive features of the hive environment. *Ecology and Evolution, 14*, e10918. https://doi.org/10.1002/ece3.10918

Cajka, T., Hajslova, J., Pudil, F., & Riddellova, K. (2009). Traceability of honey origin based on volatiles pattern processing by artificial neural networks. *Journal of Chromatography A, 1216*(9), 1458–1462. https://doi.org/10.1016/j.chroma.2008.12.066

Calatayud-Vernich, P., Calatayud, F., Simó, E., & Picó, Y. (2018). Pesticide residues in honey bees, pollen and beeswax: Assessing beehive exposure. *Environmental pollution (Barking, Essex: 1987), 241*, 106–114. https://doi.org/10.1016/j.envpol.2018.05.062

Castro-Vázquez, L. M., Díaz-Maroto, M. C., Guchu, E., & Pérez-Coello, M. S. (2006). Analysis of volatile compounds of eucalypt honey by solid phase extraction followed by gas chromatography coupled to mass spectrometry. *European Food Research and Technology, 224*, 27–31. https://doi.org/10.1007/s00217-006-0284-2

Castro-Vázquez, L. M., Díaz-Maroto, M. C., González-Viñas, M. A., & Pérez-Coello, M. S. (2009). Differentiation of monofloral citrus, rosemary, eucalyptus, lavender, thyme and heather honeys based on volatile composition and sensory descriptive analysis. *Food Chemistry, 112*(4), 1022–1030. https://doi.org/10.1016/j.foodchem.2008.06.036

Catalano, P., Della Sala, F., Cavaliere, M., Caputo, C., Pecoraro, D., Crispino, G., Lettera, S., Caioni, G., Esposito, M., Verre, A., Castellone, L., Bianco, E., & Amorena, M. (2024). Use of honey bees and hive products as bioindicators to assess environmental contamination in targeted areas of the Campania region (Italy). *Animals: An Open Access Journal from MDPI, 14*(10), 1446. https://doi.org/10.3390/ani14101446

Cebi, N., Bozkurt, F., Yilmaz, M. T., & Sagdic, O. (2020). An evaluation of FTIR spectroscopy for prediction of royal jelly content in hive products. *Journal of Apicultural Research, 59*(2), 146–155. https://doi.org/10.1080/00218839.2019.1707009

Charnock, H. M., Pickering, G. J., & Kemp, B. S. (2022). The Maillard reaction in traditional method sparkling wine. *Frontiers in Microbiology, 13*, 979866. https://doi.org/10.3389/fmicb.2022.979866

Chen, L., Mehta, A., Berenbaum, M., Zangerl, A. R., & Engeseth, N. J. (2000). Honeys from different floral sources as inhibitors of enzymatic browning in fruit and vegetable homogenates. *Journal of Agricultural and Food Chemistry, 48*(10), 4997–5000. https://doi.org/10.1021/jf000373j

Chen, M., Xiao, J., El-Seedi, H. R., Woźniak, K. S., Daglia, M., Little, P. J., & Xu, S. (2022). Kaempferol and atherosclerosis: From mechanism to medicine. *Critical Reviews, Food Science and Nutrition, 64*(8), 2157–2175. https://doi.org/10.1080/10408398.2022.2121261

Chernetsova, E. S., & Morlock, G. E. (2012). Assessing the capabilities of direct analysis in real time mass spectrometry for 5-hydroxymethylfurfural quantitation in honey. *International Journal of Mass Spectrometry, 314*, 22–32. https://doi.org/10.1002/9781119113324.ch3

Chudzinska, M., & Baralkiewicz, D. (2010). Estimation of honey authenticity by multielements characteristics using inductively coupled plasma-mass spectrometry (ICP-MS) combined with chemometrics. *Food and Chemical Toxicology, 48*(1), 284–290. https://doi.org/10.1016/j.fct.2009.10.011

Cianciosi, D., Forbes-Hernández, T. Y., Afrin, S., Gasparrini, M., Reboredo-Rodriguez, P., Manna, P. P., Zhang, J., Bravo Lamas, L., Martínez Flórez, S., Agudo Toyos, P., Quiles, J. L., Giampieri, F., & Battino, M. (2018). Phenolic compounds in honey and their associated health benefits: A review. *Molecules (Basel, Switzerland), 23*(9), 2322. https://doi.org/10.3390/molecules23092322

Ciulu, M., Oertel, E., Serra, R., Farre, R., Spano, N., Caredda, M., Malfatti, L., & Sanna, G. (2020). Classification of unifloral honeys from SARDINIA (Italy) by ATR-FTIR spectroscopy and random forest. *Molecules, 26*(1), 88. https://doi.org/10.3390/molecules26010088

Cotte, J. F., Casabianca, H., Giroud, B., Albert, M., Lheritier, J., & Grenier-Loustalot, M. F. (2004). Characterization of honey amino acid profiles using high-pressure liquid chromatography to control authenticity. *Analytical and Bioanalytical Chemistry, 378*(5), 1342–1350. https://doi.org/10.1007/s00216-003-2430-z

Coury, C., & Dillner, A. M. (2009). ATR-FTIR characterization of organic functional groups and inorganic ions in ambient aerosols at a rural site. *Atmospheric Environment, 43*(4), 940–948. https://doi.org/10.1016/j.atmosenv.2008.10.056

Cruz-Arreola, O., Orduña-Diaz, A., Domínguez, F., Reyes-Leyva, J., Vallejo-Ruiz, V., Domínguez-Ramírez, L., & Santos-López, G. (2022). *In silico* testing of flavonoids as potential inhibitors of protease and helicase domains of dengue and Zika viruses. *PeerJ, 10*, e13650. https://doi.org/10.7717/peerj.13650

Cunningham, M. M., Tran, L., McKee, C. G., Polo, R. O., Newman, T., Lansing, L., Griffiths, J. S., Bilodeau, G. J., Rott, M., & Guarna, M. M. (2022). Honey bees as biomonitors of environmental contaminants, pathogens, and climate change. *Ecological Indicators, 134*, 108457. https://doi.org/10.1016/j.ecolind.2021.108457

Daniels, B. J., Prijic, G., Meidinger, S., Loomes, K. M., Stephens, J. M., Schlothauer, R. C., ... & Brimble, M. A. (2016). Isolation, structural elucidation, and synthesis of lepteridine from manuka (Leptospermum scoparium) honey. *Journal of Agricultural and Food Chemistry, 64*(24), 5079–5084

da Silva, P. M., Gonzaga, L. V., de Azevedo, M. S., Biluca, F. C., Schulz, M., Costa, A. C. O., & Fett, R. (2020). Stability of volatile compounds of honey during prolonged storage. *Journal of Food Science and Technology, 57*, 1167–1182. https://doi.org/10.1007/s13197-019-04163-0

Dabeek, W. M., & Marra, M. V. (2019). Dietary quercetin and kaempferol: Bioavailability and potential cardiovascular-related bioactivity in humans. *Nutrients, 11*, 2288. https://doi.org/10.3390/nu11102288

Damto, T., Zewdu, A., & Birhanu, T. (2023). Application of Fourier transform infrared (FT-IR) spectroscopy and multivariate analysis for detection of adulteration in honey markets in Ethiopia. *Current Research in Food Science, 7*, 100565. https://doi.org/10.1016/j.crfs.2023.100565. PMID: 37664005; PMCID: PMC10470187.

Das, S., Sultana, K. W., Ndhlala, A. R., Mondal, M., & Chandra, I. (2023). Heavy metal pollution in the environment and its impact on health: Exploring green technology for remediation. *Environmental Health Insights, 17.1*(10), 11786302231201259. https://doi.org/10.1177/11786302231201259

De La Fuente, E., Ruiz-Matute, A. I., Valencia-Barrera, R. M., Sanz, J., & Castro, I. M. (2011). Carbohydrate composition of Spanish unifloral honeys. *Food Chemistry, 129*(4), 1483–1489. https://doi.org/10.1016/j.foodchem.2011.05.121

Desmouliere, A., Bonte, F., Couquet, Y., Rigal, M. L. (2013). Honey, which benefit in healing? (Le miel, quel intérêt en cicatrisation?). *Actualites Pharmaceutiques, 52*, 17–35. (in French)

de Souza, R. R., de Sousa Fernandes, D. D., & Diniz, P. H. G. D. (2021). Honey authentication in terms of its adulteration with sugar syrups using UV–vis spectroscopy and one-class classifiers. *Food Chemistry, 365*, 130467. https://doi.org/10.1016/j.foodchem.2021.130467

De Toledo, V. D. A. A., & DechechiChambó, E. (2020). Honey analysis – New advances and challenges. *Intech Open*. https://doi.org/10.5772/intechopen.77854

del Campo, G., Zuriarrain, J., Zuriarrain, A., & Berregi, I. (2016). Quantitative determination of carboxylic acids, amino acids, carbohydrates, ethanol and hydroxymethylfurfural in honey by $_1$H NMR. *Food Chemistry, 196*, 1031–1039. https://doi.org/10.1016/j.foodchem.2021.130467

Dembinski, J. L., Hungnes, O., Hauge, A. G., Kristoffersen, A. C., Haneberg, B., & Mjaaland, S. (2014). Hydrogen peroxide inactivation of influenza virus preserves antigenic structure and immunogenicity. *Journal of Virological Methods, 207*, 232–237. https://doi.org/10.1016/j.jviromet.2014.07.003

De-Melo, A. A. M., & de Almeida-Muradian, L. B. (2017). Chemical composition of bee pollen. In *Bee products-chemical and biological properties* (pp. 221–259). https://doi.org/10.1007/978-3-319-59689-1_11

Di Petrillo, A., Orrù, G., Fais, A., & Fantini, M. C. (2022). Quercetin and its derivates as antiviral potentials: A comprehensive review. *Phytotherapy Research: PTR, 36*(1), 266–278. https://doi.org/10.1002/ptr.7309

Díaz, M. A., O'Connell, D. P., Jordan, S., O'Connor, C., Martin, P., Jones, J. C., & Garvey, J. (2023). Analysis of pesticide levels in honey and pollen from Irish honey bee colonies using a modified Dutch mini-luke method with gas and liquid chromatography–Tandem mass spectrometry detection. *Journal of Agricultural and Food Chemistry, 71*(34), 12657–12667. https://doi.org/10.1021/acs.jafc.3c02250

Díaz-Galiano, F. J., Heinzen, H., Gómez-Ramos, M. J., Murcia-Morales, M., & Fernández-Alba, A. R. (2023). Identification of novel unique mānuka honey markers using high-resolution mass spectrometry-based metabolomics. *Talanta, 260*, 124647. https://doi.org/10.1016/j.talanta.2023.124647. Epub 2023 May 6.

Dixon, R. A., & Pasinetti, G. M. (2010). Flavonoids and isoflavonoids: From plant biology to agriculture and neuroscience. *Plant Physiology, 154*, 453–457. https://doi.org/10.1104/pp.110.161430. PMID: 20921162; PMCID: PMC2948995.

Dzobo, K. (2022). The role of natural products as sources of therapeutic agents for innovative drug discovery. In *Comprehensive pharmacology* (pp. 408–422). https://doi.org/10.1016/B978-0-12-820472-6.00041-4

Eisele, F. (2000). *Synthesis of lapidated peptide structures of viral and cellular proteins* (pp. 107–108). Faculty of Chemistry, University of Dortmund. https://d-nb.info/961588497/34

El Hawari, K., Al Iskandarani, M., Jaber, F., Ezzeddine, R., Ziller, L., Perini, M., Bontempo, L., Pellegrini, M., & Camin, F. (2021). Evaluation of honey authenticity in Lebanon by analysis of carbon stable isotope ratio using elemental analyzer and liquid chromatography coupled to isotope ratio mass spectrometry. *Journal of Mass Spectrometry, 56*(6), e4730. https://doi.org/10.1002/jms.4730

El Sohaimy, S. A., Masry, S. H. D., & Shehata, M. G. (2015). Physicochemical characteristics of honey from different origins. *Annals of Agricultural Sciences, 60*(2), 279–287. https://doi.org/10.1016/j.aoas.2015.10.015

El-Bondkly, E. A. M., Shammari, A. B., El-Gendy, M. M. A. A., Alsafari, I. A., El-Bondkly, A. A. M. A., El-Shenawy, F. S., & El-Bondkly, A. M. (2022). Phytochemical screening, antifungal, and anticancer activities of medicinal plants ThymelaeaHirsuta, Urginea Maritima, and Plantago Albicans. *BioMed Research International, 2022*, 9544915. https://doi.org/10.1155/2022/9544915

Ellegaard, K. M., Suenami, S., Miyazaki, R., & Engel, P. (2020). Vast differences in strain-level diversity in the gut microbiota of two closely related honey bee species. *Current Biology, 30*(13), 2520–2531. https://doi.org/10.1016/j.cub.2020.04.070

Erejuwa, O. O., Sulaiman, S. A., Wahab, M. S., Sirajudeen, K. N., Salleh, M. S., & Gurtu, S. (2010). Antioxidant protection of Malaysian tualang honey in pancreas of normal and streptozotocin-induced diabetic rats. *Annales d'Endocrinologie, 71*(4), 291–296. https://doi.org/10.1016/j.ando.2010.03.003

Fadzil, M. A. M., Mustar, S., & Rashed, A. A. (2023). The potential use of honey as a neuroprotective agent for the Management of Neurodegenerative Diseases. *Nutrients, 15*(7), 1558. https://doi.org/10.3390/nu15071558

Felsner, M. L., Cano, C. B., Bruns, R. E., Watanabe, H. M., Almeida-Muradian, L. B. D., & Matos, J. D. R. (2004). Characterization of monofloral honeys by ash contents through a hierarchical design. *Journal of Food Composition and Analysis, 17*(6), 737–747. https://doi.org/10.1016/j.jfca.2003.11.001

Fernandes, L., Ribeiro, H., Oliveira, A., Silva, A. S., Freitas, A., Henriques, M., & Rodrigues, M. E. (2020). Portuguese honeys as antimicrobial agents against *Candida* species. *Journal of Traditional and Complementary Medicine, 11*, 130–136. https://doi.org/10.1016/j.jtcme.2020.02.007. PMID: 33728273; PMCID: PMC7936102.

Fujieda, N. (2020). His-Cys and Trp-Cys cross-links generated by post-translational chemical modification. *Bioscience, Biotechnology, and Biochemistry, 84*(3), 445–454. https://doi.org/10.1080/09168451.2019.1696178

Gopalan, C., Rama Sastri, B. V., & Balasubramanian, S. C. (1989). Nutritive Value of Indian Foods. National Institute of Nutrition, ICMR, Hyderabad, 59–93.

Garcia, J. M., Chambers, E., Matta, Z., & Clark, M. (2005). Viscosity measurements of nectar-and honey-thick liquids: Product, liquid, and time comparisons. *Dysphagia, 20*, 325–335. https://doi.org/10.1007/s00455-005-0034-9

Gerginova, D., Simova, S., Popova, M., Stefova, M., Stanoeva, J. P., & Bankova, V. (2020). NMR profiling of north Macedonian and Bulgarian honeys for detection of botanical and geographical origin. *Molecules, 25*(20), 4687. https://doi.org/10.3390/molecules25204687

Gok, S., Severcan, M., Goormaghtigh, E., Kandemir, I., & Severcan, F. (2015). Differentiation of Anatolian honey samples from different botanical origins by ATR-FTIR spectroscopy using multivariate analysis. *Food Chemistry, 170*, 234–240. https://doi.org/10.1016/j.foodchem.2014.08.040

Gonçalves, W. B., Teixeira, W. S. R., Cervantes, E. P., Mioni, M. D. S. R., Sampaio, A. N. D. C. E., Martins, O. A., Gruber, J., & Pereira, J. G. (2023). Application of an electronic nose as a new technology for rapid detection of adulteration in honey. *Applied Sciences, 13*(8), 4881. https://doi.org/10.3390/app13084881

Gonzalez, A. C., Costa, T. F., Andrade, Z. A., & Medrado, A. R. (2016). Wound healing – A literature review. *Anais Brasileiros de Dermatologia, 91*(5), 614–620. https://doi.org/10.1590/abd1806-4841.20164741

Gooding, J. J., Hibbert, D. B., & Yang, Y. (2001). Electrochemical metal ion sensors. Exploiting amino acids and peptides as recognition elements. *Sensors, 1*, 75–90. https://doi.org/10.3390/s10300075

Gręda, K., Leśniewicz, A., Kowalczyk, K., & Pohl, P. (2024). Analysis of diluted honey samples from the province of lower Silesia (Poland) by inductively coupled plasma mass spectrometry. *Spectrochimica Acta Part B: Atomic Spectroscopy, 217*, 106978. https://doi.org/10.1016/j.sab.2024.106978

Guo, M., Ni, J., Yu, J., Jin, J., Ma, L., Zhang, H., Wang, D., Zhang, X., Dou, J., & Zhou, C. (2019). Antiviral activity of benzoic acid derivative NC-5 against influenza a virus and its neuraminidase inhibition. *International Journal of Molecular Sciences, 20*(24), 6261. https://doi.org/10.3390/ijms20246261

Guttentag, A., Krishnakumar, K., Cokcetin, N., Hainsworth, S., Harry, E., & Carter, D. (2021a). Inhibition of dermatophyte fungi by Australian jarrah honey. *Pathogens (Basel, Switzerland), 10*(2), 194. https://doi.org/10.3390/pathogens10020194

Guttentag, A., Krishnakumar, K., Cokcetin, N., Harry, E., & Carter, D. (2021b). Factors affecting the production and measurement of hydrogen peroxide in honey samples. *Access Microbiology, 3*(3), 000198. https://doi.org/10.1099/acmi.0.000198

Hassapidou, M., Fotiadou, E., Maglara, E., & Papadopoulou, S. K. (2006). Energy intake, diet composition, energy expenditure, and body fatness of adolescents in northern Greece. *Obesity (Silver Spring), 14*, 855–862. https://doi.org/10.1038/oby.2006.99

Hossain, K. S., Hossain, M. G., Moni, A., Rahman, M. M., Rahman, U. H., Alam, M., Kundu, S., Rahman, M. M., Hannan, M. A., & Uddin, M. J. (2020). Prospects of honey in fighting against COVID-19: Pharmacological insights and therapeutic promises. *Heliyon, 6*(12), e05798. https://doi.org/10.1016/j.heliyon.2020.e05798

Huang, Z., Liu, L., Li, G., Li, H., Ye, D., & Li, X. (2019). Nondestructive determination of diastase activity of honey based on visible and near-infrared spectroscopy. *Molecules, 24*(7), 1244. https://doi.org/10.3390/molecules24071244

Hunter, M., Kellett, J., Toohey, K., & Naumovski, N. (2021). Sensory and compositional properties affecting the likeability of commercially available Australian honeys. *Foods (Basel, Switzerland), 10*(8), 1842. https://doi.org/10.3390/foods10081842

Iglesias, M. T., Martín-Álvarez, P. J., Polo, M. C., de Lorenzo, C., González, M., & Pueyo, E. (2006). Changes in the free amino acid contents of honeys during storage at ambient temperature. *Journal of Agricultural and Food Chemistry, 54*(24), 9099–9104. https://doi.org/10.1021/jf061712x

Iorizzo, M., Letizia, F., Ganassi, S., Testa, B., Petrarca, S., Albanese, G., Di Criscio, D., & De Cristofaro, A. (2022). Functional properties and antimicrobial activity from lactic acid bacteria as resources to improve the health and welfare of honey bees. *Insects, 13*(3), 308. https://doi.org/10.3390/insects13030308

Jakkielska, D., Frankowski, M., & Zioła-Frankowska, A. (2024). Speciation analysis of arsenic in honey using HPLC-ICP-MS and health risk assessment of water-soluble arsenic. *Journal of Hazardous Materials, 471*, 134364. https://doi.org/10.1016/j.jhazmat.2024.134364

Jerković, I., Marijanović, Z., Kranjac, M., & Radonić, A. (2015). Comparison of different methodologies for detailed screening of *Taraxacum officinale* honey volatiles. *Natural Product Communications, 10*(2), 1934578X1501000238.

Jones, A. J., Grkovic, T., Sykes, M. L., & Avery, V. M. (2013). Trypanocidal activity of marine natural products. *Marine Drugs, 11*(10), 4058–4082. https://doi.org/10.3390/md11104058

Jones, L., Lowe, A., Ford, C. R., Christie, L., Creer, S., & de Vere, N. (2022). Temporal patterns of honeybee foraging in a diverse floral landscape revealed using pollen DNA metabarcoding of honey. *Integrative and Comparative Biology, 62*(2), 199–210. https://doi.org/10.1093/icb/icac029

Joo, S. H. (2012). Cyclic peptides as therapeutic agents and biochemical tools. *Biomolecules & Therapeutics, 20*(1), 19–26. https://doi.org/10.4062/biomolther.2012.20.1.019

Kamboj, R., Sandhu, R. S., Kaler, R. S. S., Bera, M. B., & Nanda, V. (2019). Optimization of process parameters on hydroxymethylfurfural content, diastase and invertase activity of coriander honey. *Journal of Food Science and Technology, 56*, 3205–3214. https://doi.org/10.1016/j.focha.2024.100766

Karami, Z., & Akbari-Adergani, B. (2019). Bioactive food derived peptides: A review on correlation between structure of bioactive peptides and their functional properties. *Journal of Food Science and Technology, 56*(2), 535–547. https://doi.org/10.1007/s13197-018-3549-4

Kaškonienė, V., Venskutonis, P. R., & Čeksterytė, V. (2010). Carbohydrate composition and electrical conductivity of different origin honeys from Lithuania. *LWT-Food Science and Technology, 43*(5), 801–807. https://doi.org/10.1016/j.lwt.2010.01.007

Kawashima, H., Suto, M., & Suto, N. (2019). Stable carbon isotope ratios for organic acids in commercial honey samples. *Food Chemistry, 289*, 49–55. https://doi.org/10.1016/j.foodchem.2019.03.053

Kędzierska-Matysek, M., Matwijczuk, A., Florek, M., Barłowska, J., Wolanciuk, A., Matwijczuk, A., Chruściel, E., Walkowiak, R., Karcz, D., & Gładyszewska, B. (2018). Application of FTIR spectroscopy for analysis of the quality of honey. In *BIO web of conferences* (Vol. 10, p. 02008). EDP sciences. https://doi.org/10.1051/bioconf/20181002008

Kešnerová, L., Emery, O., Troilo, M., Liberti, J., Erkosar, B., & Engel, P. (2020). Gut microbiota structure differs between honeybees in winter and summer. *The ISME Journal, 14*(3), 801–814. https://doi.org/10.1038/s41396-019-0568-8

Khan, S. U., Anjum, S. I., Rahman, K., Ansari, M. J., Khan, W. U., Kamal, S., Khattak, B., Muhammad, A., & Khan, H. U. (2018). Honey: Single food stuff comprises many drugs. *Saudi Journal of Biological Sciences, 25*(2), 320–325. https://doi.org/10.1016/j.sjbs.2017.08.004

Khansaritoreh, E., Salmaki, Y., Ramezani, E., Azirani, T. A., Keller, A., Neumann, K., Alizadeh, K., Zarre, S., Beckh, G., & Behling, H. (2020). Employing DNA metabarcoding to determine the geographical origin of honey. *Heliyon, 6*(11). https://doi.org/10.1016/j.heliyon.2020.e05596

Kim, S. R., Jeong, M. S., Mun, S. H., Cho, J., Seo, M. D., Kim, H., Lee, J., Song, J. H., & Ko, H. J. (2021). Antiviral activity of Chrysin against influenza virus replication via inhibition of autophagy. *Viruses, 13*(7), 1350. https://doi.org/10.3390/v13071350

Kot, B., Sytykiewicz, H., Sprawka, I., & Witeska, M. (2020). Effect of manuka honey on biofilm-associated genes expression during methicillin resistant *Staphylococcus aureus* biofilm formation. *Scientific Reports, 10*(1), 13552. https://doi.org/10.1038/s41598-020-70666-y

Kowalski, S., Kopuncová, M., Ciesarová, Z., & Kukurová, K. (2017). Free amino acids profile of polish and Slovak honeys based on LC–MS/MS method without the prior derivatisation. *Journal of Food Science and Technology, 54*, 3716–3723. https://doi.org/10.1007/s13197-017-2838-7

Krüger, N., Kronenberger, T., Xie, H., Rocha, C., Pöhlmann, S., Su, H., Xu, Y., Laufer, S. A., & Pillaiyar, T. (2023). Discovery of polyphenolic natural products as SARS-CoV-2 M^{pro} inhibitors for COVID-19. *Pharmaceuticals (Basel, Switzerland), 16*(2), 190. https://doi.org/10.3390/ph16020190

Kus, P. M., Szweda, P., Jerkovic, I., & Tuberoso, C. I. G. (2016). Activity of polish unifloral honeys against pathogenic bacteria and its correlation with colour, phenolic content, antioxidant capacity and other parameters. *Letters in Applied Microbiology, 62*, 269–276. https://doi.org/10.1111/lam.12541

Kwakman, P. H. S., & Zaat, S. A. J. (2012). Antibacterial components of honey. *IUBMB Life, 64*, 48–55. https://doi.org/10.1002/iub.578

Kwon, M. J., Shin, H. M., Perumalsamy, H., Wang, X., & Ahn, Y. J. (2019). Antiviral effects and possible mechanisms of action of constituents from Brazilian propolis and related compounds. *Journal of Apicultural Research, 59*(4), 413–425. https://doi.org/10.1080/00218839.2019.1695715

Kwon, E. B., Kim, Y. S., Kim, B., Kim, S. G., Na, S. J., Go, Y., Choi, H. M., Lee, H. J., Han, S. M., & Choi, J. G. (2023). Korean chestnut honey suppresses HSV-1 infection by regulating the ROS–NLRP3 Inflammasome pathway. *Antioxidants., 12*(11), 1935. https://doi.org/10.3390/antiox12111935

Laallam, H., Boughediri, L., Bissati, S., Menasria, T., Mouzaoui, M. S., Hadjadj, S., Hammoudi, R., & Chenchouni, H. (2015). Modeling the synergistic antibacterial effects of honey characteristics of different botanical origins from the Sahara Desert of Algeria. *Frontiers in Microbiology, 6*, 1–12. https://doi.org/10.3389/fmicb.2015.01239

Låg, M., Øvrevik, J., Refsnes, M., et al. (2020). Potential role of polycyclic aromatic hydrocarbons in air pollution-induced non-malignant respiratory diseases. *Respiratory Research, 21*, 299. https://doi.org/10.1186/s12931-020-01563-1

Lao, M. R., Bautista VII, A. T., Mendoza, N. D., & Cervancia, C. R. (2021). Stable carbon isotope ratio analysis of Philippine honeys for the determination of adulteration with C4 sugars. *Food Analytical Methods, 14*(7), 1443–1455. https://doi.org/10.1007/s12161-021-01968-8

Laubert, M., Bonifacius, A., Dragon, A. C., Mangare, C., Blasczyk, R., Huehn, J., & Eiz-Vesper, B. (2022). Enhancement of antiviral T-cell responses by vitamin C suggests new strategies to improve manufacturing of virus-specific T cells for adoptive immunotherapy. *Biology, 11*(4), 536. https://doi.org/10.3390/biology11040536

Lee, J. L., Loe, M. W. C., Lee, R. C. H., & Chu, J. J. H. (2019). Antiviral activity of pinocembrin against Zika virus replication. *Antiviral Research, 167*, 13–24. https://doi.org/10.1016/j.antiviral.2019.04.003

Lee, S., Kalcic, F., Duarte, I. F., Titera, D., Kamler, M., Mrna, P., Hyrsl, P., Danihlik, J., Dobes, P., Kunc, M., & Pudlo, A. (2022). 1H NMR profiling of honey bee bodies revealed metabolic differences between summer and winter bees. *Insects, 13*(2), 193. https://doi.org/10.3390/insects13020193

Leponiemi, M., Freitak, D., Moreno-Torres, M., Pferschy-Wenzig, E. M., Becker-Scarpitta, A., Tiusanen, M., Vesterinen, E. J., & Wirta, H. (2023). Honeybees' foraging choices for nectar and pollen revealed by DNA metabarcoding. *Scientific Reports, 13*(1), 14753. https://doi.org/10.1038/s41598-023-42102-4

Leska, A., Nowak, A., Nowak, I., & Górczyńska, A. (2021). Effects of insecticides and microbiological contaminants on *Apis mellifera* health. *Molecules (Basel, Switzerland), 26*(16), 5080. https://doi.org/10.3390/molecules26165080

Li, J., Zhi, A., Jia, G., Nie, H., & Ai, L. (2018). Development of an ic-ELISA and immunochromatographic strip for detection of sparfloxacin in honey. *Food and Agricultural Immunology, 29*(1), 147–158. https://doi.org/10.1080/09540105.2017.1364706

Li, Y., Li, J., Huang, H., Jian, D., Shan, Y., Wang, S., & Liu, F. (2021). Rapid quantitative detection for multiple antibiotics in honey using a quantum dot microsphere immunochromatographic strip. *Food Control, 130*, 108256. https://doi.org/10.1016/j.foodcont.2021.108256

Li, C., Tang, M., Li, X., & Zhou, X. (2022). Community dynamics in structure and function of honey bee gut bacteria in response to winter dietary shift. *MBio, 13*(5), e01131–e01122. https://doi.org/10.1128/mbio.01131-22

Ling, C. N., & Sowndhararajan, K. (2020). *A review on analytical methods for honey classification, identification and authentication*. Intech Open. https://doi.org/10.5772/intechopen.90232

Liu, L., Xu, L., Suryoprabowo, S., Song, S., & Kuang, H. (2018). Rapid detection of tulathromycin in pure milk and honey with an immunochromatographic test strip. *Food and Agricultural Immunology, 29*(1), 358–368. https://doi.org/10.1080/09540105.2017.1376040

Luong, D. V., Tam, N. Q., Xuan, D. T. T., & Tai, N. T. (2019). NMR based metabolomic approach for evaluation of Vietnamese honey. *Vietnam Journal of Chemistry, 57*(6), 712–716. https://doi.org/10.1002/vjch.2019000101

Ma, W., Zheng, X., Li, L., Shen, J., Li, W., & Gao, Y. (2020). Changes in the gut microbiota of honey bees associated with jujube flower disease. *Ecotoxicology and Environmental Safety, 198*, 110616. https://doi.org/10.1016/j.ecoenv.2020.110616

Machado, A. M., Miguel, M. G., Vilas-Boas, M., & Figueiredo, A. C. (2020). Honey volatiles as a fingerprint for botanical origin – A review on their occurrence on Monofloral honeys. *Molecules, 25*(2), 374. https://doi.org/10.3390/molecules25020374

Machado, A. M., Antunes, M., Miguel, M. G., Vilas-Boas, M., & Figueiredo, A. C. (2021). Volatile profile of Portuguese monofloral honeys: Significance in botanical origin determination. *Molecules, 26*(16), 4970. https://doi.org/10.3390/molecules26164970

Mair, K. S., Irrgeher, J., & Haluza, D. (2023). Elucidating the role of honey bees as biomonitors in environmental health research. *Insects, 14*(11), 874. https://doi.org/10.3390/insects14110874

Manisalidis, I., Stavropoulou, E., Stavropoulos, A., & Bezirtzoglou, E. (2020). Environmental and health impacts of air pollution: A review. *Frontiers in Public Health, 8*, 14. https://doi.org/10.3389/fpubh.2020.00014

Manyi-Loh, C. E., Ndip, R. N., & Clarke, A. M. (2011). Volatile compounds in honey: A review on their involvement in aroma, botanical origin determination and potential biomedical activities. *International Journal of Molecular Sciences, 12*(12), 9514–9532. https://doi.org/10.3390/ijms12129514

Margaoan, R., Papa, G., Nicolescu, A., et al. (2024). Environmental pollution effect on honey bees and their derived products: A comprehensive analysis. *Environmental Science and Pollution Research, 32*, 10370. https://doi.org/10.1007/s11356-024-33754-4

Marín-García, P. J., Peyre, Y., Ahuir-Baraja, A. E., Garijo, M. M., & Llobat, L. (2022). The role of *Nosemaceranae* (microsporidia: Nosematidae) in Honey Bee Colony losses and current insights on treatment. *Veterinary Sciences, 9*(3), 130. https://doi.org/10.3390/vetsci9030130

Martinello, M., Manzinello, C., Dainese, N., Giuliato, I., Gallina, A., & Mutinelli, F. (2021). The honey bee: An active biosampler of environmental pollution and a possible warning biomarker for human health. *Applied Sciences., 11*(14), 6481. https://doi.org/10.3390/app11146481

Masad, R. J., Haneefa, S. M., Mohamed, Y. A., Al-Sbiei, A., Bashir, G., Fernandez-Cabezudo, M. J., & Al-Ramadi, B. K. (2021). The immunomodulatory effects of honey and associated flavonoids in cancer. *Nutrients, 13*(4), 1269. https://doi.org/10.3390/nu13041269

Mato, I., Huidobro, J. F., Simal-Lozano, J., & Sancho, M. T. (2006). Rapid determination of non-aromatic organic acids in honey by capillary zone electrophoresis with direct ultraviolet detection. *Journal of Agricultural and Food Chemistry, 54*, 1541–1550. https://doi.org/10.1021/jf051757i

Matuszewska, E., Klupczynska, A., Maciołek, K., Kokot, Z. J., & Matysiak, J. (2021). Multielemental analysis of bee pollen, propolis, and royal jelly collected in west-central Poland. *Molecules (Basel, Switzerland), 26*(9), 2415. https://doi.org/10.3390/molecules26092415

Mazarakioti, E. C., Zotos, A., Thomatou, A. A., Kontogeorgos, A., Patakas, A., & Ladavos, A. (2022). Inductively coupled plasma-mass spectrometry (ICP-MS), a useful tool in authenticity of agricultural products' and foods' origin. *Food, 11*(22), 3705. https://doi.org/10.3390/foods11223705

Mbikay, M., & Chrétien, M. (2022). Isoquercetin as an anti-Covid-19 medication: A potential to realize. *Frontiers in Pharmacology, 13*, 830205. https://doi.org/10.3389/fphar.2022.830205

Mehrabi, A., Mahmoudi, R., Norian, R., Mosavi, S., Ahmadi, Z., Kazemi, M., & Alizadeh, A. (2022). Evaluation of sulfonamide antibiotic residues of honey samples produced in different regions of Qazvin Province by ELISA. *Journal of Chemical Health Risks*. https://doi.org/10.22034/jchr.2021.1905026.1159

Mendis, E., Rajapakse, N., Byun, H. G., & Kim, S. K. (2005). Investigation of jumbo squid (*Dosidicus gigas*) skin gelatin peptides for their in vitro antioxidant effects. *Life Sciences, 77*(17), 2166–2178. https://doi.org/10.1016/j.lfs.2005.03.016

Meo, S. A., Al-Asiri, S. A., Mahesar, A. L., & Ansari, M. J. (2017). Role of honey in modern medicine. *Saudi Journal of Biological Sciences, 24*(5), 975–978. https://doi.org/10.1016/j.sjbs.2016.12.010

Miguel, M. G., Antunes, M. D., & Faleiro, M. L. (2017). Honey as a complementary medicine. *Integrative Medicine Insights, 12*, 1178633717702869. https://doi.org/10.1177/1178633717702869

Milla, L., Sniderman, K., Lines, R., Mousavi-Derazmahalleh, M., & Encinas-Viso, F. (2021). Pollen DNA metabarcoding identifies regional provenance and high plant diversity in Australian honey. *Ecology and Evolution, 11*(13), 8683–8698.

Milla, L., Schmidt-Lebuhn, A., Bovill, J., & Encinas-Viso, F. (2022). Monitoring of honey bee floral resources with pollen DNA metabarcoding as a complementary tool to vegetation surveys. *Ecological Solutions and Evidence, 3*(1), 12120. https://doi.org/10.1002/2688-8319.12120

Mohamadzade, N. S., Kim, M. J., Son, M., & Jung, C. (2022). Honey DNA metabarcoding revealed foraging resource partitioning between Korean native and introduced honey bees (hymenoptera: Apidae). *Scientific Reports, 12*(1), 14394. https://doi.org/10.1038/s41598-022-18465-5

Molan, P. C. (1992). The Antibacterial Activity of Honey: 2. Variation in the potency of the antibacterial activity. *Bee World, 73*(2), 59–76. https://doi.org/10.1080/0005772X.1992.11099118

Molan, P. C. (1999). Why honey is effective as a medicine: 1. Its use in modern medicine. *Bee World, 80*, 80–92.

Molan, P. C. (2001). Why honey is effective as a medicine: 2. The scientific explanation of its effects. *Bee World, 82*, 22–40.

MOLBASE. (2021). MOLBASE compound summary. Accessed May 10. CAS no: 748808-93-7, L-Aspartic acid, L-seryl-L-seryl-L-threonyl-. http://www.molbase.com/moldata/36272906.html

Molnar, C. M., Berghian-Grosan, C., & Magdas, D. A. (2020). An optimized green preparation method for the successful application of Raman spectroscopy in honey studies. *Talanta, 208*, 120432. https://doi.org/10.1016/j.talanta.2019.120432

Montano, L., Pironti, C., PintoG, R. M., BuonoA, B. C., Venier, M., Piscopo, M., Amoresano, A., & Motta, O. (2022). Polychlorinated biphenyls (PCBs) in the environment: Occupational and exposure events, effects on human health and fertility. *Toxics, 10*(7), 365. https://doi.org/10.3390/toxics10070365

Montenegro, G., Gómez, M., Casaubon, G., Belancic, A., Mujica, A. M., & Peña, R. C. (2009). Analysis of volatile compounds in three unifloral native Chilean honeys. *Phyton (Buenos Aires), 78*(1), 61–65. ISSN 1851-5657.

Mráz, P., Hýbl, M., Kopecký, M., Bohatá, A., Konopická, J., Hoštičková, I., Konvalina, P., Šipoš, J., Rost, M., & Čurn, V. (2021). The effect of artificial media and temperature on the growth and development of the honey bee brood pathogen *Ascosphaera apis*. *Biology, 10*(5), 431. https://doi.org/10.3390/biology10050431

Mulu, A., Kassu, A., Anagaw, B., et al. (2013). Frequent detection of *'azole'* resistant *Candida* species among late presenting AIDS patients in northwest Ethiopia. *BMC Infectious Diseases, 13*, 82. https://doi.org/10.1186/1471-2334-13-82

Murcia-Morales, M., Tzanetou, E. N., García-Gallego, G., Kasiotis, K. M., Vejsnaes, F., Brodschneider, R., Hatjina, F., Machera, K., & Van der Steen, J. J. M. (2023). Environmental assessment of PAHs through honey bee colonies – A matrix selection study. *Heliyon, 10*(1), 23564. https://doi.org/10.1016/j.heliyon.2023.e23564

Mureşan, C. I., Cornea-Cipcigan, M., Suharoschi, R., Erler, S., & Mărgăoan, R. (2022). Honey botanical origin and honey-specific protein pattern: Characterization of some European honeys. *LWT, 154*, 112883. https://doi.org/10.1016/j.lwt.2021.112883

Newman, D. J., & Cragg, G. M. (2017). Current status of marine-derived compounds as warheads in anti-tumor drug candidates. *Marine Drugs, 15*(4), 99. https://doi.org/10.3390/md15040099

Nguyen, H. T. L., Panyoyai, N., Kasapis, S., Pang, E., & Mantri, N. (2019). Honey and its role in relieving multiple facets of atherosclerosis. *Nutrients, 11*(1), 167. https://doi.org/10.3390/nu11010167

Namsi, A., Nury, T., Hamdouni, H., Yammine, A., Vejux, A., Vervandier-Fasseur, D., Latruffe, N., Masmoudi-Kouki, O., & Lizard, G. (2018). Induction of Neuronal Differentiation of Murine N2a Cells by Two Polyphenols Present in the Mediterranean Diet Mimicking Neurotrophins Activities: Resveratrol and Apigenin. *Diseases (Basel, Switzerland), 6*(3), 67. https://doi.org/10.3390/diseases6030067

National Center for Biotechnology Information (NCBI) (2021). PubChem Compound Summary for CID 145455020, Gln-Cys-His. https://pubchem.ncbi.nlm.nih.gov/compound/Gln-Cys-His. Accessed May 10, 2021

Nikhat, S., & Fazil, M. (2022). History, phytochemistry, experimental pharmacology and clinical uses of honey: A comprehensive review with special reference to Unani medicine. *Journal of Ethnopharmacology, 282*, 114614. https://doi.org/10.1016/j.jep.2021.114614

Nolan, V. C., Harrison, J., & Cox, J. A. G. (2019). Dissecting the antimicrobial composition of honey. *Antibiotics (Basel, Switzerland), 8*(4), 251. https://doi.org/10.3390/antibiotics8040251

Odeh, I., Abu-Lafi, S., & Al-Najjar, I. (2013). Determination of potential volatiles markers from citrus, eucalyptus, cotton, and wild flower Palestinian honeys using SPME followed by GCMS analysis. https://dspace.alquds.edu/handle/20.500.12213/995

Olaitan, P. B., Adeleke, O. E., & Ola, I. O. (2007). Honey: a reservoir for microorganisms and an inhibitory agent for microbes. *African Health Sciences, 7*(3), 159–165. https://doi.org/10.5555/afhs.2007.7.3.159

Olas, B. (2020). Honey and its phenolic compounds as an effective natural medicine for cardiovascular diseases in humans? *Nutrients, 12*(2), 283. https://doi.org/10.3390/nu12020283

Olofsson, T. C., Butler, È., Markowicz, P., Lindholm, C., Larsson, L., & Vásquez, A. (2016). Lactic acid bacterial symbionts in honeybees – An unknown key to honey's antimicrobial and therapeutic activities. *International Wound Journal, 13*(5), 668–679. https://doi.org/10.1111/iwj.12345

Oroian, M., Ropciuc, S., & Paduret, S. (2018). Honey adulteration detection using Raman spectroscopy. *Food Analytical Methods, 11*, 959–968. https://doi.org/10.1007/s12161-017-1072-2

Orsolić, N., Knezević, A., Sver, L., Terzić, S., Hackenberger, B. K., & Basić, I. (2003). Influence of honey bee products on transplantable murine tumours. *Veterinary and Comparative Oncology, 1*(4), 216–226. https://doi.org/10.1111/j.1476-5810.2003.00029.x

Osés, S. M., Rodríguez, C., Valencia, O., Fernández-Muiño, M. A., & Sancho, M. T. (2024). Relationships among hydrogen peroxide concentration, catalase, glucose oxidase, and antimicrobial activities of honeys. *Food, 13*(9), 1344–1351. https://doi.org/10.3390/foods13091344

Palma-Morales, M., Huertas, J. R., & Rodríguez-Pérez, C. (2023). A comprehensive review of the effect of honey on human health. *Nutrients, 15*(13), 3056. https://doi.org/10.3390/nu15133056

Panseri, S., Manzo, A., Chiesa, L. M., & Giorgi, A. (2013). Melissopalynological and volatile compounds analysis of buckwheat honey from different geographical origins and their role in botanical determination. *Journal of Chemistry, 2013*(1), 904202.v. https://doi.org/10.1155/2013/904202

Papa, G., Maier, R., Durazzo, A., Lucarini, M., Karabagias, I. K., Plutino, M., Bianchetto, E., Aromolo, R., Pignatti, G., Ambrogio, A., Pellecchia, M., & Negri, I. (2022). The honey bee *Apis mellifera*: An insect at the Interface between human and ecosystem health. *Biology, 11*(2), 233. https://doi.org/10.3390/biology11020233

Pascual-Mate, A., Oses, S. M., Fernandez-Muino, M. A., & Sancho, M. T. (2018). Methods of analysis of honey. *Journal of Apicultural Research, 57*(1), 38–74. https://doi.org/10.1080/00218839.2017.1411178

Pasupuleti, V. R., Sammugam, L., Ramesh, N., & Gan, S. H. (2017). Honey, propolis, and royal jelly: A comprehensive review of their biological actions and health benefits. *Oxidative Medicine and Cellular Longevity, 2017*, 1259510. https://doi.org/10.1155/2017/1259510

Pataca, L. C., Neto, W. B., Marcucci, M. C., & Poppi, R. J. (2007). Determination of apparent reducing sugars, moisture and acidity in honey by attenuated total reflectance-Fourier transform infrared spectrometry. *Talanta, 71*(5), 1926–1931. https://doi.org/10.1080/00218839.2017.1411178

Patel, A. B., Shaikh, S., Jain, K. R., Desai, C., & Madamwar, D. (2020). Polycyclic aromatic hydrocarbons: Sources, toxicity, and remediation approaches. *Frontiers in Microbiology, 11*, 562813. https://doi.org/10.3389/fmicb.2020.562813

Pathiraja, D., Cho, J., Kim, J., & Choi, I. G. (2023). Metabarcoding of eDNA for tracking the floral and geographical origins of bee honey. *Food Research International, 164*, 112413. https://doi.org/10.1016/j.foodres.2022.112413

Peña, R. M., Barciela, J., Herrero, C., & García-Martín, S. (2004). Solid-phase microextraction gas chromatography-mass spectrometry determination of monoterpenes in honeys. *Journal of Separation Science, 27*, 1540–1544. https://doi.org/10.1002/jssc.200301705

Pohl, P., Stecka, H., Sergiel, I., & Jamroz, P. (2012). Different aspects of the elemental analysis of honey by flame atomic absorption and emission spectrometry: A review. *Food Analytical Methods, 5*, 737–751. https://doi.org/10.1007/s12161-011-9309-y

Pomastowski, P., Złoch, M., Rodzik, A., Ligor, M., Kostrzewa, M., et al. (2019). Analysis of bacteria associated with honeys of different geographical and botanical origin using two different identification approaches: MALDI-TOF MS and 16S rDNA PCR technique. *PLoS One, 14*(5), e0217078. https://doi.org/10.1371/journal.pone.0217078

Poppinga, L., & Genersch, E. (2015). Molecular pathogenesis of American foulbrood: How *Paenibacillus* larvae kills honey bee larvae. *Current Opinion in Insect Science, 10*, 29–36. https://doi.org/10.1016/j.cois.2015.04.013

Poungmalai, P., Buakeaw, A., Puthong, S., & Khongchareonporn, N. (2021). A specific monoclonal antibody for chlortetracycline detection in milk and honey samples based on ELISA. *Food and Agricultural Immunology, 32*(1), 163–173. https://doi.org/10.1080/09540105.2021.1897531

Prasanth, J., Vincy, M. V., & Brilliant, R. (2022). Screening and quantitation of pesticide residues in Indian honey samples by LC-MS/MS and GC-MS/MS. *Indian Journal of Science and Technology, 15*, 1112–1123. https://doi.org/10.17485/IJST/v15i23.120

Prata, J. C., & da Costa, P. M. (2024). Fourier transform infrared spectroscopy use in honey characterization and authentication: A systematic review. *ACS Food Science & Technology, 4*(8), 1817–1828. https://doi.org/10.1021/acsfoodscitech.4c00377

Ptaszyńska, A. A., Latoch, P., Hurd, P. J., Polaszek, A., Michalska-Madej, J., Grochowalski, Ł., Starosta, A. L., et al. (2021). Amplicon sequencing of variable 16S rRNA from bacteria and ITS2 regions from fungi and plants, reveals honey bee susceptibility to diseases results from their forage availability under anthropogenic landscapes. *Pathogens (Basel, Switzerland), 10*(3), 381. https://doi.org/10.3390/pathogens10030381

Pujol, C., Danoun, S., Biasini, G., Retailleau, E., Masson, J., Balayssac, S., & Gilard, V. (2024). Benchtop NMR coupled with Chemometrics: A workflow for unveiling hidden drug ingredients in honey-based supplements. *Molecules, 29*(9), 2086. https://doi.org/10.3390/molecules29092086

Purać, J., Nikolić, T. V., Kojić, D., Ćelić, A. S., Plavša, J. J., Blagojević, D. P., & Petri, E. T. (2019). Identification of a metallothionein gene in honey bee *Apis mellifera* and its expression profile in response to Cd, Cu and Pb exposure. *Molecular Ecology, 28*, 731–745. https://doi.org/10.1111/mec.14984

Qaid, E. Y. A., Zakaria, R., Mohd Yusof, N. A., Sulaiman, S. F., Shafin, N., Othman, Z., Ahmad, A. H., Abd Aziz, C. B., & Muthuraju, S. (2020). Tualang honey ameliorates hypoxia-induced memory deficits by reducing neuronal damage in the *Hippocampus* of adult male Sprague Dawley rats. *Turkish Journal of Pharmaceutical Sciences, 17*(5), 555–564. https://doi.org/10.4274/tjps.galenos.2019.32704

Rabie, E., Serem, J. C., Oberholzer, H. M., Gaspar, A. R. M., & Bester, M. J. (2016). How methylglyoxal kills bacteria: An ultrastructural study. *Ultrastructural Pathology, 40*(2), 107–111. https://doi.org/10.3109/01913123.2016.115491

Rahman, M. M., Gan, S. H., & Khalil, M. I. (2014). Neurological effects of honey: Current and future prospects. *Evidence-based Complementary and Alternative Medicine, 2014*, 958721. https://doi.org/10.1155/2014/958721

Rahman, F., Tabrez, S., Ali, R., Alqahtani, A. S., Ahmed, M. Z., & Rub, A. (2017). Molecular docking analysis of rutin reveals possible inhibition of SARS-CoV-2 vital proteins. *Journal of Traditional and Complementary Medicine, 11*(2), 173–179. https://doi.org/10.1016/j.jtcme.2021.01.006

Rahman, M. M., Alam, M. N., Fatima, N., Shahjalal, H. M., Gan, S. H., & Khalil, M. I. (2021). Chemical composition and biological properties of aromatic compounds in honey: An overview. *Journal of Food Biochemistry, 41*, e12405. https://doi.org/10.1111/jfbc.12405

Ranneh, Y., Akim, A. M., Hamid, H. A., Khazaai, H., Fadel, A., Zakaria, Z. A., Albujja, M., & Bakar, M. F. A. (2021). Honey and its nutritional and anti-inflammatory value. *BMC Complementary Medicine and Therapies, 21*(1), 30. https://doi.org/10.1186/s12906-020-03170-5

Ratiu, I. A., Al-Suod, H., Bukowska, M., Ligor, M., & Buszewski, B. (2020). Correlation study of honey regarding their physicochemical properties and sugars and cyclitols content. *Molecules, 25*(1), 34. https://doi.org/10.3390/molecules25010034

Reams, T., & Rangel, J. (2022). Understanding the enemy: A review of the genetics, behavior and chemical ecology of Varroa destructor, the parasitic mite of *Apis mellifera. Journal of Insect Science (Online), 22*(1), 18. https://doi.org/10.1093/jisesa/ieab101

Ren, H., Yue, J., Wang, D., Fan, J., & An, L. (2019). HPLC and 1 H-NMR combined with chemometrics analysis for rapid discrimination of floral origin of honey. *Journal of Food Measurement and Characterization, 13*, 1195–1204. https://doi.org/10.1007/s11694-019-00035-5

Rimkus, G. G., Huth, T., & Harms, D. (2020). Screening of stereoisomeric chloramphenicol residues in honey by ELISA and CHARM® II test–the potential risk of systematically false-compliant (false negative) results. *Food Additives & Contaminants: Part A, 37*(1), 94–103.10.1080/19440049.2019.1682685.

Rosiak, E., Madras-Majewska, B., Teper, D., Łepecka, A., & Zielińska, D. (2021). Cluster analysis classification of honey from two different climatic zones based on selected physicochemical and of microbiological parameters. *Molecules, 26*(8), 2361. https://doi.org/10.3390/molecules26082361

Rosier, B. T., Buetas, E., Moya-Gonzalvez, E. M., et al. (2020). Nitrate as a potential prebiotic for the oral microbiome. *Scientific Reports, 10*, 12895. https://doi.org/10.1038/s41598-020-69931-x

Rosli, N. H. M., Yahya, H. M., Ibrahim, F. W., Shahar, S., Ismail, I. S., Azam, A. A., & Rajab, N. F. (2020). Serum metabolomics profiling of commercially mixed functional foods-effects in Beta-amyloid induced rats measured using ^{1}H NMR spectroscopy. *Nutrients, 12*(12), 3812. https://doi.org/10.3390/nu12123812

Royal Society of Chemistry. (2021). Accessed May 9. ChemSpider compound summary for CSID: 16646752, His-Trp-Cys-Cys. http://www.chemspider.com/Chemical-Structure.16646752.html

Sahin, H., Kolayli, S., & Beykaya, M. (2020). Investigation of variations of invertase and glucose oxidase degrees against heating and timing options in raw honeys. *Journal of Chemistry, 2020*(1), 5398062. https://doi.org/10.1155/2020/5398062

Sakač, N., & Sak-Bosnar, M. (2012). A rapid method for the determination of honey diastase activity. *Talanta, 93*, 135–138. https://doi.org/10.1016/j.talanta.2012.01.063

Samarghandian, S., Farkhondeh, T., & Samini, F. (2017). Honey and health: a review of recent clinical research. *Pharm Res, 9*, 121–127. https://doi.org/10.4103/0974-8490.204647

Sampath Kumar, K. P., Bhowmik, D., Chiranjib, B., & Chandira, M. R. (2010). Medicinal uses and health benefits of honey: An overview. *Journal of Chemical and Pharmaceutical Research, 2*, 385–395. Available on line www.jocpr.com

Sanchez, L. M., Lopez, D., Vesely, B. A., Della Togna, G., Gerwick, W. H., Kyle, D. E., & Linington, R. G. (2010). Almiramides A–C: Discovery and development of a new class of leishmaniasis lead compounds. *Journal of Medicinal Chemistry, 53*(10), 4187–4197. https://doi.org/10.1021/jm100265s

Sanki, U. K., & Mandal, B. K. (2012). *In vitro-in vivo* correlation evaluation of generic alfuzosin modified release tablets. *ISRN Toxicology, 2012*, 813836. https://doi.org/10.5402/2012/813836

Scepankova, H., Combarros-Fuertes, P., Fresno, J. M., Tornadijo, M. E., Dias, M. S., Pinto, C. A., Saraiva, J. A., & Estevinho, L. M. (2021). Role of honey in advanced wound care. *Molecules (Basel, Switzerland), 26*(16), 4784. https://doi.org/10.3390/molecules26164784

Schramm, D. D., Karim, M., Schrader, H. R., Holt, R. R., Cardetti, M., & Keen, C. L. (2003). Honey with high levels of antioxidants can provide protection to healthy human subjects. *Journal of Agricultural and Food Chemistry, 51*(6), 1732–1735. https://doi.org/10.1021/jf025928k

Seisonen, S., Kivima, E., & Vene, K. (2015). Characterisation of the aroma profiles of different honeys and corresponding flowers using solid-phase microextraction and gas chromatography–mass spectrometry/olfactometry. *Food Chemistry, 169*, 34–40. https://doi.org/10.1016/j.foodchem.2014.07.125

Serra Bonvehí, J., Soliva Torrentó, M., & Muntané Raich, J. (2000). Invertase activity in fresh and processed honeys. *Journal of the Science of Food and Agriculture, 80*(4), 507–512. https://doi.org/10.1002/(SICI)1097-0010(200003)80:4%3C507::AID-JSFA558%3E3.0.CO;2-5.

Shapla, U. M., Solayman, M., Alam, N., et al. (2018). 5-Hydroxymethylfurfural (HMF) levels in honey and other food products: Effects on bees and human health. *Chemistry Central Journal, 12*, 35. https://doi.org/10.1186/s13065-018-0408-3

Sharma, N., & Singhvi, R. (2017). Effects of chemical fertilizers and pesticides on human health and environment: A review. *International Journal of Agriculture, Environment and Biotechnology, 10*(6), 675–679.

Shi, N., Liu, Y., Li, W., Yan, S., Ma, L., Xu, X., & Chen, D. (2024). One-pot derivatization/magnetic solid-phase extraction coupled with liquid chromatography-fluorescence detection for the rapid determination of sulfonamide residues in honey. *Food Chemistry: X, 21*, 101090. https://doi.org/10.1016/j.fochx.2023.101090

Silva, B., Biluca, F. C., Gonzaga, L. V., Fett, R., Dalmarco, E. M., Caon, T., & Costa, A. C. O. (2021). *In vitro* anti-inflammatory properties of honey flavonoids: A review. *Food Research International (Ottawa, Ont.), 141*, 110086. https://doi.org/10.1016/j.foodres.2020.110086

Silici, S. (2011). Determination of volatile compounds of pine honeys. *Turkish Journal of Biology, 35*(5), 641–645. https://doi.org/10.3906/biy-1009-112

Simonin, A., Montalbetti, N., Gyimesi, G., Pujol-Giménez, J., & Hediger, M. A. (2015). The hydroxyl side chain of a highly conserved serine residue is required for cation selectivity and substrate transport in the glial glutamate transporter GLT-1/SLC1A2. *The Journal of Biological Chemistry, 290*(51), 30464–30474. https://doi.org/10.1074/jbc.M115.689836

Singh, I., & Singh, S. (2018). Honey moisture reduction and its quality. *Journal of Food Science and Technology, 55*, 3861–3871. https://doi.org/10.17977/um016v4i22020p153

Smetanska, I., Alharthi, S. S., & Selim, K. A. (2021). Physicochemical, antioxidant capacity and color analysis of six honeys from different origin. *Journal of King Saud University-Science, 33*(5), 101447. https://doi.org/10.1016/j.jksus.2021.101447

Soares, S., Bornet, M., Grosso, C., Ramalhosa, M. J., Gouvinhas, I., Garcia, J., Rodrigues, F., & Delerue-Matos, C. (2022). Honey as an adjuvant in the treatment of COVID-19 infection: A review. *Applied Sciences, 12*(15), 7800. https://doi.org/10.3390/app12157800

Solayman, M., Islam, M. A., Paul, S., Ali, Y., Khalil, M. I., Alam, N., & Gan, S. H. (2016). Physicochemical properties, minerals, trace elements, and heavy metals in honey of different origins: A comprehensive review. *Comprehensive Reviews in Food Science and Food Safety, 15*(1), 219–233. https://doi.org/10.1111/1541-4337.12182

Song, Y. Q., Milne, R. I., Zhou, H. X., Ma, X. L., Fang, J. Y., & Zha, H. G. (2019). Floral nectar chitinase is a potential marker for monofloral honey botanical origin authentication: A case study from loquat (Eriobotrya japonica Lindl.). *Food Chemistry, 282*, 76–83. https://doi.org/10.1016/j.foodchem.2018.12.107

Soria, A. C., Sanz, J., & Martínez-Castro, I. (2009). SPME followed by GC–MS: A powerful technique for qualitative analysis of honey volatiles. *European Food Research and Technology, 228*, 579–590. https://doi.org/10.1007/s00217-008-0966-z

Souza, A. P. F., Petrarca, M. H., de Campos Braga, P. A., Rodrigues, N. R., & Reyes, F. G. R. (2021). Analysis of insecticide residues in honey by liquid chromatography tandem mass spectrometry using QuEChERS optimized by the Plackett Burman design. *CyTA-Journal of Food, 19*(1), 326–332. https://doi.org/10.1080/19476337.2021.1901785

Spiteri, C., Lia, F., & Farrugia, C. (2020). Determination of the geographical origin of Maltese honey using 1H NMR fingerprinting. *Food, 9*(10), 1455. https://doi.org/10.3390/foods9101455

Stanković, M., Bartolić, D., Šikoparij, B., Spasojević, D., Mutavdžić, D., Natić, M., & Radotić, K. (2019). Variability estimation of the protein and phenol total content in honey using front face fluorescence spectroscopy coupled with MCR–ALS analysis. *Journal of Applied Spectroscopy, 86*, 256–263. https://doi.org/10.1007/s10812-019-00809-1

Stavropoulou, E., Remmas, N., Voidarou, C. C., Vrioni, G., Konstantinidis, T., Ntougias, S., & Tsakris, A. (2023). Microbial community structure among honey samples of different pollen origin. *Antibiotics (Basel, Switzerland), 12*(1), 101. https://doi.org/10.3390/antibiotics12010101

Su, Q., Tang, M., Hu, J., Tang, J., Zhang, X., Li, X., Niu, Q., Zhou, X., Luo, S., & Zhou, X. (2022). Significant compositional and functional variation reveals the patterns of gut microbiota evolution among the widespread Asian honeybee populations. *Frontiers in Microbiology, 13*, 934459. https://doi.org/10.3389/fmicb.2022.934459

Suchart, C. (2023). Confectionery and bakery products; including honey. In V. Andersen, H. Lelieveld, & Y. Motarjemi (Eds.), *Food safety management* (2nd ed., pp. 291–303, ISBN 9780128200131). Academic. https://doi.org/10.1016/B978-0-12-820013-1.00045-0

Suetsuna, K., & Nakano, T. (2000). Identification of an antihypertensive peptide from peptic digest of wakame (Undaria pinnatifida). *The Journal of Nutritional Biochemistry, 11*(9), 450–454. https://doi.org/10.1016/s0955-2863(00)00110-8

Suhandy, D., & Yulia, M. (2021). The use of UV spectroscopy and SIMCA for the authentication of Indonesian honeys according to botanical, entomological and geographical origins. *Molecules, 26*(4), 915. https://doi.org/10.3390/molecules26040915

Suhandy, D., Kuncoro, S., Rahmawati, W., & Yulia, M. (2022). Authentication of pure and adulterated Sumbawa monofloral honey using ultraviolet-visible spectroscopy. In *IOP conference series: Earth and environmental science* (Vol. 1027, 1, p. 012014). IOP Publishing. https://doi.org/10.1088/1755-1315/1027/1/012014

Suhandy, D., Yulia, M., & Al Riza, D. F. (2024). Discrimination between real and fake honey using portable fluorescence spectroscopy and simca. In *IOP conference series: Earth and environmental science* (Vol. 1290, No. 1, p. 012003). IOP Publishing. https://doi.org/10.1088/1755-1315/1290/1/012003

Sun, X., Sarteshnizi, R. A., Boachie, R. T., Okagu, O. D., et al. (2020). Peptide-mineral complexes: Understanding their chemical interactions, bioavailability, and potential application in mitigating micronutrient deficiency. *Foods (Basel, Switzerland), 9*(10), 1402. https://doi.org/10.3390/foods9101402

Suto, M., Kawashima, H., & Suto, N. (2019). Heart-cutting two-dimensional liquid chromatography combined with isotope ratio mass spectrometry for the determination of stable carbon isotope ratios of gluconic acid in honey. *Journal of Chromatography A, 1608*, 460421. https://doi.org/10.1016/j.chroma.2019.460421

Suto, M., Kawashima, H., & Nakamura, Y. (2020). Determination of organic acids in honey by liquid chromatography with tandem mass spectrometry. *Food Analytical Methods, 13*(12), 2249–2257. https://doi.org/10.1007/s12161-020-01845-w

Swellam, T., Miyanaga, N., Onozawa, M., Hattori, K., Kawai, K., Shimazui, T., & Akaza, H. (2003). Antineoplastic activity of honey in an experimental bladder cancer implantation model: *In vivo* and *in vitro* studies. *International Journal of Urology: Official Journal of the Japanese Urological Association, 10*(4), 213–219. https://doi.org/10.1046/j.0919-8172.2003.00602.x

Szwajgier, D., Baranowska-Wójcik, E., Winiarska-Mieczan, A., & Gajowniczek-Ałasa, D. (2022). Honeys as Possible Sources of Cholinesterase Inhibitors. *Nutrients, 14*(14), 2969. https://doi.org/10.3390/nu14142969

Takahashi, T., Vo Ngo, B. C., Xiao, L., Arya, G., & Heller, M. J. (2016). Molecular mechanical properties of short-sequence peptide enzyme mimics. *Journal of Biomolecular Structure & Dynamics, 34*(3), 463–474. https://doi.org/10.1080/07391102.2015.1039586

Takamatsu, D. (2023). A typical *Melissococcusplutonius* strains: Their characteristics, virulence, epidemiology and mysteries. *The Journal of Veterinary Medical Science, 85*(9), 880–894. https://doi.org/10.1292/jvms.23-0180

Tananaki, C. H., Thrasyvoulou, A., Giraudel, J. L., & Montury, M. (2007). Determination of volatile characteristics of Greek and Turkish pine honey samples and their classification by using Kohonen self organising maps. *Food Chemistry, 101*(4), 1687–1693. https://doi.org/10.1016/j.foodchem.2006.04.042

Tananaki, C., Liolios, V., Kanelis, D., & Rodopoulou, M. A. (2021). Investigation of volatile compounds in combination with multivariate analysis for the characterization of monofloral honeys. *Applied Sciences, 12*(1), 264. https://doi.org/10.3390/app12010264

Tashkandi, H. (2021). Honey in wound healing: An updated review. *Open Life Sciences, 16*(1), 1091–1100. https://doi.org/10.1515/biol-2021-0084

Taysi, S., Algburi, F. S., Taysi, M. E., & Caglayan, C. (2023). Caffeic acid phenethyl ester: A review on its pharmacological importance, and its association with free radicals, COVID-19, and radiotherapy. *Phytotherapy Research: PTR, 37*(3), 1115–1135. https://doi.org/10.1002/ptr.7707

Teeters, B. S., Johnson, R. M., Ellis, M. D., & Siegfried, B. D. (2012). Using video-tracking to assess sublethal effects of pesticides on honey bees (Apis mellifera L.). *Environmental Toxicology and Chemistry, 31*(6), 1349–1354

Tedesco, R., Barbaro, E., Zangrando, R., Rizzoli, A., Malagnini, V., Gambaro, A., Fontana, P., & Capodaglio, G. (2020). Carbohydrate determination in honey samples by ion chromatography–

mass spectrometry (HPAEC-MS). *Analytical and Bioanalytical Chemistry, 412*, 5217–5227. https://doi.org/10.1007/s00216-020-02732-3
Tian, H., Shen, Y., Yu, H., & Chen, C. (2018). Aroma features of honey measured by sensory evaluation, gas chromatography mass spectrometry, and electronic nose. *International Journal of Food Properties, 21*(1), 1755–1768. https://doi.org/10.1080/10942912.2016.1213744
Tran, Q., Pham, D. T., & Phan, V. (2017). Using 16S rRNA gene as marker to detect unknown bacteria in microbial communities. *BMC Bioinformatics, 18*(Suppl 14), 499. https://doi.org/10.1186/s12859-017-1901-8
Trautvetter, S., Koelling-Speer, I., & Speer, K. (2009). Confirmation of phenolic acids and flavonoids in honeys by UPLC-MS. *Apidologie, 40*(2), 140–150. https://doi.org/10.1051/apido/2008072
Uçar, M., Kemal, M., Kanbur, E. D., Kara, Y., Özcelik, A. E., & Kolaylı, S. (2023). The botanical, physicochemical, and biochemical characteristics of Northern Cyprus honeys. *European Food Research and Technology, 249*(6), 1531–1541. https://doi.org/10.1007/s00217-023-04233-w
Veloso, A. C., Sousa, M. E., Estevinho, L., Dias, L. G., & Peres, A. M. (2018). Honey evaluation using electronic tongues: An overview. *Chem, 6*(3), 28. https://doi.org/10.3390/chemosensors6030028
Vincēviča-Gaile, Z. (2010). Macro- and trace elements in honey. *LLU-Raksti, 25*, 54–66. https://www.researchgate.net/publication/303881217_Macro-_and_trace_elements_in_honey
Vit, P., Fernandez-Maeso, M. C., & Ortiz-Valbuena, A. (1998). Potential use of the three frequently occurring sugars in honey to predict stingless bee entomological origin. *Journal of Applied Entomology, 122*, 5–8. https://doi.org/10.1111/j.1439-0418.1998.tb01452.x
Vit, P., Medina, M., & Enriquez, M. E. (2004). Quality standards for medicinal uses of Meliponinae honey in Guatemala, Mexico and Venezuela. *Bee World, 85*, 2–5. https://doi.org/10.1080/0005772X.2004.11099603
Vit, P., Pedro, S. R., & Roubik, D. (2013). *Pot-honey: A legacy of stingless bees*. Springer. https://doi.org/10.1007/978-1-4614-4960-7
Voica, C., Iordache, A. M., & Ionete, R. E. (2020). Multi-elemental characterization of honey using inductively coupled plasma mass spectrometry fused with chemometrics. *Journal of Mass Spectrometry, 55*(7), e4512. https://doi.org/10.1002/jms.4512
Vorlova, L., & Pridal, A. (2002). Invertase and diastase activity in honeys of Czech provenience. *Acta Universitatis Agriculturae et Silviculturae Mendelianae Brunensis, 5*, 57–66.
Wang, X., Li, Y., Chen, L., & Zhou, J. (2022). Analytical strategies for LC–MS-based untargeted and targeted metabolomics approaches reveal the entomological origins of honey. *Journal of Agricultural and Food Chemistry, 70*(4), 1358–1366. https://doi.org/10.1021/acs.jafc.1c07153
Wirta, H., Abrego, N., Miller, K., Roslin, T., & Vesterinen, E. (2021). DNA traces the origin of honey by identifying plants, bacteria and fungi. *Scientific Reports, 11*(1), 4798. https://doi.org/10.1038/s41598-021-84174-0
Wolski, T., Tambor, K., Rybak-Chmielewska, H., & Kedzia, B. (2006). Identification of honey volatile components by solid phase microextraction (SPME) and gas chromatography/mass spectrometry (GC/MS). *Journal of Apicultural Science, 50*(2), 115–126.
Wu, F., Zhao, H., Sun, J., Guo, J., Wu, L., Xue, X., & Cao, W. (2021). ICP-MS-based ionomics method for discriminating the geographical origin of honey of *Apis cerana* Fabricius. *Food Chemistry, 354*, 129568. https://doi.org/10.1016/j.foodchem.2021.129568
Wu, Y., Zheng, Y., Wang, S., Chen, Y., Tao, J., Chen, Y., Chen, G., Zhao, H., Wang, K., Dong, K., & Hu, F. (2022). Genetic divergence and functional convergence of gut bacteria between the eastern honey bee Apis cerana and the Western honey bee *Apis mellifera*. *Journal of Advanced Research, 37*, 19–31. https://doi.org/10.1016/j.jare.2021.08.002
Xagoraris, M., Revelou, P. K., Alissandrakis, E., Tarantilis, P. A., & Pappas, C. S. (2021). The use of right angle fluorescence spectroscopy to distinguish the botanical origin of greek common honey varieties. *Applied Sciences, 11*(9), 4047. https://doi.org/10.3390/app11094047

Xia, Y., Lu, T., Wang, L., Mo, J., Jin, Y., Zhang, L., & Du, S. (2021). Intrinsic Raman signal amplification for rapid identification and detection of methylglyoxal in manuka honey. *Analytica Chimica Acta, 1181*, 338902. https://doi.org/10.1016/j.aca.2021.338902

Xu, X., Liu, L., Cui, G., Wu, X., & Kuang, H. (2019). Development of an immunochromatography assay for salinomycin and methyl salinomycin in honey. *Food and Agricultural Immunology, 30*(1), 995–1006. https://doi.org/10.1080/09540105.2019.1649370

Xu, J., Liu, X., Wu, B., & Cao, Y. (2020a). A comprehensive analysis of 13C isotope ratios data of authentic honey types produced in China using the EA-IRMS and LC-IRMS. *Journal of Food Science and Technology, 57*(4), 1216–1232. https://doi.org/10.1007/s13197-019-04153-2

Xu, Y., Zhong, P., Jiang, A., Shen, X., Li, X., Xu, Z., Shen, Y., Sun, Y., & Lei, H. (2020b). Raman spectroscopy coupled with chemometrics for food authentication: A review. *TrAC Trends in Analytical Chemistry, 131*, 116017. https://doi.org/10.1016/j.trac.2020.116017

Yamagata, K. (2019). Polyphenols regulate endothelial functions and reduce the risk of cardiovascular disease. *Current Pharmaceutical Design, 25*, 2443–2458. https://doi.org/10.2174/1381612825666190722100504

Yang, Y., Albanyan, H., Lee, S., Aloysius, H., Liang, J. J., Kholodovych, V., Sahota, A., & Hu, L. (2018). Design, synthesis, and evaluation of l-cystine diamides as l-cystine crystallization inhibitors for cystinuria. *Bioorganic & Medicinal Chemistry Letters, 28*(8), 1303–1308. https://doi.org/10.1016/j.bmcl.2018.03.024

Yang, R., Roshani, D., Gao, B., Li, P., & Shang, N. (2024). Metallothionein: A comprehensive review of its classification, structure, biological functions, and applications. *Antioxidants, 13*(7), 825. https://doi.org/10.3390/antiox13070825

Yong, C. H., Muhammad, S. A., Nasir, F. I., Mustafa, M. Z., Ibrahim, B., Kelly, S. D., Cannavan, A., & Seow, E. K. (2022). Detecting adulteration of stingless bee honey using untargeted 1H NMR metabolomics with chemometrics. *Food Chemistry, 368*, 130808. https://doi.org/10.1016/j.foodchem.2021.130808

Yu, W., Zhang, G., Wu, D., Guo, L., Huang, X., Ning, F., Liu, Y., & Luo, L. (2023). Identification of the botanical origins of honey based on nanoliter electrospray ionization mass spectrometry. *Food Chemistry, 418*, 135976. https://doi.org/10.1016/j.foodchem.2023.135976

Zaky, A. A., Simal-Gandara, J., Eun, J. B., Shim, J. H., & Abd El-Aty, A. M. (2022). Bioactivities, applications, safety, and health benefits of bioactive peptides from food and by-products: A review. *Frontiers in Nutrition, 8*, 815640. https://doi.org/10.3389/fnut.2021.815640

Zalpoor, H., Bakhtiyari, M., Shapourian, H., Rostampour, P., Tavakol, C., & Nabi-Afjadi, M. (2022). Hesperetin as an anti-SARS-CoV-2 agent can inhibit COVID-19-associated cancer progression by suppressing intracellular signaling pathways. *Inflammopharmacology, 30*(5), 1533–1539. https://doi.org/10.1007/s10787-022-01054-3

Zammit Young, G. W., & Blundell, R. (2023). A review on the phytochemical composition and health applications of honey. *Heliyon, 9*(2), 12507. https://doi.org/10.1016/j.heliyon.2022.e12507

Zhang, Z., Mu, X., Cao, Q., Shi, Y., Hu, X., & Zheng, H. (2022). Honeybee gut lactobacillus modulates host learning and memory behaviors via regulating tryptophan metabolism. *Nature Communications, 13*(1), 2037. https://doi.org/10.1038/s41467-022-29760-0

Zhang, W., Zheng, Y., Yan, F., Dong, M., & Ren, Y. (2023). Research progress of quercetin in cardiovascular disease. *Frontiers in Cardiovascular Medicine, 10*, 1203713. https://doi.org/10.3389/fcvm.2023.1203713

Zheng, L., Zhao, Y., Dong, H., Su, G., & Zhao, M. (2016). Structure-activity relationship of antioxidant dipeptides: Dominant role of Tyr, Trp, Cys and met residues. *Journal of Functional Foods, 21*, 485–496. https://doi.org/10.1016/2015.12.003

Zhu, G., Wang, Q., Lu, S., & Niu, Y. (2017). Hydrogen peroxide: A potential wound therapeutic target? *Medical Principles and Practice: International Journal of the Kuwait University, Health Science Centre, 26*(4), 301–308. https://doi.org/10.1159/000475501

Zhu, W., Zhang, J., Zhang, Y., Zhang, H., Miao, K., Luo, J., & Yang, M. (2024). Establishment of a rapid and sensitive ic-ELISA for the detection of thiacloprid residues in honey and medicinal

herbs using a novel highly specific monoclonal antibody. *Ecotoxicology and Environmental Safety, 284*, 116911. https://doi.org/10.1016/j.ecoenv.2024.116911
Zielinski, L., Deja, S., Jasicka-Misiak, I., & Kafarski, P. (2014). Chemometrics as a tool of origin determination of polish monofloral and multifloral honeys. *Journal of Agricultural and Food Chemistry, 62*, 2973–2981. https://doi.org/10.1021/jf4056715
Zugravu, C. A., Parvu, M., Patrascu, D., & Stoian, A. (2009). Correlations between lead and cadmium pollution of honey and environmental heavy metal presence in two Romanian counties. Bulletin of University of Agricultural Sciences and Veterinary Medicine Cluj-Napoca. *Agriculture, 66*(2).

Chapter 7
Scientific Methods for Assessing Honey Quality

Sidra Nisar Ahmed, Abida Aziz, Muhammad Farooq Azhar, Sara Aimen, and Qurat Ul Ain Hyder

1 Methods for Honey Adulteration

1.1 Direct Adulteration

Honey adulteration is a significant apprehension in the food industry, affecting the quality, safety, and authenticity of honey (Fakhlaei et al., 2020). Honey adulteration refers to the intentional or unintentional alteration of honey's composition, affecting its quality, authenticity, and safety. Honey adulteration compromises the reliability and value of honey, affecting consumers, producers, and the industry as a whole (Ntakoulas et al., 2022). Adding sugar syrups to honey is a familiar adulteration practice that diminishes the quality, nutritional value, and authenticity of honey. Sugar syrups are often less than honey, so adding them to honey increases profit margins. The addition of sugar syrups increases the volume of honey, allowing the supplier to sell more products. Sugar syrups can conceal the flavors, aromas, or textures in low-quality honey (Brar et al., 2023). The consequences of adding high-fructose corn syrup (HFCS) to honey are plentiful and significant, affecting the product's quality, nutritional value, and consumer health (Mohamat et al., 2023). HFCS changes honey's distinct taste and smell and compromises honey's natural preservative qualities. It can lead to fermentation and spoilage. HFCS denatures enzymes, reducing honey's nutritional value. HFCS adds empty calories, reducing

S. N. Ahmed (✉)
Department of Plant Sciences, Quaid-i-Azam University, Islamabad, Pakistan

A. Aziz · S. Aimen · Q. U. A. Hyder
Department of Botany, The Women University Multan, Multan, Pakistan

M. F. Azhar
Department of Forestry and Range Management, Bahauddin Zakariya University, Multan, Pakistan

M. Zafar et al. (eds.), *Pure Honey: Assurance & Authentication*,
https://doi.org/10.1007/978-3-031-98913-1_7

honey's nutritional density, and can cause a rapid spike in blood sugar levels. It dilutes honey's natural mineral profile and can destroy or degrade honey's natural antioxidants (da Silva et al., 2016). Beet sugar adulteration of honey involves adding beet sugar syrups to honey to increase volume, reduce costs, and mask defects. It occurs due to meeting high demand without sufficient honey supply. Beet syrup can enhance honey's viscosity (Nhlapho, 2023). Cane sugar is characterized by high sucrose content, a higher C4/C3 sugar ratio, lower conductivity, different pH levels, and a distinct spectroscopic signature. Adulteration employing cane sugar at the level of 5–10% is difficult to detect. Medium-level adulteration, 10–20%, is detected by advanced methods (Sonawane, 2016). Rice syrup is another potential adulterant in honey. Rice syrup is characterized by high carbohydrates (70–80%), low glycemic index, a mild and sweet flavor, and a viscous, syrupy texture (Anwar et al., 2023). It is used for adulteration due to its low production cost, similar viscosity to honey, and ease of blending with honey, which masks the natural flavor of honey. The consequences of adulteration include an altered nutritional profile, changed flavor and texture, reduced antioxidant activity, a negative economic impact on legitimate beekeepers and producers, and the potential for regulatory action (Soares et al., 2017).

Invert sugar, also known as inverted sugar syrup, is a mixture of glucose and fructose. It is produced by hydrolyzing sucrose (table sugar) into its two component sugars. Invert sugar is added to honey for several reasons. Invert sugar helps prevent honey crystallization, maintaining its smooth, liquid texture (Nhlapho, 2023). Its humectant properties help retain moisture, thereby extending the shelf life of honey. Some producers believe invert sugar balances out the honey flavor profile. Adding invert sugar compromises honey's natural, pure state, potentially violating labeling regulations. It alters honey's nutritional content, adding more fructose and glucose. Many countries regulate or prohibit adding invert sugar to honey (Castro-Muñoz et al., 2022).

The addition of molasses to honey involves blending molasses, a thick, dark liquid derived from refining sugar cane or sugar beets, into honey to enhance its properties. It improves color and texture, enhances the flavor profile, increases shelf life, and reduces crystallization. Molasses can be added in three forms—sugarcane molasses, sugar beet molasses, and blackstrap molasses, which is thick, dark, and robust (Adbhai et al., 2022). These are added by batch mixing, continuous blending, and ultrasonic mixing. The rate of molasses addition is typically 5–20% depending on desired properties and regulations. Manufacturers can make informed decisions about enhancing their products while maintaining quality and regulatory standards by understanding the addition of molasses to honey (Ajala et al., 2021).

The addition of corn syrup to honey involves blending corn syrup, a sweetener derived from cornstarch, into honey to alter its properties. It prevents crystallization, improves texture and flow, enhances shelf life, and reduces cost. There are three types of corn syrup—light corn syrup (33% solids), heavy corn syrup (42% solids), and high-fructose corn syrup (HFCS). Corn syrup is typically added at a 5–20% rate (Helstad, 2019). With an informed understanding of the effects and processes of adding corn syrup to honey, manufacturers can make informed decisions about modifying their products while maintaining quality and regulatory standards. Some countries or regions may prohibit or restrict the addition of corn syrup to honey (Fakhlaei et al., 2020).

Saccharin, aspartame, sucralose, acesulfame potassium, and neotame are artificial sweeteners used for honey adulteration. They are used to reduce calorie content, enhance sweetness, improve shelf life, and decrease production costs. Typically, they are added at the rate of 0.1–1.0% of total weight, depending upon desired sweetness and regulations (Wu et al., 2022). Considerations regarding uses of sweeteners include impact on the flavor profile, potential health risks, regulatory compliance, and consumer acceptance. The drawbacks of using sweeteners include alteration of natural characteristics, potential health risks, regulatory noncompliance, and negative consumer perception (Saraiva et al., 2020). Artificial sweeteners may compromise honey's status as a natural food. Bose & Padmavati (2024) encourage maintaining honey's natural integrity by avoiding artificial sweeteners, focusing on quality control and beekeeping practices, and labeling products transparently. By understanding the effect of artificial sweeteners on honey, manufacturers can make informed decisions about product modification, while maintaining quality and regulatory standards (Drewnowski et al., 2019).

1.2 Indirect Adulteration

Indirect adulteration of honey refers to the unintentional or intentional addition of non-honey substances or contaminants during processing, storage, or handling, compromising a honey's quality and legitimacy (Soares, 2017). Feeding methods for bees provide supplemental nutrition to support colony health and honey production. There are different feeding methods. Dry sugar feeding involves providing bees with pure, granulated sugar to supplement their diet during times of low nectar flow (Tsuruda et al., 2021). Liquid sugar feeding methods refer to the various techniques used to provide sugar solutions to animals, particularly bees, birds, and other wildlife, as well as in agricultural settings. Liquid sugar feeding methods include: syrup feeding, mixing 1:1 sugar-water in solution; fondant feeding, using premade or homemade fondant; and candy boards, involving soaking in sugar water (Medrzycki et al., 2013).

Pollen supplements are patties composed of pollen, sugar, and water, while pollen substitutes utilize soy flour or other alternatives, incorporating sugar and water (Haefeker, 2021). Nectar replacement with sugar or sugar syrups is a common practice in beekeeping to support bee colonies, especially during times of low nectar flow or when natural nectar sources are scarce. It provides bees with essential carbohydrates for energy, helps maintain a healthy population, promotes brood production, and supplements bees' diet during periods of low nectar flow or drought. The types of sugar used are sucrose, fructose, and glucose. Sugar syrup may be thin syrup (1:1 ratio) or thick syrup (2:1 ratio; Kavimugilan et al., 2024).

Honey blending involves merging honey with other sweeteners or honey from unknown sources to produce a uniform product. This practice is common in the honey industry to guarantee uniform taste, texture, and color to address supply shortages or seasonal variations. It also decreases production costs by incorporating cheaper sweeteners, and the shelf life can be extended by adding preservatives or other sweeteners (Petros, 2016).

Ultrafiltration (UF) in honey processing requires the use of semipermeable membranes to filter out impurities, pollen, and other particles. While UF improves appearance and shelf life, it can also eliminate beneficial natural compounds (Abu-Reidah, 2020). Benefits of ultrafiltration include improved clarity, increased shelf life, consistent quality, and a lower risk of contamination. Drawbacks of UF are linked to the removal of natural enzymes, vitamins, minerals, and beneficial pollen, thus reducing potential health benefits (Ghosh et al., 2019). UF can change honey's natural flavor and aroma, and the elimination of beneficial compounds may reduce honey's nutritional value. Microfiltration, ultrafiltration, and nanofiltration are types of ultrafiltration (Samborska et al., 2019). Enzyme treatment in honey processing involves using enzymes to break down complex sugars into simpler ones, altering honey's natural composition (Baglio, 2017). This process can enhance honey's properties, but also raises concerns. The enzymes used are diastase—which breaks down starches into simpler sugars such as amylase, which converts complex carbohydrates into simpler ones—and invertase, which converts sucrose into glucose and fructose. Methods of enzyme treatment include batch treatment, continuous treatment, and immobilized enzyme treatment (Palma et al., 2021).

Heat treatment in honey processing involves exposing honey to high temperatures to achieve various benefits, but excessive heat can damage honey's natural properties. Heat treatment kills bacteria, yeast, and other microorganisms. Heat also dissolves glucose crystals, reduces granulation, inhibits fermentation and spoilage, and reduces viscosity, making filtering more efficient (Eshete and Eshete, 2019). Heat treatment methods are pasteurization (heating to 63 °C, 145 °F, for 30 min), flash pasteurization (heating to 72 °C, 162 °F, for 15 s), and heat-and-hold (heating to 60 °C, 140 °F, for 30 min; Manzoor et al., 2022). Heat damages natural enzymes, destroys vitamins and minerals, and impacts flavor, aroma, and caramelization (Kathuria et al., 2023).

Moisture adjustment in honey processing involves controlling the water content to prevent fermentation, spoilage, and crystallization (Bako et al., 2019). Moisture content can be adjusted by drying, humidification, and blending. Moisture levels can be measured by refractometry, Karl Fischer titration, and oven drying (Mauer, 2024).

2 Scientific Methods for Assessment of Honey Quality

2.1 *Physical and Chemical Methods*

2.1.1 Refractometry

Refractometry is a widely used technique for honey identification and quality control. Refractometry measures the refractive index (RI) of honey, which is affected by its sugar content, moisture, and temperature. In honey analysis, producers and quality control specialists can efficiently verify honey's authenticity, classify its type, and ensure compliance with international standards by applying refractometry. Measuring the refractive index (RI) provides a rapid, noninvasive method for differentiating pure honey from adulterated samples. Singh and Singh (2018) investigated the refractive index of 50 honey samples from various regions and botanical origins. Results show significant variations in RI among honey types, with a range of 1.38–1.42. Correlation analysis reveals strong relationships between RI and sugar content, moisture, and pollen count. Díaz-Galiano et al. (2023) reported that the use of liquid chromatography and high-resolution mass spectrometry, combined with metabolomics, has led to the identification of potential markers for authenticating manuka honey. The ability to detect adulteration, even at low levels (75% purity), is significant.

2.1.2 Polarimetry

Imaging polarimetry has broad applications across modern scientific technologies. Given the global challenge of honey adulteration in the honey industry, this research aimed to assess honey quality by examining various polarization parameters derived from imaging polarimetry techniques. These techniques were applied to honey samples mixed with different adulterants (glucose, fructose, sucrose) to distinguish polarization characteristics between pure and adulterated samples (Brickel et al., 1982) marked the beginning of polarimetry applications in biomedical detection by measuring polarization effects of light scattered from bacterial suspensions and planned polarized light as a categorization tool for biological tissue. The optical polarization method has revolutionized imaging and microscopy. Techniques such as polarized light microscopy (PLM), polarization-sensitive optical coherence tomography (PS-OCT), second-harmonic generation (SHG) Microscopy, and fluorescence polarization microscopy are employed, utilizing various light sources like lamps, LEDs, and lasers (Huynh, 2020). The method offers significant advantages over conventional optical microscopy techniques, particularly in cellular and tissue imaging, as well as material characterization (Tepeli et al., 2022). Exciting advancements have been made in biomedical imaging using polarization principles.

2.1.3 High-Performance Liquid Chromatography (HPLC)

High-performance liquid chromatography (HPLC) is a powerful analytical approach. HPLC applications in honey analyses are used for sugar estimation (e.g., fructose, glucose, and sucrose), amino acid profiling (e.g., proline, tyrosine), pollen content analysis (e.g., botanical origin), uncovering of additives and adulterants (e.g., sugars, syrups), and quantification of bioactive compounds (e.g., flavonoids, phenolic acids). HPLC modalities offer diverse chromatographic separation mechanisms—reversed-phase (RP-HPLC), normal-phase (NP-HPLC), size-exclusion (SEC), and ion-exchange (IEC). Detection methods used in HPLC are ultraviolet (UV) detection, fluorescence detection, mass spectrometry (MS) detection, and evaporative light-scattering detection (ELSD).

Combined HPLC and gas chromatography-mass spectrometry (GC-MS) comprehensively analyzed 14 honey samples. Total sugar content was determined to range from 50.26–74.74 g/100 g. HPLC analysis revealed the presence of sucrose in two samples, albeit below regulatory thresholds. The average fructose-to-glucose ratio in these honey samples was 1.3 (Alghamdi et al., 2020). Researchers (Ouchemoukh et al., 2010) conducted an inclusive study on sugar profiles of Algerian honey. 50 honey samples were analyzed (25 multifloral, 25 unifloral), in which 11 sugars were quantified (2 monosaccharides, 9 oligosaccharides). Ranges of fructose (35.99–42.57%) and glucose (24.63–35.06%) were observed. Regional variations were also observed in sugar profiles. A high-performance liquid chromatography (HPLC) system, incorporating refractive index and diode array detectors, was implemented to quantify the sugar profile and hydroxymethylfurfural (HMF) levels in Saudi honey samples. The average values of fructose (33.10–44.77%) and glucose (26.68%–37.91%) were observed. Maltose was ubiquitously present in all honey types, with average concentrations ranging from 0.37% to 2.97%. Sucrose, however, was absent in six honey types, present at 0.25% in one unifloral honey, and reached 3.25% in one multifloral honey (Ghramh et al., 2020). High-performance liquid chromatography with electrochemical detection (HPLC-ECD) combined with chemometric analysis has been shown to effectively detect resin-based adulteration in acacia honey (Wang et al., 2020). Additionally, García-Seval et al. (2022) demonstrated that HPLC-UV fingerprinting, particularly following off-line SPE polyphenolic isolation and extraction, can be valuable for identifying and classifying honey based on botanical origin, serving as chemical descriptors for honey. Jamal et al. (2020) found that all 40 honey samples in their study were free of the acaricide flumethrin, as confirmed through HPLC residue analysis. This HPLC-based method for detecting flumethrin demonstrated high sensitivity, presenting a viable alternative to gas chromatography for accurately assessing flumethrin residues in honey and beeswax. In related work, Pang et al. (2021) employed high-performance liquid chromatography in tandem with mass spectrometry (HPLC–MS/MS) to determine oxytetracycline, tetracycline, chlortetracycline, and doxycycline, achieving detection limits (LOD) of 0.073 to 0.435 ng/g and quantitation limits (LOQ) from 0.239 to 1.449 mg/g across four honey types, with recovery rates from 88.1% to 126.2%. Furthermore, Das et al. (2022) analyzed the HMF

(Hydroxymethylfurfural) content and sugar profiles of 70 honey samples collected from various regions of Bangladesh using high-performance liquid chromatography (HPLC). Their findings conclude that HMF content ranging from 1.41 to 2063.90 mg/kg indicates varying levels of honey quality and processing conditions. The limit of detection (LOD) was 0.10 mg/kg. The limit of quantification (LOQ) was 0.33 mg/kg, and the coefficient of determination (R2) was 0.9994, which presents excellent linearity. They measured fructose at 14.75% to 52.44%, glucose at 8.19% to 42.63%, and sucrose at 0.10% to 21.12%. Kara et al. (2019) investigated the phenolic profile and total phenolic content of oak honey. Methanolic extracts of the honey were enriched via liquid-liquid extraction using diethyl ether and ethyl acetate, followed by analysis with reverse-phase high-performance liquid chromatography with ultraviolet detection (RP-HPLC-UV) using an acetonitrile-water mobile phase. Calibration curves, constructed with 19 phenolic standards, quantified total phenolic contents ranging from 54 to 88 mg gallic acid equivalent (GAE) per 100 g of sample.

The study concluded that oak honey possesses significant apitherapeutic potential due to its diverse polyphenol composition. Phenolics contribute to honey's antioxidant activity, beneficial for human health. Oak honey's high phenolic content suggests potential therapeutic applications. Phenolic profile can serve as a marker for oak honey authenticity and quality.

Hernanz et al. (2023) analyzed 58 Spanish oak honeydew honey samples using ultra-high-performance liquid chromatography-mass spectrometry (UHPLC-MS), identifying 23 phenolic compounds. Salicylic acid (mean value = 24.31 mg/100 g) and five other compounds are potential markers—p-hydroxybenzoic acid, p-coumaric acid, naringenin, syringic acid, and galangin. These compounds distinguish oak honey from European honeydew honey and samples from other regions. There are correlations between phenolic compounds, antioxidant activity, and color parameters as measured by CIELAB—a color measurement system (Lab* color space) used to quantify color attributes in samples, such as lightness (L*), red-green axis (a*), and yellow-blue axis (b*). Wang et al. (2023a, b) developed a selective, sensitive, and universal HPLC-MS/MS method for analyzing endogenous phenolic compounds in honey. A two-stage extraction using ethyl acetate and 0.01 M hydrochloric acid was employed. Clean-up was performed using hydrophilic-lipophilic balanced (HLB) columns. Linearity (R2) was recorded at greater than 0.99, limits of detection (LOD) at 0.2 to 1.0 μg/kg, limits of quantitation (LOQ) at 0.6 to 3.0 μg/kg, recoveries at 71.5 to 115.2%, and Relative Standard Deviations (RSD) of less than 10%. In Xinjiang honey samples, nine compounds were identified and quantified, such as 4-Hydroxybenzoic acid, rutin, vanillic acid, p-coumaric acid, syringic acid, caffeic acid, ferulic acid, and gallic acid. Nasab et al. (2020) investigated the utility of Fourier transform near-infrared spectroscopy (FT-NIR) and high-performance liquid chromatography with diode-array detection (HPLC-DAD), combined with multivariate data analysis, classifying honey samples based on their botanical origin. Seventy honey samples from seven different varieties were analyzed, and data were collected using FT-NIR and HPLC-DAD. Partial least squares-discriminant analysis (PLS-DA) and data fusion techniques were applied to the collected data. The results obtained from PLS-DA were promising, demonstrating comparable or superior performance to individual matrix analysis. These findings confirm the feasibility of this methodology for accurate honey classification.

2.1.4 Gas Chromatography-Mass Spectrometry (GC-MS)

GC-MS combines gas chromatography (separation of compounds) with mass spectrometry (identification of compounds). Applications in honey analysis include identification and quantification of volatile compounds contributing to honey's aroma and flavor, detecting adulteration with sugar syrups or other sweeteners, monitoring presence of contaminants, such as pesticides or heavy metals, and identifying markers for specific plant species (Sotiropoulou et al., 2021). GC-MS parameters are usually DB-5 or DB-35 ms columns, steady temperature increase (e.g., 50–250 °C), electron ionization (EI) or chemical ionization (CI), and mass range was 50–500 m/z. Commonly detected compounds in honey were unpredictable—terpenes (e.g., limonene, α-pinene), aldehydes (e.g., benzaldehyde, hexanal), esters (e.g., ethyl acetate, methyl benzoate), and alcohols (e.g., phenylethyl alcohol), sugar profiles (including fructose, glucose, and sucrose), contaminants (pesticides like chlorpyrifos, carbaryl), and heavy metals (e.g., lead, mercury). There are different types of analysis: GC-MS-based metabolomics for honey's chemical fingerprinting (metabolomics), GC-MS combined with chemometrics for adulteration detection (authenticity), and GC-MS analysis of volatile compounds for honey's botanical origin identification (botanical origin). Kozłowicz et al. (2020) conducted a comprehensive investigation into the chemical and structural characterization of powders derived from chestnut flower honey (HFCh) and honey with inca berry (HBlu). Utilizing a spray-drying technique at temperatures of 80 to 50 °C and maltodextrin (DE 15) as a coating agent, they employed gas chromatography–mass spectrometry (GC-MS) to analyze phenolic compounds and sugars. Additionally, scanning electron microscopy (SEM) was used to examine morphology, Fourier-transform infrared (FTIR) spectroscopy for assessing molecular structure, and x-ray diffraction (XRD) to evaluate crystallinity. They found simple sugars ranged from 72.4 to 100 g (HFCh), 90.2 to 100 g (HBlu), 41.3 to 100 g (glucose-dominant HFCh), and 51.6 to 100 g (HBlu). Phenolic compounds were found, including 3-acid and ferulic acid in HFCh (high-fructose chestnut honey), as well as m-coumaric acid, benzoic acid, and cinnamic acid in HBlu (blueberry honey).

Sun et al. (2021) investigated the molecular mechanism of mature honey (MH) formation using metabolomics. They used UPLC-QTOF-MS-based metabolomics analysis, comparison of stomach honey (SH), immature honey (IMH), and MH samples, and verification of findings using GC-MS. Study provides a molecular basis for assessing honey maturity. Decenedioic acid can serve as a marker for MH authentication, revealing the importance of fatty acids in honey maturation. Bazaid et al. (2022) investigated the phytochemical composition and biological activity of medical-grade manuka honey (MH). GC-MS analysis was utilized to identify and quantify a diverse range of bioactive compounds, including fatty acids, phenols, and esters. Zhou et al. (2021) identified volatile compounds and aroma features specific to five unifloral honeys (Apocynum, Jujube, Amorpha, Elaeagnus, and Lavender) for authentication. There were 92 volatile compounds identified in honey samples using HS-SPME/GC-MS, and 51 volatiles correlated with floral origin; 27 contributed to aroma. El-Wahed et al. (2023) undertook a multianalytical approach to characterize the chemical composition, physicochemical properties, bioactive compounds, and sensory attributes of Sidr honey from Egypt and Saudi Arabia. High-performance liquid chromatography (HPLC), mass spectrometry (LC-MS/MS), nuclear magnetic resonance ($HNMR^1$), and solid-phase micro-extraction coupled with gas chromatography (SPME-GC-MS) were utilized to comprehensively analyze the honey samples. Moisture content measures were 18.03% (Saudi Arabia) and 19.03% (Egypt). Electrical conductivity, 1.18 mS/cm (Saudi Arabia), and 1.16 mS/cm (Egypt). The analysis recorded pH at 4.87 (Saudi Arabia) and 5.10 (Egypt), and free acidity was found at 37.50 meq/kg (Saudi Arabia) and 36.50 meq/kg (Egypt). Sugar profiling analysis reported glucose at 22.51% (Saudi Arabia) and 26.62% (Egypt), fructose at 40.33% (Saudi Arabia) and 35.28% (Egypt), sucrose at 8.94% (Saudi Arabia) and 8.87% (Egypt), and maltose at 8.22% (Saudi Arabia) and 8.13% (Egypt). The study aimed to characterize the chemical composition and physicochemical properties of Sidr honey from Egypt and Saudi Arabia. LC-MS/MS analysis was utilized to identify and quantify a diverse range of bioactive compounds, including flavonoids, sugars, vitamins, and nitrogenous compounds. GC-MS analysis was employed to identify volatile compounds such as aldehydes, ketones, phenols, acids, esters, anthraquinones, hydrocarbons, and nitrogenous compounds. The results of this study highlight the significant impact of geographical origin on the chemical composition and physicochemical properties of Sidr honey. The findings also demonstrate that the honey samples analyzed meet the quality standards established by the International Honey Commission (IHC) and the GCC Standardization Organization (GSO). Oymen et al. (2022) developed and validated a modified QuEChERS (quick, easy, cheap, effective, rugged, and safe) method for the multiresidue analysis of 346 pesticides in honey samples. The method involved an acetate-buffered acetonitrile extraction followed by dispersive solid-phase extraction (d-SPE) cleanup, using a combination of C18-E, PSA, and MgSO4 sorbents. The method was validated according to EU SANTE/2019/12682 guidelines. A total of 100 honey samples from Northern Cyprus were analyzed using the developed method. Four pesticide residues were detected: coumaphos, thiamethoxam, N-(2,4-dimethylphenyl) formamide, and piperonyl butoxide. Contamination with at least one pesticide residue was detected in 42 samples. Coumaphos and N-(2,4-dimethylphenyl) formamide were the most frequently detected residues, with two samples exceeding the maximum residue limit (MRL) for N-(2,4-dimethylphenyl)formamide. The study highlights the potential contamination of honey in Northern Cyprus with acaricides and veterinary drugs used in hive treatments. Peñalver et al. (2021) develop a method for detecting plastic additives and contaminants in honey using gas chromatography-mass spectrometry (GC-MS) and dispersive liquid-liquid microextraction (DLLME). In targeted analysis 15 contaminants were quantified, including styrene, phthalates, fatty acids, alkylphenols, bisphenol A; in untargeted analysis, additional contaminants were identified, including two phthalates, three esters, one aldehyde, one hydrocarbon, two alkyl phenol compounds. The proposed method is a valuable tool for monitoring plastic contaminants in honey, ensuring food safety and consumer health. The study highlights the potential risks associated with plastic packaging and the importance of regular testing. Kasiotis et al. (2023) identified chemical markers distinguishing orange blossom honey from Egypt, Italy, and Greece using UHPLC-HRMS metabolomics. Their study integrated non-targeted metabolomics using UHPLC-HRMS with complementary analytical techniques such as melissopalynological analysis, GC-MS, and HPLC-PDA-ESI/MS to characterize Mediterranean orange blossom honey. Samples were collected from regions with diverse aromatic medicinal plants. The analysis identified 40 potential chemical markers, revealing distinct geographical variations. Italian honey was characterized by a higher abundance of flavonoids, while Greek honey was rich in terpenoids, iridoids, and hesperidin. Egyptian honey exhibited unique markers such as suberic acid and fatty acid ester derivatives. Furthermore, new furan derivatives were identified in Greek honey using GC-MS. The study demonstrates the power of UHPLC-HRMS metabolomics in identifying chemical markers for Mediterranean orange blossom honey, emphasizing the influence of geographical origin, citrus crop varieties, and local flora on honey composition. Wang et al. (2023a, b) employed GC-MS and GC-MS/MSc to characterize the volatile aromatic profile of *Leucosceptrum canum* Smith honey (LCH) . Four key compounds contributing to LCH's unique flavor were identified: linalool (74.22%), 3-methyl-1-butanol (18.19%), benzeneacetaldehyde (1.31%), and lilac aldehyde B (2.78%). The study revealed significantly higher linalool content (0.18 mg/kg) in LCH compared to other honey types. This detailed aromatic profile highlights LCH's distinctive flavor and potential health benefits.

2.1.5 Nuclear Magnetic Resonance (NMR) Spectroscopy

Nuclear magnetic resonance (NMR) spectroscopy is a powerful analytical technique used to elucidate the structure, dynamics, reaction state, and chemical environment. NMR spectroscopy exploits the magnetic properties of atomic nuclei. Hydrogen-1 (^{1}H) and Carbon-13 (^{13}C) are commonly studied nuclei. When exposed to a strong magnetic field, these nuclei align and resonate at specific frequencies, emitting signals that provide valuable information. In honey analysis, NMR is used for authenticity verification, botanical origin determination, quality control, and sugar profile analysis of honey samples. Song et al. (2020) utilized ^{1}H nuclear magnetic resonance (NMR) spectroscopy in conjunction with chemometric techniques to detect and quantify the adulteration of high-value acacia honey with less expensive rape honey in the Chinese market. The 1-H NMR spectroscopy analysis categorized three distinct regions: aliphatic (0.00–3.00 ppm), carbohydrate (3.00–6.00 ppm), and aromatic (6.00–9.50 ppm).

The chemometric methods applied included canonical discriminant analysis (CDA), orthogonal projection to latent structures discriminant analysis (OPLS-DA), and partial least squares (PLS) modeling to classify and predict honey quality attributes. These analyses achieved the highest prediction precision (89.7%) using CDA (0.00–6.00 ppm spectral range) and determined that OPLS-DA effectively discriminated pure acacia honey from adulterated samples.

The PLS model achieved an R2 value of 0.9996 for estimating adulterant absorption. 1-H NMR spectroscopy, integrated with chemometric techniques, effectively detects and quantifies the adulteration of acacia honey with rape honey, ensuring quality control and authenticity. Guyon et al. (2020) formed a sample preparation protocol for analyzing honey using proton nuclear magnetic resonance (1H NMR). The method involves dissolving 0.5 g of honey in 5 ml of oxalate buffer (pH 4.3) and adding a lock solution (D2O) of 3-(trimethylsilyl) propionic acid D4 sodium salt. They revealed successful discrimination between seven floral origins—acacia, chestnut, scrub, colza, lavender, sunflower, and honeydew—using soft independent modeling of class analogy (SIMCA) data treatment. The results presented a 100% accurate classification using a test set and were corroborated by sensory and pollen analyses for marketable monofloral honeys.

This protocol provides a competent and highly accurate method for honey profiling and confirmation. Nunes et al. (2023) investigated polyfloral honey samples from southeastern (São Paulo) and southern (Santa Catarina) Brazil, utilizing quantitative nuclear magnetic resonance (qNMR) spectroscopy and chemometric techniques, including statistical analysis, heatmap generation, and principal component analysis (PCA). They identified and quantified 21 metabolites, which included carbohydrates, amino acids, organic acids, ketones, alcohols, esters, and alkaloids, successfully distinguishing the geographical and botanical origins of the samples. They detected *Hovenia dulcis* (Santa Catarina), *Citrus spp*. (São Paulo), and *Eucalyptus spp*. (both regions) botanical clusters.

The combination of qNMR spectroscopy and chemometrics faithfully profiled Brazilian polyfloral honey, enabling identification of geographical and botanical origins, and demonstrating the potential for honey authentication and quality control. Nunes et al. (2022) utilized ultraviolet-visible (UV-Vis) spectroscopy, near-infrared (NIR) spectroscopy, and nuclear magnetic resonance (NMR) spectroscopy to discriminate between floral honey samples from Santa Catarina (SC) and São Paulo (SP) in Brazil. This study offers an inclusive analytical approach for identifying the influential geographical origin of Brazilian floral honey, potentially detecting adulteration, and ensuring honey authenticity.

Cagliani et al. (2022) investigated the detection of honey adulteration using targeted and untargeted nuclear magnetic resonance (NMR) spectroscopy, statistical analysis (OPLS models), and calibration curves. This approach successfully detected and quantified the presence of inulin, corn/malt syrup, and invert sugar syrup adulterants in Italian chestnut, polyfloral, and acacia honey samples, estimating the adulteration percentage range at 10% to 40%. Zhou et al. (2022) developed a method to identify honey's botanical source using 1H nuclear magnetic resonance (1H NMR) spectroscopy and support vector machine (SVM) categorization. Spectral data were collected from 122 honey samples across four floral categories—vitex, rape, acacia, and sunflower—to characterize their chemical profiles.

Classification models were established using four integration intervals—full spectrum (δ 0.10–δ 9.50), fat zone (δ 0.10–δ 3.00), carbohydrate zone (δ 3.00–δ 6.00), and aromatic zone (δ 6.00–δ 9.50). The combination of 1H NMR spectroscopy and support vector machine (SVM) successfully identifies the botanical origin.

Consonni et al. (2019) differentiate the organic and conventional honeys by pairing 1H nuclear magnetic resonance (1H NMR) spectroscopy with chemometrics (Orthogonal Signal Correction, OSC). This study fills a knowledge gap in distinguishing organic and conventional honey, demonstrates the promise of 1H NMR spectroscopy and chemometrics combination in honey authentication, provides a method for ensuring honey quality and traceability, and supports the development of organic honey production and documentation. Hao et al. (2022) detected adulteration in acacia honey using alternating current (AC) impedance spectroscopy and 1H nuclear magnetic resonance (1H NMR) profiling. The 1H NMR profile reveals the adulterant identity and quantity. The combination of AC impedance spectroscopy with 1H NMR profiling offers a rapid, nondestructive method for verifying honey authenticity and effectively detecting adulteration in acacia honey.

Yong et al. (2022) investigated stingless bee honey (SBH) adulteration in the Malaysian market using untargeted 1H nuclear magnetic resonance (NMR) metabolomics, Principal Component Analysis (PCA), and discriminant analysis. SBH's attractive price leads to adulteration with extraordinary substances (vinegar, floral honey). They found that the 1H NMR fingerprints differentiate authentic SBH from adulterated samples down to 1% adulteration. 1H NMR metabolomics provides an efficient method for detecting SBH adulteration, improving consumer trust in SBH authenticity, and reducing economic losses due to adulteration. Schievano et al. (2020) used nuclear magnetic resonance (NMR) spectroscopy to determine the sugar profile of 46 European acacia honeys. The study identified minor saccharides (20) as a fingerprint for authenticity, and compared them with Chinese acacia honeys of similar botanical origin. The comparative analysis of European and Chinese acacia honeys revealed significant differences in sugar composition. The presence of specific sugar profiles in Chinese honey samples suggested potential adulteration with industrial sugar syrups. This practice involving the overfeeding of bee colonies with sugar syrups during the main nectar flow period was suspected to be the cause of the observed adulteration.

2.2 *Technological Advances*

2.2.1 Near-Infrared Spectroscopy (NIRS)

Near-infrared (NIR) spectroscopy offers a rapid, nondestructive method for the qualitative and semiquantitative analysis of various physical and chemical properties of materials, including honey. However, the quantitative analysis of sucrose in honey using NIR spectroscopy is limited to low levels of adulteration, typically up to 3% sucrose (Garcia-Alvarez et al., 2002). To address this limitation,

Aliaño-González et al. (2019) employed visible and near-infrared spectroscopy (Vis-NIRS) in conjunction with chemometric tools to detect and quantify adulteration in high-quality Granada honey. The study investigated adulteration levels ranging from 5% to 50% using inverted sugar, rice syrup, brown cane sugar, and fructose syrup as adulterants. The results demonstrated the effectiveness of Vis-NIRS in discriminating between unadulterated and adulterated honey samples, highlighting the potential of this technique for the detection of honey adulteration.

Raypah et al. (2022a, b) developed a rapid, nondestructive method combining near-infrared spectroscopy (NIRS) in the 700–1100 nm range, with chemometrics (PCA, HCA, PCA-LDA, PLSR). The study evaluated the impact of water (W) and apple cider vinegar (AC) adulteration on honey samples at adulteration levels ranging from 4.76%–50%. Use of PCA-LDA model was able to determine the complete distinction between pure, adulterated, and adulterant samples with 100% classification accuracy. NIRS combined with chemometrics effectively detects adulteration in stingless bee honey and quantifies adulteration levels.

Bisutti et al. (2019) utilized near-infrared spectroscopy (NIRS) coupled with chemometric techniques to classify 119 honey samples into four botanical origins—polyfloral, acacia, chestnut, and linden. Furthermore, the study sought to characterize the physical-chemical properties of these honey samples. NIRS combined with chemometrics effectively predicts botanical origin, authenticates acacia honeys, and shows potential for improving the classification of other varieties.

Escuredo et al. (2021) created prediction models utilizing portable near-infrared spectroscopy (MicroNIR) alongside multivariate data analysis techniques, including principal component analysis (PCA), partial least squares regression (PLSR), and linear discriminant analysis (LDA). This approach was employed to authenticate honeys from northwest Spain and to predict their physicochemical properties. The study achieved strong predictive capabilities for moisture, hydroxymethylfurfural (HMF), color (Pfund, L, b* coordinates), and flavonoid content (with RSQ values exceeding 0.75 and RPD values greater than 2.0). Additionally, acceptable predictions were obtained for electrical conductivity (EC), pH, and phenolic content (with RSQ values over 0.61 and RPD values above 1.5).

The LDA method successfully classified 88.1% of the honey samples according to their botanical origins, which included heather, chestnut, eucalyptus, blackberry, honeydew, and multifloral types. Portable MicroNIR Spectroscopy combined with chemometrics effectively authenticates honey and predicts physicochemical properties. Valinger et al. (2021) employed UV-VIS and NIR spectroscopy, coupled with partial least squares (PLS) and artificial neural network (ANN) modeling, to detect adulteration and quantify physical and chemical properties in pure acacia honey samples from Krapina-Zagorje County, Croatia, and adulterated samples containing 10–90% corn syrup. The study demonstrated the efficacy of these techniques in discriminating between pure and adulterated honey samples, as well as in quantifying key honey parameters.

Raypah et al. (2022a, b) utilized visible and near-infrared (Vis-NIR) absorbance spectroscopy, along with aquaphotomics analysis, to profile, classify, and quantify adulterated stingless bee honey (SBH). The study explored the potential of these techniques for the quality assessment and authenticity verification of SBH. The adulterants examined in this study included distilled water, apple cider vinegar, and high fructose syrup, with concentrations ranging from 10% to 90%. The researchers found that the Vis-NIR region (400–1100 nm) was effective for both discrimination and quantification of these adulterants. Vis-NIR spectroscopy combined with aqua photomics effectively authenticates SBH, detects adulteration, and quantifies contamination levels.

Bodor et al. (2023) evaluated NIRS (950–1650 nm) for detecting the addition of sugar syrup to Hungarian honeys (acacia, rape, forest, sunflower, linden) for high fructose to content ratios of sugar syrup, rice syrup, and homemade glucose-fructose syrup at adulteration levels of 3%, 5%, and 10% (weight to weight ratios). NIRS performance depends on syrup and honey types. In this study, NIRS effectively detects honey adulteration, especially involving rice syrup and F40 syrup.

Skaff et al. (2022) evaluated infrared spectroscopy coupled with chemometrics for detecting honey adulteration of six authentic Lebanese honeys and 18 intentionally adulterated samples (10%, 30%, 50%). Adulterants included glucose, fructose, sucrose, and high fructose corn syrup (HFCS). They successfully classified samples by adulterant sugars, and defined spectral bands attributed to specific adulterant sugars. The detection limit was around 10%. Mid-infrared (MIR) spectroscopy outperformed near-infrared spectroscopy (NIRS) in detecting and quantifying fructose and HFCS adulteration. Infrared spectroscopy combined with chemometrics effectively authenticates Lebanese honey and offers a quick, cost-efficient method. Calle et al. (2023) established a method that integrates visible near-infrared spectroscopy (Vis-NIRs) with machine learning (ML) algorithms to detect adulterants in honey and quantify the levels of adulteration in orange blossom and sunflower honey samples, with adulteration levels ranging from 5%–50%. Combining Vis-NIR spectroscopy with machine learning (ML) effectively enables automated quality control. Tan et al. (2021) collected 30 samples of stingless bee honey sourced from Malaysian beekeepers and suppliers and evaluated quality against Malaysian standards and implemented near-infrared (NIR) spectroscopy for adulteration detection. This study concluded that 1 of 30 samples failed moisture content and hydroxymethylfurfural (HMF) requirements, and only 5 of 30 samples met pH value requirements. This is effective for rapid, nondestructive adulteration detection. NIR spectroscopy combined with PCA and Logistic Regression effectively evaluates stingless bee honey quality and offers a rapid, nondestructive testing method. Benković et al. (2022) used NIR spectroscopy for adulteration detection and analysis of 15 acacia honey samples mixed with glucose syrup (10–90%). Principal component analysis (PCA) effectively distinguished between pure and adulterated honey and accurately identified adulterants using near-infrared (NIR) spectroscopy. When combined with machine learning, NIR spectroscopy effectively detects honey adulteration, quantifies adulteration levels, and analyzes physical and chemical properties. Zhu et al. (2022) analyzed acacia, red jujube, and rape honey using near-infrared (NIR) spectroscopy, extracted spectral features from preprocessed spectral reflectance curves, and developed machine learning models for honey variety classification and adulteration detection. In this study, honey variety classification accuracy was 95.83%, and adulteration detection accuracy was 97.92%. NIR spectroscopy combined with machine learning effectively classifies honey varieties, detects adulteration, and ensures honey quality and authenticity.

2.2.2 Raman Spectroscopy

Raman spectroscopy is a vibrational spectroscopic technique that involves the inelastic scattering of monochromatic light to provide information about the vibrational modes of molecules within a sample, including honey. This type of spectroscopy works via inelastic scattering of light by molecules and measurement of Raman shifts, or vibrational frequency changes. Applications in honey analysis include authentication and detection of adulteration, botanical origin identification, quality control (measurement of moisture and sugar content), and detection of additives and contaminants. Magdas et al. (2021) investigated Raman spectroscopy for geographical origin (Romania, France) and botanical origin for mono-varietal (unifloral) honeys. This method utilizes Soft Independent Modeling Class Analogy (SIMCA) and Machine Learning (ML) algorithms to analyze the data. Correlation between SIMCA and ML results are more distinct for honey variety discrimination and varietal composition than for geographical origin. Magdas and Berghian-Grosan (2023) developed new models for rapid botanical differentiation of honey varieties using Raman spectroscopy and machine learning algorithms. The key findings of this study were a correct prediction rate of > 81%, accurate detection of honey mixtures, and estimation of mixture component percentages. Hu et al. (2022) developed a method to detect low-concentration adulteration in Suichang native honey (China's National Geographical Indication Agricultural Product) using Raman spectroscopy and machine learning algorithms. This study utilized support vector machine (SVM), Probabilistic Neural Network (PNN), and convolutional neural network (CNN) algorithms for classification and prediction. They found accuracies of SVM at 100%, PNN at 100%, and CNN at 99.75%, which is an effective detection of low-concentration adulteration.

Molnar et al. (2020) developed a simple, eco-friendly sample preparation protocol, involving a 1:1 (w/v) honey-to-distilled water dilution to overcome crystallization and fluorescence in Raman spectroscopy for honey analysis. They obtained reliable and reproducible Raman spectra protocol, effectively eliminating crystallization and fluorescence interference, making it suitable for various honey types. Wu et al. (2020) integrated Raman spectroscopy with convolutional neural networks (CNN) and chemometric techniques to detect the adulteration of honey with high fructose corn syrup, rice syrup, maltose syrup, and blended syrup, as well as to quantify the levels of adulteration. This study achieved >97% accuracy in classifying single-syrup adulteration, and 94.79% accuracy in detecting multiple-syrup adulteration. The convolutional neural network (CNN) demonstrated superior performance compared to chemometric classification algorithms, while partial least squares Regression (PLS) effectively predicted the honey purity levels with $R2 > 0.98$ and root mean square error of prediction (RMSEP) < 3.50.

David and Magdas (2024) combined FT-Raman spectroscopy and partial least squares discriminant analysis (PLS-DA) to differentiate four types of honey botanical sources and identify two periods of harvesting years. The study achieved over 97% accuracy in identifying botanical origins and harvesting years of honey samples, with Raman spectral markers strongly correlated with their chemical composition. Anjos et al. (2021) conducted a comparative analysis utilizing Fourier transform infrared spectroscopy with attenuated total reflectance (FTIR-ATR) and Fourier transform Raman spectroscopy (FT-Raman) to assess the quality of *Lavandula spp.* (lavender) honey. The study focused on various quality parameters, including acidity, presence of reducing sugars, hydroxymethylfurfural (HMF), electrical conductivity, ash content, proline levels, diastase activity, apparent sucrose, total flavonoids, and total phenolic contents. Both FTIR-ATR and FT-Raman spectroscopy proved effective for evaluating the honey's quality, and FT-Raman yielded more precise models.

Tahir et al. (2020) combined near-infrared spectroscopy (NIR) and Raman spectroscopy (RS) to identify 16 phenolic compounds in 100 honey samples and develop prediction models for *p*-hydroxybenzoic acid (major compound). RS outperformed NIR in prediction accuracy (RS: $R^2 = 0.9500$, RMSEP = 6.83 and NIR: $R^2 = 0.8147$, RMSEP = 13.80). Data fusion methodologies improved NIR models, low-level fusion: $R^2 = 0.9553$, RMSEP = 6.59, and mid-level fusion: $R^2 = 0.9563$, RMSEP = 7.95 were found. NIR, RS, and data fusion with partial least squares (PLS) effectively quantify *p*-hydroxybenzoic acid in honey and offer alternative methods to HPLC analysis. Huzortey et al. (2022) used 532-nm laser-induced Raman spectroscopy to analyze Iranian honey varieties, develop classification models, and identify critical wavelengths for honey characterization. The study identified unique Raman spectral profiles for different honey varieties, developed classification and prediction models, and established key wavelengths for honey characterization.

Cherigui et al. (2023) employed Raman spectroscopy in conjunction with principal component analysis (PCA) and hierarchical cluster analysis (HCA) to identify pure Algerian honey and assess adulteration. For this purpose, 16 authentic honey samples from west and southwest Algeria were collected, and 72 adulterated samples—mixed with fructose, glucose, sucrose, syrup—at adulteration ratios of 5% to 50% were established. The study successfully identified pure and adulterated honey samples. Węglińska et al. (2020) used vibrational spectroscopy (Raman, ATR, DR) to determine macronutrient levels (protein, reducing sugars, and fat), quantify total polyphenolic (TPs) content, measure antioxidant activity, and analyze pH and color parameters. They attained accurate predictions for macronutrients (RSEP: 1.7–3.6%, and TPs content (RSEP: 3.1–3.5%).

Ikedi (2022) used laser Raman spectroscopy to detect honey adulteration with molasses and develop predictive models for authenticity verification. Raman bands characteristic of authentic honey and molasses, and high classification accuracies were found. Song et al. (2021) introduced a novel surface-enhanced Raman scattering (SERS) substrate, Ag@WS_2QD, composed of monodispersed tungsten disulfide quantum dots modified onto silver nanospheres. This substrate exhibited uniform and stable SERS signals within 2 minutes and demonstrated a fourfold increase in analytical enhancement compared to unmodified colloidal silver nanoparticles. The method facilitated rapid *mixing and reading* detection while improving charge transfer and chemical extraction efficiency. SERS was used to detect thiram (toxic agrochemical) in honey and four types of juice.

2.2.3 Electronic Nose (E-Nose)

Electronic nose (E-nose) and electronic tongue (E-tongue) are advanced technologies designed to replicate human sensory perception, enabling the analysis and evaluation of food quality. These systems can identify the distinct characteristics of honey, including its floral origin, overall quality, and potential adulteration. An electronic nose (E-nose) is a device that imitates the human sense of smell using sensors and machine-learning algorithms to sense and identify specific odors or volatile organic compounds (VOCs). This technology can perceive adulteration, identify honey adulterated with sugar syrups or other sweeteners, determine floral origin, monitor quality, and authenticate geographical origin (Tan et al., 2021).

Gonçalves et al. (2023) produced an E-nose using electrodes covered with ionogel films (ionic liquid + gelatin + Fe3O4 nanoparticles) and successfully discriminated between authentic and adulterated honey samples. Chen et al. (2023) examined the variation in the flavor profile of royal jelly (RJ) during storage, utilizing advanced analytical techniques. They concluded that storage at 25°C (77°F) significantly altered the water-soluble protein content, moisture, and whiteness of royal jelly (RJ). E-nose and E-tongue analyses revealed distinct flavor variations—37 volatile compounds were identified, with octanoic acid being the most abundant (47.61% at 25°C after 21 days). There are seven key compounds that contribute to RJ's characteristic flavor, such as 5-butyldihydro, 2(3H)-furanone, trans-β-ocimene, 2-heptanone, 2-nonen-4-one, 2-nonanone, 2-octenoic acid (E), and methyl benzoate. Borgonovo et al. (2021) used an E-nose and chemical analyses to determine the botanical and geographical provenance of honey samples of Italian sainfoin (*Onobrychis viciifolia*) and eucalyptus (*Eucalyptus camaldulensis*).

A combination of ten metal oxide semiconductor sensors (MOS) and headspace solid phase microextraction-gas chromatography-mass spectrometry (HS-SPME-GC-MS) was used to analyze the aromatic profiles and volatile species present in the samples. E-nose and chemical analysis combined effectively to classify and discriminate between honey types. Zaki Dizaji et al. (2021) developed an E-nose device to rapidly assess sugarcane syrup quality, specifically sucrose purity, in the agroindustry. The syrup samples analyzed in this study were derived from four sugarcane varieties: CP57, CP69, IRC99–02, and CP48.

Chemical tests were conducted to assess purity (PTY) and refined sugar (RS) content. They used an E-nose with eight sensors to detect odor compounds. E-Nose sensors MQ3, MQ5, and MQ9 responded powerfully to PTY. Sipos et al. (2020) investigated the efficiency of various methods for classifying bee pollen and predicting sensory and colorimetric attributes. The study showed the effectiveness of a multisensory approach—E-nose and E-tongue—combined with NIR spectroscopy for classifying bee pollen and estimating sensory and colorimetric attributes, offering a precious tool for quality control and authentication in the food industry.

Viejo et al. (2020) developed a transportable, economical E-nose with nine gas sensors to detect volatile compounds in beer coupled with artificial neural network (ANN) models for prediction for accurate aroma detection. The approach surpassed traditional sensory methods by integrating gas chromatography-mass spectrometry (GC-MS) for precise volatile analysis and validation by 12 trained panelists. This study investigated the effectiveness of combining E-nose technology with machine learning for precise beer aroma prediction, offering a valuable tool for the industry.

Szczurek et al. (2020) developed a method to review honeybee colony invasion by *Varroa destructor* using a gas sensor array to analyze beehive air volatiles in conjunction with machine learning for classification. The study found that gas sensor responses to beehive air volatiles fluctuate over 24 hours, with the accuracy of *Varroa destructor* infestation assessment (*low, medium, high*) varying by measurement duration and optimal detection times differing by infestation category. The Szczurek study highlights the impending of gas sensor arrays for detecting *Varroa destructor* infestation in honeybee colonies and underscores the significance of timing in measurement data collection.

Rasekh et al. (2021) produced an E-nose system, MAU-9, to authenticate plant-derived products by analyzing volatile essential oils. The E-nose is equipped with nine metal oxide semiconductor (MOS) gas sensors, and chemometric data analysis was conducted using principal component analysis (PCA), linear discriminant analysis (LDA), quadratic discriminant analysis (QDA), and support vector machine (SVM) methods. Classifications were 100% correct using PCA, LDA, QDA, and SVM. The Rasekh E-nose successfully differentiated between essential oils from herbal and fruit plant sources, and found unique sensor array responses to distinct volatile organic compounds (VOCs). This study concluded the effectiveness of the MAU-9 e-nose system in authenticating plant-derived products, offering a priceless tool for quality control and regulatory fulfillment.

Rasekh and Karami (2021) developed an E-nose system by means of nine metal oxide semiconductor (MOS) sensors to categorize natural and industrial fruit juices. They identified six of the nine sensors that showed effective discrimination power, and employed artificial neural networks (ANN) to analyze sensor response patterns. Principal component regression (PCR) was used to evaluate sensor presentation, allowing the use of a small number of effective sensors, thereby reducing system cost while increasing accuracy. This method offers fast, nondestructive quality control for fruit juice and improved classification accuracy. María et al. (2023) used a low-cost E-nose prototype to characterize Iberian honeys based on floral origin. They distinguish honey types from local beekeepers in Madrid, Spain, and evaluate E-nose measurements with physicochemical and pollen analyses. This study confirmed the potential of E-nose technology for characterizing honey based on floral origin, which introduces a helpful tool for quality control and authentication in the honey industry.

2.2.4 Artificial Intelligence (AI) and Machine Learning (ML)

Artificial intelligence (AI) and machine learning (ML) are essential in sensing applications, especially for pattern recognition and predictive analysis. Chenchouni and Laallam (2024) are working on self-organizing maps (SOMs) and principal component analysis (PCA) to analyze honey quality in Algeria. They used 62 honey samples and measured 8 parameters—water content 15.14%, electrical conductivity 0.5 μS/cm, pH 4.20, total sugar content 83%, content of reducing sugar 63.83%, proline concentration 382.7 mg/kg, hydroxymethylfurfural level 77.4 mg/kg, and pollen grain density 4.56×10^5 grains/10 g. They offer advanced food quality evaluation through data-driven techniques and supply insights into honey quality variations across North Africa.

Mateo et al. (2021) compared machine learning (ML) algorithms to classify seven types of Spanish unifloral kinds of honey—rosemary, citrus, lavender, sunflower, eucalyptus, heather, and forest honeydew—based on physicochemical parameters. They concluded that penalized discriminant analysis (PDA) achieved the highest accuracy (92.5%) for assessing beehive or food quality, while artificial neural networks (ANN), extremely randomized trees (ET), random forest (RF), and extreme gradient boosting trees (XGBoost) also demonstrated strong performance. The strategy efficiently classified unifloral honey using ML algorithms, reduced the need for expert personnel in palynological analysis, and provided potential.

Bernardes et al. (2022) conducted a systematic review using AI-aided meta-analysis, finding that combined stressors (e.g. pesticides, habitat loss) drive pollinator decline. Toxicological studies overly focus on honey bees (*Apis mellifera*), neglecting other pollinators—bumble bees, solitary bees, and stingless bees—as well as the significant risks that agrochemicals pose to pollinators, particularly *Hymenoptera*, thereby effecting pollinator decline. They estimate that 70% of studies have focused on honey bees.

The limited research on other bee species (bumble bees, solitary bees, stingless bees) amid increased publication trends on pollinator decline and agrochemicals is a concern. Liu et al. (2022) used machine learning and high-throughput sequencing (HTS)–based metabarcoding to determine the geographical origins of 28 honey samples. The study concluded that machine learning accurately identified honey's geographical origin with a >99% success rate and biological components in honey provided distinctive traits for geographic tracing.

Ahmed (2024) developed a machine learning-based approach using hyperspectral imaging to detect honey adulteration, achieving 98% classification accuracy in detecting adulterated honey, effective identification of botanical origins using feature reduction techniques and improved performance through hyperparameter tuning.

Machuca et al. (2022) developed a system combining a confocal optical microscope with a tunable optical filter and artificial intelligence networks to identify honey samples artificially adulterated with syrups. His article discusses a study on using hyperspectral microscopy technology and machine learning tools to detect adulteration in Guindo Santo and Quillay honey, which are endemic to Chile. The study aims to provide a method for certifying the origin and authenticity of Chilean honey, supporting local beekeepers, and enhancing competitiveness in the global market.

Karaboga et al. (2020) reviews the integration of the artificial bee colony (ABC) algorithm, a nature-inspired optimization technique, with machine learning (ML) algorithms to enhance its performance. The ABC algorithm mimics honeybees' intelligent foraging behavior. The study was conducted to provide valuable insights for researchers working with ML and nature-inspired optimization (NIO) algorithms, predominantly ABC. Othman et al. (2023) highlights the effective incorporation of artificial intelligence (AI) techniques with sensing devices for detecting food adulteration and agricultural product defects. They confirmed that AI sensor integration achieves 81.2–100% accuracy in adulteration and defect detection, whereas case-specific AI techniques improve high-dimensional dataset understanding, classification, and prediction. The study demonstrates AI's potential in improving food and agriculture quality assessment, emphasizing the need for continued innovation and validation.

2.3 *Biological Methods*

2.3.1 Pollen Analysis

Pollen analysis is a critical technique for determining the botanical origin, authenticity, and quality of honey. It discovers honey's botanical origin, authenticity verification, detects added sugars or artificial sweeteners, determines geographical origin, and ensures fulfillment with labeling claims (Ulberth, 2016). Özler (2018) analyzed pollen in 19 honey samples from Konya, Karaman, and Ulukışla regions in Turkey. They recognized 55 plant taxa, 21 families, 30 genera, one tribe, and three species, along with multifloral honey samples with secondary and minor pollen groups. They predict dominant pollen from Fabaceae (two samples), *Scrophulariaceae* (samples), *Helianthus annuus* (one sample), and *Zea mays* (one sample), as well as secondary pollen from *Amaranthaceae*, *Brassicaceae*, *Fabaceae*, *Poaceae*, *Rosaceae*, and *Scrophulariaceae*.

Sniderman et al. (2018) used 173 unblended honey samples from southern Australia for pollen analysis using microscopy to verify the geographic origin of Australian honey. They demonstrate Eucalyptus (*Myrtaceae*) dominates Australian honey pollen, and *Corymbia/Angophora*, *Leptospermeae*, *Brassicaceae*, *Echium*, *Macadamia*, and *Acacia* are significant components. This study demonstrates the potential of pollen analysis for ensuring the authenticity and quality of Australian honey.

Rašić et al. (2018) studied the pollen content of honey samples from northeastern Croatia. They analyzed 60 honey samples from different parts of northeastern Croatia, identified 84 different pollen grains belonging to 40 families. The assortment of pollen types indicates the rich flora of northeastern Croatia. The supremacy of certain pollen types (*Brassicaceae* and *Robinia pseudacacia*) suggests the prevalence of specific plant species in the region. Unifloral honeys from black locust, rapeseed, and lime tree can be marketed for their exclusive characteristics and potential health benefits. This research contributes to information in the field of melissopalynology (the study of pollen in honey) and highlights the magnitude of pollen analysis in determining the origin, quality, and characteristics of honey.

Şik et al. (2017) researched pollen analysis of honey samples from Ardahan in northeast Anatolia. Eleven honey samples were analyzed from different localities in Ardahan. 23 different pollen taxa from 13 families were recognized. They found dominant pollen on *Astragalus spp.* (14,019 grains), *Apiaceae* (7050 grains), and *Brassicaceae* (6379 grains). All 11 honey samples are multifloral, representing a varied flora in the region. The high taxon richness in some samples (up to 23 taxa) suggests a diverse and complex plant community. *Astragalus spp.*, *Apiaceae*, and *Brassicaceae* are the primary nectar sources for bees in the region. The presence of *Fabaceae* and *Lamiaceae* pollen in inconsequential quantities highlights the secondary importance of these plant families. This study contributes important information to the field of melissopalynology and provides insights into the botanical origin and diversity of honey from Ardahan, Northeast Anatolia.

El-Sofany et al. (2020) conducted a melissopalynological analysis of seven types of Egyptian honey to authenticate their botanical origin and labeling. The study aimed to identify the pollen sources utilized by honey bees in Egypt, providing valuable insights into the floral diversity and foraging behavior of bees in the region. They discovered pollen from 20 plant taxa, belonging to 14 families, botanical origin and labeling confirmed for clover honey, sidr honey, citrus honey, and banana honey. Mislabeling is a significant issue in the Egyptian honey market. Pollen analysis is an effective tool for authenticating honey origin and labeling. This study highlights the significance of pollen analysis in ensuring the authenticity and quality of honey products, as well as its potential to address mislabeling problems in the honey market.

Alkan (2020) worked on 16 chestnut honey samples from 12 localities in 2018 from Samsun, Sinop, Kastamonu, and Giresun in the Black Sea Region of Turkey in an elevation range of 10 to 859 meters. The study confirms the Black Sea Region's potential for producing high-quality, unifloral chestnut honey. These findings maintain the region's reputation for chestnut honey production. Microscopical pollen analysis proves efficient in verifying the botanical origin and quality of honey. This study proved the importance of pollen analysis in verifying the authenticity and quality of chestnut honey from the Black Sea Region of Turkey, supporting the region's standing as a significant producer of high-quality honey.

2.3.2 DNA Analysis

DNA analysis through polymerase chain reaction (PCR) is a powerful technique for detecting both plant and animal DNA in honey, enabling the identification of specific plant species and revealing region-specific plant DNA as well as animal-derived compounds. Mohamadzade Namin et al. (2022) presented a PCR-based approach to determine the entomological origin of honey derived from three honeybee species—*Apis cerana* (ACH), *Apis dorsata* (ADH), and *Apis mellifera* (AMH). The researchers designed three species-specific primers targeting the NADH dehydrogenase 2 (ND2) region of the mitochondrial DNA of honeybees. Their results indicated that each primer set effectively amplified the ND2 region from the respective honeybee DNA, with distinct amplicon sizes for each species: 224 bp for *Apis cerana*, 302 bp for *Apis dorsata*, and 377 bp for *Apis mellifera*. The method detects 1% AMH in combination with ACH or ADH. Real-time PCR and melting curve analysis can differentiate honey origins. This study provides a valuable tool for detecting honey adulteration and ensuring authenticity, addressing concerns about mislabeling and mixing.

Wirta et al. (2021) demonstrate the potential of DNA-based methods (metabarcoding and metagenomics) for determining the geographical origin of honey. DNA-based methods do better than traditional morphological pollen identification. Plant and fungal DNA provide clearer distinctions between countries. Joint species distribution modeling efficiently links taxonomic content to its country of origin. The study affirms the ability of DNA-based methods for authentication and origin tracing in food products, particularly honey.

Indahsari and Purwanto (2024) used DNA analysis to detect honey fraud in the market. They identified the entomological origin of honey claimed as forest honey and stingless bee honey and detected potential honey fraud using 16S rRNA gene amplicon analysis. They tested 60 honey samples (30 forest honey, 30 stingless bee honey), isolated DNA from honey samples, amplified 16S rRNA gene using PCR, and sequenced and analyzed the amplicons. The study showed the effectiveness of 16S rRNA gene sequencing, enabling authentication of its botanical or geographical origin. DNA analysis can resourcefully detect honey fraud, ensuring authenticity and protecting consumers. This method can be functional to monitor honey quality in the Indonesian market.

Honrado et al. (2022) studied a novel, modernized method for authenticating honey's entomological origin using real-time PCR together with high-resolution melting (HRM) analysis. They tested 44 commercial honeys from various countries, investigating their test results with the declared geographical origin and recommended mixture of honeys from different mitotypes in Slovenian samples. This study concluded the potential of real-time PCR-HRM analysis for rapid and accurate honey authentication, benefiting consumers, beekeepers, and the honey industry.

Chavan et al. (2022) concluded with a breakthrough in honey authentication, overcoming the restrictions of traditional pollen microscopy. The study successfully isolated pollen-free DNA from filtered honey samples using three adsorbents: anti-dsDNA antibodies coupled to magnetic microspheres, an anion-exchange adsorbent, and a ceramic hydroxyapatite. The study demonstrated that an anion-exchange adsorbent efficiently captured trace pollen-free DNA, enabling precise identification of the botanical or geographical origin. A metagenomic approach utilizing next-generation sequencing (NGS) of the internal transcribed spacer 2 (ITS2) region was employed to identify the botanical origins of honey samples. This inventive approach revolutionizes honey authentication, ensuring authenticity and reliability in the universal honey market.

Regina and Gallo (2024) worked on DNA-based methods for authenticating chestnut monofloral honey in Calabria, South Italy. DNA barcoding, PCR-RFLP, and LAMP (loop-mediated isothermal amplification) approaches effectively identified botanical composition. All methods correctly identified *Castanea sativa* (chestnut) as the dominant plant species. This study shows the potential of DNA-based methods, predominantly LAMP, for ensuring the authenticity and quality of honey, promoting consumer reliance, and protecting the integrity of the food supply chain.

2.3.3 Enzyme-Linked Immunosorbent Assay (ELISA)

Enzyme-linked immunosorbent assay (ELISA) is a sensitive and specific immunoassay technique widely utilized for the detection and quantification of various analytes in honey. ELISA employs antibodies that bind to target molecules, enabling the identification of botanical origin (e.g., pollen, nectar), adulterants (e.g., added sugars, syrups), contaminants (e.g., pesticides, heavy metals), and allergens (e.g., bee proteins). The four primary types of ELISA for honey analysis are direct ELISA (D-ELISA), indirect ELISA (I-ELISA), sandwich ELISA, and competitive ELISA.

Hendrickson et al. (2024) showed a highly sensitive immunoenzyme assay (EIA) for detecting flumequine (FLU), a fluoroquinolone antibiotic, in honey and developed an indirect competitive EIA with colorimetric detection. This study concluded the potential of EIA for detecting antibiotic residues in honey, sustaining the honey industry's promise to quality and safety.

Al-Eisa et al. (2023) demonstrated that nigella, moringa, sidr, and pumpkin honey varieties enhance immune response and survival rates in mice with invasive aspergillosis, a fungal infection in individuals with weakened immune systems. Honey demonstrates potential as a natural immunostimulator, enhancing natural immune responses and survival rates against invasive aspergillosis. The study concluded that honey may be used as a prophylactic or beneficial agent for various diseases.

Wu et al. (2020) developed and validated two immunoanalytical methods, direct competitive enzyme-linked immunosorbent assay (dcELISA) and nanoparticle immunochromatographic strip (Immunostrip), for the detection of chloramphenicol, a prohibited antibiotic, in food samples such as honey and milk. Ten honey samples and six milk samples were analyzed using both methods, and all samples tested negative for chloramphenicol (CAP). The developed dcELISA and Immunostrip methods demonstrated high sensitivity and appropriateness for CAP detection in food products, making them valuable tools for ensuring food safety and compliance with regulatory standards.

In a related study, Zhu et al. (2024) focused on detecting thiacloprid residues in environmental and food samples. Thiacloprid, a neonicotinoid insecticide, raises concerns due to its toxicity to the environment and nontarget organisms. Although its use is banned in several countries, it continues to be detected in various samples.

The study reported recovery rates of 87.50–116.11% in honey and medicinal herbs (*Crataegi fructus* and *Citri reticulatae pericarpium*). The multi-immunogen strategy proved effective in preparing sensitive and specific monoclonal antibodies. The established ELISA shows promising potential for detecting thiacloprid residues in complex matrices, contributing to the advancement of detection methods essential for safeguarding human health and environmental safety.

Li et al. (2021) developed the honey antibiotic detection system (HADS), demonstrating a critical concern in food safety, providing a rapid, sensitive, and user-friendly solution for detecting antibiotic residues in honey. HADS ensures honey quality, protects consumers from antibiotic residue exposure, and enables honey producers and regulators to monitor antibiotic residues, while adhering to international standards. It reduces the danger of honey rejection or recall due to antibiotic contamination and helps prevent antibiotic pollution in the environment, thereby mitigating antimicrobial resistance development.

2.4 *Isotope and Elemental Analysis*

2.4.1 Stable Isotope Ratio Analysis (SIRA) and Inductively Coupled Plasma Mass Spectrometry (ICP-MS)

IRMS deals with the natural variations in stable isotopic ratios of elements like C, N, O, H, and S in honey, providing an exclusive fingerprint for authentication and adulteration recognition. Gas chromatography-combustion-IRMS (GC-C-IRMS) measures δ13C, δ15N, and δ18O in honey components, liquid chromatography-IRMS (LC-IRMS) analyzes δ13C and δ15N in sugars and other compounds, and elemental analyzer-IRMS (EA-IRMS) determines δ13C, δ15N, δ18O, and δ2H in bulk honey. The isotopic ratios used are δ13C, representing C3/C4 plant origin. The value δ15N is the nitrogen isotopic ratio, helpful for nitrogen source, while δ18O is the oxygen isotopic ratio, reflecting climate and geographical origin. The value δ2H is the hydrogen isotopic ratio, demonstrating botanical and geographical origin. And finally, δ34S is the sulfur isotopic ratio, which identifies the sulfur source. They are used for authenticity verification, adulteration detection, quality control, origin tracing, and food. By using IRMS technology, the honey industry can ensure authenticity, detect adulteration, and verify quality, ultimately protecting consumers and maintaining the integrity of this valuable food product.

Bonini et al. (2024) examined the impact of botanical origin (monofloral and polyfloral) on stable isotope ratios within the same geographic region. Their study reviewed the effectiveness of stable isotope ratios in distinguishing honeys from the boundary regions of Concordia (Argentina) and Salto (Uruguay), as well as from Canelones and Maldonado. Analyzing 82 monofloral and polyfloral honey samples, they found that the botanical origin did not significantly influence stable isotope ratios, except for δ15N. They concluded that stable isotope ratio analysis effectively discriminates citrus honey from different geographical origins in Argentina and

Uruguay. Stable isotope analysis of sulfur in proteins and carbon in honey provides valuable information and is considered a key parameter. This study provides a valuable tool for authenticity verification and protection of citrus honey's geographical indication.

Bontempo et al. (2017) aimed to develop methods for authenticating the geographical and botanical origins of Italian honey. The researchers examined 265 honey samples from various botanical sources, including polyfloral, citrus, rhododendron, eucalyptus, acacia, chestnut, and honeydew. They identified specific ranges of variability associated with authentic Italian honey. This study advanced honey authenticity methods by analyzing the stable isotope ratios (SIRs) and mineral element content, ensuring regulatory compliance and safeguarding consumer interests.

Kropf et al. (2010) employed nitrogen stable isotopes to authenticate honey using a dataset of 271 samples from four distinct geographical regions in Slovenia. Similarly, Schellenberg et al. (2010) analyzed multiple stable isotope ratios of hydrogen, carbon, nitrogen, and sulfur in 516 authentic honey samples sourced from 20 European regions. Daniele et al. (2018) developed a method for differentiating honeys from seven botanical origins through organic acid analysis. Their findings indicated that by combining various organic acid concentrations with isotopic ratio values and applying statistical methods such as principal component analysis (PCA), it is feasible to distinguish honey samples based on their botanical origin.

Cotte et al. (2007) utilized site-specific natural isotopic fractionation assessed by NMR to evaluate its potential for characterizing honey and detecting adulteration. They concluded that this method was inadequate for identifying syrup adulteration at levels below 20%. ICP-MS is an influential tool for honey analysis, providing valuable insights into its composition, origin, and authenticity by analyzing elemental composition (e.g., Sr, Ba, Pb, Cd, Hg, As), isotopic ratios (e.g., Sr, Pb, Nd), and trace elements (e.g., Cu, Zn, Fe, Mn).

Voica et al. (2020) studies the elemental composition of four Romanian honey varieties (linden, acacia, rape, and sunflower) using inductively coupled plasma quadrupole mass spectrometry (ICP-Q-MS). It analyzed 18 elements—Al, As, Ba, Ca, Cd, Co, Cr, Cu, Mg, Mn, Na, Ni, K, Pb, Sr, Ti, V, and Zn. The most abundant minerals were K (248.70 mg/kg), Ca (59.97 mg/kg), Mg (20.54 mg/kg), and Na (11.92 mg/kg). Mineral content differed considerably among honey samples of diverse botanical origins. This study showed the potential of ICP-Q-MS and PCA for honey authentication and traceability, highlighting the magnitude of elemental composition in distinguishing honey varieties.

Gohar and Shakeel (2021) demonstrated the environmental impact on honey quality in Pakistan and Turkey by analyzing 11 essential and 17 toxic elements in 24 honey samples. They found noteworthy variations in elemental composition among honey samples from various geographical regions. The honeybee species (*Apis dorsata, Apis mellifera, Apis flora*) affected phosphorus (P) content, and geographical regions significantly impacted concentrations of essential elements—K, Mn, Fe, Cu, P, Mo, and toxic elements. Environmental factors, honeybee species, and botanical flora significantly affect honey's elemental composition, disturbing its quality and safety.

Luccas et al. (2022) optimized sample preparation techniques for the elemental analysis of Brazilian honey samples using triple quadrupole inductively coupled plasma mass spectrometry (TQ-ICP-MS). The study employed machine learning algorithms to classify the honey samples based on their entomological origin.

2.5 *Sensory Evaluation*

Visual inspection is a decisive aspect of sensory evaluation for honey, offering imperative insights into its color and texture, both of which are closely linked to its botanical origin and on the whole quality. While traditional methods have long been used, recent advancements in analytical techniques have enhanced the trustworthiness and objectivity of these assessments. The color of honey serves as a central visual indicator linked to its botanical source. For example, lighter honeys are classically derived from clover or acacia, whereas darker varieties are often connected with buckwheat or chestnut (Gün & Karaoğlu, 2024). Modern analytical techniques, including UV-VIS spectroscopy and chemometric approaches, have been effectively applied to detect adulteration in honey. By measuring the color spectrum, these methods can successfully distinguish between pure honey and those adulterated with sugar syrups (Yan et al., 2022). Texture, which includes factors such as thickness and consistency, plays a major role in the sensory perception of consumers. Trained evaluators visually judge these attributes, which can also point out possible adulteration (Damto et al., 2024). Advanced techniques, such as thermal imaging by convolutional neural networks, offer improved capabilities for detecting adulterants. By identifying temperature changes linked to sugar composition, these methods can offer important insights into honey's authenticity (Boulbarj et al., 2024).

2.6 *Other Methods for Detecting Honey Adulteration*

Except for the above highlighted analytical methods, there are other methods that are used in the detection of honey adulteration. Two of them are thermal analysis and ultraviolet spectrum analysis, also known as UV spectroscopy.

2.6.1 Thermal Analysis

The consideration of the thermal properties of honey, especially through differential scanning calorimetry (DSC), is key in identifying any adulteration in the product. This method demonstrates how the sugar composition and moisture content of honey affect its melting and crystallization points, showing that adulterated samples have different thermal properties than pure honey (Damto et al., 2024). The crystallization process is considerably influenced by the sugar profile of honey; adulterated samples often show altered thermal characteristics, indicating the presence of inauthentic ingredients (Gün & Karaoğlu, 2024). Recent findings showed that the introduction of sugar syrups alters the glass transition temperature (Tg) values of honey, serving as useful indicators of adulteration (Boulbarj et al., 2024).

2.6.2 Ultraviolet (UV) Spectroscopy

UV spectroscopy is proving to be an authoritative tool in the determination of contaminants in honey and is special for adulterants like HMF and the different sweeteners. This method employs the characteristic absorption spectra of honey samples to ensure food safety and verify authenticity, detecting adulteration in honey products. UV spectroscopy efficiently distinguishes pure honey from adulterated samples, as concluded by studies employing principal component analysis (PCA) and canonical discrimination analysis (CDA; Damto et al., 2024). Research showed that UV absorbance patterns can accurately identify specific contaminants, such as glucose syrup (Nunes et al., 2023).

Recent advancements in UV spectroscopy, combined with chemometric methods, provide highly sensitive and selective detection for verifying honey authenticity and quality. This combination of methods enables analyses to be done faster and at a cheaper price, thus facilitating adulteration detection in honey (Ameer et al., 2024). It has also made the testing process easier to undertake, since the use of multivariate analysis in combination with UV spectral data has also been proven to increase the reliability of the results (Nunes et al., 2023). However, as it has also been seen, UV spectroscopy has its problems, mainly in terms of sensitivity and data analysis. There is a continuous need for such research to improve these methods and to extend their application to food safety.

In general, UV spectroscopy is a useful method to ensure the preservation of any honey product (Rygula et al., 2024). Another one of the most important techniques used in detecting honey adulteration, apart from UV spectroscopy, is thermal analysis. Both melting and crystallization points combined with UV absorbance reveal significant characteristics of honey and its authenticity (Valinger et al., 2021). In the future, with new advanced technology becoming a part of people's daily lives, both thermal analysis and UV spectroscopy will be of particular significance in enhancing consumer protection and establishing trust in honey products (Shruti et al., 2024).

References

Abu-Reidah, I. M. (2020). Carbonated beverages. In *Trends in non-alcoholic beverages* (pp. 1–36).

Adbhai, A. R., Dewanjee, S., Patel, K. G., & Karmakar, N. (2022). Sugar beet molasses production and utilization. In *Sugar beet cultivation, management and processing* (pp. 885–904). Springer Nature.

Ahmed, E. (2024). Detection of honey adulteration using machine learning. *PLOS Digital Health, 3*(6), e0000536.

Ajala, E. O., Ighalo, J. O., Ajala, M. A., Adeniyi, A. G., & Ayanshola, A. M. (2021). Sugarcane bagasse: A biomass sufficiently applied for improving global energy, environment and economic sustainability. *Bioresources and Bioprocessing, 8*, 1–25.

Al-Eisa, R. A., Helal, M., Aljahani, A. H., Sami, R., Aljaadi, A. M., Alshahrani, M. Y., Banjer, H. J., & Algehainy, N. A. (2023). The immunological effects against invasive aspergillosis disease on inbred mice after the dietary intake of honey varieties with the determination of diastase and invertase enzyme activities. *Materials Express, 13*(6), 1088–1094.

Alghamdi, B. A., Alshumrani, E. S., Saeed, M. S. B., Rawas, G. M., Alharthi, N. T., Baeshen, M. N., Helmi, N. M., Alam, M. Z., & Suhail, M. (2020). Analysis of sugar composition and pesticides using HPLC and GC–MS techniques in honey samples collected from Saudi Arabian markets. *Saudi Journal of Biological Sciences, 27*(12), 3720–3726.

Aliaño-González, M. J., Ferreiro-González, M., Espada-Bellido, E., Palma, M., & Barbero, G. F. (2019). A screening method based on visible-NIR spectroscopy for the identification and quantification of different adulterants in high-quality honey. *Talanta, 203*, 235–241.

Alkan, P. E. (2020). Pollen analysis of chestnut honey in some provinces of the black sea region, Turkey. *Mellifera, 20*(2), 18–31.

Ameer, K., Murtaza, M. A., Jiang, G., Zhao, C. C., Siddique, F., Kausar, T., Mueen-ud-Din, G., & Mahmood, S. (2024). Fluorescence and ultraviolet–visible spectroscopy in the honey analysis. In *Advanced techniques of honey analysis* (pp. 153–191). Academic Press.

Anjos, O., Guiné, R. P., Santos, A. J., Paula, V. B., Pereira, H., & Estevinho, L. M. (2021). Evaluation of FT-Raman and FTIR-ATR spectroscopy for the quality evaluation of Lavandula spp. honey. *Open Agriculture, 6*(1), 47–56.

Anwar, S., Syed, Q. A., Munawar, F., Arshad, M., Ahmad, W., Rehman, M. A., & Arshad, M. K. (2023). Inclusive overview of sweeteners trends: Nutritional safety and commercialization. *ACS Food Science & Technology, 3*(2), 245–258.

Baglio, E. (2017). *Chemistry and technology of honey production*. Springer.

Bako, T., Mamai, E. A., & Bature, B. J. (2019). Determination of quality parameters of honey from Taraba state, Nigeria. *Chemical and Biomolecular Engineering, 4*(1), 1–9.

Bazaid, A. S., Alamri, A., Almashjary, M. N., Qanash, H., Almishaal, A. A., Amin, J., Binsaleh, N. K., Kraiem, J., Aldarhami, A., & Alafnan, A. (2022). Antioxidant, anticancer, antibacterial, antibiofilm properties and gas chromatography and mass spectrometry analysis of manuka honey: A nature's bioactive honey. *Applied Sciences, 12*(19), 9928.

Benković, M., Jurina, T., Longin, L., Grbeš, F., Valinger, D., Jurinjak Tušek, A., & Gajdoš Kljusurić, J. (2022). Qualitative and quantitative detection of acacia honey adulteration with glucose syrup using near-infrared spectroscopy. *Separations, 9*(10), 312.

Bernardes, R. C., Botina, L. L., Araujo, R. D. S., Guedes, R. N. C., Martins, G. F., & Lima, M. A. P. (2022). Artificial intelligence-aided meta-analysis of toxicological assessment of agrochemicals in bees. *Frontiers in Ecology and Evolution, 10*, 845608.

Bisutti, V., Merlanti, R., Serva, L., Lucatello, L., Mirisola, M., Balzan, S., Tenti, S., Fontana, F., Trevisan, G., Montanucci, L., & Contiero, B. (2019). Multivariate and machine learning approaches for honey botanical origin authentication using near infrared spectroscopy. *Journal of Near Infrared Spectroscopy, 27*(1), 65–74.

Bodor, Z., Majadi, M., Benedek, C., Zaukuu, J. L. Z., Veresné Bálint, M., Csajbókné Csobod, É., & Kovacs, Z. (2023). Detection of low-level adulteration of Hungarian honey using near infrared spectroscopy. *Chem, 11*(2), 89.

Bonini, A., Tamano, G., Farina, L., Boido, E., Dellacassa, E., Pianezze, S., & Perini, M. (2024). Improving honey identification: Stable isotope ratios variability of mono and polyfloral honeys from the citrus growing area of Salto/Concordia and from Uruguayan coastal areas. *Journal of Food Composition and Analysis, 129*, 106132.

Bontempo, L., Camin, F., Ziller, L., Perini, M., Nicolini, G., & Larcher, R. (2017). Isotopic and elemental composition of selected types of Italian honey. *Measurement, 98*, 283–289.

Borgonovo, F., Marzorati, S., Piana, L., Rizzi, R., Panseri, S., Verotta, L., & Guarino, M. (2021). Electronic nose for the classification of honeys of different floral origins. *Chemical Engineering Transactions, 85*, 205–210.

Bose, D., & Padmavati, M. (2024). Honey authentication: A review of the issues and challenges associated with honey adulteration. *Food Bioscience, 61*, 105004.

Boulbarj, I., Abdelaziz, B., Alami, Y.E., Samira, D., & Hassan, D. (2024). Unmasking honey adulteration: a breakthrough in quality assurance through cutting-edge convolutional neural network analysis of thermal images. *arXiv preprint arXiv:2402.08122*.

Brar, D. S., Nayik, G. A., Aggarwal, A. K., Kaur, S., Nanda, V., Saxena, S., Gautam, S., Ramniwas, S., & Tolcha, T. D. (2023). Chemical and functional characteristics to detect sugar syrup adulteration in honey from different botanical origins. *International Journal of Food Properties, 26*(1), 1390–1413.

Brickel, D., Bolduan, J., & Farah, R. (1982). The effect of vasectomy-vasovasostomy on normal physiologic function of the vas deferens. *Fertility and Sterility, 37*(6), 807–810.

Cagliani, L. R., Maestri, G., & Consonni, R. (2022). Detection and evaluation of saccharide adulteration in Italian honey by NMR spectroscopy. *Food Control, 133*, 108574.

Calle, J. L. P., Punta-Sánchez, I., González-de-Peredo, A. V., Ruiz-Rodríguez, A., Ferreiro-González, M., & Palma, M. (2023). Rapid and automated method for detecting and quantifying adulterations in high-quality honey using Vis-NIRs in combination with machine learning. *Food, 12*(13), 2491.

Castro-Muñoz, R., Correa-Delgado, M., Córdova-Almeida, R., Lara-Nava, D., Chávez-Muñoz, M., Velásquez-Chávez, V. F., Hernández-Torres, C. E., Gontarek-Castro, E., & Ahmad, M. Z. (2022). Natural sweeteners: Sources, extraction and current uses in foods and food industries. *Food Chemistry, 370*, 130991.

Chavan, D., Adolacion, J. R. T., Crum, M., Nandy, S., Lee, K. H., Vu, B., Kourentzi, K., Sabo, A., & Willson, R. C. (2022). Isolation and barcoding of trace pollen-free DNA for authentication of honey. *Journal of Agricultural and Food Chemistry, 70*(43), 14084–14095.

Chen, L., Ning, F., Zhao, L., Ming, H., Zhang, J., Yu, W., Yi, S., & Luo, L. (2023). Quality assessment of royal jelly based on physicochemical properties and flavor profiles using HS-SPME-GC/MS combined with electronic nose and electronic tongue analyses. *Food Chemistry, 403*, 134392.

Chenchouni, H., & Laallam, H. (2024). Revolutionizing food quality assessment: Unleashing the potential of artificial intelligence for enhancing honey physicochemical, biochemical, and melissopalynological insights. *Journal of the Saudi Society of Agricultural Sciences, 23*(4), 312–325.

Cherigui, S., Dergal, F., Chikhi, I., Chaker, H., & Chabane, N. (2023). Detection of Algerian honey adulteration by Raman spectroscopy and Chemometrics methods. *Agriculturae Conspectus Scientificus, 88*(3), 231–242.

Consonni, R., Bernareggi, F., & Cagliani, L. R. (2019). NMR-based metabolomic approach to differentiate organic and conventional Italian honey. *Food Control, 98*, 133–140.

Cotte, J. F., Casabianca, H., Lhéritier, J., Perrucchietti, C., Sanglar, C., Waton, H., & Grenier-Loustalot, M. F. (2007). Study and validity of 13C stable carbon isotopic ratio analysis by mass spectrometry and 2H site-specific natural isotopic fractionation by nuclear magnetic resonance isotopic measurements to characterize and control the authenticity of honey. *Analytica Chimica Acta, 582*(1), 125–136.

da Silva, P. M., Gauche, C., Gonzaga, L. V., Costa, A. C. O., & Fett, R. (2016). Honey: Chemical composition, stability and authenticity. *Food Chemistry, 196*, 309–323.

Damto, T., Zewdu, A., & Birhanu, T. (2024). Impact of different adulterants on honey quality properties and evaluating different analytical approaches for adulteration detection. *Journal of Food Protection, 87*(4), 100241.

Daniele, G., Giroud, B., Jabot, C., & Vulliet, E. (2018). Exposure assessment of honeybees through study of hive matrices: Analysis of selected pesticide residues in honeybees, beebread, and beeswax from French beehives by LC-MS/MS. *Environmental Science and Pollution Research, 25*, 6145–6153.

Das, S., Uddin, M. N., Haque, M. S., Chakraborty, D., Mostafa, M., Hasnaine, A., Das, S. K., & Uddin, M. (2022). Hydroxymethylfurfural content and sugar profile of honey available in Bangladesh using validated HPLC-PDA and HPLC-RID. *Journal of Food Quality and Hazards Control*.
David, M., & Magdas, D. A. (2024). Authentication of honey origin and harvesting year based on Raman spectroscopy and chemometrics. *Talanta Open, 10*, 100342.
Díaz-Galiano, F. J., Heinzen, H., Gómez-Ramos, M. J., Murcia-Morales, M., & Fernández-Alba, A. R. (2023). Identification of novel unique mānuka honey markers using high-resolution mass spectrometry-based metabolomics. *Talanta, 260*, 124647.
Drewnowski, A., Tappy, L., Forde, C. G., McCrickerd, K., Tee, E. S., Chan, P., Amin, L., Trinidad, T. P., & Amarra, M. S. (2019). Sugars and sweeteners: Science, innovations, and consumer guidance for Asia. *Asia Pacific Journal of Clinical Nutrition, 28*(3), 645–663.
El-Sofany, A., Naggar, Y. A., Naiem, E., Giesy, J. P., & Seif, A. (2020). Authentication of the botanical and geographic origin of Egyptian honey using pollen analysis methods. *Journal of Apicultural Research, 59*(5), 946–955.
El-Wahed, A. A. A., Rashwan, E. H., AlAjmi, M. F., Khalifa, S. A., Saeed, A., Zhao, C., Naggar, Y. A., Guo, Z., Musharraf, S. G., Wang, K., & El-Seedi, H. R. (2023). Sidr honeys physical and chemical characterization, a comprehensive approach through LC-MS/MS, NMR, and GC-MS analysis. *Separations, 10*(7), 372.
Escuredo, O., Rodríguez-Flores, M. S., Meno, L., & Seijo, M. C. (2021). Prediction of physico-chemical properties in honeys with portable near-infrared (Micronir) spectroscopy combined with multivariate data processing. *Food, 10*(2), 317.
Eshete, Y., & Eshete, T. (2019). A review on the effect of processing temperature and time duration on commercial honey quality. *Madridge Journal of Food Technology, 4*(1), 158–162.
Fakhlaei, R., Selamat, J., Khatib, A., Razis, A. F. A., Sukor, R., Ahmad, S., & Babadi, A. A. (2020). The toxic impact of honey adulteration: A review. *Food, 9*(11), 1538.
Garcia-Alvarez, M., Ceresuela, S., Huidobro, J. F., Hermida, M., & Rodríguez-Otero, J. L. (2002). Determination of polarimetric parameters of honey by near-infrared transflectance spectroscopy. *Journal of Agricultural and Food Chemistry, 50*(3), 419–425.
García-Seval, V., Martínez-Alfaro, C., Saurina, J., Núñez, O., & Sentellas, S. (2022). Characterization, classification and authentication of Spanish blossom and honeydew honeys by non-targeted HPLC-UV and off-line SPE HPLC-UV polyphenolic fingerprinting strategies. *Food, 11*(15), 2345.
Ghosh, P., Garg, S., Mohanty, I., Sahoo, D., & Pradhan, R. C. (2019). Comparison and storage study of ultra-filtered clarified jamun (Syzygium cumini) juice. *Journal of Food Science and Technology, 56*, 1877–1889.
Ghramh, H. A., Khan, K. A., Zubair, A., & Ansari, M. J. (2020). Quality evaluation of Saudi honey harvested from the Asir province by using high-performance liquid chromatography (HPLC). *Saudi Journal of Biological Sciences, 27*(8), 2097–2105.
Gohar, A., & Shakeel, M. (2021). Assessment of environmental impact on essential and toxic elements composition in natural honeys by using inductively coupled plasma mass spectrometry. *Environmental Science and Pollution Research, 28*, 15794–15805.
Gonçalves, W. B., Teixeira, W. S. R., Cervantes, E. P., Mioni, M. D. S. R., Sampaio, A. N. D. C. E., Martins, O. A., Gruber, J., & Pereira, J. G. (2023). Application of an electronic nose as a new technology for rapid detection of adulteration in honey. *Applied Sciences, 13*(8), 4881.
Gün, R., & Karaoğlu, M. M. (2024). Detection of honey adulteration by characterization of the physico-chemical properties of honey adulterated with the addition of glucose–fructose and maltose corn syrups. *European Food Research and Technology, 250*, 1–18.
Guyon, F., et al. (2020). Metabolomics applied to proton nuclear magnetic resonance profile for the identification of seven floral origin of French honeys. *Journal of Food & Nutrition Research, 59*(2).
Haefeker, W. (2021). Pollen supplements and substitutes in the EU feed market: A product/market survey for bees and other animal species. *EFSA Supporting Publications, 18*(2), 6461E.

Hao, S., Yuan, J., Cui, J., Yuan, W., Zhang, H., & Xuan, H. (2022). The rapid detection of acacia honey adulteration by alternating current impedance spectroscopy combined with 1H NMR profile. *LWT, 161*, 113377.

Helstad, S. (2019). Corn sweeteners. In *Corn* (pp. 551–591). AACC International Press.

Hendrickson, O. D., Zherdev, A. V., & Dzantiev, B. B. (2024). Sensitive immunoenzyme assay for the detection of antibiotic flumequine in honey. *Analytical Methods, 16*(5), 721–730.

Hernanz, D., Jara-Palacios, M. J., Santos, J. L., Pajuelo, A. G., Heredia, F. J., & Terrab, A. (2023). The profile of phenolic compounds by HPLC-MS in Spanish oak (Quercus) honeydew honey and their relationships with color and antioxidant activity. *LWT, 180*, 114724.

Honrado, M., Lopes, A. R., Pinto, M. A., & Amaral, J. S. (2022). A novel real-time PCR coupled with high resolution melting analysis as a simple and fast tool for the entomological authentication of honey by targeting Apis mellifera mitochondrial DNA. *Food Research International, 161*, 111761.

Hu, S., Li, H., Chen, C., Chen, C., Zhao, D., Dong, B., Lv, X., Zhang, K., & Xie, Y. (2022). Raman spectroscopy combined with machine learning algorithms to detect adulterated Suichang native honey. *Scientific Reports, 12*(1), 3456.

Huynh, D. N. (2020). *Development of optical indices of articular cartilage damage and microstructure*. The Catholic University of America.

Huzortey, A. A., Arefi, A., Anderson, B., Khadem, H., Sackey, S. S., Mahmoodi-Khaledi, E., & Tavassoli, S. H. (2022). 532-nm laser-excited Raman spectroscopic evaluation of Iranian honey. *Food Analytical Methods, 15*(3), 772–782.

Ikedi, R. I. (2022). *Laser Raman spectroscopic assessment of honey adulteration by molasses*. Doctoral dissertation, University of Nairobi.

Indahsari, A. N., & Purwanto, H. (2024). Detection of entomological origin of honey sold in Indonesia based on 16S rRNA gene analysis. *Journal of Tropical Biodiversity and Biotechnology, 9*(3), 86256.

Jamal, M., Aziz, M. A., Naeem, M., Iqbal, Z., Khalid, A., Siddique, F., Khan, K. A., & Ghramh, H. A. (2020). Detection of flumethrin acaricide residues from honey and beeswax using high performance liquid chromatography (HPLC) technique. *Journal of King Saud University-Science, 32*(3), 2229–2235.

Kara, Y., Can, Z., & Kolaylı, S. (2019). HPLC analyses of polyphenolic compounds in oak (Querces frainetto) honey from Kırklareli region of Turkey. *Turkish Journal of Analytical Chemistry, 1*(1).

Karaboga, D., Akay, B., & Karaboga, N. (2020). A survey on the studies employing machine learning (ML) for enhancing artificial bee colony (ABC) optimization algorithm. *Cogent Engineering, 7*(1), 1855741.

Kasiotis, K. M., Baira, E., Iosifidou, S., Manea-Karga, E., Tsipi, D., Gounari, S., Theologidis, I., Barmpouni, T., Danieli, P. P., Lazzari, F., & Dipasquale, D. (2023). Fingerprinting chemical markers in the mediterranean orange blossom honey: UHPLC-HRMS metabolomics study integrating melissopalynological analysis, GC-MS and HPLC-PDA-ESI/MS. *Molecules, 28*(9), 3967.

Kathuria, D., Gautam, S., & Thakur, A. (2023). Maillard reaction in different food products: Effect on product quality, human health and mitigation strategies. *Food Control, 153*, 109911.

Kavimugilan, S., Kalyanasundaram, A., & Ambethgar, V. (2024). Managing bee's health during dearth periods: A review of pollen and nectar substitutes for tropical climates. *Uttar Pradesh Journal Of Zoology, 45*(16), 472–489.

Kozłowicz, K., Różyło, R., Gładyszewska, B., Matwijczuk, A., Gładyszewski, G., Chocyk, D., Samborska, K., Piekut, J., & Smolewska, M. (2020). Identification of sugars and phenolic compounds in honey powders with the use of GC–MS, FTIR spectroscopy, and X-ray diffraction. *Scientific Reports, 10*(1), 16269.

Kropf, U., Golob, T., Necemer, M., Kump, P., Korosec, M., Bertoncelj, J., & Ogrinc, N. (2010). Carbon and nitrogen natural stable isotopes in Slovene honey: Adulteration and botanical and geographical aspects. *Journal of Agricultural and Food Chemistry, 58*(24), 12794–12803.

Li, Y., Li, J., Huang, H., Jian, D., Shan, Y., Wang, S., & Liu, F. (2021). Rapid quantitative detection for multiple antibiotics in honey using a quantum dot microsphere immunochromatographic strip. *Food Control, 130*, 108256.

Liu, S., Lang, D., Meng, G., Hu, J., Tang, M., & Zhou, X. (2022). Tracing the origin of honey products based on metagenomics and machine learning. Food chemistry, 371, 131066.

Luccas, F. S., Fernandes, E. A. D. N., Mazola, Y. T., Bacchi, M. A., & Sarriés, G. A. (2022). Optimization of sample preparation of Brazilian honeys for TQ-ICP-MS analysis. *Talanta Open, 5*, 100117.

Machuca, G., Staforelli, J., Rondanelli-Reyes, M., Garces, R., Contreras-Trigo, B., Tapia, J., Sanhueza, I., Jara, A., Lamas, I., Troncoso, J. M., & Coelho, P. (2022). Hyperspectral microscopy technology to detect syrups adulteration of endemic Guindo Santo and Quillay honey using machine-learning tools. *Food, 11*(23), 3868.

Magdas, D. A., & Berghian-Grosan, C. (2023). Botanical honey recognition and quantitative mixture detection based on Raman spectroscopy and machine learning. *Spectrochimica Acta Part A: Molecular and Biomolecular Spectroscopy, 293*, 122433.

Magdas, D. A., Guyon, F., Berghian-Grosan, C., & Molnar, C. M. (2021). Challenges and a step forward in honey classification based on Raman spectroscopy. Food Control, 123, 107769.

Manzoor, A., Jan, B., Zahoor, I., Anjum, N., Nabi, A., Allai, F. M., Rizvi, Q. U. E. H., Shiekh, R. A., Sheikh, M. A., & Ahmad, S. (2022). Thermal treatment of foods: Science, shelf life, and quality. In *Shelf life and food safety* (pp. 165–180). CRC Press.

María, E. G., Luna, A. M., Celdrán, A. C., Muñoz, G. M., Oates, M. J., & Ruiz-Canales, A. (2023). Classification of Monofloral honeys by measuring a low-cost electronic nose prototype based on resistive metal oxide sensors. *Agronomy, 13*(8), 2183.

Mateo, F., Tarazona, A., & Mateo, E. M. (2021). Comparative study of several machine learning algorithms for classification of unifloral honeys. *Food, 10*(7), 1543.

Mauer, L. J. (2024). Moisture and total solids analysis. In *Nielsen's food analysis* (pp. 233–260). Springer International Publishing.

Medrzycki, P., Giffard, H., Aupinel, P., Belzunces, L. P., Chauzat, M. P., Classen, C., Colin, M. E., Dupont, T., Girolami, V., Johnson, R., & Le Conte, Y. (2013). Standard methods for toxicology research in Apis mellifera. *Journal of Apicultural Research, 52*(4), 1–60.

Mohamadzade Namin, S., Yeasmin, F., Choi, H. W., & Jung, C. (2022). DNA-based method for traceability and authentication of Apis cerana and A. Dorsata honey (hymenoptera: Apidae), using the NADH dehydrogenase 2 gene. *Food, 11*(7), 928.

Mohamat, R. N., Noor, N. R. A. M., Yusof, Y. A., Sabri, S., & Zawawi, N. (2023). Differentiation of high-fructose corn syrup adulterated Kelulut honey using physicochemical, rheological, and antibacterial parameters. *Food, 12*(8), 1670.

Molnar, C. M., Berghian-Grosan, C., & Magdas, D. A. (2020). An optimized green preparation method for the successful application of Raman spectroscopy in honey studies. Talanta, 208, 120432.

Nasab, S. G., Yazd, M. J., Marini, F., Nescatelli, R., & Biancolillo, A. (2020). Classification of honey applying high performance liquid chromatography, near-infrared spectroscopy and chemometrics. *Chemometrics and Intelligent Laboratory Systems, 202*, 104037.

Nhlapho, S. (2023). *Investigation of sugars in honey for authentication and detection of adulterants using selected analytical techniques*. Doctoral dissertation, University of Johannesburg.

Ntakoulas, D. D., Pasias, I. N., Raptopoulou, K. G., Proestos, C., & Weller, P. (2022). Authenticity and Chemometrics of honey. In *Chemometrics and authenticity of foods of plant origin* (pp. 152–186). CRC Press.

Nunes, A., Azevedo, G. Z., Dos Santos, B. R., Borges, C. V., Lima, G. P. P., Crocoli, L. C., Moura, S., & Maraschin, M. (2022). Characterization of Brazilian floral honey produced in the states of Santa Catarina and São Paulo through ultraviolet–visible (UV–vis), near-infrared (NIR), and nuclear magnetic resonance (NMR) spectroscopy. *Food Research International, 162*, 111913.

Nunes, A., Sforça, M. L., Rocco, S. A., Azevedo, G. Z., dos Santos, B. R., Borges, C. V., Lima, G. P. P., Moura, S., & Maraschin, M. (2023). Characterization of Brazilian floral honey pro-

duced in the states of Santa Catarina and São Paulo using quantitative nuclear magnetic resonance (qNMR) spectroscopy. *Food Chemistry Advances, 3*, 100378.
Othman, S., Mavani, N. R., Hussain, M. A., Abd Rahman, N., & Ali, J. M. (2023). Artificial intelligence-based techniques for adulteration and defect detections in food and agricultural industry: A review. *Journal of Agriculture and Food Research, 12*, 100590.
Ouchemoukh, S., Schweitzer, P., Bey, M. B., Djoudad-Kadji, H., & Louaileche, H. (2010). HPLC sugar profiles of Algerian honeys. *Food Chemistry, 121*(2), 561–568.
Oymen, B., Aşır, S., Türkmen, D., & Denizli, A. (2022). Determination of multi-pesticide residues in honey with a modified QuEChERS procedure followed by LC-MS/MS and GC-MS/MS. *Journal of Apicultural Research, 61*(4), 530–542.
Özler, H. (2018). Pollen analysis of the honey from South Anatolia. *Uludağ Arıcılık Dergisi, 18*(2), 73–86.
Palma, B. G., Leão, R. A., de Souza, R. O., & Pandoli, O. G. (2021). Immobilization of lipases on lignocellulosic bamboo powder for biocatalytic transformations in batch and continuous flow. *Catalysis Today, 381*, 280–287.
Pang, Y. H., Lv, Z. Y., Sun, J. C., Yang, C., & Shen, X. F. (2021). Collaborative compounding of metal–organic frameworks for dispersive solid-phase extraction HPLC–MS/MS determination of tetracyclines in honey. *Food Chemistry, 355*, 129411.
Peñalver, R., Arroyo-Manzanares, N., Campillo, N., & Viñas, P. (2021). Targeted and untargeted gas chromatography-mass spectrometry analysis of honey samples for determination of migrants from plastic packages. *Food Chemistry, 334*, 127547.
Petros, M. (2016). *Physico-chemical properties and consumer acceptance of honey flavored Ethiopian curd.* Doctoral dissertation, Haramaya University.
Rasekh, M., & Karami, H. (2021). E-nose coupled with an artificial neural network to detection of fraud in pure and industrial fruit juices. *International Journal of Food Properties, 24*(1), 592–602.
Rasekh, M., Karami, H., Wilson, A. D., & Gancarz, M. (2021). Classification and identification of essential oils from herbs and fruits based on a MOS electronic-nose technology. *Chem, 9*(6), 142.
Rašić, S., Štefanić, E., Antunović, S., Jović, J., & Kristek, S. (2018). Pollen analysis of honey from North-Eastern Croatia. *Poljoprivreda, 24*(2), 43–49.
Raypah, M. E., Omar, A. F., Muncan, J., Zulkurnain, M., & Abdul Najib, A. R. (2022a). Identification of stingless bee honey adulteration using visible-near infrared spectroscopy combined with aquaphotomics. *Molecules, 27*(7), 2324.
Raypah, M. E., Zhi, L. J., Loon, L. Z., & Omar, A. F. (2022b). Near-infrared spectroscopy with chemometrics for identification and quantification of adulteration in high-quality stingless bee honey. *Chemometrics and Intelligent Laboratory Systems, 224*, 104540.
Regina, T. M. R., & Gallo, F. (2024). Evaluating the effectiveness of DNA-based methods to detect mislabelling and adulteration of chestnut honey. *Food Bioscience, 61*, 104798.
Rygula, A., Klisińska-Kopacz, A., Matosz, M., Wilkosz, T., Obarzanowski, M., Krupska-Wolas, P., Skóra, K., Kopyciak, A., Harasim-Grym, J., & del Hoyo-Meléndez, J. M. (2024). Noninvasive multitechnique analysis of baroque Amber artworks from Gdansk: The benefits of using Raman spectroscopy. *Journal of Raman Spectroscopy, 55*, 1372.
Samborska, K., Barańska, A., Bodel, D., & Jedlińska, A. (2019). Investigation on the possibility to enhancing honey spray drying process and powders properties by ultrafiltration pre-treatment. *Chemical and Process Engineering, 40*(3).
Saraiva, A., Carrascosa, C., Raheem, D., Ramos, F., & Raposo, A. (2020). Natural sweeteners: The relevance of food naturalness for consumers, food security aspects, sustainability and health impacts. *International Journal of Environmental Research and Public Health, 17*(17), 6285.
Schellenberg, A., Chmielus, S., Schlicht, C., Camin, F., Perini, M., Bontempo, L., Heinrich, K., Kelly, S. D., Rossmann, A., Thomas, F., & Jamin, E. (2010). Multielement stable isotope ratios (H, C, N, S) of honey from different European regions. *Food Chemistry, 121*(3), 770–777.

Schievano, E., Sbrizza, M., Zuccato, V., Piana, L., & Tessari, M. (2020). NMR carbohydrate profile in tracing acacia honey authenticity. *Food Chemistry, 309*, 125788.

Shruti, A., Bage, N., & Kar, P. (2024). Nanomaterials based sensors for analysis of food safety. *Food Chemistry, 433*, 137284.

Şik, L., Güvensen, A., Durmuşkahya, C., & Erol, O. (2017). Pollen analysis of honeys from Ardahan/Turkey. *Biological Diversity and Conservation, 10*(2), 12–19.

Singh, I., & Singh, S. (2018). Honey moisture reduction and its quality. *Journal of Food Science and Technology, 55*, 3861–3871.

Sipos, L., Végh, R., Bodor, Z., Zaukuu, J. L. Z., Hitka, G., Bázár, G., & Kovacs, Z. (2020). Classification of bee pollen and prediction of sensory and colorimetric attributes – A sensometric fusion approach by e-nose, e-tongue and NIR. *Sensors, 20*(23), 6768.

Skaff, W., El Hajj, R., Hanna-Wakim, L., & Estephan, N. (2022). Detection of adulteration in honey by infrared spectroscopy and chemometrics: Effect on human health. *Journal of Food Processing and Preservation, 46*(10), e15438.

Sniderman, J. K., Matley, K. A., Haberle, S. G., & Cantrill, D. J. (2018). Pollen analysis of Australian honey. *PLoS One, 13*(5), e0197545.

Soares, S. E. F. (2017). A contribution to assess the quality and authenticity oh honey: Study of chemical parameters and genomic-based markers.

Soares, S., Amaral, J. S., Oliveira, M. B. P., & Mafra, I. (2017). A comprehensive review on the main honey authentication issues: Production and origin. *Comprehensive Reviews in Food Science and Food Safety, 16*(5), 1072–1100.

Sonawane, B. V. (2016). *Environmental regulation of CO2 concentrating mechanisms in C4 grasses with different biochemical subtypes*. Doctoral dissertation, Western Sydney University

Song, X., She, S., Xin, M., Chen, L., Li, Y., Vander Heyden, Y., Rogers, K. M., & Chen, L. (2020). Detection of adulteration in Chinese monofloral honey using 1H nuclear magnetic resonance and chemometrics. *Journal of Food Composition and Analysis, 86*, 103390.

Song, Y., Huang, H. C., Lu, W., Li, N., Su, J., Cheng, S. B., Lai, Y., Chen, J., & Zhan, J. (2021). Ag@ WS2 quantum dots for surface enhanced Raman spectroscopy: Enhanced charge transfer induced highly sensitive detection of thiram from honey and beverages. *Food Chemistry, 344*, 128570.

Sotiropoulou, N. S., Xagoraris, M., Revelou, P. K., Kaparakou, E., Kanakis, C., Pappas, C., & Tarantilis, P. (2021). The use of SPME-GC-MS IR and Raman techniques for botanical and geographical authentication and detection of adulteration of honey. *Food, 10*(7), 1671.

Sun, J., Zhao, H., Wu, F., Zhu, M., Zhang, Y., Cheng, N., Xue, X., Wu, L., & Cao, W. (2021). Molecular mechanism of mature honey formation by GC-MS-and LC-MS-based metabolomics. *Journal of Agricultural and Food Chemistry, 69*(11), 3362–3370.

Szczurek, A., Maciejewska, M., Zajiczek, Ż., Bąk, B., Wilk, J., Wilde, J., & Siuda, M. (2020). The effectiveness of Varroa destructor infestation classification using an e-nose depending on the time of day. *Sensors, 20*(9), 2532.

Tahir, H. E., Arslan, M., Mahunu, G. K., Shi, J., Zou, X., Gasmalla, M. A. A., & Mariod, A. A. (2020). Data fusion approach improves the prediction of single phenolic compounds in honey: A study of NIR and Raman spectroscopies. *eFood, 1*(2), 173–180.

Tan, S. H., Pui, L. P., Solihin, M. I., Keat, K. S., Lim, W. H., & Ang, C. K. (2021). Physicochemical analysis and adulteration detection in Malaysia stingless bee honey using a handheld near-infrared spectrometer. *Journal of Food Processing and Preservation, 45*(7), e15576.

Tepeli, A., Sanzu, L., Charles, P. (2022). Assessment of the honey purity by using imaging polarimetry technique.

Tsuruda, J. M., Chakrabarti, P., & Sagili, R. R. (2021). Honey bee nutrition. *Veterinary Clinics: Food Animal Practice, 37*(3), 505–519.

Ulberth, F. (2016). Advances in testing for adulteration in honey. In *Advances in food authenticity testing* (pp. 729–753). Woodhead Publishing.

Valinger, D., Longin, L., Grbeš, F., Benković, M., Jurina, T., Kljusurić, J. G., & Tušek, A. J. (2021). Detection of honey adulteration–the potential of UV-VIS and NIR spectroscopy coupled with multivariate analysis. *LWT, 145*, 111316.

Viejo, C. G., Fuentes, S., Godbole, A., Widdicombe, B., & Unnithan, R. R. (2020). Development of a low-cost e-nose to assess aroma profiles: An artificial intelligence application to assess beer quality. *Sensors and Actuators B: Chemical, 308*, 127688.

Voica, C., Iordache, A. M., & Ionete, R. E. (2020). Multielemental characterization of honey using inductively coupled plasma mass spectrometry fused with chemometrics. *Journal of Mass Spectrometry, 55*(7), e4512.

Wang, Q., Zhao, H., Zhu, M., Zhang, J., Cheng, N., & Cao, W. (2020). Method for identifying acacia honey adulterated by resin absorption: HPLC-ECD coupled with chemometrics. *LWT, 118*, 108863.

Wang, X., Yan, S., Zhao, W., Wu, L., Tian, W., & Xue, X. (2023a). Comprehensive study of volatile compounds of rare Leucosceptrum canum smith honey: Aroma profiling and characteristic compound screening via GC–MS and GC–MS/MS. *Food Research International, 169*, 112799.

Wang, Y., Xing, L., Zhang, J., Ma, X., & Weng, R. (2023b). Determination of endogenous phenolic compounds in honey by HPLC-MS/MS. *LWT, 183*, 114951.

Węglińska, M., Szostak, R., Kita, A., Nemś, A., & Mazurek, S. (2020). Determination of nutritional parameters of bee pollen by Raman and infrared spectroscopy. *Talanta, 212*, 120790.

Wirta, H., Abrego, N., Miller, K., Roslin, T., & Vesterinen, E. (2021). DNA traces the origin of honey by identifying plants, bacteria and fungi. *Scientific Reports, 11*(1), 4798.

Wu, S. W., Wang, M. Y., Liu, B. H., & Yu, F. Y. (2020). Sensitive enzyme-linked immunosorbent assay and gold nanoparticle immunochromatocgraphic strip for rapid detecting chloramphenicol in food. *Journal of Food Safety, 40*(2), e12759.

Wu, X., Xu, B., Ma, R., Niu, Y., Gao, S., Liu, H., & Zhang, Y. (2022). Identification and quantification of adulterated honey by Raman spectroscopy combined with convolutional neural network and chemometrics. *Spectrochimica Acta Part A: Molecular and Biomolecular Spectroscopy, 274*, 121133.

Yan, S., Sun, M., Wang, X., Shan, J., & Xue, X. (2022). A novel, rapid screening technique for sugar syrup adulteration in honey using fluorescence spectroscopy. *Food, 11*(15), 2316.

Yong, C. H., Muhammad, S. A., Nasir, F. I., Mustafa, M. Z., Ibrahim, B., Kelly, S. D., Cannavan, A., & Seow, E. K. (2022). Detecting adulteration of stingless bee honey using untargeted 1H NMR metabolomics with chemometrics. *Food Chemistry, 368*, 130808.

Zaki Dizaji, H., Adibzadeh, A., & Aghili Nategh, N. (2021). Application of E-nose technique to predict sugarcane syrup quality based on purity and refined sugar percentage. *Journal of Food Science and Technology, 58*, 4149–4156.

Zhou, M., Wang, X. H., Zhu, Q., Jiang, F., Wang, H. X. (2021). Identification of the botanic source of honey by 1H nuclear magnetic resonance and support vector machine.

Zhou, M., Long, T., Zhao, Z., Chen, J., Wu, Q., Wang, Y., & Zou, Z. (2022). Honey quality detection based on near-infrared spectroscopy. *Food Science and Technology, 43*, e98822.

Zhu, M., Sun, J., Zhao, H., Wu, F., Xue, X., Wu, L., & Cao, W. (2022). Volatile compounds of five types of unifloral honey in Northwest China: Correlation with aroma and floral origin based on HS-SPME/GC–MS combined with chemometrics. *Food Chemistry, 384*, 132461.

Zhu, W., Zhang, J., Zhang, Y., Zhang, H., Miao, K., Luo, J., & Yang, M. (2024). Establishment of a rapid and sensitive ic-ELISA for the detection of thiacloprid residues in honey and medicinal herbs using a novel highly specific monoclonal antibody. *Ecotoxicology and Environmental Safety, 284*, 116911.

Chapter 8
The Influence of Honey's Botanical and Geographical Origins on Its Therapeutic Potential

İlginç Kızılpınar Temizer

Abbreviations

B. subtilis	*Bacillus subtilis*
S. aureus	*Staphylococcus aureus*
S. epidermidis	*Staphylococcus epidermidis*
S. pyogenes	*Staphylococcus pyogenes*

1 Introduction

Honey is a natural sweet substance produced by honeybees from flower nectar, secretions from parts of living plants, or excretions from plant-sucking insects (Hemiptera) found on plant surfaces. Bees gather these materials, mix them with their enzymes, deposit the mixture into honeycombs, dehydrate it, and allow it to ripen into honey. Blossom honey, also known as nectar honey, is derived from the nectar of flowers, while honeydew honey is primarily produced from the secretions left on plant surfaces by plant-sucking insects or directly from plant secretions (Codex Alimentarius, 1981).

Beyond being a viscous, supersaturated sugar solution, honey contains over 200 components, including proteins, enzymes, phenolic acids, flavonoids, vitamins, minerals, and other phytochemicals, making it a rich and diverse source of bioactive

İ. K. Temizer (✉)
Department of Crop and Animal Production, Vocational School of Food, Agriculture and Livestock, Bingol University, Bingol, Turkiye

Department of Medical Services and Techniques, High School of Health Services, Giresun University, Giresun, Turkiye
e-mail: ilginc.kizilpinar@giresun.edu.tr

M. Zafar et al. (eds.), *Pure Honey: Assurance & Authentication*,
https://doi.org/10.1007/978-3-031-98913-1_8

compounds (Çobanoğlu et al., 2023; Dai et al., 2020; Kızılpınar Temizer, 2024; Pita-Calvo & Vázquez, 2017). The composition and chemical diversity of honey are largely influenced by the nectar and pollen sources of the plants, the plant species, and the geographical origin of the honey (Çobanoğlu et al., 2023). Many of these compounds have demonstrated significant health benefits.

The therapeutic use of bee products is known as "apitherapy," with honey being one of the most significant products in this field. For centuries, honey has been widely used in traditional medicine across various cultures, and today it continues to be a focal point of modern scientific research worldwide. The phenolic compounds, enzymes, and other bioactive substances in honey provide it with antioxidant, antibacterial (effective against approximately 60 types of bacteria), antibiofilm, anti-inflammatory, anti-swarming, antiviral, anticancer, and wound-healing properties (Eteraf-Oskouei & Najafi, 2013).

Recent studies have confirmed that the bioactive components in honey work synergistically to provide comprehensive health benefits, solidifying its position in modern medicine as more than just a sweetener but as a holistic therapeutic agent. The ongoing exploration of honey's chemical diversity and its impact on human health continues to uncover new applications, reinforcing honey's status as a valuable adjunct in the prevention and treatment of various health conditions.

2 Historical, Cultural, and Religious Insights into the Use of Honey

Honey has played a vital role across various cultures throughout history, being valued not only as a food but also for its medicinal, symbolic, and ritualistic significance. In ancient Egypt, particularly in Lower Egypt, honey had been a key element in royal ceremonies, symbolizing the power, health, and prosperity of gods and pharaohs (Bowie, 2020). Egyptian papyri had revealed numerous recipes that utilized honey's therapeutic properties, highlighting its esteemed status in Egyptian society (Kuropatnicki et al., 2018). Similarly, ancient Greek scholars had praised honey for its health benefits, using it as both a protective and healing substance (Eteraf-Oskouei & Najafi, 2013). However, after the fall of the Roman Empire, honey's medicinal use had declined in medieval Europe, where it had not been widely adopted by practitioners of the time (Kuropatnicki et al., 2018).

In the Middle East and Asia, honey had been both a staple food and a revered remedy. Known as ʿasl in Arabic, angabīn in Persian, shehed in Urdu, and madhu in Hindi, this amber-colored, thick, and aromatic liquid has been highly valued across diverse cultures (Nikhat & Fazil, 2022). Although the history of beekeeping in Iran had not been well documented, evidence had suggested that honey was widely used in ancient times. During the Persian Empire (circa 500 BCE), the honey trade had been lucrative and subject to taxation. The Greek historian Aeschylus had noted honey among substances used in rituals, and reliefs at Persepolis had depicted "honey-filled beehives" offered to the king by the Ionians (Komeili, 1990).

The discovery of a large-scale beehive from the tenth to early ninth century BCE at Tel Rehov in the Jordan Valley, northern Israel, has underscored the economic and cultural importance of honey and beeswax production in the ancient Near East (Mazar et al., 2008). In India, honey has been an essential component of Ayurveda, used to balance the body and treat various ailments, with more than 634 therapeutic applications involving honey have been developed within this medical system (Khan et al., 2014). Traditional Chinese Medicine has also utilized honey for thousands of years and has continued to value it for its health-promoting properties (Khan et al., 2014). In Jewish visual culture, especially in traditional art and crafts, honey and bees have served as powerful symbols of harmony, prosperity, and community (Rodov, 2020).

The health benefits of honey had been highlighted in the Quran over 1400 years ago, where it had not only emphasized honey's therapeutic properties but had also described the process by which bees produced it (Hasam et al., 2020). In the Quran's Surah An-Nahl, it had detailed how bees collected nectar from various plants to create honey, highlighting its healing qualities (Eker and Kök, 2017; Hasam et al., 2020). Similarly, in Christianity, honey and bees had been frequently mentioned in the Bible, notably in the books of Exodus, Judges, Matthew, and Proverbs, symbolizing abundance and divine blessing. Honey's significance had extended across numerous religions and traditions, including Buddhism, Judaism, Hinduism, and Vedic cultures, each of which had recognized its sacred and healing nature (Ajibola et al., 2012).

3 Therapeutic Potential of Different Honey Types

There are over 300 types of monofloral honey worldwide, varying in color from water white to dark brown or black (Singh, 2021). Today, numerous detailed studies are conducted to explore the therapeutic properties of these honey types, derived from different botanical sources and based on their chemical compositions. These studies provide an in-depth examination of how the distinctive components of each honey type influence health. By exploring these characteristics, researchers gain valuable insights into honey's therapeutic potential and its wide range of health benefits and practical applications.

3.1 Honeydew Honey Types

Cedar Honey

It is produced from the nectar of trees belonging to the *Cedrus* genus, commonly known as cedar trees. These trees are part of the Pinaceae family and include species such as *Cedrus libani* A. Rich, *Cedrus atlantica* Manetti, and *Cedrus brevifolia* (Hook.f.) Elwes & A. Henry. In regions like Egypt and the Eastern Mediterranean, cedar trees are prevalent and play a significant role in the production of honey. El-Sofany et al. (2018) investigated the antimicrobial activity of cedar honey against various pathogens, including the bacteria *Bacillus subtilis*, *Staphylococcus aureus*, *Bacillus cereus var. toyonensis*, *Pseudomonas aeruginosa*, *Escherichia coli*, and *Salmonella typhimurium*, as well as the fungi *Candida albicans* and *Aspergillus niger*. The study identified cedar honey as one of the most potent Egyptian honeys in terms of antibacterial activity. Seyhan et al. (2017) found that cedar honey exhibits some of the highest cytotoxic effects among honey types, particularly against breast cancer cell lines (MCF7, SKBR3, MDAMB-231) and the fibroblastic breast epithelial cell line used as a control (MCF10A).

Fir Honey

It is a specific type of honeydew honey produced from the secretions of insects, primarily aphids, that feed on fir trees (*Abies* L.) (Bacandritsos et al., 2006). Fir trees are coniferous trees found in temperate regions across Europe, Asia, and North America. Stavropoulou et al. (2022) reported that fir honey exhibits effective anti-inflammatory and antibacterial properties by reducing TNF-α and iNOS production and preventing the migration of bacteria into tissues, particularly in inflammatory conditions such as acute endotoxemia and polymicrobial peritonitis. Majtan et al. (2013) demonstrated that fir honey inhibits MMP-9 production induced by TNF-α in keratinocytes, thereby preventing proteolytic activity associated with skin inflammation. They identified flavonoids such as apigenin and kaempferol as key components responsible for these effects, suggesting that fir honey may offer a natural remedy for chronic inflammatory dermatological issues. Ramić et al. (2023) found that fir honeydew honey possesses strong antibacterial activity against *Campylobacter jejuni*. Additionally, Dżugan et al. (2024) highlighted that fir honeydew honey effectively inhibits the migration of breast cancer cells while only mildly affecting the movement of normal fibroblasts. Broznić et al. (2018) demonstrated that *Abies alba* Mill. honeydew honey possesses both antiproliferative potential in cancer treatment and antimicrobial effects against antibiotic-resistant bacteria.

Oak Honey

It is a type of honeydew honey produced from the secretions of oak aphids, insects that feed on species of the *Quercus* genus, including *Quercus frainetto* Ten., *Quercus ilex* L., *Quercus cerris* L., *Quercus robur* L., and *Quercus pyrenaica* Willd. (Hernanz et al., 2023; Özkök et al., 2019). Oak trees are widespread and found in various temperate regions, including parts of Europe, North America, and Asia (Hernanz et al., 2023; Özkök et al., 2019). Sahin (2016) reported that oak honey is a strong source of urease and xanthine oxidase (XO) enzyme inhibition due to its high phenolic content. Kolayli et al. (2016) identified oak honey as one of the types with high anti-hyaluronidase activity. Kocyigit et al. (2019) found that honey from *Q. pyrenaica* in the Ida Mountains exhibits cytotoxic, genotoxic, and apoptotic effects due to its high phenolic and antioxidant content, which contributes to pro-oxidant activities. Dinkov (2013) noted that oak honey has significant antioxidant and antibacterial effects.

Pine Honey

Produced in the Eastern Mediterranean region, such as Greece and Turkey, pine honey is derived from the excretions of the *Marchalina hellenica* Gennadius insect, which is specific to the *Pinus brutia* Ten. and *Pinus halepensis* Miller species (Kafantaris et al., 2021; Tananaki et al., 2007; Tsavea et al., 2022). Tsavea et al. (2022) reported that Greek pine honey exhibits high antioxidant activity and substantial antimicrobial capacity. Eraslan et al. (2010) suggested that due to its phenolic compounds and amino acids, pine honey alleviates oxidative stress and recommended its daily inclusion in the diet to reduce the risk of trichlorfon poisoning. Ögretmen and İnanan (2014) highlighted that pine honey is an effective method in cryomedicine, particularly in improving the success rate of fertilization with frozen-thawed sperm. Sıcak et al. (2021) noted that pine honey contains antioxidant, anticholinesterase, and urease-inhibiting compounds, offering numerous beneficial effects and significant antiproliferative activity against C6 (brain) and PC3 (prostate) cancer cell lines. Kafantaris et al. (2021) demonstrated that pine honey inhibits the genomic expression of P. aeruginosa and is effective in wound healing when used in wound dressings. Spilioti et al. (2014) emphasized that pine honey is rich in protocatechuic acid and p-hydroxybenzoic acid, and the high concentrations of these phenolic acids are associated with the honey's strong antioxidant, anticancer (in PC-3 and MCF-7 cells), and antiatherogenic effects.

3.2 *Blossom Honey Types*

3.2.1 Polyfloral Honey

Anzer Honey

It is a world-renowned type of honey produced in the Anzer Plateau of İkizdere district in Rize province, located in Turkey's Eastern Black Sea Region. This special honey is obtained from a diverse range of endemic plant species that grow at an altitude of approximately 2000–3000 m. Korkmaz and Kolankaya (2009) reported that a decrease in glutathione levels contributes to liver damage caused by N-ethylmaleimide and suggested that the hepatoprotective effect of Anzer honey may be related to its role in sulfhydryl-sensitive processes. Bagatir et al. (2022) demonstrated that Anzer honey's antioxidant and protective properties could reduce DNA damage induced by X-rays. Doğan and Kolankaya (2005) found that Anzer honey has a protective effect against ethanol-induced increased vascular permeability, which is likely related to its content of ascorbic acid. Özdemir et al. (2023) highlighted the importance of addressing the genotoxicity and cytotoxicity of antidiabetic drugs due to the increasing number of T2DM patients needing long-term treatment. Their study showed that VM exhibited cytotoxic and genotoxic effects on hTERT-HPNE cells, but supplementation with Anzer honey extract increased cell viability, and no cytotoxicity was observed.

3.2.2 Monofloral Honey

Buckwheat Honey

It, a unifloral honey derived from the nectar and pollen of the buckwheat plant (*Fagopyrum esculentum* L.), is commonly produced in regions such as North America (Canada and California), China, and various European countries including Poland, Russia, the Netherlands, and Germany (Pasini et al., 2013). According to Deng et al. (2018), the primary phenolic compounds in buckwheat honey are p-hydroxybenzoic acid, chlorogenic acid, and p-coumaric acid. Their research showed that buckwheat honey exhibits antibacterial activity comparable to manuka honey against *S. aureus* and *P. aeruginosa* and superior cellular antioxidant activity compared to manuka honey. Dżugan et al. (2020) reported that buckwheat honey is the darkest honey in Poland and possesses the strongest antibacterial and antioxidant properties. They also found that its polyphenol content provides protective effects against hydroxyl radicals in *Saccharomyces cerevisiae* yeast cells. Van Den Berg et al. (2008) highlighted buckwheat honey's effectiveness in wound-healing products due to its ability to reduce reactive oxygen species (ROS) levels, attributing this effect to the honey's high phenolic content, low pH, and high free acid levels, which enhance its antibacterial and wound-healing properties. Hammond et al. (2016) demonstrated that buckwheat honey has a predominantly bactericidal effect against various pathogens, including *Clostridium difficile*, Methicillin-Resistant *S. aureus* (MRSA), *P. aeruginosa*, Methicillin-Susceptible *S. aureus* (MSSA), Vancomycin-Resistant *Enterococcus faecalis*, and *Klebsiella pneumoniae*. Jiang et al. (2020) found that buckwheat honey supports gut health by selectively promoting beneficial Bifidobacteria while inhibiting pathogenic bacteria in the gastrointestinal tract. Cheng et al. (2015) showed that buckwheat honey enhances antioxidant capacity and protects against liver and DNA damage induced by carbon tetrachloride (CCl_4) in mice. The honey displayed high total phenolic content, free radical scavenging capacity, and ferric reducing antioxidant properties. A 10-week treatment with buckwheat honey improved serum antioxidant levels and provided hepatoprotective effects, shielding DNA from CCl_4-induced oxidative damage. Salonen et al. (2017) further confirmed buckwheat honey's potent antibacterial effects against *P. aeruginosa* and *S. aureus*, as well as its high antioxidant activity.

Carob Honey

It is an under-represented unifloral honey produced from the nectar and pollen of carob trees (*Ceratonia siliqua* L.) (Terrab et al., 2003). Carob trees are found in regions with a Mediterranean basin (Terrab et al., 2003; Tous et al., 2013). The honey is derived from the sweet secretions produced by certain insects that feed on the leaves and stems of these trees. Carob honey typically consists of insect secretions or the extracts from the carob plant itself. Carob honey has been shown to possess high antioxidant content and antioxidant properties. El-haskoury et al. (2018) reported that carob honey provides protective effects against kidney and liver toxicity induced by CCl4, demonstrating its effectiveness by maintaining the activity of the antioxidant defence system. Fihri et al. (2016) found that carob honey has therapeutic effects against lead-induced liver and kidney toxicity and anemia. El-Haskoury et al. (2019) observed that the ethyl acetate extract of carob honey is particularly effective in reducing hyperglycemia and improving lipid and liver function disorders. Furthermore, El Haskoury et al. (2015) reported that carob honey exhibits diuretic, natriuretic, and kaliuretic activities without the side effects commonly associated with furosemide-induced hypokalemia. Bouhlali et al. (2019) highlighted that carob honey, rich in malic, citric, succinic, and oxalic acids, demonstrates antioxidant properties due to its ability to bind with metals.

Eucalyptus Honey
It is produced from the pollen and nectar of *Eucalyptus* trees (genus *Eucalyptus*). These trees are native to Australia but have been introduced to various parts of the world with suitable climates, including regions of the Americas, Africa, and parts of Europe (Bobis et al., 2020). Valdés-Silverio et al. (2018) reported that eucalyptus honey effectively inhibits the growth of *S. aureus*, *S. pyogenes*, *P. aeruginosa*, and *E. coli*. Gweirif et al. (2015) found that Libyan eucalyptus honey exhibited a significantly larger inhibition zone against multidrug-resistant bacteria (MDR) compared to conventional antibiotics. Cianciosi et al. (2021) demonstrated that eucalyptus honey protects against inflammation by modulating key inflammatory biomarkers, such as TNF-α, IL-10, and iNOS, and reducing NF-κB expression in macrophages stressed with bacterial endotoxin lipopolysaccharide, while also improving mitochondrial respiration and related parameters. Proaño et al. (2021) found that eucalyptus honey effectively inhibited biofilm formation and disrupted pre-formed biofilms of two multidrug-resistant human pathogens, *S. aureus* MRSA S21 and *P. aeruginosa* P28.

Fireweed Honey
The plant *Epilobium angustifolium* L., which is the source of fireweed honey, is a circum-boreal species commonly found in the temperate zones of North America and Eurasia (Adamczak et al., 2019). Salonen et al. (2017) reported that fireweed honey exhibits both antibacterial activity against *P. aeruginosa* and *S. aureus* and antioxidant properties.

Heather Honey
It is produced from the nectar of heather plants, specifically from the species *Calluna vulgaris* (L.) Hull and various *Erica* species. In Portugal, heather honey is primarily derived from *Erica* sp., while in Spain and France, it comes from both *Calluna* and *Erica* species (Feás et al., 2013). *C. vulgaris* (heather), a significant plant for honey production, yields honey that is highly valued by beekeepers (Rodrigues da). This honey, derived from the nectar of *C. vulgaris*, is characterized by its high content of catechin derivatives and abscisic acids. These components are associated with the honey's strong antioxidant capacity and biological activity. Rodrigues da Silva et al. (2021) demonstrated that heather honey effectively prevents oxidative damage in human erythrocytes induced by ROO• radicals by inhibiting hemolysis, lipid peroxidation, and hemoglobin oxidation. Kolayli et al. (2016)

also noted that heather honey is one of the varieties with high anti-hyaluronidase activity.

Lavender Honey

It is typically produced in the Mediterranean region, and species of the *Lavandula* genus, such as *Lavandula latifolia* Medik. LC, *L. stoechas* L., L. *penduculata* (Mill.) Cav., *L. viridis* L'Hér., and *L. x allardii*, are important sources of this honey (Castro-Vázquez et al., 2014; Lusby et al., 2006; Soares et al., 2018). Lusby et al. (2006) have shown that *L. x allardii* honey has beneficial effects on wound healing. Aliyu and Aliyu (2022) have found that monofloral lavender honey could be effective against vulvovaginal candidiasis due to its high fluconazole resistance, and it inhibits the growth of all Candida species, showing the lowest minimum inhibitory concentration. Additionally, Estevinho et al. (2011) tested lavender honey samples against *Candida albicans*, *C. krusei*, and *Cryptococcus neoformans* and found that its non-sugar components reduced the growth of all yeast species.

Linden Honey

Linden honey, also known as lime honey, is derived from the nectar of *Tilia* species, commonly known as linden or lime trees. These trees are widespread in temperate regions of the Northern Hemisphere, including Europe, Asia, and North America (Jones, 1968). Stojiljković et al. (2024) studied the use of linden honey in reducing oxidative damage caused by radiation in the blood of irradiated rats. Sakač et al. (2022) reported that linden honey exhibits strong antioxidant, antibacterial (*E. faecalis* (ATCC 29212), *S. epidermidis*, *S. aureus* (ATCC 25923), *Proteus mirabilis*, *E. coli* (ATCC 8739)), and antiproliferative properties. Hulea et al. (2022) highlighted that linden honey has a high antioxidant profile and antimicrobial activity, showing effectiveness against certain bacterial/fungal ATCC strains isolated from chronic otitis in dogs, as well as multidrug-resistant strains. Balázs et al. (2021) found that linden honey is effective against common biofilm-forming respiratory pathogens, including *Haemophilus* spp., *P. aeruginosa*, and *S. pneumoniae*, demonstrating both antibacterial and antibiofilm properties.

Manuka Honey

It is a monofloral honey derived from the Manuka tree (*Leptospermum scoparium* J. R. Forst & G. Forst.), which belongs to the Myrtaceae family (Johnston et al., 2018). This tree primarily grows as a shrub or small tree in New Zealand and Australia (Ahmed & Othman, 2013; Johnston et al., 2018). Manuka honey is known for its antioxidant, antimicrobial, and antiproliferative properties (Ahmed & Othman, 2013; Carter et al., 2016). In addition to inhibiting bacterial growth, compounds like glyoxal and methylglyoxal found in Manuka honey have been reported to enhance wound healing and tissue regeneration due to their immunomodulatory effects (Sinha et al., 2023). Manuka honey is used in the treatment of various types of open wounds, including diabetic foot ulcers, pressure sores, burns, and surgical wounds. The product helps prevent bacterial infections, maintain wound moisture, and speed up natural healing processes (Lusby et al., 2006). Hillitt et al. (2017) investigated the effect of Manuka honey on the cell-wall-free bacteria *Ureaplasma urealyticum* and *U. parvum*, finding it effective against all isolates, including antibiotic-resistant strains. They also determined that the bacteria were unable to develop resistance to this honey. Al-Kafaween and Nagi Al-Jamal (2022) found that Manuka honey exhibited a stronger effect on *P. aeruginosa* and *S. pyogenes* in both planktonic and biofilm cultures compared to other tested honeys. They reported that MH demonstrated potent antibacterial, antibiofilm, and anti-virulence activities against all isolates, including antibiotic-resistant bacteria.

Rapeseed Honey
Rapeseed honey is produced from the nectar of the rapeseed plant (*Brassica napus* L.), which is a member of the Brassicaceae family. The rapeseed plant is widely cultivated in temperate regions around the world, including Europe, North America, and parts of Asia. Bucekova et al. (2019) and Farkasovska et al. (2019) reported that rapeseed honey exhibits antibacterial activity against *S. aureus* and *P. aeruginosa.* Hulea et al. (2022) found that rapeseed honey has a high antioxidant profile and is effective against certain ATCC bacterial/fungal strains isolated from chronic otitis in dogs, as well as against multidrug-resistant strains.

Raspberry Honey
Raspberry honey, also known as red raspberry honey, is produced from the plant *Rubus idaeus* L., a major agricultural crop primarily cultivated in Eastern Europe and Northern Asia (Leoni et al., 2024; Saftić Martinović et al., 2024). Studies have demonstrated its therapeutic potential; Oinaala et al. (2015) found that Rubus honey is effective against *Clostridium perfringens*, while Salonen et al. (2017) reported its strong antibacterial activity against *P. aeruginosa* and *Staphylococcus aureus*, along with notable antioxidant properties. Additionally, Bucekova et al. (2019) confirmed its potent antibacterial effects against *S. aureus* and *P. aeruginosa.*

Sunflower Honey
It is a type of honey produced from the nectar and pollen of the sunflower (*Helianthus annuus* L.) plant. Although the sunflower plant is originally native to the Americas, it is now widely cultivated around the world. Sunflower honey has been the subject of several studies highlighting its beneficial properties. Şahin (2021a) reported that fresh sunflower honey samples exhibit anticholinesterase activity by inhibiting the enzymes acetylcholinesterase and butyrylcholinesterase. Milosavljević et al. (2021) found that organic sunflower honey has antimicrobial effects, showing significant inhibition zones ranging from 40% to 100% against five bacterial strains: *S. aureus*, *E. coli*, *Salmonella enterica*, *Yersinia enterocolitica*, and *Bacillus subtilis*. Şahin (2021b) observed that sunflower honey samples demonstrate excellent antioxidant properties and effectively inhibit lipid peroxidation. Additionally, Balázs et al. (2021) identified that sunflower honey has antibacterial and antibiofilm effects against common respiratory pathogens that form biofilms, such as *Haemophilus* spp., *P. aeruginosa*, and *Streptococcus pneumoniae.*

Sidr Honey
It is a high-quality monofloral honey derived from the nectar and pollen of *Ziziphus* species, which are predominantly found in regions such as the Middle East, North

Africa, India, and Pakistan (Saied et al., 2008). Ghramh et al. (2020) investigated the anticancer, immunomodulatory, and antimicrobial effects of Sidr honey sourced from Saudi Arabia and Pakistan. The honey was diluted to 20%, and it demonstrated the ability to produce spherical silver nanoparticles, modulate normal splenic cell growth, and exhibit antimicrobial activity against various pathogens, including *Bacillus subtilis*, *E. coli*, *P. aeruginosa,* and *Candida albicans*. When assessing its anticancer properties, Sidr honey showed effectiveness against HepG2 liver cancer cells but did not exhibit the same effect on HeLa cervical cancer cells. Manal Mohamed Elhassan et al. (2015) investigated the anti-ulcer effects of Sidr honey (*Ziziphus spina-christi* L. Wild.) and its role in preventing oxidative damage and apoptosis during ethanol-induced gastric ulceration. Experimental animals were treated with Sidr honey at doses of 100, 250, and 500 mg/kg or omeprazole before exposure to 95% ethanol. Sidr honey significantly inhibited gastric lesions at a 100 mg/kg dose, reduced lipid peroxidation, prevented glutathione depletion, and modulated Bax protein immunoexpression. These findings suggest that Sidr honey acts as a potent antioxidant and antiapoptotic agent, offering gastroprotective effects. Othman (2014) reported that Sidr honey from Yemen and Egypt exhibited antibacterial activity against various bacteria, including *Salmonella typhi*, *Neisseria meningitides*, *E. coli*, *Klebsiella pneumoniae*, *S. aureus*, *P. aeruginosa*, *Haemophilus influenzae*, *Shigella flexneri*, and *Proteus vulgaris*. Similarly, Al-Kafaween et al. (2021) demonstrated Sidr honey's therapeutic potential against Pseudomonas aeruginosa and Streptococcus pyogenes. Alzubier and Okechukwu (2011) evaluated the effects of Yemeni Sidr honey on acetic acid and formalin-induced writhing, carrageenan and histamine-induced paw edema, and yeast-induced hyperthermia. They found that at doses of 100, 200, and 500 mg/kg, Sidr honey exhibited anti-inflammatory, analgesic, and antipyretic effects. Djebli et al. (2021) confirmed the significant gastroprotective effects of Saharan Sidr honey against HCl/alcohol-induced gastric ulcers through histological evaluations, demonstrating its ability to significantly reduce plasma TNF-α, IL-6, and PGE2 levels.

References

Adamczak, A., Dreger, M., Seidler-Łożykowska, K., & Wielgus, K. (2019). Fireweed (Epilobium angustifolium L.): Botany, phytochemistry and traditional uses. A review. *Herba Polonica, 65*(3), 51–63.

Ahmed, S., & Othman, N. H. (2013). Review of the medicinal effects of tualang honey and a comparison with manuka honey. *The Malaysian Journal of Medical Sciences, 20*(3), 6.

Ajibola, A., Chamunorwa, J. P., & Erlwanger, K. H. (2012). Nutraceutical values of natural honey and its contribution to human health and wealth. *Nutrition & Metabolism, 9*(1), 61.

Aliyu, A., & Aliyu, A. (2022). A scoping review: Efficacy of medical-grade honey as an alternative therapy against vulvovaginal candidiasis. *EC Microbiology, 18*, 08–24.

Al-kafaween, M. A., Hilmi, A. B. M., Al-Jamal, H. A. N., Al-Groom, R. M., Elsahoryi, N. A., & Al-Sayyed, H. (2021). Potential antibacterial activity of Yemeni Sidr honey against Pseudomonas aeruginosa and Streptococcus pyogenes. *Anti-Infective Agents, 19*(4), 51–65.

Al-Kafaween, M. A., & Nagi Al-Jamal, H. A. (2022). A comparative study of antibacterial and anti-virulence activities of four selected honeys to Manuka honey. *Iranian Journal of Microbiology, 14*(2), 238–251.

Alzubier, A. A., & Okechukwu, P. N. (2011). Investigation of anti-inflammatory, antipyretic and analgesic effect of Yemeni Sidr honey. *World Academy of Science, Engineering and Technology, 80*, 47–52.

Bacandritsos, N., Sabatini, A., Papanastasiou, I., & Saitanis, C. (2006). Physico-chemical characteristics of Greek fir honeydew honey from Marchalina hellenica (Gen.) in comparison to other Mediterranean honeydew honeys. *Italian Journal of Food Science/Rivista Italiana di Scienza degli Alimenti, 18*(1), 21–31.

Bagatir, G., Kaya, M., Suer, I., Cefle, K., Palanduz, A., Palanduz, S., Becerir, H. B., Koçyiğit, M., & Ozturk, S. (2022). The effect of Anzer honey on X-ray induced genotoxicity in human lymphocytes: An in vitro study. *Microscopy Research and Technique, 85*(6), 2241–2250.

Balázs, V. L., Nagy-Radványi, L., Filep, R., Kerekes, E., Kocsis, B., Kocsis, M., & Farkas, Á. (2021). In vitro antibacterial and antibiofilm activity of Hungarian honeys against respiratory tract bacteria. *Food, 10*(7), 1632.

Bobis, O., Moise, A. R., Ballesteros, I., Reyes, E. S., Durán, S. S., Sánchez-Sánchez, J., Cruz-Quintana, S., Giampieri, F., Battino, M., & Alvarez-Suarez, J. M. (2020). Eucalyptus honey: Quality parameters, chemical composition and health-promoting properties. *Food Chemistry, 325*, 126870.

Bouhlali, E. D. T., Bammou, M., Sellam, K., El Midaoui, A., Bourkhis, B., Ennassir, J., Alem, C., & Filali-Zegzouti, Y. (2019). Physicochemical properties of eleven monofloral honey samples produced in Morocco. *Arab Journal of Basic and Applied Sciences, 26*(1), 476–487.

Bowie, A. (2020). The ritual role of honey in ancient Egypt, Hatti and Greece. *Истраживања, 31*, 7–23.

Broznić, D., Ratkaj, I., Malenica Staver, M., Kraljević Pavelić, S., Žurga, P., Bubalo, D., & Gobin, I. (2018). Evaluation of the antioxidant capacity, antimicrobial and antiproliferative potential of fir (Abies alba Mill.) honeydew honey collected from Gorski kotar (Croatia). *Food Technology and Biotechnology, 56*(4), 533–545.

Bucekova, M., Jardekova, L., Juricova, V., Bugarova, V., Di Marco, G., Gismondi, A., Leonardi, D., Farkasovska, J., Godocikova, J., Laho, M., Klaudiny, J., Majtan, V., Canini, A., & Majtan, J. (2019). Antibacterial activity of different blossom honeys: New findings. *Molecules, 24*(8), 1573.

Carter, D. A., Blair, S. E., Cokcetin, N. N., Bouzo, D., Brooks, P., Schothauer, R., & Harry, E. J. (2016). Therapeutic manuka honey: No longer so alternative. *Frontiers in Microbiology, 7*, 194754.

Castro-Vázquez, L., Leon-Ruiz, V., Alañon, M. E., Pérez-Coello, M. S., & González-Porto, A. V. (2014). Floral origin markers for authenticating Lavandin honey (Lavandula angustifolia x latifolia). Discrimination from lavender honey (Lavandula latifolia). *Food Control, 37*, 362–370.

Cheng, N., Wu, L., Zheng, J., & Cao, W. (2015). Buckwheat honey attenuates carbon tetrachloride-induced liver and DNA damage in mice. *Evidence-based Complementary and Alternative Medicine, 2015*(1), 987385.

Cianciosi, D., Forbes-Hernandez, T. Y., Alvarez-Suarez, J. M., Ansary, J., Quinzi, D., Amici, A., Navarro-Hortal, M. D., Esteban-Muñoz, A., Quiles, J. L., Battino, M., & Giampieri, F. (2021). Anti-inflammatory activities of Italian chestnut and eucalyptus honeys on murine RAW 264.7 macrophages. *Journal of Functional Foods, 87*, 104752.

Çobanoğlu, D. N., Akyıldız, İ. E., Kızılpınar Temizer, İ., Damarlı, E., & Çelik, Ş. (2023). Phenolic compound, organic acid, mineral, and carbohydrate profiles of pine and blossom honeys. *European Food Research and Technology, 249*, 1503–1515.

Codex-Alimentarius. (1981). *Revised codex standard for honey (Codex Stan 12–1981)*. Food and Agriculture Organization of the United Nations & World Health Organization.

Dai, Y., Jin, R., Verpoorte, R., Lam, W., Cheng, Y.-C., Xiao, Y., Xu, J., Zhang, L., Qin, X.-M., & Chen, S. (2020). Natural deep eutectic characteristics of honey improve the bioactivity and safety of traditional medicines. *Journal of Ethnopharmacology, 250*, 112460.

Deng, J., Liu, R., Lu, Q., Hao, P., Xu, A., Zhang, J., & Tan, J. (2018). Biochemical properties, antibacterial and cellular antioxidant activities of buckwheat honey in comparison to manuka honey. *Food Chemistry, 252*, 243–249.

Dinkov, D. (2013). *Correlation between antibacterial and antioxidant activity in oak honeydew and Acacia (Robinia pseudoacacia l.) bee honeys*. In Proceedings of the V. international scientific and practical conference on "current state and perspectives of food industry and catering development (pp. 1–5).

Djebli, N., Mustafa, M. R., Keskin, M., & Kolayli, S. (2021). Anti-Ulcerogenic and cytoprotective effects of Saharian (Sidr) honey from Algeria. *Combinatorial Chemistry & High Throughput Screening, 24*(10), 1664–1670.

Doğan, A., & Kolankaya, D. (2005). Protective effect of Anzer honey against ethanol-induced increased vascular permeability in the rat stomach. *Experimental and Toxicologic Pathology, 57*(2), 173–178.

Dżugan, M., Grabek-Lejko, D., Swacha, S., Tomczyk, M., Bednarska, S., & Kapusta, I. (2020). Physicochemical quality parameters, antibacterial properties and cellular antioxidant activity of Polish buckwheat honey. *Food Bioscience, 34*, 100538.

Dżugan, M., Ciszkowicz, E., Tomczyk, M., Miłek, M., & Lecka-Szlachta, K. (2024). Coniferous honeydew honey: Antibacterial activity and anti-migration properties against breast cancer cell line (MCF-7). *Applied Sciences, 14*(2), 710.

Eker, Ö., & Kök, A. (2017). Divân-I Hikmet ve Atabetü'l Hakâyik'taki Bal ve Söz Metaforları. *Türk Dünyası Araştırmaları, 115*(227), 103–114.

El Haskoury, R., Zizi, S., Touzani, S., Al Waili, N., AlGhamdi, A., & Lyoussi, B. (2015). Diuretic activity of carob (Ceratonia silique L.) honey: Comparison with furosemide. *African Journal of Traditional, Complementary and Alternative Medicines, 12*(4), 128–133.

El-haskoury, R., Al-Waili, N., Kamoun, Z., Makni, M., Al-Waili, H., & Lyoussi, B. (2018). Antioxidant activity and protective effect of carob honey in CCl4-induced kidney and liver injury. *Archives of Medical Research, 49*(5), 306–313.

El-Haskoury, R., Al-Waili, N., El-Hilaly, J., Al-Waili, W., & Lyoussi, B. (2019). Antioxidant, hypoglycemic, and hepatoprotective effect of aqueous and ethyl acetate extract of carob honey in streptozotocin-induced diabetic rats. *Veterinary World, 12*(12), 1916–1923.

El-Sofany, A., Naggar, Y., Naiem, E., & Seif, A. (2018). Characterization of Apis mellifera honey of different botanical and geographical origins in Egypt. *Journal of Experimental Biology, 14*, 75.

Eraslan, G., Kanbur, M., Silici, S., & Karabacak, M. (2010). Beneficial effect of pine honey on trichlorfon induced some biochemical alterations in mice. *Ecotoxicology and Environmental Safety, 73*(5), 1084–1091.

Estevinho, M. L., Afonso, S. E., & Feás, X. (2011). Antifungal effect of lavender honey against Candida albicans, Candida krusei and Cryptococcus neoformans. *Journal of Food Science and Technology, 48*(5), 640–643.

Eteraf-Oskouei, T., & Najafi, M. (2013). Traditional and modern uses of natural honey in human diseases: A review. *Iranian Journal of Basic Medical Sciences, 16*(6), 731–742.

Farkasovska, J., Bugarova, V., Godocikova, J., Majtan, V., & Majtan, J. (2019). The role of hydrogen peroxide in the antibacterial activity of different floral honeys. *European Food Research and Technology, 245*, 2739–2744.

Feás, X., Iglesias, A., Rodrigues, S., & Estevinho, L. M. (2013). Effect of Erica sp. honey against microorganisms of clinical importance: Study of the factors underlying this biological activity. *Molecules, 18*(4), 4233–4246.

Fihri, A. F., Al-Waili, N. S., El-Haskoury, R., Bakour, M., Amarti, A., Ansari, M. J., & Lyoussi, B. (2016). Protective Effect of Morocco Carob Honey Against Lead-Induced Anemia and Hepato-Renal Toxicity. *Cellular Physiology and Biochemistry, 39*(1), 115–122. https://doi.org/10.1159/000445610

Ghramh, H. A., Ibrahim, E. H., & Kilany, M. (2020). Study of anticancer, antimicrobial, immunomodulatory, and silver nanoparticles production by Sidr honey from three different sources. *Food Science & Nutrition, 8*(1), 445–455.

Gweirif, S., Ahmed, N., Elmhdwi, M., Altayar, M., & Attitalla, I. (2015). Antibacterial activity of eucalyptus honey of Libyan against multi drug resistant bacteria (MDR). *International Journal of Microbiological Research, 6*, 240–244.

Hammond, E. N., Duster, M., Musuuza, J. S., & Safdar, N. (2016). Effect of United States buckwheat honey on antibiotic-resistant hospital acquired pathogens. *Pan African Medical Journal, 25*, 212.

Hasam, S., Qarizada, D., & Azizi, M. (2020). A review: Honey and its nutritional composition. *Asian Journal of Research in Biochemistry, 7*(3), 34–43.

Hernanz, D., Jara-Palacios, M. J., Santos, J. L., Gómez Pajuelo, A., Heredia, F. J., & Terrab, A. (2023). The profile of phenolic compounds by HPLC-MS in Spanish oak (Quercus) honeydew honey and their relationships with color and antioxidant activity. *LWT – Food Science and Technology, 180*, 114724.

Hillitt, K. L., Jenkins, R. E., Spiller, O. B., & Beeton, M. L. (2017). Antimicrobial activity of Manuka honey against antibiotic-resistant strains of the cell wall-free bacteria Ureaplasma parvum and Ureaplasma urealyticum. *Letters in Applied Microbiology, 64*(3), 198–202.

Hulea, A., Obiştioiu, D., Cocan, I., Alexa, E., Negrea, M., Neacşu, A.-G., Hulea, C., Pascu, C., Costinar, L., Iancu, I., Tîrziu, E., & Herman, V. (2022). Diversity of Monofloral honey based on the antimicrobial and antioxidant potential. *Antibiotics, 11*(5), 595.

Jiang, L., Xie, M., Chen, G., Qiao, J., Zhang, H., & Zeng, X. (2020). Phenolics and carbohydrates in buckwheat honey regulate the human intestinal microbiota. *Evidence-based Complementary and Alternative Medicine, 2020*(1), 6432942.

Johnston, M., McBride, M., Dahiya, D., Owusu-Apenten, R., & Nigam, P. S. (2018). Antibacterial activity of Manuka honey and its components: An overview. *AIMS Microbiology, 4*(4), 655–664.

Jones, G. N. (1968). Taxonomy of American species of linden (Tilia) 39. *Illinois Biological Monographs, 39*.

Kafantaris, I., Tsadila, C., Nikolaidis, M., Tsavea, E., Dimitriou, T. G., Iliopoulos, I., Amoutzias, G. D., & Mossialos, D. (2021). Transcriptomic analysis of Pseudomonas aeruginosa response to pine honey via RNA sequencing indicates multiple mechanisms of antibacterial activity. *Food, 10*(5), 936.

Khan, I. U., Dubey, W., & Gupta, V. (2014). Medicinal properties of honey: A review. *Indian Journal of Pure & Applied Biosciences, 2*(5), 149–156.

Kızılpınar Temizer, İ. (2024). Botanical origin and elemental content of Turkish honey: Implications for health risks from essential and non-essential elements. *International Journal of Environmental Health Research, 34*(3), 1737–1750.

Kocyigit, A., Aydogdu, G., Balkan, E., Yenigun, V. B., Guler, E. M., Bulut, H., Koktasoglu, F., Gören, A. C., & Atayoglu, A. T. (2019). Quercus pyrenaica honeydew honey with high phenolic contents cause DNA damage, apoptosis, and cell death through generation of reactive oxygen species in gastric adenocarcinoma cells. *Integrative Cancer Therapies, 18*, 1534735419876334.

Kolayli, S., Sahin, H., Can, Z., Yildiz, O., & Sahin, K. (2016). Honey shows potent inhibitory activity against the bovine testes hyaluronidase. *Journal of Enzyme Inhibition and Medicinal Chemistry, 31*(4), 599–602.

Komeili, A. B. (1990). Beekeeping in Iran. *Bee World, 71*(1), 12–24.

Korkmaz, A., & Kolankaya, D. (2009). Anzer honey prevents N-ethylmaleimide-induced liver damage in rats. *Experimental and Toxicologic Pathology, 61*(4), 333–337.

Kuropatnicki, A. K., Kłósek, M., & Kucharzewski, M. (2018). Honey as medicine: Historical perspectives. *Journal of Apicultural Research, 57*(1), 113–118.

Leoni, V., Panseri, S., Giupponi, L., Pavlovic, R., Gianoncelli, C., Coatti, G., Beretta, G., & Giorgi, A. (2024). Phytochemical profiling of red raspberry (L.) honey and investigation of compounds related to its pollen occurrence. *Journal of the Science of Food and Agriculture, 104*(9), 5391–5406.

Lusby, P. E., Coombes, A. L., & Wilkinson, J. M. (2006). A comparison of wound healing following treatment with Lavandula x allardii honey or essential oil. *Phytotherapy Research, 20*(9), 755–757.
Majtan, J., Bohova, J., Garcia-Villalba, R., Tomas-Barberan, F. A., Madakova, Z., Majtan, T., Majtan, V., & Klaudiny, J. (2013). Fir honeydew honey flavonoids inhibit TNF-α-induced MMP-9 expression in human keratinocytes: A new action of honey in wound healing. *Archives of Dermatological Research, 305*(7), 619–627.
Manal Mohamed Elhassan, T., Siddig Ibrahim, A., Rashad, E., Bassem, Y. S., Mahmood Ameen, A., Saif Eldeen, B., Husham, E., & Eldaw, M. (2015). Effectiveness of Sidr honey on the prevention of ethanol-induced gatroulcerogenesis: Role of antioxidant and antiapoptotic mechanism. *Pharmacognosy Journal, 7*(3), 157–164.
Mazar, A., Namdar, D., Panitz-Cohen, N., Neumann, R., & Weiner, S. (2008). Iron age beehives at Tel Reḥov in the Jordan valley. *Antiquity, 82*(317), 629–639.
Milosavljević, V., Zugic Petrovic, T., Mladenovic, K., Grujović, M., Kolašinac, R., & Orović, D. (2021). Quality assessment, antimicrobial activity organic sunflower honey and use of Maldi-tof mass spectrometry for the identification bacteria isolated from honey. *Progress in Nutrition, 23*(2), e2021182.
Nikhat, S., & Fazil, M. (2022). History, phytochemistry, experimental pharmacology and clinical uses of honey: A comprehensive review with special reference to Unani medicine. *Journal of Ethnopharmacology, 282*, 114614.
Ögretmen, F., & İnanan, B. E. (2014). Evaluation of cryoprotective effect ofturkish pine honey on common carp (Cyprinus Carpio) spermatozoa. *CryoLetters, 35*(5), 427–437.
Oinaala, D., Lehesvaara, M., Lyhs, U., & Tikkanen-Kaukanen, C. (2015). Antimicrobial activity of organic honeys against food pathogenic bacterium Clostridium perfringens. *Organic Agriculture, 5*(2), 153–159.
Othman, A. S. (2014). Antibacterial activity of bee and Yemeni sidr honey against some pathogenic bacterial species. *International Journal of Current Microbiology and Applied Sciences, 3*(10), 1015–1025.
Özdemir, Ö., Çöl, Z., & Ertürk, Ö. (2023). Efficacy of bee products (Anzer honey, pollen and Propolis) in detection and healing of damage induced by antidiabetic drug Vildagliptin/metformin hydrochloride in healthy human pancreatic cells: Cytotoxic, genotoxic and biochemical studies. *Current Medical Science, 43*(6), 1173–1182.
Özkök, A., Ipek, A., & Sorkun, K. (2019). The origin of the oak honeydew honey produced from Quercus frainetto ten. *Fresenius Environmental Bulletin and Advances in Food Sciences, 28*(4), 2650–2655.
Pasini, F., Gardini, S., Marcazzan, G. L., & Caboni, M. F. (2013). Buckwheat honeys: Screening of composition and properties. *Food Chemistry, 141*(3), 2802–2811.
Pita-Calvo, C., & Vázquez, M. (2017). Differences between honeydew and blossom honeys: A review. *Trends in Food Science & Technology, 59*, 79–87.
Proaño, A., Coello, D., Villacrés-Granda, I., Ballesteros, I., Debut, A., Vizuete, K., Brenciani, A., & Álvarez-Suarez, J. M. (2021). The osmotic action of sugar combined with hydrogen peroxide and bee-derived antibacterial peptide Defensin-1 is crucial for the antibiofilm activity of eucalyptus honey. *LWT – Food Science and Technology, 136*, 110379.
Ramić, D., Gobin, I., Vučković, D., Planinić, A., Brčić Karačonji, I., & Smole Možina, S. (2023). Fir (Abies alba Mill.) honeydew honey inhibits growth and adhesion of campylobacter jejuni in vitro. *Applied Sciences, 13*(23), 12735.
Rodov, I. (2020). A taste of honey: Metaphorizing nature in traditional Jewish art. *Journal for the Study of Religion, Nature & Culture, 14*(3), 370–394.
Rodrigues da Silva, L., Campos Chisté, R., & Fernandes, E. (2021). Chemical and antioxidant characterization of the portuguese heather honey from Calluna vulgaris. *Separations, 8*(10), 177.
Saftić Martinović, L., Birkic, N., Pavlešić, T., Planinić, A., Gobin, I., Mišetić Ostojić, D., & Pedisić, S. (2024). Chemical characterization of rare Unifloral honeys of ailanthus (Ailanthus

altissima), fennel (Foenicum vulgare), and raspberry (Rubus idaeus) and their antimicrobial and antioxidant activity. *Agricultural Research*. Early access.

Sahin, H. (2016). Honey as an apitherapic product: Its inhibitory effect on urease and xanthine oxidase. *Journal of Enzyme Inhibition and Medicinal Chemistry, 31*(3), 490–494.

Şahin, B. (2021a). Can sunflower honey have a protective effect against Alzheimer's disease? *Journal of Ongoing Chemical Research, 6*(1), 6–9.

Şahin, B. (2021b). Sunflower honey produced in Turkey scavenges singlet oxygen and inhibits lipid peroxidation via antiradical activity. *Journal of Ongoing Chemical Research, 6*(1), 1–5.

Saied, A. S., Gebauer, J., Hammer, K., & Buerkert, A. (2008). Ziziphus spina-christi (L.) Willd.: A multipurpose fruit tree. *Genetic Resources and Crop Evolution, 55*, 929–937.

Sakač, M., Jovanov, P., Marić, A., Četojević-Simin, D., Novaković, A., Plavšić, D., Škrobot, D., & Kovač, R. (2022). Antioxidative, antibacterial and Antiproliferative properties of honey types from the Western Balkans. *Antioxidants, 11*(6), 1120.

Salonen, A., Virjamo, V., Tammela, P., Fauch, L., & Julkunen-Tiitto, R. (2017). Screening bioactivity and bioactive constituents of Nordic unifloral honeys. *Food Chemistry, 237*, 214–224.

Seyhan, M. F., Yılmaz, E., Timirci-Kahraman, Ö., Saygılı, N., Kısakesen, H. İ., Eronat, A. P., Ceviz, A. B., Bilgiç Gazioğlu, S., Yılmaz-Aydoğan, H., & Öztürk, O. (2017). Anatolian honey is not only sweet but can also protect from breast cancer: Elixir for women from Artemis to present. *International Union of Biochemistry and Molecular Biology Life, 69*(9), 677–688.

Sıcak, Y., Şahin-Yağlıoğlu, A., & Öztürk, M. (2021). Bioactivities and phenolic constituents relationship of Muğla thyme and pine honey of Turkey with the chemometric approach. *Journal of Food Measurement and Characterization, 15*(4), 3694–3707.

Singh, R. (2021). Current honey market in India-volume and value. *International Journal of Ayurveda and Pharma Research, 9*(8), 82–88.

Sinha, S., Sehgal, A., Ray, S., & Sehgal, R. (2023). Benefits of Manuka honey in the management of infectious diseases: Recent advances and prospects. *Mini Reviews in Medicinal Chemistry, 23*(20), 1928–1941.

Soares, S., Grazina, L., Costa, J., Amaral, J. S., Oliveira, M. B. P. P., & Mafra, I. (2018). Botanical authentication of lavender (Lavandula spp.) honey by a novel DNA-barcoding approach coupled to high resolution melting analysis. *Food Control, 86*, 367–373.

Spilioti, E., Jaakkola, M., Tolonen, T., Lipponen, M., Virtanen, V., Chinou, I., Kassi, E., Karabournioti, S., & Moutsatsou, P. (2014). Phenolic acid composition, antiatherogenic and anticancer potential of honeys derived from various regions in Greece. *PLoS One, 9*(4), e94860.

Stavropoulou, E., Ieronymaki, E., Dimitroulia, E., Constantinidis, T. C., Vrioni, G., Tsatsanis, C., & Tsakris, A. (2022). Anti-inflammatory and antibacterial effects and mode of action of Greek arbutus, chestnut, and fir honey in mouse models of inflammation and sepsis. *Microorganisms, 10*(12), 2374.

Stojiljković, V. R., Gavrilović, L. V., Stanić, V. D., Stanković, S. J., Nikolić, D. M., Pejić, S. A., & Pajović, S. B. (2024). Radioprotective effects of linden honey in rat peripheral blood. *Nuclear Technology and Radiation Protection, 39*(1), 81–87.

Tananaki, C., Thrasyvoulou, A., Giraudel, J. L., & Montury, M. (2007). Determination of volatile characteristics of Greek and Turkish pine honey samples and their classification by using Kohonen self organising maps. *Food Chemistry, 101*(4), 1687–1693.

Terrab, A., Díez, M. J., & Heredia, F. J. (2003). Palynological, physico-chemical and colour characterization of Moroccan honeys: III. Other unifloral honey types. *International Journal of Food Science & Technology, 38*(4), 395–402.

Tous, J., Romero, A., & Batlle, I. (2013). The carob tree: Botany, horticulture, and genetic resources. *Horticultural Reviews, 41*, 385–456.

Tsavea, E., Vardaka, F.-P., Savvidaki, E., Kellil, A., Kanelis, D., Bucekova, M., Grigorakis, S., Godocikova, J., Gotsiou, P., Dimou, M., Loupassaki, S., Remoundou, I., Tsadila, C., Dimitriou, T. G., Majtan, J., Tananaki, C., Alissandrakis, E., & Mossialos, D. (2022). Physicochemical characterization and biological properties of pine honey produced across Greece. *Food, 11*(7), 943.

Valdés-Silverio, L. A., Iturralde, G., García-Tenesaca, M., Paredes-Moreta, J., Narváez-Narváez, D. A., Rojas-Carrillo, M., Tejera, E., Beltrán-Ayala, P., Giampieri, F., & Alvarez-Suarez, J. M. (2018). Physicochemical parameters, chemical composition, antioxidant capacity, microbial contamination and antimicrobial activity of eucalyptus honey from the Andean region of Ecuador. *Journal of Apicultural Research, 57*(3), 382–394.

Van Den Berg, A. J. J., Van Den Worm, E., Quarles van Ufford, H. C., Halkes, S. B. A., Hoekstra, M. J., & Beukelman, C. J. (2008). An in vitro examination of the antioxidant and anti-inflammatory properties of buckwheat honey. *Journal of Wound Care, 17*(4), 172–178.

Chapter 9
Innovations in Honey Quality Control

Asif Ahmad, Shiza Atif, and Zain Ul Iman

Abbreviations

AI	Artificial intelligence
GC-MS	Gas chromatography-mass spectrometry
HMF	Hydroxymethylfurfural
HPLC	High-performance liquid chromatography
IoT	Internet of things
NIR	Near infrared

1 Introduction

Honey is a natural sweetener and has been used for thousands of years in history. It is well known for its sweet taste and its medicinal purposes, making it more popular for consumption. The health-friendly implications of honey are also available in numerous historic transcripts of Muslim historians (Khalid et al. 2014; Shafiq et al., 2012). However, its popularity and the complexity of the composition make it a main target for adulteration. Ensuring the safety and integrity of honey has emerged as a critical responsibility for food scientists. This urgency stems from growing concerns among consumers and regulators over economic fraud, which involves

A. Ahmad (✉) · S. Atif
Institute of Food and Nutritional Sciences, PMAS-Arid Agriculture University, Rawalpindi, Pakistan
e-mail: asifahmad@uaar.edu.pk

Z. U. Iman
University Institute of Information Technology, PMAS-Arid Agriculture University, Rawalpindi, Pakistan

M. Zafar et al. (eds.), *Pure Honey: Assurance & Authentication*,
https://doi.org/10.1007/978-3-031-98913-1_9

adulterating honey with syrup sweeteners, artificial additives, pollutants, colorants, pesticide residues, and even antibiotics, all of which compromise the authenticity of this honey (Siddiqui et al., 2017). Significant research in recent years has been carried out for the quantification and detection of various adulterants, pollutants, and chemicals in honey. Determining the honey quality underlines the importance that honey samples fulfill the needed standards for consumer safety and trade purposes (Haderbache & Mohammedi, 2015).

Ensuring quality control of honey is important from the consumer perspective, and modern analytical techniques play an important role in determining the composition and ensuring quality. Conventionally, many methods are in use to determine the presence of adulterants in the honey, but with growing knowledge, numerous advanced techniques have been evolved for the determination of adulterants and ultimately for honey quality (Chin & Sowndhararajan, 2020; Ohmenhaeuser et al., 2013). As honey is used as a natural sweetener with endowed nutritional and medicinal effects, there have been increasing challenges for the quality control, authenticity, and safety of honey in recent years. Consumers often face issues of adulteration, contamination, and the use of poor production procedures.

In the search for answers, over the past years, new methods of analysis of the quality of honey have been developed by researchers and experts from all around the world. Such measures are aimed at assuring that honey is fit for consumption and meets all specifications required. Modern, innovative analytical methods can trace honey adulteration or improper storage. One of the more creative solutions at this point would be the analysis of honey by GC-MS and HPLC, a recent technique able to realize measurements to identify and quantify quite a large number of compounds in honey with very high accuracy. The techniques will therefore work out the adulterants or contaminants that one could find in honey, including pesticides, antibiotics, and heavy metals. An application of infrared spectroscopy is a new approach in the management of honey quality, positioned in chemo-profiling methods. These techniques offer a fast and reliable tool for the chemical composition of honey to be analyzed, and in that way, deviations or abnormalities from the expected profile can thus be sorted out. The other perspective in which developments in DNA-based techniques have greatly influenced is in control of honey quality. Honey quality, considering its complex phytochemical composition and heterogeneous structure, has been discussed here in the global honey industry.

In this way, potential adulterants or contaminants like pesticides, antibiotics, and heavy metals are identified in honey. Another such new perspective in honey quality management is characterized by the application of chemo-profiling methods, such as infrared spectroscopy. These techniques provide a quick and safe way of chemical composition analysis of honey for a possible deviation or abnormality from the usual profile. Moreover, DNA-based techniques have been developed that play an important role in controlling the quality of honey. The aims of this chapter are focused on the various aspects of the problem of honey quality: on one side, the intricate phytochemical composition; and on the other, the heterogeneous structure of the global honey industry.

1.1 Importance of Honey Quality Control

Traditionally honey has been analyzed using many specific methods, including pollen and sensory analysis, but all these methods have constraints regarding their specificity, capacity, and sensitivity to identify the novel and unfamiliar adulterants (Tsagkaris et al., 2021; Araujo et al., 2020). The quality and authenticity of the honey are crucial to preserving its intrinsic characteristics for fit consumption and better acceptability among consumers (Chin & Sowndhararajan, 2020). Several factors play an important role in the preservation of honey, and variations in those factors can endanger the quality of the honey. That is the reason monitoring those factors is critical, like if only the moisture control is not regulated correctly, it impacts the vulnerability of honey to fermentation and preservation. Similarly, the color of honey plays a vital role too. If the hue of the honey can vary from nearly transparent to deep brown, it serves as an indicator of both variety and caliber and impacts consumer inclination and acceptance (Prica et al., 2014). The presence of reducing and nonreducing sugars, including glucose, fructose, and sucrose, in the composition of honey can provide important information on its processing and purity. Moreover, pollen count can also provide the basis for the distinctive botanical identifier, offering the opportunity to trace and verify the origin in terms of both location and plant species Juan-Borrás et al. (2015).

In honey quality control management, hydroxymethylfurfural (HMF) plays a significant role and becomes apparent when honey is either aged or heated beyond limits. It also serves as a gauge to check the honey's freshness and its storage conditions, various quality parameters are presented in Table 9.1. Elevated levels of HMF could refer to prolonged storage and/or the presence of the contaminants (Escriche et al., 2008). Viscosity, scent, electrical conductivity, and the activity of particular enzymes are the extra elements that improve the all-encompassing method of controlling honey quality and offer a more in-depth analysis of its characteristics.

Since advanced analytical methods can give us more definitive information about nearly anything, they are essential for the regulation of honey quality. This allows easy detection and quantification of numerous honey compounds, contaminants, impurities as well as chemical elements (Trifković et al., 2017). Various methods are developed to perform the analyses necessary for the quality control of honey. These methods, which are recognized for their innovation and efficacy, can be used to detect and quantify a large number of compounds in honey, such as sugars, organic acids, phenolic compounds, and amino acids (Schievano et al., 2020) (Table 9.1).

2 Honey Authenticity Issues

For the food industry, quality control and the pureness of the honey have become crucial concerns. Honey loses its purity when adulterants are added and causes adverse health effects. Moreover, it decreases customer acceptance and trust (Soares

Table 9.1 Honey quality control parameters in various honey across the world

Honey type	Region	Compositional quality control measures	Reference values	Reference citations
Acacia honey	Europe	HMF content, pollen analysis	HMF: <10 mg/kg (Codex Alimentarius) pollen: < 15 mg/kg (EU Directive)	Pauliuc et al. (2021)
Tupelo honey	USA	Glucose-fructose ratio, moisture content	Glucose-fructose ratio: 0.8–1.0 Moisture content: <20% (EU Directive)	Gardiner, (2015)
Manuka honey	New Zealand	MGO content, diastase activity	MGO: 100–800 mg/kg diastase activity: >8 Schade (Schade units) (EU Directive)	Machado (2022)
Sidr honey	Middle East	Electrical conductivity, pH	Electrical conductivity: 0.8–1.3 mS/cm pH: 3.2–4.5	Smetanska et al. (2021)
Eucalyptus honey	Australia	Total phenolic content, antioxidant activity	Total phenolic content: 50–100 mg/kg Antioxidant activity: 5–15 μmol Trolox equivalent/kg	Ciappini and Calviño (2020)
Lychee honey	China	Fructose-glucose ratio, moisture content	Fructose-glucose ratio: 1.3–1.5 Moisture content: <20% (EU Directive)	Yingying et al. (2021)
Thyme honey	Mediterranean	Hydroxymethylfurfural (HMF) content, pH	HMF: <10 mg/kg (Codex Alimentarius) pH: 3.5–4.5	Ailli et al. (2024)
Multifloral honey	Pakistan	Diastase activity, color intensity	Diastase activity: >10 Schade (Schade units) (EU Directive) color intensity: 45–75 mm	Raweh et al. (2023)

et al., 2017). The complex compositional nature of honey, attributed to diversified floral sources, makes it difficult to validate its authentication, source, and adulterants (Palma-Morales et al., 2023).

2.1 *Methods of Honey Adulteration*

The important steps in honey authentication involve a need to identify the adulterants, impurities, and foreign matter in honey. Diversified adulterants in honey may exist, and among those, some are listed below (Brar et al., 2023a, b; Chadha et al., 2024).

1. Diluted with poor quality or inexpensive honey substitutes
2. Added either sugar syrups, substituted sweeteners, or artificial sweeteners to increase yield in honey production while reducing production costs
3. Added artificial flavorings and additives to imitate the flavor and aroma of honey in the products
4. Foreign impurities that will taint the honey and endanger the health of the consumers are added
5. Introduction of honey from doubtful or unidentified sources, as this may not adhere to safety and quality standards (some of honey adulterants are listed in Table 9.2)

Due to the above adulteration practices, the quality of honey could seriously be affected, and this could also affect the color, flavor, aroma, nutritional value, and other characteristics of honey. Besides these, when contaminated honey includes pesticides or antibiotics, it poses a serious health threat. In recent years, many significant developments have been made in connection with the management of honey's quality. According to Soares et al. (2017), these developments have introduced the latest improved techniques for assuring the safety and integrity of honey.

3 Traditional Quality Control Methods

The details of traditional quality control methods as applied to honey are described in the following subsections.

3.1 *Physicochemical Tests for Honey Quality Control*

Honey physiochemical parameters are very important in quality control, which expresses the integrity of honey between producers and users. Developments in technology are far better as compared to conventional approaches in terms of quick testing, sensitivity, and specificity. Following recent developments, the physiochemical analysis of honey is now easy and fast to undertake (Ahmed et al., 2024). The next section briefly describes physiochemical parameters for quality control that can be ascertained by conventional analytical methods. In honey, the percentage of moisture chiefly controls its predisposition to fermentation and spoilage. Near-infrared spectroscopy is a new development that allows rapid and noninvasive measurement of moisture content in honey, therefore extending the shelf life and avoiding fermentation. Moreover, high-performance liquid chromatography (HPLC) and ultra-performance liquid chromatography are new techniques for the correct quantification of sugars and adulterants' detection instead of the conventional approach like qualitative tests and Fehling's test for sugar estimation. Otherwise, infrared spectroscopy combined with robust procedures has the

Table 9.2 Common adulterants of honey in different regions

Adulterant	Purpose	Regions	References
Cane or beet-based sugar syrup	Increase volume, mimic natural sweetness	Global, particularly in Asia and Africa	Soares et al. (2017)
High fructose corn syrup	Increase volume, cheaper than honey	North America, Europe, Asia	Brar et al. (2023b)
Glucose syrup	Increase volume, mimic honey viscosity	Global	Ciursa and Oroian (2021)
Rice syrup	Increase volume, mimic natural sweetness	Asia (especially China)	Fakhlaei et al. (2020)
Corn syrup	Increase volume, reduce cost	North America	Chen et al. (2011)
Inverted sugar syrup	Increase volume, maintain similar sweetness	Europe, North America	Tura and Seboka (2019)
Maltose syrup	Increase volume, mimic natural sweetness	Asia, especially Southeast Asia	Ismail (2019)
Molasses	Alter color and flavor, increase volume	North America, Europe	Tura and Seboka (2019)
Fructose	Increase sweetness, mimic honey properties	Global	Downey et al. (2003)
Starch-based syrups	Increase volume, reduce production cost	Global	Brar and Nanda (2024)
Artificial flavors	Mimic natural honey flavors	Global	Zakaria et al. (2011)
Water	Increase volume, reduce cost	Global	Tura and Seboka (2019)
Jaggery	Increase volume, add sweetness and color	South Asia (India, Pakistan)	Brar et al. (2023a)
Industrial glucose	Increase volume, mimic honey properties	Asia, especially China	Soares et al. (2017)
Antibiotics	Prevent bacterial contamination, increase shelf life	Asia (especially China and India)	Al-Waili et al. (2012)
Commercial honey	Blend different types to standardize quality	Global	Soares et al. (2017)
Ascorbic acid (vitamin C)	Mimic antioxidant properties of honey, preserve color	Global	Damto et al. (2024)

potential for the rapid evaluation of enzyme activity that can critically determine the freshness of honey (Hussain et al., 2023).

When honey ages or is overheated, the levels of HMF rise; therefore, low levels of HMF depict that the honey is fresh and less processed. On the other hand, high levels of HMF reflect the overheating of the honey or the old-aged honey (Ohmenhaeuser et al., 2013). Thus, HMF acts as an indicator of honey's age and heating level and can be assessed by using many recently developed methods, including capillary electrophoresis and nuclear magnetic resonance, which offer fast and more reliable detection (Hussain et al., 2023). For the pH and acidity analysis of honey, modern ion-selective electrodes have been developed, as they are crucial parameters for the honey's stability and resistance against microbes. Electrical conductivity is also an important characteristic in honey quality monitoring since it indicates the mineral content and purity of the honey, and it can be measured using a conductivity meter or a bridge. It is now simpler to assess the quality of honey on-site and during industrial operations owing to recent advancements in portable and inexpensive technologies, such as ion-selective electrodes, which provide accurate electrical conductivity measurements (Pascual-Mate et al., 2018).

3.2 Sensory Evaluation

The sensory parameters of honey are as important as the physiochemical parameters. This goes in line with the fact that honey quality or quality assessment is significant through sensing evaluation since it detects the characteristics of flavor, odor, color, and texture. Therefore, based on the differences among the types of kinds of honey, we need to identify and evaluate these characteristics to keep their uniqueness and check the quality of honey. These intrinsic sensory properties are relevant for consumer choice and perception and thus count a lot. These are the characteristics that impress the sensory evaluation, which include color, aroma, flavor, taste, texture, and overall appeal. Traditionally sensory evaluation is carried out by trained panelists using different scales such as, the 9- point Hedonic scale (Akhtar et al., 2010), Yousaf et al. (2013). Notably, new approaches for evaluating its sensory characteristics have been pronounced over the past years. This section considers the standards developed for analyzing honey, with an emphasis on such new approaches (Singh-Ackbarali & Maharaj, 2014).

The very first thing consumers see and observe is the color of the honey that depends on the flower source from which it was derived; nectar color ranges from brown to orange and further to transparent (Insha et al., 2024). Sometimes, we need quantitative data using colorimeter for color analysis. Similarly, honey aroma originates through numerous volatile compounds and could be quantified (Mahmoud et al., 2024; Kang et al., 2023). Advanced techniques based on mass spectrophotometry and gas chromatography provide a complete flavor profile of honey with basic composition. Flavor not only defining the aroma and taste characteristics but also to customer acceptance of honey. As such, sweetness, sourness, bitterness, and individual sensations like floral, fruity, or woody are identified in honey. For an accurate determination of the flavors' profile, sensory panels and electronic tongues were used (Noor et al., 2024; Zhang et al., 2024). The texture of honey also comprises viscosity, graininess, and mouthfeel among others. The properties around the texture can be investigated concerning consistency and presence of crystals by the adoption of either rheometers or sensory evaluation panels Tiwari et al. (2013). The overall acceptability by trained panels or consumer segments of a product is based on its sensory attributes (Cavdan & Drewing, 2024). Sensory evaluation data may therefore give valuable insights into customer preference by statistical analyses conducted for this purpose.

3.3 *Limitations of Traditional Analysis and Quality Control Methods*

Conventional approaches to honey analysis and quality control are highly problematic. One disadvantage is the subjective nature of the sensory evaluation panels, which are diverse in their composition (Puōcion-Jakubik et al., 2020). Besides, traditional sensory testing methods are also time-consuming since trained panelists must physically inspect and score honey products (Puōcion-Jakubik et al., 2020). In addition, the limited capacity of human sensory systems might not allow conventional processes to capture all possible sensations. This is countered by contemporary quality control methods, like AI-driven analyses and sensory tools, with faster, more objective means to evaluate honey characteristics (Shafiee et al., 2013).

The physicochemical studies provided a foundation for the conventional methods of quality control of honey. Nonetheless, these protocols/techniques have some limitations, including special equipment requirements, repeated damage to the sample, and time consumption. Hence, to ensure that only pure and unadulterated honey reaches the market, a full range of physiochemical assays are performed. Although standard approaches provide plenty of essential information, they have their limitations (Pascual-Maté et al., 2018).

4 Melissopalynological Analysis

Melissopalynology is an analysis of pollen and its counting. It is crucial in the determination of honey purity. It protects the interests and trust of honest producers and consumers by ensuring the authenticity of the honey samples and products in the market. There have been various advances in pollen analysis too, number of novel ways, including liquid chromatography-mass spectrometry along with chemometric data analysis. In addition to this, developments in DNA meta-barcoding have also been made to enable a more reliable and complete means of floral source identification of honey (Hawkins et al., 2015; Richardson et al., 2015). This method helps in the accurate identification of pollen numbers and kinds in the honey.

5 Advanced Analytical Techniques

The limitations of traditional methods provide a basis for innovative analytical methods for the examination and control of honey quality. Novel methods have several advantages over conventional methods. They need less processing time, and they also do not degrade the sample. These rapid techniques provide more accurate analysis of adulterants in honey. Rapid advancements in the techniques like HPLC, NIR, FTIR, and GC-MS provide more viable solutions to determine adulterants in honey. With these techniques, use of machine learning and artificial intelligence can augment the process of adulterant determination with increased efficiency, thus offer reliable analysis of moisture, total sugars, reducing sugars, contaminants, and pesticide residues in honey (Chin & Sowndhararajan, 2020).

5.1 High-Performance Liquid Chromatography (HPLC)

The HPLC is another modern technique used for detailed analysis of honey. This technique is based on separation, identification, and quantification of the honey components. The HPLC data can be further used for quality assurance and to maintain the standards of identity in honey (Fatima et al., 2022). The working principle for HPLC involves the passing of a liquid sample through a column filled with a solid stationary phase, and as a result, different mixture components interact differently with these phases, causing the mixture to separate and flow out of the column at different times. These times are known as retention times. For detection purposes, specialized detectors, such as mass spectrometers, UV-Vis, and refractive index detectors, are used to analyze the molecules. The sample preparation steps are very crucial and involve homogenization and extraction using methods like liquid–liquid or solid-phase extraction. After extraction, the extract is cleaned to remove any remaining matrix components. Once the sample is inserted into the column, the chemicals (for instance, the pesticide) interact with the stationary phase Prencipe et al. (2014). Common detectors are tandem MS/MS and photodiode array detectors, which have higher sensitivities in analysis. Repeatability, robustness, sensitivity, specificity, accuracy, and precision of the method should be checked because of regulatory requirements (Puścion-Jakubik et al., 2020; Tsagkaris et al., 2021). The diagram illustrating the working principle of HPLC is provided in Fig. 9.1.

In a 2018 study, HPLC was shown to detect and analyze 24 fragrance allergens in honey. Notably, the test even explains how HPLC can check various types of toxins using just a single analytical run. A recent article describes the use of electrospray ionization-mass spectrometry and HPLC coupled to diode array detection for metabolite fingerprinting of bioactive compounds in *Humulus lupulus L.* This technique is therefore very suitable for investigating such a complex plant as hop, due to the detailed multiple-component analysis of secondary metabolites possible with it (Kaleem et al., 2016; Soo Lim et al., 2018).

The introduction of automation systems capable of handling more than one sample at a time has boosted consistency and quality control in several years due to progressions in HPLC. With the application of sophisticated software and tools meant for the analysis of data from HPLC, it is possible to automatically identify and quantify compounds within the honey. These tools enable complex manipulation and understanding of the information generated by chromatograms. For example, there are innovative developments lately that have resulted in a reduction in the size of sample preparation equipment and an improvement in the extraction techniques used when preparing honey samples for HPLC analysis. This combination improves the ability to distinguish complex adulterants and verify whether honey is authentic or not (Fig. 9.1).

5.2 Gas Chromatography-Mass Spectrometry

Another helpful quality assessment tool is gas chromatography-mass spectrometry (GC-MS). It possesses a mass spectrometry identification of the spectra property and m/z of the compounds, while gas chromatography is done for the separation of molecules based on their volatile character (Pastor et al., 2019). For this purpose, the analysis of the volatile compounds found in honey through GC-MS analysis may be a way. This would be helpful for the identification and quantification of a

Fig. 9.1 Working principle of HPLC

wide range of volatile compounds, such as pollutants, pesticides, and odorants. The GC-MS is a quite powerful technique for determining trace compounds within complex matrices, such as honey, because it has high sensitivity and selectivity (Sotiropoulou et al., 2021). With this system, the chemical composition of honey and chemotaxonomic markers of ammonia can be uniquely determined with gas chromatography-mass spectroscopy, which verifies intangible signs for its authenticity and/or identifies the origin. One of the variants included is multi-dimensional gas chromatography, which has greatly advanced the analysis and separation of complex honey (Karabagias et al., 2019). The other aspect is the sample preparation techniques, such as solid-phase microextraction, that have improved the sensitivity of gas-chromatograph-mass spectrometry on honey samples. In this respect, the GC-MS technique of analyzing volatile chemicals could be used as an experimental method to ensure the quality control of honey and obtain more characterization of information about the chemical composition, authenticity, or geographical origin (Chin & Sowndhararajan, 2020; Puścion-Jakubik et al., 2020; Lehotay & Hajšlová, 2002).

The GC- and GC-MS-based techniques can be used for multiple purposes that may include the analysis of sugars, volatile and flavoring compounds in honey samples. Based on these results, inferences about quality and freshness of honey can be reported. Furthermore, the probability of occurrence for adulterants in honey samples can be inferred through these techniques. The chromatograms that originated from these techniques have valuable information about important quantitative and qualitative parameters. Software and libraries come with GC and GC-MS systems to read the chromatogram peaks and peak areas and identify the substances and their concentrations in the honey sample. Gas chromatography is specifically helpful for identifying nonpolar-moderately polar pesticides to meet regulatory requirements for the safety and quality of honey (Jansson, 2000). Since this method allows separation efficiency to be high enough, it becomes possible to measure and identify such substances at low concentrations. Because pesticide residues in honey are often difficult to identify due to its complex matrix and numerous potential pesticides that might be present. Gas chromatography with a mass spectrometric detector, among other advanced detectors, can be used successfully on several such problems (Jansson, 2000; Hamadamin & Hassan, 2020).

To determine the quality and purity of honey, relevant metabolomics were carried out by GC-MS to quantify specific compounds, i.e., sugars, amino acids, and organic acids, as galveledoscetin markers. This technique assists in discriminating between adulterants and certifies consumer acceptance while fulfilling legislative constraints on honey and is under targeted metabolomics (Kasiotis et al., 2023). The use of biomarkers and metabolomics data obtained through GC-MS has been employed for determining the purity levels of honey samples. GC-MS is a useful technique for the detection of a range of metabolites for adulterant detection purposes. In modern quality control, this technique is vital to ensure traceability and maintain the purity of honey (Shamsudin et al., 2022). A comparative flow chart of targeted versus non-targeted metabolomics is presented in Fig. 9.2.

Another viable technique used with GC-MS is for computation of stable isotope ratio. This technique is valuable to provide indirect information about the source-based origin of honey, and adulteration with distinct isotopic signatures can be identified. For this technique, carbon and oxygen stable isotopes can be used for measuring ratios of stable isotopes (Tsagkaris et al., 2021). The methodology strengthens quality control by guaranteeing authenticity and traceability for kinds of honey and supports any origin or purity claims that exist. The other is referred to as volatile organic compounds (VOCs) profiling, which takes into consideration the volatile compounds responsible for honey taste, aroma, and overall flavor. Each kind of honey has a unique profile of VOCs that originate from floral sources in a specific location. Some researchers indicated that comparison between profiles may help in characterizing the botanical origin of honey, together with recording notes of bad-smelling compounds or even adulterants containing artificial/foreign elements that can be identified (Mahmoud et al., 2024). It is necessary to control the sensory properties of honey and verify its authenticity. On the other hand, GC-MS and atomic absorption spectrophotometry (AAS) analyze antioxidants and trace elements in another way. The AAS is instrumental to identify and quantify minerals in honey. The amounts of minerals can be varied in honey samples according to the soil and environmental conditions; this also helps to detect its place of origin with pollution level and the presence of heavy metal contamination (Pavlin et al., 2023), Khalid et al. (2017). It also helps in quality control by confirming the provenance and purity of the honey and making sure it satisfies safety requirements (Fig. 9.2).

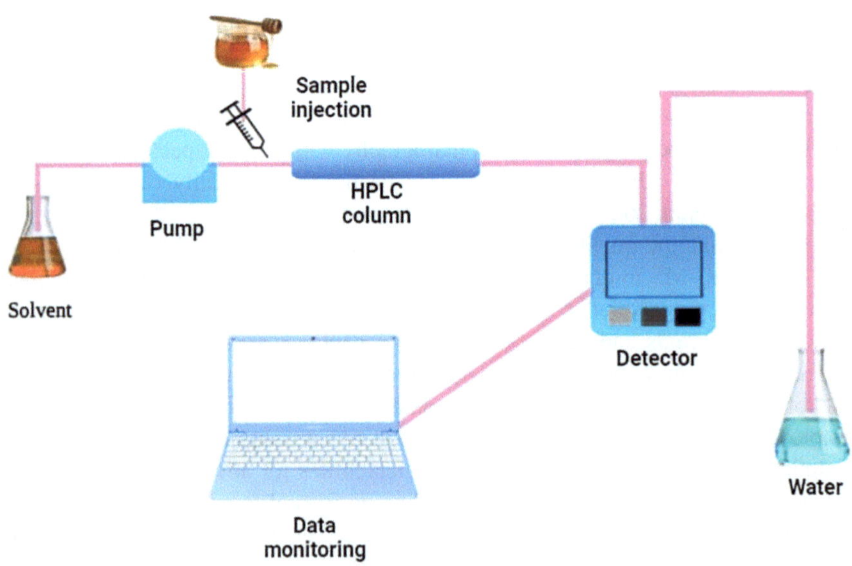

Fig. 9.2 Targeted vs non-targeted metabolomics

Advanced GC-MS methods have immensely improved the detection of honey quality by removing a lot of challenges and mitigating them through various means. Brief details are mentioned as under:

(a) **Complex Matrix**

A variety of components, such as sugars, proteins, moisture, enzymes, vitamins, organic acids, and aromatic compounds, are all present in honey. In addition to these, certain impurities in the form of pesticides also exist. Isolation of these is a complex and cumbersome process, but it is possible to separate volatile and semi-volatile substances with greater efficiency through the use of the GC technique, which simplifies the separation process and yields accurate results with great precision and accuracy along with the identification of compounds (Drabińska et al., 2020).

(b) **Low Pesticide Concentrations**

Pesticide residues in honey often exist in minute quantities, often reported in parts per billion (ppb), but GC techniques can identify these impurities. When combined with sensitive detectors like the electron capture detector (ECD), the technique identifies and quantifies notorious pesticides, like organochlorine and organophosphorus. GC-MS is very sensitive in the identification of small amounts of diversified pesticides (Hamadamin & Hassan, 2020).

(c) **Co-elution with Other Compounds**

Co-elution of pesticide residues with honey matrix substances increases challenges in profiling pesticides while using GC, but the high resolution of this technique allows separating pesticide peaks apart from other chemicals. Furthermore, a diversified range of columns are available that provide support to resolve the issue (Pan et al., 2023).

(d) **Variability of Pesticide Properties**

Certain parameters of pesticide residues vary considerably; these include stability, volatility, and polarity of pesticides, which guide how these can be extracted and made detectable. This technique has the advantage of operating at multiple temperatures and has compatibility with several stationary phases; this property makes it suitable for testing a broad range of pesticides. The thermal stability of the system also allows it to handle a range of volatile compounds and pesticides (Janta et al., 2022; Pico et al., 2020).

(e) **Need for Confirmatory Analysis**

While developing methods for pesticide analysis on GC, the positive pesticide residue detection needs to be verified to prevent false positives. During analysis, a wide range of mass spectral data may sometimes appear. GC-MS offers a viable and sensitive technique for confirmatory analysis in such cases. The unique feature of this technique is the availability of a mass spectrum library search, which offers chemical matching for confirmation (Gavage et al., 2021).

5.3 Near-Infrared Spectroscopy

Infrared spectroscopy is one of the analytical methods in the range used to evaluate honey quality by measuring the interaction between the infrared radiation and the honey, giving information about its chemical composition. The technique works based on the principle that different molecules absorb infrared light at distinct wavelengths; hence it is simple to identify and measure many types of chemical compounds that are found in honey. The working principle of NIR is illustrated in Fig. 9.3.

Honey quality control has seen a lot of progress lately, and more emphasis is further given to innovative state-of-the-art technologies for guaranteeing authenticity and purity. One such advanced method that has proved to be an important tool in monitoring and controlling honey quality is near-infrared spectroscopy (Al-Awadhi & Deshmukh, 2023). The assessment of the interaction between the molecules of honey and near-infrared light is a fast, nondestructive tool in near-IR spectroscopy. This can be applied to obtain further information related to honey's chemical composition: sugar profile, humidity content, and presence of impurities or extraneous matter. Finally, NIR is also applicable in chemo-profiling and fingerprinting honeys

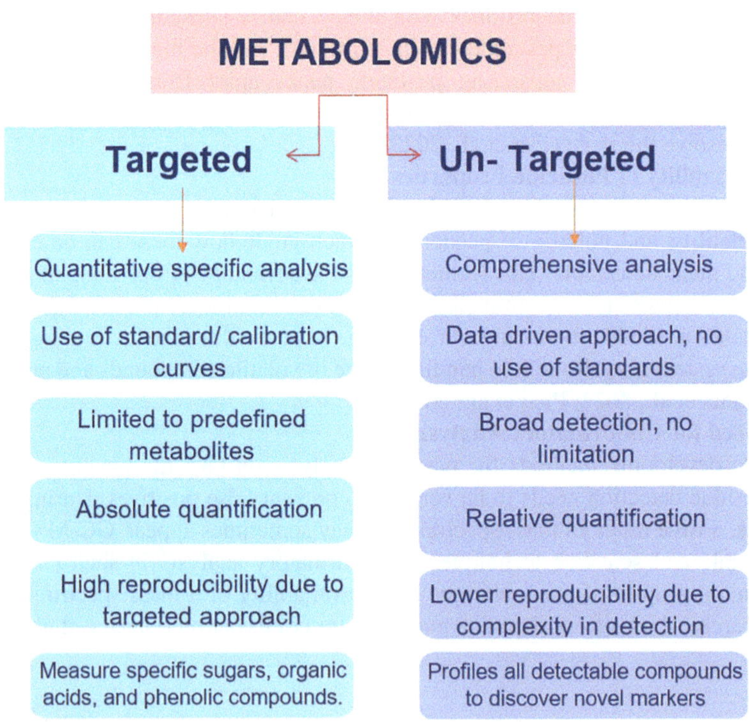

Fig. 9.3 Working principle of NIR – spectroscopy

(Nada et al., 2022). Such methods are mainly applied to investigate various types of food products, among them honey (Fig. 9.3).

Near-infrared spectroscopy has given promising results for honey quality control. A study indicated that the chemical composition of honey could be differentiated according to various kinds of honey, such as mono-floral and polyfloral honey, using NIR. Besides, the application of NIR spectroscopy in detecting and quantifying honey adulterants has also been reported. Because of its nondestructive nature, near-infrared spectroscopy in honey quality management has many benefits as compared to the traditional methods (Huck, 2017). It is a fast technique that allows for the simultaneous measurement of multiple factors. Near-infrared spectroscopy, therefore, eliminates the need for sample preparation, saving both time and resources. Additionally, it offers an inexpensive and environmentally friendly method for honey quality control, as it does not require chemical reagents (Wang et al., 2022). Near-infrared spectroscopy advancement has brought a revolution into how the quality assurance of authenticity and purity of honey is done to be reliable, useful, and innovative in so doing.

Recent advances in NIR techniques benefit the quality control of honey since these techniques are more complex and improvised. FT-NIR spectroscopy is a technique that improves the resolution and sensitivity of NIR measurements by recording a wide spectrum of light simultaneously FTIR (Ahmad & Ayub, 2022, Beć et al., 2022). It finds applications in many food analysis fields, especially invaluable for the identification of minute differences in honey chemical composition regarding sugars, moisture, and amino acids, and provides deep estimation-based quality assessment. Also, very large spatial and spectral information can be obtained about honey samples by combining NIR spectroscopy and imaging in a technique called hyperspectral imaging. This kind of information can be able to help spot heterogeneous areas arising from contamination, adulteration, and crystallization (Shao et al., 2022). This method vividly brings out the quality and homogeneity aspects of honey, so it is a very effective tool in quality control.

In addition, chemometric analysis uses mathematical and statistical techniques to get information about complex NIR spectra in the identification and quantification of particular compounds in honey that allows for the detection of adulteration, verification of botanical and geographical origin, and assessment of the overall quality more accurately (Biswas & Chaudhari, 2024). Chemometric models can be trained to detect patterns related to quality honey, thus allowing for the discrimination between genuine and adulterated samples. Another high-end technique is multipoint NIR spectroscopy, which requires several NIR readings to be taken at different locations in a honey sample. This approach would allow for data obtained to properly represent the whole sample by accounting for sampling heterogeneity (Hong & Chia, 2021). This technique allows one to more reliably assess the quality of honey since it averages spectra from several sites and thus reduces the effect of local oscillations. To this end, multipoint NIR spectroscopy becomes very useful in large-scale quality control procedures wherein precise and consistent measurements are required to uphold product standards.

Yet another advancement in near-infrared technology is diffuse reflectance spectroscopy. It detects scattered light from the sample of honey. Therefore, it gives information about its composition and internal structure. Application of this technique to turbid or opaque samples, for example, crystallized honey, gives good results (Abbas et al., 2020). Honey can be investigated by capturing diffuse reflectance to check on the quality of honey and irregularities or adulteration. This technique offers a nondestructive mechanism for honey sample evaluation, thus perform the analysis for quality pR meters without losing the sample. Therefore, this nondestructive nature with sample preservation makes it valuable to analyze the multiple honey samples using the NIR technologies to predict accurate and reliable results on multiple parameters of honey and provide valid information on authentication, purity, and safety of honey products (Biswas & Chaudhari, 2024).

5.4 *Metabolic Fingerprinting*

The metabolic fingerprinting is an innovative technique used for monitoring and evaluation of honey composition and purity. This technique allows study of multifaceted metabolic profiles of a product. For honey samples, mass spectrometry and HPLCs are used for identification of these markers (Donno et al., 2020). These substances can be volatile: sugars, organic acids, amino acids, and phenolic compounds. Using the metabolic fingerprinting method, one also draws an examination of the low molecular weight metabolites; this information is very significant for the quality of the product and its botanical origin and authenticity and detects adulterants or pollutants, differentiating floral honey from different sources (Danieli & Lazzari, 2022). Metabolomics can be coupled with HPLC, NIR, and GC-MS as described above. The data on metabolic fingerprinting allow differentiation between kinds of honey from different botanical sources and of origins by analysis of specific metabolites. These profiles thus become a chemical marker specific to the geographic location and floral source of the honey (Chin & Sowndhararajan, 2020). One of the most powerful tools currently available for metabolic profiling is NMR spectrometry. It can rapidly provide a profile of various naturally occurring compounds in honey without requiring complex sample preparation (Ohmenhaeuser et al., 2013). Basically, in recent times, NMR spectra have been used increasingly as fingerprinting techniques because these are nondestructive techniques of detection and are highly reliable. They can identify metabolites of different kinds, which later on facilitates their quantification (Onuh & Qiu, 2021). The other typically used method for metabolic fingerprinting is LC combined with MS. LC-MS will be ideal for the sensitive and selective detection of metabolites in honey, for instance, phenolic chemicals, volatile organic compounds, or carbohydrates, since it has the power to separate complicated mixtures and offer mass spectral data (Hassan et al., 2022; Quirantes-Piné et al., 2024).

This method assures the flavor and aroma of honey originating from volatile compounds will be detected by the GC-MS. It enables the processors to maintain quality control parameters to separate the adulterated samples and categorize the honey samples according to botanical origin and geographical identities. During that process, a large dataset is created that can be further interpreted using chemometrics techniques and multivariate analysis. Metabolic profiling with procedures like principal component analysis can, therefore, actually participate in the classification and discrimination of honey samples (Tarapoulouzi et al., 2023). This scope and sensitivity of metabolic fingerprinting are expected to increase in the near future with better analytical techniques and, more importantly, bioinformatics for improved detection of counterfeiting and authentication checks that enable an overall quality assessment of honey. In other words, metabolic fingerprinting is a high-tech methodology for the integral assessment of honey quality. The non-targeted approach enables the identification of known and unknown metabolites, hence making control of the honey quality more robust (Eshawu & Ghalsasi, 2024; Wilson & Poole, 2023).

6 The Role of Technological and Managerial Innovation in Quality Management

To be able to maintain the quality of honey together with techniques of analysis, technical and administrative developments are also crucial. According to Kharbach et al. (2023), technological advancement includes improvement in packing materials as well as methods for maximum maintenance of honey quality and prevention of infection. Even more, traceability technologies, like barcode and RFID tagging, allow for honey to be traced down from hive to consumer, enforcing accountability and transparency in the supply chain. In reverse, managerial innovations that demand setting up systems of quality assurance like ISO certifications and Good Manufacturing Practices are necessary as well (Bk et al., 2023; Daoud, 2022). The development of these mechanisms offers vital methods to establish criteria for quality control at various stages of production, processing, and storage of honey. These methods are now gaining popularity in the honey business and trade with the use of smart technologies and digital platforms, as they allow for real-time data monitoring and analytics (Bose & Padmavati, 2024). These developments not only help to enhance the accuracy and correctness of honey quality control but also secure the long-term viability and competitiveness of the honey business (Hosseini et al., 2024).

In a nutshell, innovations in honey quality control have smoothed the path for detecting adulteration, authentication, and determination of honey quality (Zhang et al., 2023). New techniques are revolutionizing quality control from traditional ways by providing more accurate and full information on composition and origin. Such an integrated approach to honey quality control guarantees consumers a natural, safe, and good-quality product (Brar et al., 2023b; Żak & Wilczyńska, 2023).

7 Integration of New Technologies in Quality Control

Modern spectroscopic techniques, such as NIR, FTIR, and Raman spectroscopy, along with confocal microscopy, are promising technologies. These latest techniques ensure honey quality and safety and allow solid quality control management of honey based on chemical analysis and with the support of honey pollen identification. These techniques provide nondestructive and rapid testing along with validated data about multiple quality factors of honey, including honey composition, moisture, sugar profile, furfurals, and selected adulterants in the honey (Skaff et al., 2022). Furthermore, in this decade, DNA-based techniques are popularizing, including polymerase chain reaction (PCR) and next-generation sequencing. These techniques provide support for the identification of honey based on biological and geographical parameters (Balkanska et al., 2020; Mohamadzade Namin et al., 2023). Therefore, modern technology in quality control of honey changed the concept of honey testing.

Traditional procedures, such as melissopalynology and organoleptic analysis, are still in use in some parts of the world where the resources are limited, although they have time and operator restrictions (Tarapoulouzi et al., 2023). New technologies have emerged to address these constraints and deliver more efficient and dependable solutions. It is suggested that these techniques may be used with modern technologies (Brar et al., 2023a, b; Frigerio et al., 2024). In addition to these advances in chromatography and spectroscopy, including HPLC with tandem detectors, GC with MS detectors has also been made for the easier analysis of honey to analyze sugars and flavor profiles along with the detection and examination of specific chemical substances like flavonoids, phenolic acids, and volatile components in the honey, which give the honey its characteristic color, taste, flavor, and aroma (Ahmad et al., 2015; Tarapoulouzi et al., 2023).

8 AI in Consumer Sensory Analysis

Owing to recent advancements in artificial intelligence (AI), it is now possible to evaluate customer evaluations in great depth. Large datasets for quality parameters can be used to extract, quantify, and interpret client sentiments and preferences with machine learning and natural language processing methods. AI-based systems can assess customers' sentiment ratings, classify them as positive, neutral, or negative, and associate them with specific sensations (Yang et al., 2010). In this way, AI-powered analysis based on sensory aspects enables us to judge consumers' preferences for honey. Equally, future technology such as Standardizing Data Collection with Sensory Analysis Software may serve as a quality control tool. The sensory analysis software was developed for enhanced accuracy and reliability of sensory analysis within the honey quality control field; it could be used by the honey industry business players (Buhociu et al., 2021) (Fig. 9.4).

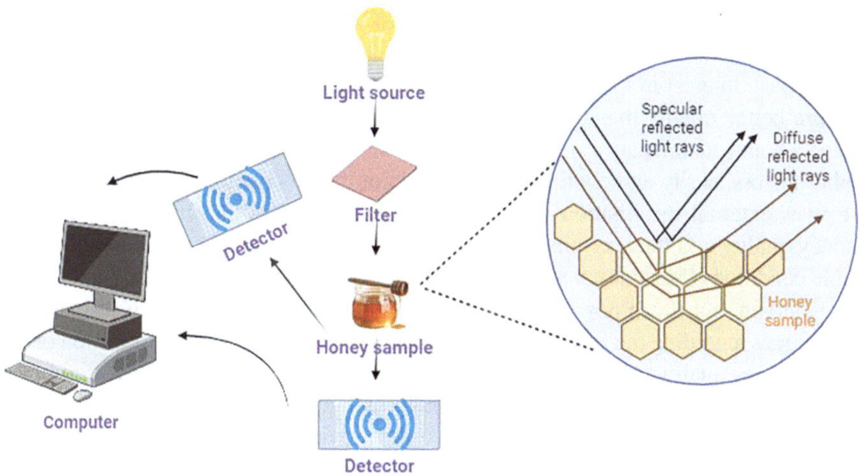

Fig. 9.4 Role of AI in the perseverance of honey integrity

The use of artificial intelligence based on user input data has made significant advances in the field of honey quality management. It helps producers identify the most important sensory aspects that customers consider when buying honey products so that they can adjust their goods accordingly. This method not only improves consumer pleasure but also allows focused marketing techniques and better product development (Fatima et al., 2022). There are many advancements made in honey quality control as a whole; these include sensory quality parameters, AI-supported customer sensory review analysis, and the use of sensory analysis software. The latest advances in sensory analysis have employed artificial intelligence (Deliza & Ares, 2018). The intelligence uses Honey-net to seek out consumer reviews that speak about honey attributes based on taste or aroma, while the tool assists in providing standardized data for collecting these attributes (sensory evaluation). There has been improvement in this area concerning precision and effectiveness in giving correct information regarding the quality of honey. In addition to sensory evaluation, there are new modern methods of assessing honey quality. The figure depicting different roles of AI in the preservation of honey integrity is presented in Fig. 9.4.

As artificial intelligence improves, it becomes ever more likely that user feedback will provide insights that can be used to improve honey production monitoring systems. Thus, the latest advances regarding honey quality control are AI-driven analysis of consumer input, which could benefit the beekeeping sector (Mădaş et al., 2020). For a beekeeper interested in meeting market demands, he or she must pay attention to preferences such as how it looks, smells, or tastes (flavor). Such AI-based tools can be applied for analyzing big datasets such as user reviews, helping us get into details, not just the general picture that is painted through simple words from one's mouth.

9 Use of IoT for Honey Quality Control

In the past, interest in Internet of Things (IoT) technologies has grown as a means to ensure honey quality management. It allows monitoring of different parameters that impact honey maintenance and storage conditions, like temperature, humidity, pH, and other factors, easily and continuously throughout the processing and storage periods. It gives external information on environmental parameters that have an impact on honey quality and thus permits previous ordering for its maintenance in the best possible conditions (Martínez et al., 2017). Honey quality control has certainly undergone a revolution, but it can now be controlled more precisely and effectively using these techniques regarding the genuineness and quality of honey. Thus, the application of the IoT in honey quality monitoring depicts that the conventional practices have combined with today's digital technologies and brought about the new concept of "Precision Apiculture." Such new quality control processes provide more accuracy and completeness in composition, origin, and authenticity management within honey. How all of this system is managed depends upon the use of an IoT network in honey production with interconnected devices and sensors for the collection of data on beehives in real-time Thilakarathne et al. (2021). Parameters of data may vary from simple ones, such as humidity and temperature, independently considered with hive weight and bee activity, to other relevant parameters. This information is then sent to a central system where it can be analyzed and interpreted by the beekeeper or an automated algorithm.

10 Future Directions in Honey Quality Control

Future research and development in honey quality control should prioritize the use of cutting-edge technology alongside conventional methods to guarantee a thorough examination. There should be more research on honey quality control that will emphasize the future use of advanced technology in addition to traditional approaches for a comprehensive analysis. As a result, there is a need for kinds of honey examination to be more accurate and straightforward. Furthermore, artificial intelligence, the Internet of Things, and other emerging technologies are set to revolutionize the future of honey quality control. It will help in making a precise analysis through machine learning or AI by detecting any patterns or irregularities that could mean adulteration or bad quality. More research is required to collect and monitor real-time data across the honey supply chain that could be facilitated by IoT, hence ensuring traceability and quality. Blockchain technology will improve transparency and trust as it securely records and verifies every step taken during the honey production and distribution process. The use of advanced sensor technologies would enable continuous monitoring without destruction, while handheld devices would allow onsite testing. Collaborative platforms can speed up innovation to maintain its purity, authenticity, and safety for human consumption only in such an environment where honey is considered unfit for trade if it is not genuine in its value-added state (El-Banna et al., 2023). In brief, future use of improved methods in controlling honey quality provides valuable ways of ensuring that it remains authentic and excellent when used for trading purposes.

11 Conclusion

The quality of honey has improved by applying new technology and using new tests to control the quality of honey better. Now it is possible to artificial things and also extremely trace back the information on when and what forms of plants these bees visited to gather the nectar so they can make the natural honey from it. By testing honey, one can say that consumer safety is protected while ensuring a safe and competitive market. In the future, research will combine new techniques with traditional approaches. By making use of these methods together, a consistent solution can be provided for proper and effective honey quality evaluation. HPLC, GCMS, Raman spectroscopy, NIR spectroscopy, and FTIR have a bright future in the honey industry for analyzing and quality control measures. The use of artificial intelligence together with machine learning algorithms can be used to control and analyze big volumes of data sets about the product, making it easier to detect counterfeit entries since they are mathematically inconsistent. These advancements offer effective mechanisms that can be used as tools in promoting honest trade-in of well-produced honey while creating accurate findings on this important research matter due to the fast pace at which these developments are changing today's world.

References

Abbas, O., Pissard, A., & Baeten, V. (2020). *Near-infrared, mid-infrared, and Raman spectroscopy chemical analysis of food Yolanda Pico* (2nd ed., pp. 77–134). Academic Press.

Ahmad, A., & Ayub, H. (2022). Fourier transform infrared spectroscopy (FTIR) technique for food analysis and authentication. In *Nondestructive quality assessment techniques for fresh fruits and vegetables* (pp. 103–142). Springer.

Ahmad, A., Kaleem, M., Ahmed, Z., & Shafiq, H. (2015). Therapeutic potential of flavonoids and their mechanism of action against microbial and viral infections – A review. *Food Research International, 77*(2), 221–235.

Ahmed, M. W., Asif, M., Ahmed, R., Khan, A. S., & Raza, R. (2024). Pharmacology, nutrition value and therapeutic potential of honey: A review. *Journal of Pharmacognosy and Phytochemistry, 13*(2), 40–47.

Ailli, A., Zibouh, K., Eddamsyry, B., Drioiche, A., Fetjah, D., Ayyad, F. Z., & Tarik, R. (2024). Physicochemical characterization of Moroccan honey varieties from the Fez-Meknes region and their antioxidant and antibacterial properties. *Metabolites, 14*(7), 364.

Akhtar, S., Riaz, M., Ahmad, A., & Nisar, A. (2010). Physico-chemical, microbiological and sensory stability of chemically preserved mango pulp. *Pakistan Journal of Botany, 42*(2), 853–862.

Al-Awadhi, M., & Deshmukh, R. (2023). Enhancing honey adulteration detection with optimal subspace wavelength reduction in Vis-NIR reflection spectroscopy. *IEEE Access, 11*.

Al-Waili, N., Salom, K., Al-Ghamdi, A., & Ansari, M. J. (2012). Antibiotic, pesticide, and microbial contaminants of honey: Human health hazards. *The Scientific World Journal, 2012*(1), 930849.

Araujo, D., Ruiz Pérez-Cacho, P., Serrano, S., Dios-Palomares, R., & Galán-Soldevilla, H. (2020). Sensory profile and physico-chemical properties of artisanal honey from Zulia, Venezuela. *Foods, 9*(3), 339.

Balkanska, R., Stefanova, K., & Stoikova-Grigorova, R. (2020). Main honey botanical components and techniques for identification: A review. *Journal of Apicultural Research, 59*(5), 852–861.
Beć, K. B., Grabska, J., & Huck, C. W. (2022). Miniaturized NIR spectroscopy in food analysis and quality control: Promises, challenges, and perspectives. *Food, 11*(10), 1465.
Biswas, A., & Chaudhari, S. R. (2024). Exploring the role of NIR spectroscopy in quantifying and verifying honey authenticity: A review. *Food Chemistry*, 138712.
Bk, B., Debnath, S., & Mudliar, S. N. (2023). Advanced computational tools for enhanced food quality and safety. In *Engineering aspects of food quality and safety* (pp. 207–247). Springer.
Brar, D. S., & Nanda, V. (2024). A comprehensive introduction to honey adulteration. In *Advanced techniques of honey analysis* (pp. 63–91). Academic Press.
Brar, D. S., Nayik, G. A., Aggarwal, A. K., Kaur, S., Nanda, V., Saxena, S., & Tolcha, T. D. (2023a). Chemical and functional characteristics to detect sugar syrup adulteration in honey from different botanical origins. *International Journal of Food Properties, 26*(1), 1390–1413.
Brar, D. S., Pant, K., Krishnan, R., Kaur, S., Rasane, P., Nanda, V., & Gautam, S. (2023b). A comprehensive review on unethical honey: Validation by emerging techniques. *Food Control, 145*, 109482.
Buhociu, F. M., Augustin, C. L., Turek Rahoveanu, M. M., Turek Rahoveanu, A., & Zugravu, G. A. (2021). Fuzzy sensory quality certification in intensive organic beekeeping. *Agriculture, 11*(7), 644.
Bose, D., & Padmavati, M. (2024). Honey Authentication: A review of the issues and challenges associated with honey adulteration. *Food Bioscience, 61*, 105004.
Cavdan, M., & Drewing, K. (2024). To touch or not to touch: The linkage between viscosity and unpleasantness. In *International conference on human haptic sensing and touch enabled computer applications* (pp. 61–73). Springer.
Chen, L., Xue, X., Ye, Z., Zhou, J., Chen, F., & Zhao, J. (2011). Determination of Chinese honey adulterated with high fructose corn syrup by near infrared spectroscopy. *Food Chemistry, 128*(4), 1110–1114.
Chin, N. L., & Sowndhararajan, K. (2020). A review on analytical methods for honey classification, identification and authentication. In *Honey analysis, new advances and challenges* (pp. 1–33).
Ciappini, M. C., & Calviño, A. (2020). A holistic view to develop descriptive sheets for argentinean clover and eucalyptus unifloral honey. *Current Nutrition and Food Science, 16*(6), 919–927.
Ciursa, P., & Oroian, M. (2021). Rheological behavior of honey adulterated with agave, maple, corn, rice and inverted sugar syrups. *Scientific Reports, 11*(1), 23408.
Chadha, D., Hamid, N., Poulain, N., Kantono, K., Janani, R., & Teo, P. S. (2024). Beyond borders: A cross-cultural analysis of consumption and purchase behaviour of sweeteners in yoghurts. Food Frontiers, 5(4), 1687–1708.
Damto, T., Zewdu, A., & Birhanu, T. (2024). Impact of different adulterants on honey quality properties and evaluating different analytical approaches for adulteration detection. *Journal of Food Protection, 87*(4), 100241.
Danieli, P. P., & Lazzari, F. (2022). Honey traceability and authenticity. Review of current methods most used to face this problem. *Journal of Apicultural Science, 66*(2), 101–119.
Daoud, B. A. S. (2022). Quality management practices and innovation: The moderating effect of ISO 9001 certification. *Journal of the Knowledge Economy, 13*(3), 2177–2202.
Deliza, R., & Ares, G. (2018). Consumer perception of novel technologies. In *Fruit preservation: Novel and conventional technologies* (pp. 1–20).
Donno, D., Mellano, M. G., Gamba, G., Riondato, I., & Beccaro, G. L. (2020). Analytical strategies for fingerprinting of antioxidants, nutritional substances, and bioactive compounds in foodstuffs based on high performance liquid chromatography–mass spectrometry: An overview. *Food, 9*(12), 1734.
Downey, G., Fouratier, V., & Kelly, J. D. (2003). Detection of honey adulteration by addition of fructose and glucose using near infrared transflectance spectroscopy. *Journal of Near Infrared Spectroscopy, 11*(6), 447–456.

Drabińska, N., Młynarz, P., de Lacy Costello, B., Jones, P., Mielko, K., Mielnik, J., & Ratcliffe, N. M. (2020). An optimization of liquid–liquid extraction of urinary volatile and semi-volatile compounds and its application for gas chromatography-mass spectrometry and proton nuclear magnetic resonance spectroscopy. *Molecules, 25*(16), 3651.

Escriche, I., Visquert, M., Carot, J. M., Domenech, E., & Fito, P. (2008). Effect of honey thermal conditions on hydroxymethylfurfural content prior to pasteurization. *Food Science and Technology International, 14*(5), 29–35.

Eshawu, A. B., & Ghalsasi, V. V. (2024). Metabolomics of natural samples: A tutorial review on the latest technologies. *Journal of Separation Science, 47*(1), 2300588.

El-Banna, B. T., Abouzeid, A. E., El-Damrawy, S. Z., & El-Rayes, T. K. (2023). Effect of bee venom on production performance and immune response of broilers. Egyptian. *Journal of Nutrition and Feeds, 26*(1), 91–100.

Fakhlaei, R., Selamat, J., Khatib, A., Razis, A. F. A., Sukor, R., Ahmad, S., & Babadi, A. A. (2020). The toxic impact of honey adulteration: A review. *Food, 9*(11), 1538.

Fatima, N., Mujtaba, G., Akhunzada, A., Ali, K., Shaikh, Z. H., Rabia, B., & Shahani, J. A. (2022). Prediction of Pakistani honey authenticity through machine learning. *IEEE Access, 10*.

Frigerio, J., Campone, L., Giustra, M. D., Buzzelli, M., Piccoli, F., Galimberti, A., & Labra, M. (2024). Convergent technologies to tackle challenges of modern food authentication. *Heliyon, 10*(11).

Gardiner, S. R. (2015). *Physicochemical and flavor characterization of Tupelo honey*. University of Illinois at Urbana-Champaign.

Gavage, M., Delahaut, P., & Gillard, N. (2021). Suitability of high-resolution mass spectrometry for routine analysis of small molecules in food, feed and water for safety and authenticity purposes: A review. *Food, 10*(3), 601.

Haderbache, L., & Mohammedi, A. (2015). Quality of imported honeys marketed in Algeria. *Journal of Fundamental and Applied Sciences, 7*(1), 139–149.

Hamadamin, A. Y., & Hassan, K. I. (2020). Gas chromatography–mass spectrometry based sensitive analytical approach to detect and quantify non-polar pesticides accumulated in the fat tissues of domestic animals. *Saudi Journal of Biological Sciences, 27*(3), 887–893.

Hassan, N. H., Cacciola, F., Chong, N. S., Arena, K., Marriott, P. J., & Wong, Y. F. (2022). An updated review of extraction and liquid chromatography techniques for analysis of phenolic compounds in honey. *Journal of Food Composition and Analysis, 114*, 104751.

Hawkins, J., De Vere, N., Griffith, A., Ford, C. R., Allainguillaume, J., Hegarty, M. J., & Adams-Groom, B. (2015). Using DNA metabarcoding to identify the floral composition of honey: A new tool for investigating honey bee foraging preferences. *PLoS One, 10*(8), e0134735.

Hong, F. W., & Chia, K. S. (2021). A review on recent near infrared spectroscopic measurement setups and their challenges. *Measurement, 171*, 108732.

Huck, C. W. (2017). Theoretical and technical advancements of near-infrared spectroscopy and its operational impact in industry. *NIR News, 28*(8), 17–21.

Hussain, A., Kausar, T., Rehman, A., Batool, A., Saleem, M., Musharraf, T. M., & Gorsi, F. I. (2023). Evaluation of the phytochemical and medicinal value of lemongrass (Cymbopogon citratus), by conversion into powders and extracts to develop a nutritional bakery product. *Future Integrative Medicine, 2*(3), 129–140.

Hosseini, S. E., Barati, A. A., & Fami, H. S. (2024). Good beekeeping farm management practices, towards sustainable development (experiences from Iran). *Environment, Development and Sustainability*, 1–27.

Insha, R. A. N., Islam, M. N., Gomasta, J., Hasan, M. N., Amin, M. R., Sarmin, N. S., & Rahman, M. M. (2024). Comprehensive honey authentication in Bangladesh: Profiling physicochemical and bioactive compounds to distinguish floral sources and detect adulteration. *Heliyon*.

Ismail, N. I. (2019). Effects of sugar adulterants on the physicochemical properties of natural honey. *Journal of Tomography System and Sensor Application, 2*(2).

Jansson, C. (2000). Multiresidue method for the gas chromatographic determination of pesticides in honey after solid-phase extraction cleanup. *Journal of AOAC International, 83*(3), 714–719.

Janta, P., Wongla, B., Phayoonhong, W., Intarapanich, O., Kokpol, S., Mahatheeranont, S., & Kulsing, C. (2022). Analysis of low-volatility pesticides in cabbage by high temperature comprehensive two-dimensional gas chromatography. *Analytical Methods, 14*(33), 3180–3187.

Juan-Borrás, M., Domenech, E., Conchado, A., & Escriche, I. (2015). Physicochemical quality parameters at the reception of the honey packaging process: Influence of type of honey, year of harvest, and beekeeper. *Journal of Chemistry, 2015*(1), 929658.

Kaleem, M., Ahmad, A., Khalid, S., & Azam, M. T. (2016). HPLC condition optimization for identification of flavonoids from Carissa opaca. *Science International (Lahore), 28*(1), 343–348.

Kang, M. J., Kim, K. R., Kim, K., Morrill, A. G., Jung, C., Sun, S., & Sung, J. (2023). Metabolomic analysis reveals linkage between chemical composition and sensory quality of different floral honey samples. *Food Research International, 173*, 113454.

Karabagias, I. K., Karabagias, V. K., & Badeka, A. V. (2019). The honey volatile code: A collective study and extended version. *Food, 8*(10), 508.

Kasiotis, K. M., Baira, E., Iosifidou, S., Manea-Karga, E., Tsipi, D., Gounari, S., & Lazzari, F. (2023). Fingerprinting chemical markers in the mediterranean orange blossom honey: UHPLC-HRMS metabolomics study integrating melissopalynological analysis, GC-MS and HPLC-PDA-ESI/MS. *Molecules, 28*(9), 3967.

Khalid, N., Ahmad, A., Khalid, S., Ahmed, A., & Irfan, M. (2014). Mineral composition and health functionality of Zamzam water: A review. *International Journal of Food Properties, 17*(3), 661–677.

Khalid, S., Ahmad, A., & Kaleem, M. (2017). Antioxidant activity and phenolic contents of Ajwa date and their effect on lipo-protein profile. *Functional Foods in Health and Disease, 7*(6), 396–410.

Kharbach, M., Alaoui Mansouri, M., Taabouz, M., & Yu, H. (2023). Current application of advancing spectroscopy techniques in food analysis: Data handling with chemometric approaches. *Food, 12*(14), 2753.

Lehotay, S. J., & Hajšlová, J. (2002). Application of gas chromatography in food analysis. *TrAC Trends in Analytical Chemistry, 21*(9–10), 686–697.

Machado, A. (2022). *Volatile chemical characterization and biological activity assessment of Portuguese honey types: Importance of botanical origin.*

Mădaş, M. N., Mărghitaş, L. A., Dezmirean, D. S., Bobiş, O., Abbas, O., Danthine, S., & Nguyen, B. K. (2020). Labeling regulations and quality control of honey origin: A review. *Food Reviews International, 36*(3), 215–240.

Mahmoud, M. A., Kilic-Buyukkurt, O., Fotouh, M. M. A., & Selli, S. (2024). Aroma active compounds of honey: Analysis with GC-MS, GC-O, and molecular sensory techniques. *Journal of Food Composition and Analysis*, 106545.

Martínez, R. A., Schvezov, N., Brumovsky, L. A., & Roman, A. B. P. (2017). Influence of temperature and packaging type on quality parameters and antimicrobial properties during Yateí honey storage. *Food Science and Technology, 38*, 196–202.

Mohamadzade Namin, S., Ghosh, S., & Jung, C. (2023). Honey quality control: Review of methodologies for determining entomological origin. *Molecules, 28*(10), 4232.

Nada, A. A., Nour, I. H., Metwally, A. M., Asaad, A. M., Eldin, S. M. S., & Ibrahim, R. S. (2022). An integrated strategy for chemical, biological and palynological standardization of bee propolis. *Microchemical Journal, 182*, 107923.

Noor, A., Kamaraj, S., Bhakti, M. A., Kumari, C., Kulshrestha, S., Pandohee, J., & That, L. F. L. N. (2024). Analysis of volatile organic compounds in honey using gas chromatography and mass spectrometry. In *Advanced techniques of honey analysis* (pp. 287–307). Academic Press.

Ohmenhaeuser, M., Monakhova, Y. B., Kuballa, T., & Lachenmeier, D. W. (2013). Qualitative and quantitative control of honeys using NMR spectroscopy and chemometrics. *International Scholarly Research Notices, 2013*(1), 825318.

Onuh, J. O., & Qiu, H. (2021). Metabolic profiling and metabolites fingerprints in human hypertension: Discovery and potential. *Metabolites, 11*(10), 687.

Palma-Morales, M., Huertas, J. R., & Rodríguez-Pérez, C. (2023). A comprehensive review of the effect of honey on human health. *Nutrients, 15*(13), 3056.

Pan, L., Qin, Y., Chen, M., Liu, H., Liu, S., Sun, X., & Cui, H. (2023). Accurate identification and quantification of ultra-multi-target flavours in essential oils through a combination of retention index distribution-based parallel dual-column GC-MS/MS and analyte protectants. *Industrial Crops and Products, 193*, 116186.

Pascual-Mate, A., Oses, S. M., Fernandez-Muino, M. A., & Sancho, M. T. (2018). Methods of analysis of honey. *Journal of Apicultural Research, 57*(1), 38–74.

Pauliuc, D., Oroian, M., & Ciursă, P. (2021). Organic acids content, sugars content and physicochemical parameters of Romanian acacia honey. *Ukrainian Food Journal, 10*(1), 158–170.

Pavlin, A., Kočar, D., Imperl, J., Kolar, M., Marolt, G., & Petrova, P. (2023). Honey origin authentication via mineral profiling combined with chemometric approaches. *Food, 12*(15), 2826.

Pico, Y., Alfarhan, A. H., & Barcelo, D. (2020). How recent innovations in gas chromatography-mass spectrometry have improved pesticide residue determination: An alternative technique to be in your radar. *TrAC Trends in Analytical Chemistry, 122*, 115720.

Prencipe, F. P., Brighenti, V., Rodolfi, M., Mongelli, A., dall'Asta, C., Ganino, T., & Pellati, F. (2014). Development of a new high-performance liquid chromatography method with diode array and electrospray ionization-mass spectrometry detection for the metabolite fingerprinting of bioactive compounds in Humulus lupulus L. *Journal of Chromatography A, 1349*, 50–59.

Prica, N., Baloš, M. Ž., Jakšić, S., Mihaljev, Ž., Kartalović, B., Babić, J., & Savić, S. (2014). Moisture and acidity as indicators of the quality of honey originating from Vojvodina region. *Archives of Veterinary Medicine, 7*(2), 99–109.

Puścion-Jakubik, A., Borawska, M. H., & Socha, K. (2020). Modern methods for assessing the quality of bee honey and botanical origin identification. *Food, 9*(8), 1028.

Quirantes-Piné, R., Sanna, G., Mara, A., Borrás-Linares, I., Mainente, F., Picó, Y., & Ciulu, M. (2024). Mass spectrometry characterization of honeydew honey: A critical review. *Food, 13*(14), 2229.

Raweh, H. S., Badjah-Hadj-Ahmed, A. Y., Iqbal, J., & Alqarni, A. S. (2023). Physicochemical composition of local and imported honeys associated with quality standards. *Food, 12*(11), 2181.

Richardson, R. T., Lin, C. H., Quijia, J. O., Riusech, N. S., Goodell, K., & Johnson, R. M. (2015). Rank-based characterization of pollen assemblages collected by honey bees using a multi-locus metabarcoding approach. *Applications in Plant Sciences, 3*(11), 1500043.

Schievano, E., Sbrizza, M., Zuccato, V., Piana, L., & Tessari, M. (2020). NMR carbohydrate profile in tracing acacia honey authenticity. *Food Chemistry, 309*, 125788.

Shafiee, S., Minaei, S., Moghaddam-Charkari, N., Ghasemi-Varnamkhasti, M., & Barzegar, M. (2013). Potential application of machine vision to honey characterization. *Trends in Food Science and Technology, 30*(2), 174–177.

Shamsudin, S., Selamat, J., Sanny, M., Jambari, N. N., Sukor, R., Salleh, N. A., & Khatib, A. (2022). Integrated gas chromatography–mass spectrometry and liquid chromatography-quadruple time of flight-mass spectrometry-based untargeted metabolomics reveal possible metabolites related to antioxidant activity in stingless bee honey. *Food Analytical Methods, 15*(11), 3209–3224.

Shao, Y., Shi, Y., Xuan, G., Li, Q., Wang, F., Shi, C., & Hu, Z. (2022). Hyperspectral imaging for non-destructive detection of honey adulteration. *Vibrational Spectroscopy, 118*, 103340.

Siddiqui, A. J., Musharraf, S. G., & Choudhary, M. I. (2017). Application of analytical methods in authentication and adulteration of honey. *Food Chemistry, 217*, 687–698.

Singh-Ackbarali, D., & Maharaj, R. (2014). Sensory evaluation as a tool in determining acceptability of innovative products developed by undergraduate students in food science and technology at the University of Trinidad and Tobago. *Journal of Curriculum and Teaching, 3*(1), 10–27.

Skaff, W., El Hajj, R., Hanna-Wakim, L., & Estephan, N. (2022). Detection of adulteration in honey by infrared spectroscopy and chemometrics: Effect on human health. *Journal of Food Processing and Preservation, 46*(10), e15438.

Smetanska, I., Alharthi, S. S., & Selim, K. A. (2021). Physicochemical, antioxidant capacity and color analysis of six honeys from different origin. *Journal of King Saud University-Science, 33*(5), 101447.

Soares, S., Amaral, J. S., Oliveira, M. B. P., & Mafra, I. (2017). A comprehensive review on the main honey authentication issues: Production and origin. *Comprehensive Reviews in Food Science and Food Safety, 16*(5), 1072–1100.

Soo Lim, D., Min Choi, S., Kim, K.-B., Yoon, K., Kacew, S., Sik Kim, H., & Lee, B.-M. (2018). Determination of fragrance allergens and their dermal sensitization quantitative risk assessment (QRA) in 107 spray perfumes. *Journal of Toxicology and Environmental Health, Part A, 81*(22), 1173–1185.

Sotiropoulou, N. S., Xagoraris, M., Revelou, P. K., Kaparakou, E., Kanakis, C., Pappas, C., & Tarantilis, P. (2021). The use of SPME-GC-MS IR and Raman techniques for botanical and geographical authentication and detection of adulteration of honey. *Food, 10*(7), 1671.

Shafiq, H., Iftikhar, F., Ahmad, A., Kaleem, M., & Sair, A. T. (2012). Effect of crystallization on the water activity of honey. *International Journal of Food and Nutritional Sciences, 3*(3), 1–6.

Tarapoulouzi, M., Mironescu, M., Drouza, C., Mironescu, I. D., & Agriopoulou, S. (2023). Insight into the recent application of Chemometrics in quality analysis and characterization of bee honey during processing and storage. *Food, 12*(3), 473.

Thilakarathne, N. N., Yassin, H., Bakar, M. S. A., & Abas, P. E. (2021). *Internet of things in smart agriculture: Challenges, opportunities and future directions*. Paper presented at the 2021 IEEE Asia-Pacific conference on computer science and data engineering (CSDE).

Tiwari, K., Tudu, B., Bandyopadhyay, R., & Chatterjee, A. (2013). Identification of monofloral honey using voltammetric electronic tongue. *Journal of Food Engineering, 117*(2), 205–210.

Trifković, J., Andrić, F., Ristivojević, P., Guzelmeric, E., & Yesilada, E. (2017). Analytical methods in tracing honey authenticity. *Journal of AOAC International, 100*(4), 827–839.

Tsagkaris, A. S., Koulis, G. A., Danezis, G. P., Martakos, I., Dasenaki, M., Georgiou, C. A., & Thomaidis, N. S. (2021). Honey authenticity: Analytical techniques, state of the art and challenges.*Research. Advances, 11*(19), 11273–11294.

Tura, A. G., & Seboka, D. B. (2019). Review on honey adulteration and detection of adulterants in honey. *Gastroenterology, 4*(1), 1–6.

Wang, Y., Xiang, J., Tang, Y., Chen, W., & Xu, Y. (2022). A review of the application of near-infrared spectroscopy (NIRS) in forestry. *Applied Spectroscopy Reviews, 57*(4), 300–317.

Wilson, I. D., & Poole, C. F. (2023). Planar chromatography–current practice and future prospects. *Journal of Chromatography B, 1214*, 123553.

Yang, C. C., Tang, X., Wong, Y., & Wei, C.-P. (2010). Understanding online consumer review opinions with sentiment analysis using machine learning. *Pacific Asia Journal of the Association for Information Systems, 2*(3), 7.

Yingying, D., Zengzhao, C., & Yusen, C. (2021). Comparative study on the quality of sugar-derived honey, flower-derived honey and commercial honey. *Food Research and Development, 42*(1).

Żak, N., & Wilczyńska, A. (2023). The importance of testing the quality and authenticity of food products: The example of honey. *Food, 12*(17), 3210.

Zakaria, A., Shakaff, A. Y. M., Masnan, M. J., Ahmad, M. N., Adom, A. H., Jaafar, M. N., & Kamarudin, L. M. (2011). A biomimetic sensor for the classification of honeys of different floral origin and the detection of adulteration. *Sensors, 11*(8), 7799–7822.

Zhang, X.-H., Gu, H.-W., Liu, R.-J., Qing, X.-D., & Nie, J.-F. (2023). A comprehensive review of the current trends and recent advancements on the authenticity of honey. *Food Chemistry: X, 19*, 100850.

Zhang, M., Fang, F., Chen, M., Kan, J., Qian, C., Liu, J., & Jin, C. (2024). Comparison of flavor profiles of honey peach cultivars by Gas Chromatography–Mass Spectrometry (GC–MS) with Odor Activity Values (OAVs), E-tongue, and sensory evaluation combined with chemometrics. *Analytical Letters*, 1–22.

Chapter 10
Local Flora Influence on Honey: A Focus on Pollen Analysis

Duygu Nur Çobanoğlu and İlginç Kızılpınar Temizer

1 Introduction

There are various reasons why it is desirable to know the floral origin of a honey. For example, for quality control in marketing, due to fair trade legislation requiring accurate descriptions on labels, and where there are regulatory concerns about the country of origin of honeys (Molan, 1998). Currently, probably due to a lack of alternative methods, palynological analysis of honey is used to determine the origin of honeys.

The scientific branch that studies pollen is called "palynology," and this term was introduced into the literature by Hyde and Williams in 1944 (Stephen, 2014). Pollen grains exhibit considerable variability in their morphological and biochemical characteristics, influenced by factors such as species, environmental conditions, and pollination types. Pollen grains differ from each other in terms of features such as symmetry, polarity, shape, size, structure, sculpture, and apertures, depending on pollination patterns (Álvarez & Ramírez, 2000). These features allow for the identification and classification of pollen, making them useful in many scientific and applied fields. Palynological studies are not only used to identify plant species but have also led to the emergence of subdisciplines such as paleopalynology,

D. N. Çobanoğlu (✉)
Department of Crop and Animal Production, Vocational School of Food, Agriculture and Livestock, Bingol University, Bingol, Turkiye
e-mail: dncobanoglu@bingol.edu.tr

İ. K. Temizer
Department of Crop and Animal Production, Vocational School of Food, Agriculture and Livestock, Bingol University, Bingol, Turkiye

Department of Medical Services and Techniques, High School of Health Services, Giresun University, Giresun, Turkiye
e-mail: ilginc.kizilpinar@giresun.edu.tr

M. Zafar et al. (eds.), *Pure Honey: Assurance & Authentication*,
https://doi.org/10.1007/978-3-031-98913-1_10

iatropalynology, entomopalynology, pharmacopalynology, forensic palynology, quaternary palynology, and aeropalynology, which are used to understand rock and soil compositions, assess air pollution levels, resolve plant-insect relationships, etc. (Stephen, 2014). One of these subdisciplines, melissopalynology, helps to understand the environmental interactions and feeding habitats of honey bees while also focusing on determining the purity, botanical, and geographical origins of honey with palynological analysis (Koelzer et al., 2024).

Pollen grains play the role of male gametophytes in the reproduction of Spermatophyta (gymnosperms and angiosperms/flowering plants) (Liu & Wang, 2021). For reproduction in plants, pollen must be transferred from the anther to the stigma, a process known as pollination. Wind, water, birds, and insects are known as pollination agents (Faheem et al., 2004). Among insect pollinators, bees are particularly prominent. The honeybee (*Apis mellifera* L.) is responsible for much of the pollination workforce (Altieri et al., 2015). *Apis mellifera* has been intentionally transported and domesticated by humans, which has led to its spread across all continents, with the only exception being Antarctica and many oceanic islands. This intentional intervention has solidified its role as the dominant plant pollinator in natural habitats worldwide (Hung et al., 2018). Bees generally collect pollen and nectar during their foraging activities. Pollen analysis of some bee products (honey, bee pollen, perga, propolis, and royal jelly) has been used for many years to understand the foraging behavior of honey bee colonies and their preferences among available floral resources.

The nectar source is one of the key factors determining the commercial and therapeutic value of honey; therefore, verifying its botanical origin is of great importance in terms of quality control. For determining the nectar source of honey, melissopalynology has been used (Islam et al., 2022). Microscopic analysis is conducted to identify the types and proportions of various pollen grains in honey, with the dominant pollen being recognized as an indicator of the honey's nectar source (Borkowska et al., 2024). The pollen composition in honey varies qualitatively and quantitatively depending on the plants preferred by the bees. This diversity causes each honey to have unique characteristics reflecting the specific floral sources visited by the bees.

The experience and knowledge of pollen experts play a major role in the correct identification of pollen. The experience and knowledge of pollen experts play a major role in the correct identification of pollen. Technological advances, especially the development of light microscopy and other microscopic techniques, allow for more detailed and precise identification of pollen morphology. These advances minimize errors in the separation and nomenclature of pollen, significantly reducing the possibility of mislabeling. Therefore, the combination of technological innovations and expert opinion is of great importance to improve the accuracy of the botanical origin of honey and to contribute to the establishment of international standards. Technological advances, especially the development of light microscopy and other microscopic techniques, allow for more detailed and precise identification of pollen morphology. These advances minimize errors in the separation and nomenclature of pollen, significantly reducing the possibility of mislabeling. Therefore, the combination of technological innovations and expert opinion is of great importance to improve the accuracy of the botanical origin of honey and to contribute to the establishment of international standards.

2 Melissopalynology: From Historical Foundations to Global Studies

Melissopalynology is a branch of palynology which studies the geographical and botanical origin of honey by evaluating honey sediment and therefore pollen and the other honeydew elements contained in it (Ricciardelli D'Albore, 1997). Melissopalynological analysis is used not only in honey but also in plant-based bee products such as pollen and bee bread to gain insights into the botanical and geographical origins of these products (Çobanoğlu, 2024; Çobanoğlu et al., 2023b, c; Deveza et al., 2015; Dursun et al., 2024). Additionally, pollen grains that unintentionally contaminate other bee products, such as propolis, royal jelly, and bee venom, are also analyzed, providing valuable information about the bees' feeding preferences and geographical origin (Ayna et al., 2021; Kekecoglu et al., 2021; Moreira et al., 2008; Pyrgioti et al., 2022). Briefly, melissopalynology is the study of pollen collected by bees and provides information about plants visited by foraging honeybees (Nabila et al., 2023).

Honey is a natural food and sweetener produced by honeybees (*Apis mellifera*) from nectar and secretions of living parts of plants or also from the secretions of sucking insects living on trees (Bodor et al., 2021). According to its botanical origin, honey is typically classified into two categories: blossom honey and honeydew. While honeydew honey primarily comes from plant exudations or insect excretions, blossom honey is derived from the nectar of flowers (Hernanz et al., 2023). It should also be mentioned that unifloral and polyfloral. When most of the nectar or honeydew in honey comes from a single plant species, it is referred to as unifloral honey. If a honey is unifloral, it is named after the plant from which it originates, for example, mustard, niger, astragalus, thyme, sunflower, and litchi honeys. Diverse nectar sources are present in polyfloral honey (Ruoff et al., 2006). The physicochemical, microscopical, and organoleptical properties of unifloral honey can be used to identify it (Soria et al., 2004). The polyfloral honeys are particularly difficult to authenticate since, in contrast to unifloral honeys, they exhibit a great deal of diversity regarding all measurands rather than specific physical or chemical properties (Ruoff et al., 2006).

Pollen is present in honey in trace amounts, primarily from the stigma of a flower where bees foraged with the help of wind and may have added pollen from bee bread when extracting honey (Molan, 1998). The pollen inadvertently collected by bees while gathering nectar serves as a key indicator of the botanical, and particularly the geographical origin of the nectar (Painkra et al., 2024). In addition to being essential to the bee's food, pollen is also critical to its growth, development, and reproduction (Ghouizi et al., 2020). Melissopalynological data can guide the development of plants to help pollinators (Nabila et al., 2023). By examining the types and quantities of pollen in honey samples, the plant species that bees have visited can be identified. These analyses contribute to assessing the authenticity and quality of honey while also being effective in detecting cases of adulteration (Soares et al., 2017). Melissopalynology also enables the determination of a honey's geographical origin, as the pollen spectrum comprising all the pollen in the sediment mirrors the floral composition of the region where the honey was produced (Jiang et al., 2024). The identification of a honey's geographical origin typically relies on the presence of a specific combination of pollens that are unique to a particular region and not found elsewhere (*Rhododendron*, pine, manuka, acacia, lavender, ulmo, etc.). These marker pollens, when present, are sufficient to determine the origin of the honey variety in which they are found (Ricciardelli D'Albore, 1997). Different geographical areas have distinct floral associations, which are reflected in the pollen found in the honey. As well as the variance in floral relationships becomes more noticeable with increasing climate difference. The pollen preserved in honey provides valuable information about plant diversity, land use, and climate change, making melissopalynology significant for ecological and historical research.

Pollen analysis in honey is a crucial method used to determine the botanical and geographical origin of honey, with the earliest studies in this field dating back to the early ninteenth century. The roots of melissopalynology can be traced to Pfister's work in 1895, which initiated the systematic analysis of pollen grains in honey. Pfister was one of the first scientists to demonstrate that pollen in honey could be systematically examined and that these pollen grains were critical for understanding the source of the honey. Since then, various researchers, including Fehlman (1911), Armbruster and Oenike (1929), Griebel (1930–1931), Armbruster and Jacobs (1934–1935), and Zander (1935, 1937, 1941, 1949, and 1951), further advanced this field (Ricciardelli D'Albore, 1997). Their contributions have significantly influenced the development of techniques that are still used today to determine the botanical and geographical origins of honey. A melissopalynological method was developed and proposed by the International Commission for Bee Botany (ICBB) and published in 1978 (Louveaux et al., 1978). To develop these methods, the International Honey Commission initiated a ring trial aimed at standardizing sample preparation and the counting of pollen grains. In these trials, two types of analyses were conducted: qualitative and quantitative. The qualitative analysis involved the ring trial group identifying plant elements on slides prepared from honey sediments. The quantitative analysis involved counting the identified plant elements on the slides and interpreting the results (Von Der Ohe et al., 2004). Von Der Ohe et al. (2004) harmonized qualitative and quantitative melissopalynological methods with the results of the ring trials. Ricciardelli D'Albore (1997) published "Textbook of Melissopalynology." These studies have made it possible today to determine the importance of different plant species for beekeeping on a rigorous scientific basis (Bodor et al., 2021; Çobanoğlu et al., 2023a; de Freitas et al., 2015; de Souza et al., 2019; Freitas et al., 2015; Islam et al., 2022; Temizer et al., 2021; Silva et al., 2013).

3 Qualitative and Quantitative Examination of Melissopalynology

Melliferous plants are divided into three groups: nectar-producing, pollen-producing, and both nectar- and pollen-producing (mixed) plants. Logically, nectar-producing plants provide pollen for honey, which is usually underrepresented. Pollen-producing plants provide over-represented pollen to honey (Bobiş et al., 2020). The number of pollen grains in honey is influenced by the morphology and physiology of the flower (Vidal et al., 2010). Furthermore, flower structure determines pollen distribution by influencing the accessibility of nectar and pollen and the way bees collect these resources (Mureşan et al., 2022). The nectar- and pollen-collecting behavior of honey bees depends not only on the plant community, but also on various factors in the hive, such as the number of larvae and the number of worker bees (Tsuruda et al., 2021).

In these methods, the pollen grains of nectariferous plants and nectarless anemophilous and entomophilous plants and honeydew elements are classified separately and counted. When identifying pollens, it is often challenging to determine the exact species, and sometimes even the genus cannot be confidently identified. In such cases, broader categories based on family, shapes, and types are used instead (Ricciardelli D'Albore, 1997). Although some pollen species are present in the composition of honey, they should be excluded from the count because they come from plant species that do not produce nectar (nectarless taxa) and therefore do not contribute to honey formation (Louveaux et al., 1978). However, many researchers have included these pollen species (nectarless taxa) for melissopalinological studies because these taxa may contribute to the composition of honey (Başoğlu et al., 1996; Çobanoğlu et al., 2023a; Ecem Bayram et al., 2019). Pollen taxa were divided into four main categories according to frequency in honey: predominant pollen ($P > 45\%$), secondary pollen (16–45%), important minor pollen (3–15%), and minor pollen (less than 3%) (Bobiş et al., 2020; Louveaux et al., 1978). The frequency of honeydew indicators is determined by comparing the number of these indicators to the pollen grains from nectar-producing plants. Honey is classified as honeydew honey if this ratio exceeds three (Çobanoğlu et al., 2023a). With melissopalynological analysis, honey is classified as honeydew or blossom honey, but melissopalynological analysis cannot determine the botanical origin of honeydew honey (Tedesco et al., 2022). Honey is classified as monofloral/unifloral and polyfloral/multifloral/heterofloral or monofloral, bifloral, and polyfloral according to the number of pollen grains that it contains. In these classifications, the frequency of the presence of pollen belonging to the plant taxa contained in honey is evaluated at different rates. Honey is often classified as unifloral and polyfloral. If the pollen grain of one species exceeds 45% of the total, honey is considered to be unifloral. Honey is considered polyfloral if there is no predominant pollen (Alvarez-Suarez et al., 2010; Bouhlali et al., 2019; Kızılpınar Temizer, 2024; Molan, 1998; Viteri et al., 2021; Zhao et al., 2016). But in some studies, the category of bifloral honey has been added to this classification. Puścion-Jakubik et al. (2020) classified honey as bifloral when the pollen grains from two plant species each constitute 22.25% of the total. It was classified as multifloral when the pollen grains from more than three botanical species each represent less than 16% of the total. In a study, multifloral honey samples were classified as "bifloral" when two pollen taxa had secondary percentages, "oligofloral" when two or more pollen taxa belonged to a single botanical family, and "polyfloral" when three or more pollen taxa belonged to secondary frequency (Ramírez-Arriaga et al., 2011). Jacinto-Pimienta et al. (2016) characterized honey as multifloral when there are three or more species with a percentage of more than 10%. Although not officially recognized, honey has been classified as "bifloral" when two species are significant in this research. García et al. (2020) classified honey types based on pollen composition, where a honey type is considered monofloral if a single species constitutes 45% or more of the pollen content. Bifloral honey types are identified when the pollen composition is predominantly derived from two species, with their combined proportion being less than 50% and the difference between the two species not exceeding 5%. Polyfloral honey types, on the other hand, are categorized as honey derived from various plant species, with none of the species contributing 45% or more to the pollen content.

Honey can be categorized numerically depending on the total number of plant elements it contains (Dobre et al., 2013). Honeys are divided into five classes on the basis of the plant elements in honey. Class I contains fewer than 20,000 pollen grains/10 g, including unifloral honeys with underrepresented pollen, such as *Robinia* and *Citrus*. Most kinds of floral honey and mixtures of flower and honeydew honey fall into class II (20,000–100,000 pollen grains/10 g honey). Including honeys with over-represented pollen and honeydew honeys are classified in class III (100,000–500,000 pollen grains/10 g). Class IV contains between 500,000 and 1,000,000 pollen grains/10 gr honey, including unifloral honeys with highly over-represented pollen and some pressed honeys. Honey obtained by pressing falls into class V also (> 1,000,0000 pollen grains/1 g honey) (Maurizio, 1955). This method has been used in many studies (Ahmad et al., 2023; Bölükbaşı, 2009; Kale Sniderman et al., 2018; Puusepp & Koff, 2014; Todd & Vansell, 1942).

4 Honey Authentication According to Melissopalynology

Pollen grains that the bees collected from the targeted plants and other particles are found in honey. Its composition helps identify its botanical and geographic origin by reflecting the types of flora that make up the honey matrix (Corvucci et al., 2015). Pollen number and taxa in honey are affected by various factors such as plant morphology and physiology (Todd & Vansell, 1942), foraging bee activity (Molan, 1998), contamination within the hive, and honey processing procedure. It is commonly recognized that microscopic examination of a honey sample can typically identify the source of the honey's nectar and the region in which it was produced with a high degree of accuracy. When researchers discovered that melissopalynology could be used to precisely identify the geographical origin of a honey sample, in addition to determining its location and whether it was unifloral or multifloral, the opinion of producing quality honeys with controlled geographical denomination started to take shape (Ricciardelli D'Albore, 1997).

There are different kinds of honey according to the geographical origin. In Malaysia and Australia, the main types of honey are tualang and gelam (Kishore et al., 2011), while New Zealand is known for its clover, manuka, and thyme honeys. Manuka honey is harvested from *Leptospermum scoparium* or *Leptospermum polygalifolium,* members of the Myrtaceae family found throughout New Zealand (Stephen, 2014) and eastern Australia. Kanuka honey, derived from *Kunzea ericoides,* a close relative of the manuka tree. In Spain, one of the most notable types of monofloral honey is that derived from the viper's-bugloss (*Echium* spp.). This honey is recognized for its high fructose and glucose content, which contributes to its sweetness and antioxidant capacity (Arroyo et al., 2017). In Greece, monofloral honeys such as thyme and pine honey are well-regarded for their unique flavors and health benefits. Greek honey has gained recognition in both national and international markets, with its distinct botanical origins contributing to its premium status (Xagoraris et al., 2021). The combination of melissopalynological analysis and sensory evaluation is essential for authenticating these honeys and ensuring their quality. Türkiye produces a variety of honeys, primarily astragalus, chestnut, euphorbia, pine, rhododendron, and thyme (Küçükaydın et al., 2023; Uçurum et al., 2023). In contrast, Bulgarian and Hungarian honeys tend to be more complex, predominantly featuring species such as coriander, thistle, buckwheat, willow, and linden, along with forest honeys. Romanian and Ukrainian monofloral honeys are characterized by canola, chestnut, linden, sunflower, and willow (Margaon et al., 2021).

In global trade and the higher economic value associated with certain honey types linked to specific origins, these products have become particularly vulnerable to fraud, including the adulteration of honey or the mislabeling of lower-value honey as a more expensive variety. The authentication of these premium honey types can be achieved through melissopalynological analysis. However, for honey to be classified as unifloral, the percentage of pollen from a specific taxon present in the honey must exceed a minimum threshold. No honey produced by free-flying bees can be considered entirely unifloral; this is only possible in special cases where bees are exposed exclusively to a particular plant species (Abdulrahaman et al., 2017; Moar, 1985). For a honey to be classified as unifloral, the major part of the nectar is derived from a single plant taxon. A unifloral honey must contain a minimum of 45% quantity of pollen grains of the predominant plant species. However, this percentage does not apply when the plant of the nectar source of honey contains a higher or lower number of pollen grains than average (Louveaux et al., 1978). It has been recognized that different plants produce pollen in greater or lesser amounts, and that because of this, due of differences in blossom structure, varying amounts of pollen may come into contact with the nectar and subsequently be recovered in honey. However, there is no consensus on the percentage of dominant pollen required for each monofloral species, as there is a lack of widely accepted international standards on this issue. Countries apply their own unifloral honey classification criteria (based on limits set by specific laboratories) (Escriche et al., 2023).

Generally, honeys with a low percentage of underrepresented pollen (in the range of 10–30%) exhibit a low pollen count (Louveaux et al., 1978). However, some plants are characterized by their highly over-represented pollen, such as *Myosotis, Castanea sativa*, and *Eucalyptus*. In honeys from these plants, the pollen percentage is typically above 70%. Underrepresented taxa—such as *Thymus, Citrus,* and *Arbutus unedo,* among others—have a relative pollen abundance that is significantly lower than 45% in honey (Rodopoulou et al., 2018). Moar (1985) stated that pollen production varies greatly among plant species. Therefore, some unifloral honeys may contain less than 20,000 pollen grains, while others may contain more than 100,000 pollen grains. Honeys with low pollen content were considered "underrepresented" and honeys with high pollen content were considered "over-represented." In Table 10.1 are given the frequency (%) value for the classification of some unifloral honeys with underrepresented, over-represented, and normal represented pollen grains.

5 Conclusion

Melissopalynological analysis, while essential for honey classification, is a complex and labor-intensive process that heavily depends on the expertise of skilled analysts. However, the lack of standardized methods, particularly in identifying, counting, and excluding specific pollens, leads to variations across laboratories and countries. The absence of a global consensus on pollen count thresholds and honey classification further complicates matters. These result in inconsistencies in honey labeling and trade. The discrepancies impact producers, consumers, and regulatory bodies alike. Despite its critical role, current methods require revision and harmonization to ensure more reliable and consistent results across the industry. The pollen profile plays an important role in the classification of honey, but the diversity of pollen sources and the proportions of pollen present in honey can create differences in the nomenclature of honey. The pollen profile of honeys named based on over-represented pollen sources usually clearly reflects the main source of this honey and contains high pollen ratios specific to a particular flower. The pollen profiles of such honeys are more homogeneous and more similar to each other than honeys named due to underrepresented pollen. However, this is not a complete reflection of the chemical properties and composition of the honey. Even honeys with the same name based on pollen content may not be chemically similar due to the different sources of other pollen and nectar contributing to the honey. The experience and knowledge of pollen experts play a major role in the correct identification of pollen. The experience of the experts is critical for the accurate determination of the morphological characteristics of pollen.

Table 10.1 Classification of some unifloral honeys according to the melissopalynological analysis

Author	Source of pollen grain	Frequency %	Honey name	Geographical origin	Class
Von Der Ohe et al. (2004)	*Arbutus*	8–20	Strawberry tree	–	Underrepresented
Oddo et al. (1995)		>8		–	
Tuberoso et al. (2010)		19.9–56.5		Italy	
Küçükaydın et al. (2023)	*Astragalus* spp.	>50	Geven	Türkiye	normally represented
Bodor et al. (2021)	Brassica	>80	Rapeseed	Hungary	Over-represented or normally represented
Szczesna et al. (2011)	*Brassica napus*	45.0–66.8		Poland	normally represented
Beck and Camps (2009)		>80		Germany	Over-represented
Persano Oddo et al. (2004)		60.7–99.2		Switzerland, Belgium,Bulgaria, CZ, Germany, France, Finland, Greece, Italy, Netherlads, Portugal, Spain	normally Represented
Vîjan et al. (2023)		46.1–81.7	Rapeseed	Romania	
Serra Bonvehí and Tarrés (1993)	*Calluna vulgaris*	10–33	Heather	Portugal	Underrepresented
Persano Oddo et al. (2004)		10–77		Germany, France, Netherlands	
Von Der Ohe et al. (2004)	*Carduus*	5–25	Thistle	–	Underrepresented
Von Der Ohe et al. (2004)	*Castanea sativa*	>86	Chestnut	–	Over-represented
Persano Oddo et al. (2004)		61–99		Switzerland, Germany, France, Greece, Italy	
Bodor et al. (2021)		60–93		Hungary	

(continued)

Table 10.1 (continued)

Author	Source of pollen grain	Frequency %	Honey name	Geographical origin	Class
Louveaux et al. (1978)	*Citrus* spp.	10–20	Citrus	–	Underrepresented
Corvucci et al. (2015)		14–76		Italy, Spain	
Persano Oddo et al. (2004), Von Der Ohe et al. (2004)		2–42		–	
Rodriguez et al. (2010)		≥5		Spain	
Arroyo et al. (2017)	*Echium* spp.	>70	Purple viper's-bugloss	Spain	
Beck and Camps (2009)	*Erica* spp.	>45	Heather	Germany	normally represented
Persano Oddo et al. (2004)	*Eucalyptus* spp.	83–100	Eucalyptus	Greece, Italy	Over-represented
Alissandrakis et al. (2011)		69–83		Greece	
Bodor et al. (2021)	*Helianthus unnuus*	3–83	Sunflower	Hungary	Underrepresented
Vîjan et al. (2023)		48.5–74.1		Romania	
Moar (1985)	*Knightia* spp.	>10	Rewarewa	New Zealand	Underrepresented pollen
Louveaux et al. (1978)	*Lavandula* spp.	10–20	Lavander	–	Underrepresented
Persano Oddo et al. (2004)		1–20		France, Spain	
Moar (1985)	*Leptospermum* spp.	>70	Manuka honey	New Zealand	Over-represented
Louveaux et al. (1978)	*Medicago* spp.	20–30	Medicago		Underrepresented
Persano Oddo et al. (2004)	*Rhododendron* spp.	15–77	Rhododendron	Switzerland, Italy	Underrepresented

Louveaux et al. (1978)	*Robinia pseudoacacia*	20–30	Black locust	–	Underrepresented
Persano Oddo et al. (2004), Von Der Ohe et al. (2004)		7–60		–	
Bodor et al. (2021)		7–43		Hungary	
Corvucci et al. (2015)		15–88		Italy, East Europe	
Beck and Camps (2009)		>20		Germany	
Vîjan et al. (2023)		24.3–53.1		Romania	
Louveaux et al. (1978)	*Rosmarinus* spp.	10–20	Rosmarinus honey	–	Underrepresented
Persano Oddo et al. (2004)		10–57		Italy, Spain	Underrepresented
Louveaux et al. (1978)	*Salvia* spp.	10–20	Salvia		Underrepresented
Persano Oddo et al. (2004)	*Taraxacum officinale*	5–40	Dandelion	Switzerland, Germany, Italy	
Moar (1985)	*Thymus*	>20	Thyme	New Zealand	Underrepresented
Persano Oddo et al. (2004)		13–68		France, Greece, Italy	
Louveaux et al. (1978)	*Tilia* spp.	20–30	Lime	–	Underrepresented
Persano Oddo et al. (2004)		1–56		Bulgaria, Switzerland, France, Germany, Italy	
Bodor et al. (2021)		5.3–66.4		Hungary	
Vîjan et al. (2023)	*Tilia tomentosa*	31.8–61.4	Linden	Romania	
Beck and Camps (2009)	*Trifolium* spp.	>70	Clover	Germany	normally represented
Persano Oddo et al. (2004)	*Taraxacum officinale*	5–40	Dandelion	Switzerland, Germany, Italy	

References

Abdulrahaman, A. A., Kolawole, S. O., Adeyemi, S. B., & Misbahudeen, H. L. (2017). Mellisopalynological analysis of some commercial honeys. *Annals. Food Science & Technology, 18*(3).

Ahmad, M., Althobaiti, A. T., Ali, W., Masood, K., Ramadan, M. F., Chaudhary, B., et al. (2023). Membrane-processed honey samples for pollen characterization with health benefits. *Chemosphere, 319*, 137994.

Alissandrakis, E., Tarantilis, P. A., Pappas, C., Harizanis, P. C., & Polissiou, M. (2011). Investigation of organic extractives from unifloral chestnut (Castanea sativa L.) and eucalyptus (Eucalyptus globulus Labill.) honeys and flowers to identification of botanical marker compounds. *LWT – Food Science and Technology, 44*(4), 1042–1051.

Altieri, M. A., Nicholls, C. I., Gillespie, M., Waterhouse, B., Wratten, S., Gbèhounou, G., & Gemmill-Herren, B. (2015). *Crops, weeds and pollinators: Understanding ecological interaction for better management*. Food and Agriculture Organization of the United Nations.

Álvarez, T., & Ramírez, R. (2000). Pollen characters and their evolutionarysignificance with special reference to the flora of Karachi. *Turkish Journal of Biology, 24*(2), 365–378.

Alvarez-Suarez, J. M., González- Paramás, A. M., Santos-Buelga, C., & Battino, M. (2010). Antioxidant characterization of native monofloral Cuban honeys. *Journal of Agricultural and Food Chemistry, 58*(17), 9817–9824.

Arroyo, T. M., González-Porto, A. V., & Esteban, C. B. (2017). Viper's bugloss (echium spp.) honey typing and establishing the pollen threshold for monofloral honey. *Plos One, 12*(10), e0185405. https://doi.org/10.1371/journal.pone.

Ayna, A., Tunç, A., Özbolat, S., Bengü, A. Ş., Aykutoğlu, G., Canli, D., Polat, R., Çiftci, M., & Darendelioğlu, E. (2021). Anticancer, and antioxidant activities of royal jelly on HT-29 colon cancer cells and melissopalynological analysis. *Turkish Journal of Botany, 45*(8), 809–819.

Başoğlu, F. N., Sorkun, K., Löker, M., Doğan, C., & Wetherilt, H. (1996). Saf ve Sahte Balların Ayırt Edilmesinde Fiziksel, Kimyasal ve Palinolojik Kriterlerin Saptanması. *The Journal of Food, 21*(2).

Beckh, G., & Camps, G. (2009). Neue Spezifikationen für Trachthonige. *Deutsche Lebensmittel-Rundschau, 105*(2), 105–110.

Bobış, O., Dezmırean, D. S., Bonta, V., Urcan, A. C., Moıse, A. R., & Mărgăoan, R. (2020). Fungal diversity and over-represented non-nectariferous plants pollen in honey. Case study on acacia honey authenticity, analyzed in APHIS Laboratory. *Bulletin of University of Agricultural Sciences and Veterinary Medicine Cluj-Napoca Animal Science and Biotechnologies, 77*(2), 54–61.

Bodor, Z., Kovacs, Z., Benedek, C., Hitka, G., & Behling, H. (2021). Origin identification of hungarian honey using melissopalynology, physicochemical analysis, and near infrared spectroscopy. *Molecules, 26*, 7274.

Bölükbaşı, D. N. (2009). Melissopalynologic analysis of packed honey. *Mellifera, 9*(18).

Borkowska, M., Burzyńska, M., Piasecka-Kwiatkowska, D., & Celińska, E. (2024). Tracking adulteration of nectar honey varieties using a high-resolution melting qPCR technique validated with melissopalinology. *Food Control, 155*, 110086.

Bouhlali, E. D. T., Bammou, M., Sellam, K., El Midaoui, A., Bourkhis, B., Ennassir, J., Alem, C., & Filali-Zegzouti, Y. (2019). Physicochemical properties of eleven monofloral honey samples produced in Morocco. *Arab Journal of Basic and Applied Sciences, 26*(1), 476–487.

Çobanoğlu, D. N. (2024). Assessing monofloral bee pollens from Türkiye: Palynological verification, phenolic profile, and antioxidant activity. *Journal of Food Science, 89*(3), 1711–1726.

Çobanoğlu, D. N., Akyıldız, İ. E., Kızılpınar Temizer, İ., Damarlı, E., & Çelik, Ş. (2023a). Phenolic compound, organic acid, mineral, and carbohydrate profiles of pine and blossom honeys. *European Food Research and Technology, 249*(6), 1503–1515.

Çobanoğlu, D. N., Kizilpinar Temizer, İ., Candan, E. D., Yolcu, U., & Güder, A. (2023b). Evaluation of the nutritional value of bee pollen by palynological, antioxidant, antimicrobial, and elemental characteristics. *European Food Research and Technology, 249*(2), 307–325.

Çobanoğlu, D. N., Şeker, M. E., Temizer, İ. K., & Erdoğan, A. (2023c). Investigation of botanical origin, phenolic compounds, carotenoids, and antioxidant properties of monofloral and multifloral bee bread. *Chemistry & Biodiversity, 20*(3), e202201124.

Corvucci, F., Nobili, L., Melucci, D., & Grillenzoni, F. V. (2015). The discrimination of honey origin using melissopalynology and Raman spectroscopy techniques coupled with multivariate analysis. *Food Chemistry, 169*, 297–304.

De Freitas, S., Augusto, J., Sattler, G., Rodrigues, B., Souza, D., Almeida-muradian, L. B., Sattler, A., Monika, O., & Barth, O. M. (2015). A melissopalynological analysis of Apis mellifera L. loads of dried bee pollen in the southern Brazilian. *Grana, 54*(4), 305–312.

de Souza, R. R., de Abreu, V. H. R., & de Novais, J. S. (2019). Melissopalynology in Brazil: A map of pollen types and published productions between 2005 and 2017. *Palynology, 43*(4), 690–700.

Deveza, M. V., Keller, K. M., Lorenzon, M. C. A., Nunes, L. M. T., Sales, É. O., & Barth, O. M. (2015). Mycotoxicological and palynological profiles of commercial brands of dried bee pollen. *Brazilian Journal of Microbiology, 46*(4), 1171–1176.

Dobre, I., Alexe, P., Escuredo, O., & Seijo, C. M. (2013). Palynological evaluation of selected honeys from Romania. *Grana, 52*(2), 113–121.

Dursun, İ., Felek, İ., & Çobanoğlu, D. N. (2024). Analyzing the antioxidant activity and fatty acid composition of monofloral mullein (Verbascum sp.) pollen oil obtained via various extraction techniques. *Chemistry & Biodiversity, 21*(3), e202400117.

Ecem Bayram, N., Yüzer, M. O., & Bayram, S. (2019). Melıssopalynology analysis, physicochemical properties, multi-element content and antimicrobial activity of honey samples collected from Bayburt, Turkey. *Uludağ Arıcılık Dergisi, 19*(2), 161–176.

Ghouizi, A. El, Menyiy, N. El, Falcão, S. I., Vilas-Boas, M., & Lyoussi, B. (2020). *Chemical composition, antioxidant activity, and diuretic effect of Moroccan fresh bee pollen in rats.*

Escriche, I., Juan-Borrás, M., Visquert, M., & Valiente, J. M. (2023). An overview of the challenges when analysing pollen for monofloral honey classification. *Food Control, 143*, 109305.

Faheem, M., Aslam, M., & Razaq, M. (2004). Pollınatıon ecology with special reference to insects-A review. *Journal of Research (Science), 15*(4), 395–409.

García, S., Troncoso, J. M., Rondanelli-Reyes, M., García, S., Troncoso, J. M., & Rondanelli-Reyes, M. (2020). Study of honey according to botanical origin and physicochemical parameters in the Biobío Region, Chile. *Chilean Journal of Agricultural Research, 80*(4), 675–685.

Hernanz, D., Jara-Palacios, M. J., Santos, J. L., Gómez Pajuelo, A., Heredia, F. J., & Terrab, A. (2023). The profile of phenolic compounds by HPLC-MS in Spanish oak (Quercus) honeydew honey and their relationships with color and antioxidant activity. *LWT, 180*, 114724.

Hung, K. L. J., Kingston, J. M., Albrecht, M., Holway, D. A., & Kohn, J. R. (2018). The worldwide importance of honey bees as pollinators in natural habitats. *Proceedings of the Royal Society B: Biological Sciences, 285*(1870).

Islam, M. K., Lawag, I. L., Green, K. J., Sostaric, T., Hammer, K. A., Lim, L. Y., & Locher, C. (2022). An investigation of the suitability of melissopalynology to authenticate Jarrah honey. *Current Research in Food Science, 5*, 506–514.

Jacinto-Pimienta, S. Y., Mendoza-Hernández, J. H. R., Zaldivar-Cruz, J. M., Sol-Sánchez, Á., Vargas-Villamil, L. M., Reyes-Sánchez, C. A., Jacinto-Pimienta, S. Y., Mendoza-Hernández, J. H. R., Zaldivar-Cruz, J. M., Sol-Sánchez, Á., Vargas-Villamil, L. M., & Reyes-Sánchez, C. A. (2016). Use of principal component in melissopalynology classification of honey from Apis mellifera L. *Revista Mexicana de Ciencias Agrícolas, 7*(SPE14), 2831–2840.

Jiang, W., Battesti, M. J., Yang, Y., Jean-Marie, É., Costa, J., Béreau, D., Paolini, J., & Robinson, J. C. (2024). Melissopalynological Analysis of Honey from French Guiana. *Food, 13*(7), 1073.

Kale Sniderman, J. M., Matley, K. A., Haberle, S. G., & Cantrill, D. J. (2018). Pollen analysis of Australian honey. *PLoS One, 13*(5), e0197545.

Kekecoglu, M., Sonmez, E., Acar, M. K., & Karaoglu, S. A. (2021). Pollen analysis, chemical composition and antibacterial activity of anatolian chestnut propolis collected from Yıgılca Region. *Biology Bulletin, 48*(6), 721–728.
Kishore, R. K., Halim, A. S., Syazana, M. N., & Sirajudeen, K. N. S. (2011). Tualang honey has higher phenolic content and greater radical scavenging activity compared with other honey sources. *Nutrition Research, 31*(4), 322–325.
Kızılpınar Temizer, İ. (2024). Botanical origin and elemental content of Turkish honey: Implications for health risks from essential and non-essential elements. *International Journal of Environmental Health Research, 34*(3), 1737–1750.
Koelzer, K., Ribarits, A., & Weber, M. (2024). Comparing the acetolysed and hydrated methods for the pollen analysis of honey. *Grana, 63*(3), 259–274.
Küçükaydın, S., Tel-Çayan, G., Çayan, F., Taş-Küçükaydın, M., Eroğlu, B., Duru, M. E., & Öztürk, M. (2023). Characterization of Turkish Astragalus honeys according to their phenolic profiles and biological activities with a chemometric approach. *Food Bioscience, 53*, 102507.
Liu, L., & Wang, T. (2021). Male gametophyte development in flowering plants: A story of quarantine and sacrifice. *Journal of Plant Physiology, 258–259*, 153365.
Louveaux, J., Maurizio, A., & Vorwohl, G. (1978). Methods of melissopalynology. *Bee World, 59*(4), 139–157.
Mărgăoan, R., Topal, E., Balkanska, R., Yücel, B., Oravecz, T., Cornea-Cipcigan, M., & Vodnar, D. C. (2021). Monofloral honeys as a potential source of natural antioxidants, minerals and medicine. *Antioxidants, 10*(7), 1023.
Maurizio, A. (1955). Beitrage zur quantitativen Pollenanalyse des Honigs. 2: Absoluter Gehalt pflanzlicher Bestandteilein Tilia- und Labiaten Honigen. *Z. F. Bienenforschung, III*(2), 32–39.
Moar, N. T. (1985). Pollen analysis of New Zealand honey. *New Zealand Journal of Agricultural Research, 28*(1), 39–70.
Molan, P. C. (1998). The limitations of the methods of identifying the floral source of honeys. *Bee World, 79*(2), 59–68.
Moreira, L., Dias, L. G., Pereira, J. A., & Estevinho, L. (2008). Antioxidant properties, total phenols and pollen analysis of propolis samples from Portugal. *Food and Chemical Toxicology, 46*(11), 3482–3485.
Mureşan, C. I., Cornea-Cipcigan, M., Suharoschi, R., Erler, S., & Mărgăoan, R. (2022). Honey botanical origin and honey-specific protein pattern: Characterization of some European honeys. *LWT, 154*.
Nabila, Ahmad, M., Althobaiti, A. T., Ali, W., Masood, K., Ramadan, M. F., Chaudhary, B., Zafar, M., Akhtar, M. S., Sultana, S., Zahmatkesh, S., Mehmood, T., Azam, M., & Asif, S. (2023). Membrane-processed honey samples for pollen characterization with health benefits. *Chemosphere, 319*, 137994.
Oddo, L. P., Piazza, M. G., Sabatini, A. G., & Accorti, M. (1995). Characterization of unifloral honeys. *Apidologie, 26*(6), 453–465.
Painkra, K., Gandhi Krishi Vishwavidyalaya Raipur, I., Mohini Devi, R., Bhagat, I. P., & Author, C. (2024). *Studies on identification of monofloral/multifloral honey by pollen morphology* (K. L. Painkra, G. P. Painkra, S. S. Shaw, V. K. Dubey, & P. K. Bhagat).
Persano Oddo, L., Piro, R., Der Ohe, V., Von Der Ohe, K., Godinho, J., Bentabol, A., Ortiz, A., Spain, V., Bogdanov, S., & Ruoff, K. (2004). Main European unifloral honeys: Descriptive sheets 1. *Apidologie, 35*, 38–81.
Puścion-Jakubik, A., Socha, K., & Borawska, M. H. (2020). Comparative study of labelled bee honey from Poland and the result of the melissopalynological analysis. *Journal of Apicultural Research, 59*(5), 928–938.
Puusepp, L., & Koff, T. (2014). Pollen analysis of honey from the Baltic region, Estonia. *Grana, 53*(1), 54–61.
Pyrgioti, E., Graikou, K., Aligiannis, N., Karabournioti, S., & Chinou, I. (2022). Qualitative analysis related to palynological characterization and biological evaluation of propolis from Prespa National Park (Greece). *Molecules, 27*(20).

Ramírez-Arriaga, E., Navarro-Calvo, L. A., & Díaz-Carbajal, E. (2011). Botanical characterisation of Mexican honeys from a subtropical region (Oaxaca) based on pollen analysis. *Grana, 50*(1), 40–54.

Ricciardelli D'Albore, G. (1997). *Textbook of melissopalynology* (p. 308).

Rodopoulou, M. A., Tananaki, C., Dimou, M., Liolios, V., Kanelis, D., Goras, G., & Thrasyvoulou, A. (2018). The determination of the botanical origin in honeys with over-represented pollen: Combination of melissopalynological, sensory and physicochemical analysis. *Journal of the Science of Food and Agriculture, 98*(7), 2705–2712.

Rodriguez, I., Salud, S., Hortensia, G., Luis, U. J., & Jodral, M. (2010). Characterisation of Sierra Morena citrus blossom honey (Citrus sp). *International Journal of Food Science and Technology, 45*(10), 2008–2015.

Ruoff, K., Luginbühl, W., Künzli, R., Iglesias, M. T., Bogdanov, S., Bosset, J. O., Von Der Ohe, K., Von Der Ohe, W., & Amadò, R. (2006). Authentication of the botanical and geographical origin of honey by mid-infrared spectroscopy. *Journal of Agricultural and Food Chemistry, 54*(18), 6873–6880.

Serra Bonvehí, J., & Tarrés, G. (1993). Physicochemical properties, composition and pollen spectrum of ling heather (Calluna vulgaris (L) Hull) honey produced in Spain. *Apidologie, 24*(6), 586–596.

Silva, T. M. S., dos Santos, F. P., Evangelista-Rodrigues, A., da Silva, E. M. S., da Silva, G. S., de Novais, J. S., dos Santos, F. d. A. R., & Camara, C. A. (2013). Phenolic compounds, melissopalynological, physicochemical analysis and antioxidant activity of jandaíra (Melipona subnitida) honey. *Journal of Food Composition and Analysis, 29*(1), 10–18.

Soares, S., Amaral, J. S., Oliveira, M. B. P. P., & Mafra, I. (2017). A comprehensive review on the main honey authentication issues: Production and origin. *Comprehensive Reviews in Food Science and Food Safety, 16*(5), 1072–1100.

Soria, A. C., González, M., De Lorenzo, C., Martínez-Castro, I., & Sanz, J. (2004). Characterization of artisanal honeys from Madrid (Central Spain) on the basis of their melissopalynological, physicochemical and volatile composition data. *Food Chemistry, 85*(1), 121–130.

Stephen, A. (2014). Pollen–A microscopic wonder of plant kingdom. *International Journal of Advanced Research in Biological Sciences, 1*(9), 45–62.

Szczesna, T., Rybak-Chmielewska, H., Was, E., Kachaniuk, K., & Teper, D. (2011). Characteristics of Polish unifloral honeys. I. Rape honey (Brassica napus L. Var. oleifera Metzger). *Journal of Apicultural Science, 55*(1), 111–119.

Tedesco, R., Scalabrin, E., Malagnini, V., Strojnik, L., Ogrinc, N., & Capodaglio, G. (2022). Characterization of botanical origin of Italian honey by carbohydrate composition and volatile organic compounds (VOCs). *Food, 11*(16), 2441.

Temizer, İ. K., Candan, E. D., Güder, A., & Çiçek, F. (2021). Floral origin, antioxidant and antimicrobial activity of some floral honey. *Türk Doğa ve Fen Dergisi, 10*(2), 95–100.

Todd, F. E., & Vansell, G. H. (1942). Pollen grains in nectar and honey. *Journal of Economic Entomology, 35*(5), 728–731.

Tsuruda, J. M., Chakrabarti, P., & Sagili, R. R. (2021). Honey bee nutrition. The veterinary clinics of North America. *Food Animal Practice, 37*(3), 505–519.

Tuberoso, C. I. G., Bifulco, E., Caboni, P., Cottiglia, F., Cabras, P., & Floris, I. (2010). Floral markers of strawberry tree (Arbutus unedo L.) honey. *Journal of Agricultural and Food Chemistry, 58*(1), 384–389.

Uçurum, H. Ö., Tepe, Ş., Yeşil, E., Güney, F., Karakuş, S., Kolayli, S., Takma, Ç., Duru, M. E., Özkök, A., Yücel, B., Karaca, Ü., Sorkun, K., Baran, A., Kilinç, A., Köseoğlu, M., Özsoy, N., Kunduracı, B. S., Türkaslan, N., Atmaca, H., Karaağaç, H. E., Topal, E., Dayıoğlu, M., Kalkan, M., Ekerbiçer, Ü., & Çakici, N. (2023). Characterization of Turkish pine honey according to their geographical origin based on physicochemical parameters and chemometrics. *European Food Research and Technology, 249*(5), 1317–1327.

Vidal, M. D. G., de Jong, D., Wien, H. C., & Morse, R. A. (2010). Pollination and fruit set in pumpkin (Cucurbita pepo) by honey bees. *Brazilian Journal of Botany, 33*(1), 106–113.

Vîjan, L. E., Mazilu, I. C., Enache, C., Enache, S., & Topală, C. M. (2023). Botanical Origin influence on some honey physicochemical characteristics and antioxidant properties. *Food, 12*(11).
Viteri, R., Zacconi, F., Montenegro, G., & Giordano, A. (2021). Bioactive compounds in Apis mellifera monofloral honeys. *Journal of Food Science, 86*(5), 1552–1582.
Von Der Ohe, W., Persano Oddo, L., Piana, M. L., Morlot, M., & Martin, P. (2004). Harmonized methods of melissopalynology. *Apidologie, 35*, 18–25.
Xagoraris, M., Revelou, P., Dedegkika, S., Kanakis, C. D., Papadopoulos, G. K., Pappas, C.,& Tarantilis, P. A. (2021). Spme-gc-ms and ftir-atr spectroscopic study as a tool for unifloral common greek honeys' botanical origin identification. *Applied Sciences, 11*(7), 3159. https://doi.org/10.3390/app11073159
Zhao, J., Du, X., Cheng, N., Chen, L., Xue, X., Zhao, J., Wu, L., & Cao, W. (2016). Identification of monofloral honeys using HPLC–ECD and chemometrics. *Food Chemistry, 194*, 167–174.

Chapter 11
Physicochemical Characterization, In Vitro Antiproliferative Effect, DNA Fragmentation, and Cell Cycle Arrest Potential of Turkish Sinop Region Chestnut Honey Against Breast Cancer Cells

Ceyda Aldag, Adem Guner, Tunay Kilic, and Nefise Ulku Karabay Yavasoglu

Abbreviations

CAM	Complementary and alternative medicine
DFF	DNA fragmentation factor
ER	Estrogen receptor
HMF	Hydroxymethyl furfural
IHC	International Honey Commission
LDH	Lactate dehydrogenase
NIH	National Institute of Health
ROS	Reactive oxygen species
TAC	Total antioxidant capacity
TOS	Total oxidant status

1 Introduction

Honey is a natural substance created from the secretions (nectar or honeydew) of plants by honeybees (*Apis mellifera* L.). Besides its sugar content, honey consists of various biologically active components such as phenolic compounds and flavonoids, amino acids, proteins, vitamins, trace elements, and enzymes such as invertase, catalase, and glucose oxidase (Hermosín et al., 2003; Moniruzzaman et al., 2013). Although honey has been a natural product used for nutritional and

C. Aldag · A. Guner · T. Kilic · N. U. K. Yavasoglu (✉)
Faculty of Science, Department of Biology, Ege University, Izmir, Turkey
e-mail: tunay.kilic@ege.edu.tr; ulku.karabay@ege.edu.tr

M. Zafar et al. (eds.), *Pure Honey: Assurance & Authentication*,
https://doi.org/10.1007/978-3-031-98913-1_11

therapeutic purposes since ancient times, in recent years, there has been a significant increase in the number of studies investigating the mechanism of action of honey for the treatment of various diseases, especially cancer. It has been revealed that the biological effects of honey on these diseases are due to the various pharmacologically active components, such as phenolic compounds and flavonoids, in it (Erejuwa et al., 2014). Some of these compounds identified in honey are apigenin, gallic acid, genistein, quercetin, ellagic acid, naringenin, chrysin, kaempferol, pinobanksin, hesperetin, pinocembrin, luteolin, p-coumaric acid, ferulic acid, caffeic acid, vanillic acid, and syringic acid (Kassim et al., 2010; Chan et al., 2013; Sergiel et al., 2014). In addition to its anticancer potential, these components have been shown to exhibit antibacterial, antioxidant, anti-inflammatory, hypoglycemic, hepatoprotective, and antihypertensive effects (Kassim et al., 2010; Samarghandian et al., 2011; Erejuwa et al., 2012; Jaganathan, 2012; Xie et al., 2013; Can et al., 2015; Živanović et al., 2019; Bayram et al., 2020; Cianciosi et al., 2021; Hunter et al., 2021; Ben Amor et al., 2022; Escuredo & Seijo, 2022). Complementary and alternative medicine (CAM) has been defined by the NIH National Cancer Institute as medical products and practices that are not part of standard medical care. Recently, it has been observed that using this term has become increasingly common, especially among cancer patients (Marijanović et al., 2021; Kanyilmaz et al., 2022). Within this scope, honey has recently attracted attention as an important component of CAM (Aljawadi et al., 2020).

Turkey's geographical location, topographic structure, and climate conditions, which make it possible to produce honey in three seasons of the year, have made it one of the richest regions in the world in terms of honey production and diversity. Since the flora and geographical differences in which honey is produced affect all its physicochemical and biologically active properties, many different types of honey varieties, both uni-floral and multi-floral, with various characteristics, are produced in Turkey, which has rich vegetation (Can et al., 2015; Bayram et al., 2020; Altun & Aydemir, 2021). In this context, Turkey hosts the world's highest quality honeys in terms of plant diversity, and it also ranks as the second honey producer in total global production.

The most important species that provide high amounts of pollen and nectar needed by bees include chestnut, willow, maple, acacia, and linden. Chestnut, one of the woody plants that can grow naturally in terms of biodiversity, has been the main food source in the mountainous regions where grains cannot grow well for thousands of years (Dag et al., 2017). While it is known that there are 10–12 species of chestnut in the world, the one found in the natural flora of Turkey is the Anatolian chestnut (*Castanea sativa* Mill.). The climate characteristics of the Black Sea Region create the ideal environment for this plant. In this region, Sinop city is one of the richest areas of chestnut tree population in Turkey. It ranks second in the region and fourth in Turkey in chestnut production. It has been observed that intensive work has been done on the production of chestnuts and especially the chestnut honey over the past decade (Derebasi et al., 2014; Can et al., 2015; Col Ayvaz et al., 2018; Silici, 2018; Aker & Nisbet, 2020; Avsar et al., 2023; Tabak et al., 2024). Considering the studies conducted with other chestnut honeys produced in the

Black Sea Region, it is seen that this honey has higher quality and richer pollen value compared to other honey varieties in the region (Tabak et al., 2024). It was noted that Sinop chestnut honey has a higher antioxidant effect and significant total phenol and proline content compared to the other honeys (Atasoy & Altingoz, 2011; Tabak et al., 2024). Due to these features, this honey has received a geographical indication certificate from the Turkish Patent and Trademark Office. Honey products obtained from different geographies of the world are the subject of many research studies for therapeutic purposes. However, no studies have been conducted on the therapeutic effectiveness of geographically indicated Sinop chestnut honey in Turkey, the world second second-largest honey producer. For this purpose, the current study aimed to reveal the anticancer effect against ER-positive breast cancer cells of Turkish Sinop region chestnut honey.

2 Materials and Methods

2.1 *Honey Samples*

Uni-floral chestnut (*Castanea sativa* Mill.) honeys of the Sinop region were used in the study. Honey samples were obtained from the SINATE brand chestnut honey produced by the Beekeeping Application Area of Sinop University Natural Products Application and Research Center (Turkey) in the 2021 harvest season. The collected honeys were kept in a cool, dark environment under 40–60% humidity levels and between 20 and 22 °C until analysis. Randomly selected samples were characterized physicochemically and investigated for their antiproliferative and genotoxic effects.

2.2 *Analytical Characterization of the Honey Samples*

Physicochemical characteristics of the honey samples were revealed according to the Harmonized Methods of the International Honey Commission (IHC) (IHC, 2009).

2.2.1 Moisture Contents

Moisture contents of the honey samples were measured using a digital refractometer (Mettler Toledo RM40, USA) (IHC, 2009). The surface of the refractometer prism was covered with the honey sample, and the refractive index was measured after 2 min at 20 °C. Referring to the standard table on refractive index and moisture content, the value corresponding to the refractive index was accepted as % moisture content (IHC, 2009). Measurements were performed in triplicate.

2.2.2 Electrical Conductivities

The electrical conductivities of the samples were measured using a conductivity meter (Mettler Toledo SevenMulti™ S47, USA) (IHC, 2009). Results are expressed in milliSiemen/centimeter (mS/cm). Measurements were performed in triplicate.

2.2.3 pH Value

pH is an important indicator during the extraction and storage of honey, as it affects its texture, stability, and shelf life (Derebasi et al., 2014). A honey solution (10% w/v) was prepared in Milli-Q water (Millipore Corporation, Billerica, Massachusetts, USA), and it was measured using a pH meter (Hanna Instruments HI 98127, Mauritius).

2.2.4 Hydroxymethyl Furfural (HMF) Analysis

The HMF level in honey samples was carried out using the HPLC system coupled with a Diode Array Detector (DAD) (Agilent 1100 Series, Germany) according to the Harmonized Methods of the IHC (IHC, 2009). Honey samples (5 g) were taken and dissolved in a volumetric flask with 25 mL of ultrapure water using a magnetic stirrer. Then it was made up to 100 mL with water, and the solution was filtered with a 0.45 μm membrane filter. The prepared analytical standards and sample solutions were injected into the HPLC system at a 5 μl volume. A flow rate was 1.2 ml/min, and a run time was 20 min. During the run, the temperature of the column was maintained at 25 °C. Chromatographic analysis was performed with a methanol and distilled water mixture (15:85 v/v) as the mobile phase. Separation was carried out using a C18 Reversed Phase Column (Agilent, Zorbax, 100 Å, 5 μm, 4.6 × 250 mm). The analysis was carried out at a 284 nm wavelength. The chromatograms were recorded for the retention times of the major peak, and the peak response was measured as peak areas. To calculate the amount of HMF in the sample, a calibration curve was drawn with the standards prepared by adding at least three different concentrations of HMF standards to the residue-free sample (matrix match). The amount of HMF was automatically calculated by the device software by comparing the peak area values of the samples with the calibration curve values.

2.2.5 Sugar Analysis

Sugar profile, which constitutes a significant part of honey, was determined using the HPLC system coupled with a Refractive Index Detector (RID) (Agilent 1100 Series, Waldbronn, Germany). The analysis method was adapted from the Harmonized Method of the IHC (IHC, 2009) and TSE 12630 standards. Standard sugar mixture solutions are prepared with fructose, glucose, sucrose, and maltose sugars at concentrations of 2%, 1%, 0.5%, 0.25%, and 0.1% for each mixture, and calibration curves were created. Sample preparation was carried out from the honey sample (5 g) by dissolving it in 40 ml ultrapure water in a beaker. Then it was transferred to a measuring tape containing 25 ml of methanol and made up to 100 ml with distilled water. The sample solution drawn into a 2 mL syringe was filtered with a 0.45 μm membrane filter. The standard and sample solutions were placed in separate vials and labeled. All samples were injected into the HPLC coupled with a refractive index detector (LC-RID). Separation was carried out using an Inersil NH_2 HLPC column (GL Sciences, 100 Å, 5 μm, 4.6 × 250 mm). Chromatographic analysis was performed with a 20:80 v/v mixture of distilled water and acetonitrile as the mobile phase. The 10 μl sample was injected with a flow rate of 1.5 ml/min. The temperature of the column and the RID detector was maintained at 30 °C during the whole run. The sugar identification in the analyzed samples was calculated by considering the calibration curve and dilution factor with the device software (Alghamdi et al., 2020).

2.2.6 C4 Sugar Ratio

C4 sugar ratio analysis was performed by $^{13}C/^{12}C$ carbon stable isotope ratio mass spectrometric analysis (ΔC13-IRMS) according to the Official Methods of Analysis of the Association of Official Analytical Chemist (AOAC 998.12, 2023) standard method. This standard specifies the method used to demonstrate C4 plant (corn or cane) sugars in honey at a concentration > 7% by stable carbon isotope ratio mass spectrometer. For this, first raw honey ΔC13 values and protein ΔC13 values were measured using the combined Elemental Analyzer-Isotope Ratio Mass Spectrometer System (EA-IRMS, Elementar Vario Pyro Cube Elemental Analyzer-EA, Elementar Vario Isotope Select-IRMS). The analysis solution was prepared using 10% tungstic acid, Na salt solution, and 0.335 M sulfuric acid solution. To calculate raw honey ΔC13 values, samples were placed in small amounts into tin capsules used in the device, placed in the automatic sampler of the device's combustion unit, and analyzed. To calculate protein ΔC13 values, 10 g of honey was taken and placed in a centrifuge tube, 4 mL of ultrapure water was added and mixed in a heated magnetic stirrer set at 80 °C until it was thoroughly dissolved. Then, 2 mL of each analysis solution was added to the honey solution in the centrifuge tube and mixed in a heated magnetic stirrer. After the first solution mixture was added, 1 mL of each analysis solution was added to the honey solution in the centrifuge tube three times with an interval of 30 min. and mixed in a heated magnetic stirrer. It was kept in an 80 °C water bath until protein precipitation was observed at the bottom of the centrifuge tube. After protein precipitation was observed, the centrifuge tube was filled with ultrapure water up to 50 mL and centrifuged at 2500–3500 rpm for 5 min. in the centrifuge device. After the process was completed, the upper phase in the tube was discarded by the transfer method. Up to 50 mL of ultrapure water was placed on the bottom sediment, mixed in a vortex, and centrifuged again in the centrifuge device at 4000 rpm for 5 min. This process was continued three times until the supernatant was clear. After the centrifugation process was completed, the precipitate (protein) remaining at the bottom was placed in a watch glass and dried in an oven at 70 °C for 1 h. The dried residue (protein) was placed in small amounts into the tin capsules used in the device and analyzed by placing it in the automatic sampler of the combustion unit of the EA-IRMS system. Analysis results were obtained directly from the device. Device parameters are set as follows: Combustion Reactor Temperature: 1040 °C, Reduction Reactor Temperature: 710 °C, CO_2 Column Temperature: 20 °C, Cooling Temperature: 100 °C.

The difference in $\Delta C13$ protein and $\Delta C13$ raw honey values was calculated from the difference in the values found. The ratio of C4 sugars was calculated with the following formula (11.1);

$$C4\,sugar\% = \left[\frac{\Delta C13\,protein - \Delta C13\,raw\,honey}{\Delta C13\,protein - (-9.7)}\right] \times 100 \tag{11.1}$$

2.2.7 Proline Content

Proline amount analyses were made using a UV-Vis Spectrophotometer (Shimadzu UV-1800 Spectrophotometer, Shimadzu Corp. Japan). The analysis method was adapted from the Harmonized Method of the IHC (IHC, 2009). The honey sample (5 g) is weighed and dissolved in approximately 50 ml of pure water in the shaker, and the volume is completed by transferring it to a 100 ml volumetric flask. 0.5 mL of sample solution was transferred to a glass tube. 0.5 mL of pure water was put into another tube, and 0.5 mL of the prepared proline solution (0.8 mg/25 mL) was put into three separate tubes. Formic acid (1 mL) and ninhydrin solution (1 mL) were added sequentially to all tubes and mixed in a vortex for 10 min, and they were kept in a boiling water bath for 15 min. The tubes removed from here were kept in a 70 °C water bath for another 15 min. Then, 5 mL of 50% 2-propanol solution was added to each tube and mixed in a vortex. After 45 min, the absorbance of the solutions was measured in 1 cm cuvettes at 510 nm with a UV-Vis spectrophotometer. The results were calculated using the following formula (11.2);

$$\text{Proline}\left(\text{mg/kg}\right) = \frac{Es}{Ea} \times \frac{E1}{E2} \times 80 \tag{11.2}$$

Es: absorption of sample solution
Ea: absorption of proline standard solution
$E1$: mg proline taken for proline standard solution
$E2$: amount of honey (g)
80: dilution factor

2.3 *Anticancer Effects*

2.3.1 Determination of Antiproliferative Effects

MCF-7 (human breast adenocarcinoma cell line, ATCC-HTB-22), A549 (human lung carcinoma, ATCC-CCL-185), HeLa (human cervical cancer cell line, ATCC-CCL-2), and healthy HEK-293 (human embryonic kidney cell line, ATCC-CRL-1573) cells were used in cell culture studies. Cells were incubated in Dulbecco's Modified Eagle's Medium (DMEM) Ham's F12 broth medium (Biological Industries, Cat: 01-170-1B) containing 10% Fetal Bovine Serum (FBS) (Biological Industries, Cat: 04-121-1), 1% L-glutamine (Biological Industries, Cat: 41-218-100), 2% $NaHCO_3$ (5.5%) (Sigma, CAS Number 144-55-8), 1 mM HEPES (Biological Industries, Cat: 03-025-1B), 100 U/ml penicillin, and 100 μg/mL streptomycin for 24 h in a 37 °C incubator under 5% CO_2. Maintenance and proliferation of the cells were ensured by passage in flasks after covering 80% of the culture vessel. For the cytotoxicity test, all cell lines at an initial concentration of 1×10^5 cells/mL were seeded with 100 μl each in 96-well cell culture plates (Mosmann, 1983). The honey sample solutions were prepared with phosphate buffered saline (PBS) pH: 7.4 to adjust different concentrations in a biological safety cabinet (Thermo Fisher Scientific™, 1500 Series A2 BSC, USA). Then, cells were treated with concentrations of 5, 50, 100, 250, 500, and 1000 μg/mL honey for 24, 48, and 72 h (Abel & Baird, 2018), and doxorubicin was used as a positive control. The percentage of viable cells was determined using a UV-visible spectrophotometer (Thermo Fisher Scientific™ Multiscan SkyHigh Microplate Spectrophotometer, USA) at 570 nm (refilter 620 nm) after 24, 48, and 72 h. Cell viability and IC_{50} values of the substances were calculated for each cell line by comparing them with the control at the applied dose.

2.3.2 Lactate Dehydrogenase (LDH) Release

LDH leakage from MCF-7 cancer cells treated with the honey samples was determined using a commercial LDH Cytotoxicity Assay Kit (Cayman, 60,117 USA) according to the manufacturer's protocol. MCF-7 cancer cells (5×10^3 cells/well) were seeded in 96-well cell culture plates and incubated overnight. Then, the cells were treated with IC_{50} values of the honey samples obtained at 72 h and incubated for 48 h. After the incubation period, 100 μL from each well was transferred into a 96-well culture plate and mixed with 100 μL of LDH reaction solution and incubated at room temperature for 30 min. The level of LDH release in the medium was measured at 490 nm using a microplate reader (BMG Labtech, Ortenberg, Germany). Mitomycin-C was used as a positive control for the assay.

2.3.3 DNA Fragmentation Analysis

DNA fragmentation potential of the honey samples was determined spectrophotometrically with the Cell Death Detection ELISA kit (Roche 11,544,675,001) according to the manufacturer's protocol. The principle of this test is the qualitative and quantitative in vitro determination of cytoplasmic histone-associated DNA fragments after induced cell death by the photometric enzyme immunoassay method. MCF-7 cells (1×10^5 cells/well) were plated in 96 wells, and then the cells were exposed to the IC_{50} values of the honey samples obtained at 72 h. At the end of the incubation period, 20 ml of the cytoplasmic fraction was transferred to a streptavidin-coated 96-well ELISA plate with anti-DNA antibodies and incubated for 2 h at room temperature. After the washing period, 2,2-azino-di-(3-ethylbenzthiazoline sulfonate) diammonium salt was added, and absorbances were measured at 405 nm using a microplate reader (BMG Labtech, Ortenberg, Germany).

2.3.4 Cell Cycle Analysis

MCF-7 cells (1×10^4 cells/well) were seeded on 96-well plates and exposed to the IC_{50} values of the honey samples obtained at 72 h. After 48-h incubation period, cell cycle analysis was performed using a Muse™ Cell Cycle Assay Kit (Merck Millipore Cat. No. MCH100106, Germany) according to the manufacturer's protocol. Briefly, cells were trypsinized with PBS and fixed with 70% cold ethanol. The cell cycle reagent was added to the resulting cell pellet and incubated for 30 min. SubG1, G0/G1, S, and G2/M percentage of cells were calculated with a Muse cell cycle analyzer (Merck Millipore, Germany).

2.4 *Antioxidant Characteristics*

Total antioxidant capacity (TAC) and total oxidant status (TOS) were determined using commercial TAC and TOS kits (Rel Assay Diagnostics, Gaziantep, Turkey) according to the manufacturer's protocol. These assays are fast and reliable automated colorimetric assays that are frequently used to determine oxidative changes (Kusano & Ferrari, 2008). The TAC assay aims to determine the antioxidant levels of the honey used by inhibiting the formation of the 2,2′-azino-bis (3-ethylbenzothi azoline-6-sulfonic acid) (ABTS) compound, which is a free radical. For this test, MCF-7 cells were treated with IC_{50} values of the honey samples obtained at the 72 h concentrations and incubated at 37 °C in a humidified atmosphere containing 5% CO_2 for 2 h. In the TAC assay, Reagent 1 (500 μL), cell lysate (30 μL), and Reagent 2 (ABTS radical cation) (75 μL) were exploited in a quartz cuvette, and the absorbance was measured at 660 nm. Trolox, a vitamin E analog, was used to calibrate the system. Ascorbic acid was used as a positive control. The results were expressed in terms of mM Trolox equivalents (TE) per liter (mmol Trolox equiv./L). To determine the TOS level, 500 µL of Reagent 1 Buffer, 75 µL of each cell lysate, and 15 µL of Reagent 2 (Prochromogen) were used, and the absorbance was measured at 530 nm. H_2O_2 was used to calibrate the system, and the results were expressed in terms of µM H_2O_2 equivalent per liter (µmol H_2O_2 equiv./L).

2.5 Genotoxic Effects

The genotoxic effects of the honey samples were determined by using the AMES test (Maron & Ames, 1983; OECD Guideline 471, 2020). In this test, the power of the honey samples to convert *Salmonella typhimurium* his- mutants into his+ was measured. The test was performed in three histidine-requiring strains of *S. typhimurium*, TA98, TA100, and TA1537 (Xenometrix AG, Switzerland). Experimental design was performed using Araclor 1254 (CAS 48586), Sigma-Aldrich (Germany)-induced rat liver S9. Sodium azide (Sigma-Aldrich Co.) at a concentration of 10 µg/mL was used as a strain-specific positive control agent.

Bacterial cultures in Xenometrix growth medium (PMM-GMOO, LOT: K05672P) were incubated at 37 °C for 12–14 h in a shaking incubator at 100 rpm continuously. Subsequently, 100 µL cultures and 100 µL honey samples (final concentrations at 5000 µg/mL) were added to 2 mL melted top agar supplemented with 0.5 mM histidine and 0.5 mM biotin. The tubes were shaken and then inoculated on the previously prepared Minimal Glucose Agar (MGA) plates. The plates were incubated at 37 °C for 48 h, and after incubation, colonies observed in the plates were counted and the number of his+ revertants was determined. These processes were repeated for all controls and honey samples in the presence and absence of S9 mixture. The number of his+ revertant colonies was expressed as the mean ± standard deviation (SD). The experiment was done in triplicate.

2.6 Data Analysis

Statistical analysis of the experimental data was performed with the GraphPad Prism 6 Package (USA). Data were expressed as the mean ± standard deviation (SD). One-way analysis of variance (ANOVA) followed by an LSD test was performed for comparison between the groups. Values of $p \leq 0.05$ were regarded as statistically significant (Snedecor & Cochran, 1968).

3 Results

3.1 Physicochemical Characterization

Physicochemical analyses showed that the refractive indexes of the honey samples ranged from 15.8% to 17.4% (mean: 16.6 ± 0.8%), which is within the limit (not more than 20%) recommended by the FAO Codex Standard for Honey (12-19811) (Codex Standard for Honey, 2001).

The analyzed honey samples were found to be acidic with a mean pH value of 4.72 ± 0.02. The pH value is a critical factor during honey extraction and preservation. The acidity of the honey depends on the presence of organic compounds and may vary according to the harvesting season. Electrical conductivities of the honey samples ranged from 0.82 to 0.91 mS/cm. The value is a characteristic that varies depending on the floral origin of honey. According to the FAO Codex Standard for Honey (12-19811) and Turkish Food Codex Honey Communiqué (2012/58), this value is expected to be not more than 0.8 mS/cm for floral honey, while it is expected to be not less than 0.8 mS/cm for honeydew, chestnut honey, and their blends (Codex Standard for Honey, 2001; Turkish Food Codex, 2012). These values of the honey samples studied were found to be appropriate with these standards.

HMF was not detected in the honey samples (Fig. 11.1). The presence of HMF occurs during the processing and aging of products and is a widely known indicator of honey freshness. The absence of HMF detection in honey samples indicates the freshness and high quality of the honey.

Fructose and glucose contents of the honey samples were found to be 38.2% and 31%, respectively, totaling 69.2%, and the fructose/glucose ratio was determined to be 1.23 (Fig. 11.2). According to the Turkish Food Codex Honey Communiqué

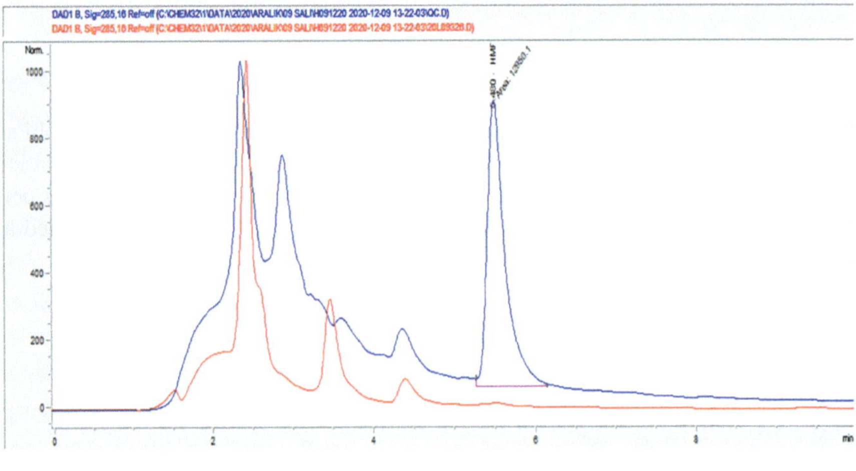

Fig. 11.1 HPLC chromatogram for HMF (blue line: HMF standard, red line: the honey sample)

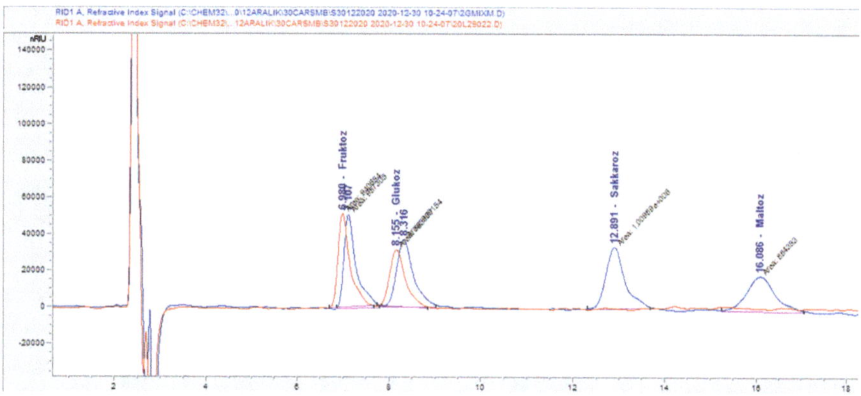

Fig. 11.2 HPLC chromatogram for sugar analysis (Blue line: standard sugar mixture, red line: the honey sample)

(2012/58), the fructose/glucose value in the honey samples should be between 1 and 1.85 (Turkish Food Codex, 2012).

The Turkish Food Codex Honey Communiqué (2012/58) recommends that the difference in ΔC13 values of honey should be (−1) or more positive, while the C4 sugar rate should be at most 7%. As a result of the analysis, the difference between the raw honey ΔC13 and protein ΔC13 values was found to be +0.51, while the % C4 sugar rate was determined to be zero.

The main amino acid found in honey is proline, and to verify the authenticity of honey, the amount of proline in samples must be determined. Proline contents of the honey samples were recorded and ranged from 832 to 1120 mg/kg. According to the Turkish Food Codex Honey Communiqué (2012/58), this value should be at least 300 mg/kg for floral honey. When these values are considered, it is revealed that the honey samples we analyzed had proline values well above the standard value.

3.2 *Cytotoxic Effects*

Graphs of % viability of the cells exposed to different concentrations of the Sinop chestnut honey samples are shown in Fig. 11.3. The honey did not show a calculable IC_{50} value on the cell lines at 24 and 48 h. The IC_{50} values at 72 h for them on cell lines are listed in Table 11.1. It showed a low cytotoxic effect at 72 h with $IC_{50} = 171.48 \pm 22.71$ μg/ml, $IC_{50} = 68.95 \pm 11.21$ μg/ml, and $IC_{50} = 62.98 \pm 18.24$ μg/ml on A549, HeLa, and HEK-293 cells, respectively. Nevertheless, it exhibited a significant cytotoxic effect on MCF-7 cells with an IC_{50} value of 10.43 ± 2.32 μg/ml. Because of this reason, to elucidate these remarkable selective cytotoxic effects of the honey samples on MCF-7 cancer cells at the molecular level, the concentrations at which the honey sample was effective on MCF-7 cells at 72 h were used in the subsequent experimental stages.

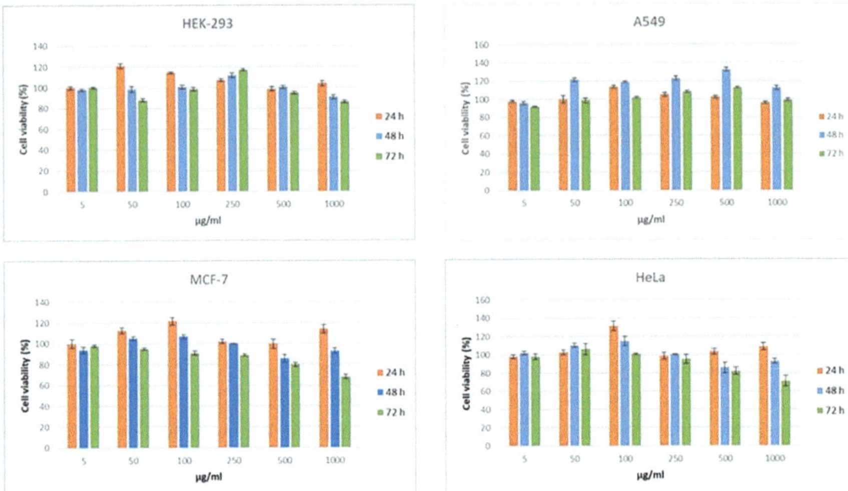

Fig. 11.3 Percentage of viability of cells treated with different honey concentrations (µg/ml) for 24, 48, and 72 h. (**a**) HEK 293, (**b**) A549, (**c**) MCF-7, (**d**) HeLa cells

Table 11.1 Cytotoxic effects (IC_{50} value) of the Sinop chestnut honey on cell lines at 72 h (mean ± SD)

	IC_{50} (µg/ml)			
Treatment	HEK-293	A549	HeLa	MCF-7
Honey	62.98 ± 18.24	171.48 ± 22.71	68.95 ± 11.21	10.43 ± 2.32
PC	4.81 ± 0.80	3.77 ± 0.33	9.06 ± 1.81	1.26 ± 0.02

PC positive control (doxorubicin)

LDH releases (U/L) on MCF-7 cells after treatments are shown in Table 11.2. Compared with the negative control, the honey samples and positive control treatments showed statistically significant increases ($p < 0.05$) in LDH release levels. Similarly to LDH results, % DNA fragmentation after treatments of honey samples and the positive control showed statistically significant increases ($p < 0.05$) compared to the negative control (Table 11.2).

Cell cycle analysis of the honey samples-treated cells showed that there is an increase in cells in G2/M phase after 72 h of treatment. The results are shown in Fig. 11.4. Compared to the untreated group, the cell population in the G2/M phase of MCF-7 cells treated with this honey increased from 16.73% to 21.9%. The number of cells in subG1 (6.52–4.22%) and G0/G1 phase (67.3–64.3%) decreased compared to untreated cells. There was no change in S phase (9.45–9.58%) in both groups. These results showed that the honey samples blocked the cell cycle in the G2/M phase.

Table 11.2 LDH levels and % DNA fragmentation in MCF-7 cells exposed to Sinop chestnut honey (mean ± SD)

Treatment	LDH (U/L)	DNA fragmentation (%)
NC	15.4 ± 2.1	12.2 ± 1.4
Honey	20.7 ± 4.4[a,b]	18.3 ± 1.9[a,b]
PC	47.6 ± 7.8[a]	29.7 ± 2.8[a]

NC negative control, *PC* positive control (10^{-7} M mitomycin C)
[a]Statistically significant difference from NC ($p < 0.05$)
[b]Statistically significant difference from PC ($p < 0.05$)

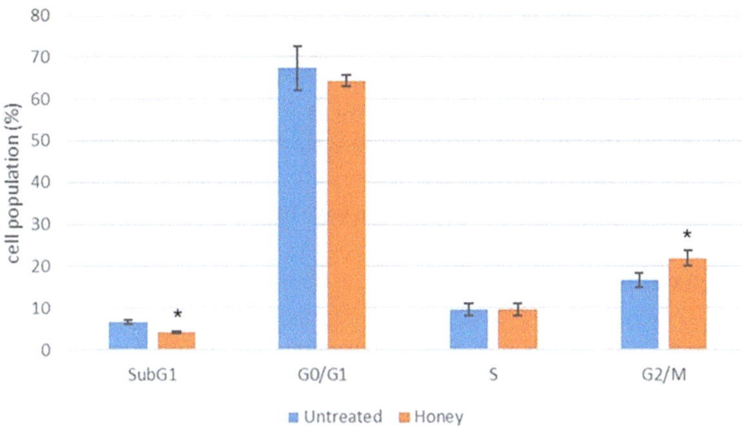

Fig. 11.4 Percentage of cell cycle progression treated with Sinop chestnut honey. Statistically significant increases in the cell population of the G2/M phase compared to the NC ($p < 0.05$). (*NC* negative control, untreated group)

3.3 Antioxidant Properties

TAC and TOS results are shown in Table 11.3. Compared to the negative control group, there are no statistically significant changes in TAC levels of cells treated with the honey ($p > 0.05$). However, TOS levels of cells treated with the honey showed a statistically significant increase when compared to the negative control ($p < 0.05$).

3.4 Genotoxic Effects

Genotoxicity data obtained from the honey samples are shown in Table 11.4. Genotoxic effects were not observed on the bacterial strains used after treatment with the honey samples.

Table 11.3 TAC and TOS levels in MCF-7 cells exposed to Sinop chestnut honey (mean ± SD)

Treatment	TAC (Trolox equivalent/mmol/L)	TOS (μmol H_2O_2 equivalent/L)
NC	14.2 ± 1.9	5.3 ± 0.8
Honey	13.9 ± 0.7	8.2 ± 1.1[a,b]
PC	35.6 ± 4.7[a]	14.6 ± 2.3[a]

NC negative control, *PC* positive control (10 μM ascorbic acid for TAC; 25 μM hydrogen peroxide for TOS)
[a]Statistically significant difference from NC ($p < 0.05$)
[b]Statistically significant difference from PC ($p < 0.05$)

Table 11.4 Genotoxic effects of Sinop chestnut honey (mean ± SD)

	Number of his+ revertant/plate					
	TA 98		TA 100		TA 1537	
Treatment	S9+	S9−	S9+	S9−	S9+	S9−
NC	27 ± 8	25 ± 6	110 ± 23	116 ± 18	18 ± 2	16 ± 4
Honey (5 mg/ml)	18 ± 9	17 ± 5	199 ± 35	206 ± 28	19 ± 8	17 ± 3
PC	600 ± 42	1750 ± 54	600 ± 101	460 ± 16	3360 ± 54	485 ± 34

NC negative control (Sterile dH_2O), *PC* positive control (10 μg/mL sodium azide)

4 Discussion

In the study, the physicochemical properties analyzed for the honey's quality and purity of the Sinop chestnut honey clearly showed that the honey produced in the Sinop region of Turkey with dense chestnut trees perfectly meets the criteria in both the FAO Codex Standard for Honey (12-19811) and the Turkish Food Codex Honey Communiqué (2012/58), besides being pure and high quality without any external processing or adulteration. Some of the physicochemical characteristics of honey are criteria for honey quality and purity recognized in international food codices; however, they do not specify the bioactive potentials of honey (Can et al., 2015). For these reasons, the current study revealed the antioxidant properties as well as the in vitro antiproliferative effect, DNA fragmentation, and cell cycle arrest potential against ER-positive breast cancer cells of geographically indicated Turkish Sinop region chestnut honey, which has not been studied for therapeutic potential before, in addition to the physicochemical properties.

In addition to the therapeutic effectiveness of compounds commonly used in cancer treatment, their toxic effects on healthy cells are their undesirable properties. Therefore, the discovery of new compounds that show cancer cell-specific cytotoxicity is of great importance. In this study, cytotoxicity analyses showed that Sinop region chestnut honey was more effective against MCF-7 cells (IC_{50} = 10.43 ± 2.32 μg/mL) than HEK-293 (IC_{50} = 62.98 ± 18.24 μg/mL), HeLa (IC_{50} = 68.95 ± 11.21 μg/mL), and A-549 (IC_{50} = 171.48 ± 22.71 μg/mL). The honey significantly inhibited viability in MCF-7 cells, depending on increasing concentrations. Moreover, it has been presented to exhibit lower cytotoxic effects on healthy cells (HEK-293). These results suggested that the honey revealed a selective cytotoxic effect on MCF-7 cells, which means it has the potential for anticancer activity. Cytotoxic effects on MCF-7 cells of Indian and Tualang honeys have also been reported (Jaganathan et al., 2010; Fauzi et al., 2011; Erejuwa et al., 2014).

In investigations aiming to clarify the molecular mechanism of the cytotoxic effects of honeys on cancer cells, it has been emphasized that the main mechanisms of action of honey are regulation of the cell cycle, modulation of oxidative stress, and induction of apoptosis (Erejuwa et al., 2014; Afrin et al., 2018; Waheed et al., 2019). In our study, the cytotoxic activity in MCF-7 cancer cells after honey applications was evidenced by leakage of LDH from the cancer cell membranes. LDH release increased significantly after the honey treatment; findings supported the MTT test. The positive correlation between both tests showed that the honey applications may trigger cytotoxicity through plasma membrane disruption in MCF-7 cells. LDH, which contributes to energy production processes, is a cytoplasmic enzyme found in all types of cells. When cells are exposed to toxic effects, their plasma membrane integrity is damaged, and the LDH enzyme leaks from the cells into their environment. It is known that the extracellular amount of this enzyme increases, since the cell integrity can be disrupted during the lipid peroxidation (the oxidative degradation of lipids) process under the condition of oxidative stress (Jovanovic et al., 2010). Thus, damaged/dead cells can be determined by measuring LDH enzyme activity after exposure (Guner, 2021). In the study, significant increases in oxidative stress levels and a decrease in the antioxidant capacity of MCF-7 cells were detected with TAC and TOS analyses after the honey application compared to the control. It is thought that lipid peroxidation caused by free radicals under oxidative stress conditions disrupted cell membrane integrity, and LDH release increases as a result. Increases in oxidative stress because of a selective effect on cancer cells are seen as an important option for destroying cancer cell viability. Although it is known that antioxidant compounds of honey can reduce the effects of oxidative reactions caused by free radicals (Liu et al., 2013), the inhibitory effect of honey on proliferation and growth of cancer is thought to be via modulation of oxidative stress, which is one of the potential cancer treatment strategies that aim to modulate levels of ROS (Gorrini et al., 2013). Similarly, studies showed that increased LDH release and decreased cell viability in MCF-7 cells after Tualang honey (*Koompassia excelsa* (Becc.) Taub.) applications. It is determined that Tualang honey induced apoptosis and reduced mitochondrial membrane potential, and the honey has a selective cytotoxic effect on breast cancer cell lines (Fauzi et al., 2011; Kadir et al., 2013).

It is known that in response to membrane lipid peroxidation, and according to specific cellular metabolic conditions and repair capacities, the cells may promote cell survival or induce cell death. Under low lipid peroxidation conditions, the cells survive through signaling pathway activation or their antioxidant defense systems, while under high lipid peroxidation conditions (toxic conditions), due to increased oxidative damage, the cells undergo apoptosis or necrosis (Ayala et al., 2014). Apoptosis describes a form of cell death in an active or inherently controlled manner, and cleavage of chromosomal DNA into oligonucleosomal-size fragments is a biochemical hallmark of apoptosis. The DNA fragmentation factor (DFF) plays a crucial role in DNA fragmentation in the presence of an activated caspase 3 in various cells upon activation of apoptosis (Liu et al., 1997; Zhang & Xu, 2000). In this study, a significant increase in DNA fragmentation levels in MCF-7 cells was detected after the honey treatments. These results clearly demonstrate that MCF-7 cells undergo apoptosis because of DNA fragmentation after the honey application.

Inhibition of the cell cycle by the honey applications may be a possible way to reduce tumor proliferation. Current studies reveal that honey's ability to halt the cell cycle is due to various phenolic compounds and flavonoids in honey. Honey and its components are demonstrated to inhibit the cell cycle of cancer cell lines such as glioma, melanoma, and colon cancers at the G1/G0 phase (Afroz et al., 2016). Also, chrysin, an important component of honey, has been shown to have an antiproliferative effect against mouse melanoma cells through cell cycle arrest at the G0/G1 phase (Pichichero et al., 2010). As well as chrysin, some other phenolics, such as quercetin and kaempferol, which are found in large amounts in honey, have also been shown to arrest the cell cycle in human melanoma at G0/G1, G1, and G2/M phases in various cancer cell lines (Zhang et al., 2008; Priyadarsini et al., 2010; Aliyu et al., 2012). Nevertheless, cytotoxicity induced by flavones is mediated by G2/M cell cycle arrest and apoptosis (Zhang et al., 2008). In our study, it was determined that the honey applications caused decreases in the cell population in the G0/G1 phase, while increases in the cell population in the G2/M phase followed inhibitions in cell viability. Cell cycle arrest most frequently occurs at the G1/S or G2/M boundaries (Shapiro & Harper, 1999). It is known that most of the antineoplastic drugs in clinical use block the cell cycle in the S or G2/M phases (Pichichero et al., 2010). Although many compounds designed to interrupt cell cycle progression or checkpoint control are known to have cytostatic effects, the presence of DNA fragmentation and apoptosis in the honey-treated cells suggests that it has a cytotoxic effect on MCF-7 cells (Shapiro & Harper, 1999).

Breast cancer is the major cause of cancer deaths among women globally. It is known that circulating estrogen levels and dysregulated estrogen signaling pathways generally play a dominant role in the development and progression of breast cancer (Germain, 2011). Therapy for breast cancer often targets the ER signaling pathway. Tsiapara et al. (2009) revealed that honey samples showed a concentration-dependent biphasic effect on MCF-7 cells: estrogenic or antiestrogenic effect depending on their concentration. Phytoestrogens are estrogen-like compounds, and four phenolic compound classes (isoflavones, stilbene, coumestan, and lignan) were categorized as phytoestrogens (Desmawati & Sulastri, 2019). Although the exact mechanism of estrogen-induced tumor regression is not fully known (Chimento et al., 2022), phytoestrogens such as quercetin have been reported to cause apoptotic effects through ER-dependent mechanisms (Bulzomi et al., 2012). Our study showed that Turkish Sinop chestnut honey can exert a cytotoxic effect against MCF-7, which are ER-positive breast cancer cells. These findings may be attributed to especially phenolic compounds categorized as phytoestrogens in honey, because these components have been shown to induce apoptosis by stimulating both ER-α/$-\beta$ subtypes.

5 Conclusion

Honey has been used as an important food source to this day, both in terms of the bioactive compounds it contains and its taste. In the current study, it was revealed that geographically labeled Sinop chestnut honey has pure and high-quality content. Moreover, our study provided evidence of the cytotoxic and anticancer activities of the honey on MCF-7 breast cancer cells, and it has been shown for the first time in this study that it reduces viability in the breast cancer cells by increasing DNA fragmentation, LDH release, and cell population in the G2/M phase through oxidative stress production. Also, the data suggest that Turkish Sinop chestnut honey exhibits anticancer effects by its antiestrogen activity and potentially induces lipid peroxidation and apoptosis in breast cancer cells. Considering its anticancer activities, it has been revealed that Sinop chestnut honey has the potential to be developed as an anticancer agent. These findings may open new avenues for research into the potential use of honey as a complementary approach to cancer treatment and prevention. It could be used alongside traditional cancer treatments to enhance their effectiveness or as a preventive measure for individuals at high risk of developing cancer.

Acknowledgment This study was supported by Sinop University, Scientific Research Fund (Project number is SGBF-1901-21-001).

References

Abel, S. D. A., & Baird, S. K. (2018). Honey is cytotoxic towards prostate cancer cells but interacts with the MTT reagent: Considerations for the choice of cell viability assay. *Food Chemistry, 241*, 70–78.

Afrin, S., Giampieri, F., Gasparrini, M., Forbes-Hernández, T. Y., Cianciosi, D., Reboredo-Rodriguez, P., Manna, P. P., Zhang, J., Quiles, J. L., & Battino, M. (2018). The inhibitory effect of Manuka honey on human colon cancer HCT-116 and LoVo cell growth. Part 2: Induction of oxidative stress, alteration of mitochondrial respiration and glycolysis, and suppression of metastatic ability. *Food & Function, 9*(4), 2158–2170.

Afroz, R., Tanvir, E. M., Zheng, W., & Little, P. J. (2016). Molecular pharmacology of honey. *Clinical and Experimental Pharmacology, 6*, 3.

Aker, D., & Nisbet, C. (2020). Antioxidant activities, total phenolic and flavonoid contents of honey collected from different botanical origins. *Ankara Üniversitesi Veteriner Fakültesi Dergisi, 67*, 133–136. https://doi.org/10.33988/auvfd.523745

Alghamdi, B. A., Alshumrani, E. S., Saeed, M. S. B., Rawas, G. M., Alharthi, N. T., Baeshen, M. N., Helmi, N. M., Alam, M. Z., & Suhail, M. (2020). Analysis of sugar composition and pesticides using HPLC and GC-MS techniques in honey samples collected from Saudi Arabian markets. *Saudi Journal of Biological Science, 27*(12), 3720–3726. https://doi.org/10.1016/j.sjbs.2020.08.018

Aliyu, M., Odunola, O. A., Farooq, A. D., Mesaik, A., Choudhary, M., Beenish, F., Qureshi, T., & Erukainure, O. (2012). Acacia honey modulates cell cycle progression, pro-inflammatory cytokines and calcium ions secretion in PC-3 cell line. *Journal of Cancer Research and Therapeutics, 4*, 401–407.

Aljawadi, M. H., Khoja, A. T., AlOtaibi, A. D., Alharbi, K. T., Alodayni, M. A., AlMetwazi, M. S., Arafah, A., Al-Shammari, S. A., & Khoja, T. A. (2020). The utilization of complementary and alternative medicine among Saudi older adults: A population-based study. *Evidence-based Complementary and Alternative Medicine*, 4357194. https://doi.org/10.1155/2020/4357194

Altun, S. K., & Aydemir, M. E. (2021). Some physical and chemical properties of raw flower-honey from Erzurum Highland. *Bee Studies, 13*(2), 57–61.

Atasoy, E., & Altingöz, Y. (2011). Dünya ve Türkiye'de Kestanenin Önemi ve Üretimi. *Coğrafya Dergisi, 1*(22), 1–13.

Avsar, C., Aslan, H., Yumak, T., Tabak, A., & Deniz, N. T. (2023). Determination of the superior quality properties of randomly selected chestnut honey samples from the Sinop region. *Spectroscopy Letters, 56*(6), 353–363. https://doi.org/10.1080/00387010.2023.2221328

Ayala, A., Muñoz, M. F., & Argüelles, S. (2014). Lipid peroxidation: Production, metabolism, and signaling mechanisms of malondialdehyde and 4-hydroxy-2-nonenal. *Oxidative Medicine and Cellular Longevity, 2014*, 360438. https://doi.org/10.1155/2014/360438

Bayram, N. E., Kara, H. H., Can, A. M., Bozkurt, F., Akman, P. K., Vardar, S. U., Çebi, N., Yılmaz, M. T., Sagdıc, O., & Dertli, E. (2020). Characterization of physicochemical and antioxidant properties of Bayburt honey from the north-east part of Turkey. *Journal of Apicultural Research, 60*(1), 46–56. https://doi.org/10.1080/00218839.2020.1812806

Ben Amor, S., Mekious, S., Allal Benfekih, L., Abdellattif, M. H., Boussebaa, W., Almalki, F. A., Ben Hadda, T., & Kawsar, S. M. A. (2022). Phytochemical characterization and bioactivity of different honey samples collected in the Pre-Saharan Region in Algeria. *Life (Basel), 12*(7), 927. https://doi.org/10.3390/life12070927

Bulzomi, P., Galluzzo, P., Bolli, A., Leone, S., Acconcia, F., & Marino, M. (2012). The pro-apoptotic effect of quercetin in cancer cell lines requires ERbeta-dependent signals. *Journal of Cellular Physiology, 227*, 1891–1898.

Can, Z., Yildiz, O., Sahin, H., Akyuz Turumtay, E., Silici, S., & Kolayli, S. (2015). An investigation of Turkish honeys: Their physico-chemical properties, antioxidant capacities and phenolic profiles. *Food Chemistry, 180*, 133–141. https://doi.org/10.1016/j.foodchem.2015.02.024

Chan, C. W., Deadman, B. J., Manley-Harris, M., Wilkins, A. L., Alber, D. G., & Harry, E. (2013). Analysis of the flavonoid component of bioactive New Zealand manuka (*Leptospermum scoparium*) honey and the isolation, characterisation and synthesis of an unusual pyrrole. *Food Chemistry, 141*, 1772–1781. https://doi.org/10.1016/j.foodchem.2013.04.092

Chimento, A., De Luca, A., Avena, P., De Amicis, F., Casaburi, I., Sirianni, R., & Pezzi, V. (2022). Estrogen receptors-mediated apoptosis in hormone-dependent cancers. *International Journal of Molecular Sciences, 23*(3), 1242. https://doi.org/10.3390/ijms23031242

Cianciosi, D., Forbes-Hernandez, T. Y., Alvarez-Suarez, J. M., Ansary, J., Quinzi, D., Amici, A., Navarro-Hortal, M. D., Esteban-Muñoz, A., Quiles, J. L., & Giampieri, M. B. F. (2021). Anti-inflammatory activities of Italian chestnut and eucalyptus honeys on murine RAW 264.7 macrophages. *Journal of Functional Foods, 87*, 104752. https://doi.org/10.1016/j.jff.2021.104752

Codex Standard for Honey (Codex Stan 12-19811). (2001). FAO. https://www.fao.org/4/w0076e/w0076e30.htm

Col Ayvaz, M., Ömür, B., Ertürk, Ö., & Kabakci, D. (2018). Phenolic profiles, antioxidant, antimicrobial, and DNA damage inhibitory activities of chestnut honeys from Black Sea region of Turkey. *Journal of Food Biochemistry, 42*(3), 1–10. https://doi.org/10.1111/jfbc.12502

Dag, B., Sıralı, R., & Tarakci, Z. (2017). Investigation of some properties of chestnut honey produced in Black Sea region of Turkey. *Batman University Journal of Life Sciences, 7*(2/2), 118–123.

Derebasi, E., Bulut, G., Col, M., Guney, F., Yasar, N., & Erturk, O. (2014). Physicochemical and residue analysis of honey from Black Sea region of Turkey. *Fresenius Environmental Bulletin, 23*(1), 10–17.

Desmawati, D., & Sulastri, D. (2019). Phytoestrogens and their health effect. *Open Access Macedonian Journal of Medical Sciences, 7*(3), 495–499. https://doi.org/10.3889/oamjms.2019.044

Erejuwa, O. O., Sulaiman, S. A., Ab Wahab, M. S., Sirajudeen, K. N., Salleh, S., & Gurtu, S. (2012). Honey supplementation in spontaneously hypertensive rats elicits antihypertensive effect via amelioration of renal oxidative stress. *Oxidative Medicine and Cellular Longevity, 2012*, 374037.

Erejuwa, O. O., Sulaiman, S. A., & Wahab, M. S. A. (2014). Effects of honey and its mechanisms of action on the development and progression of cancer. *Molecules, 19*(2), 2497–2522. https://doi.org/10.3390/molecules19022497

Escuredo, O., & Seijo, M. C. (2022). Authenticity of honey: Characterization, bioactivities and sensorial properties. *Foods, 11*(9), 1301. https://doi.org/10.3390/foods11091301

Fauzi, A.N., Norazmi, M.N., & Yaacob, N.S. (2011). Tualang honey induces apoptosis and disrupts the mitochondrial membrane potential of human breast and cervical cancer cell lines. Food and Chemical Toxicology, 49, 871–878.

Germain, D. (2011). Estrogen carcinogenesis in breast cancer. *Endocrinology and Metabolism Clinics of North America, 40*, 473–484. https://doi.org/10.1016/j.ecl.2011.05.009

Gorrini, C., Harris, I., & Mak, T. (2013). Modulation of oxidative stress as an anticancer strategy. *Nature Reviews Drug Discovery, 12*, 931–947. https://doi.org/10.1038/nrd4002

Guner, A. (2021). Toxic and irritant effects induced by zearalenone: Prevention by taurine. *Toxin Reviews, 40*(1), 25–34.

Hermosín, I., Chicón, R. M., & Dolores Cabezudo, M. (2003). Free amino acid composition and botanical origin of honey. *Food Chemistry, 83*, 263–268.

Hunter, M., Ghildyal, R., D'Cunha, N. M., Gouws, C., Georgousopoulou, E. N., & Naumovski, N. (2021). The bioactive, antioxidant, antibacterial, and physicochemical properties of a range of commercially available Australian honeys. *Current Research in Food Science, 4*, 532–542. https://doi.org/10.1016/j.crfs.2021.08.002

IHC. (2009). *Harmonised methods of The International Honey Commission (IHC)*. IHC responsible for the methods: Stefan Bogdanov. https://www.ihc-platform.net/ihcmethods2009.pdf

Jaganathan, S. K. (2012). Growth inhibition by caffeic acid, one of the phenolic constituents of honey, in HCT 15 colon cancer cells. *Scientific World Journal, 2012*, 372345. https://doi.org/10.1100/2012/372345

Jaganathan, S. K., Mandal, S. M., Jana, S. K., Das, S., & Mandal, M. (2010). Studies on the phenolic profiling, anti-oxidant and cytotoxic activity of Indian honey: In vitro evaluation. *Natural Product Research, 24*, 1295–1306. https://doi.org/10.1080/14786410903184408

Jovanovic, P., Zoric, L., Stefanovic, I., Dzunic, B., Djordjevic-Jocic, J., Radenkovic, M., & Jovanovic, M. (2010). Lactate dehydrogenase and oxidative stress activity in primary open-angle glaucoma aqueous humour. *Bosnian Journal of Basic Medical Sciences, 10*(1), 83–88. https://doi.org/10.17305/bjbms.2010.2743

Kadir, E. A., Sulaiman, S. A., Yahya, N. K., & Othman, N. H. (2013). Inhibitory effects of tualang honey on experimental breast cancer in rats: A preliminary study. *Asian Pacific Journal of Cancer Prevention, 14*, 2249–2254.

Kanyilmaz, G., Akmansu, M., Bora, H., & Yirmibeşoğlu Erkal, E. (2022). The use of complementary and alternative medicine among cancer patients treated with radiotherapy. *Turkish Journal of Oncology, 37*(1), 74–82. https://doi.org/10.5505/tjo.2021.3357

Kassim, M., Achoui, M., Mustafa, M. R., Mohd, M. A., & Yusoff, K. M. (2010). Ellagic acid, phenolic acids, and flavonoids in Malaysian honey extracts demonstrate in vitro anti-inflammatory activity. *Nutrition Research, 30*, 650–659. https://doi.org/10.1016/j.nutres.2010.08.008

Kusano, C., & Ferrari, B. (2008). Total antioxidant capacity: A biomarker in biomedical and nutritional studies. *Journal of Molecular Cell Biology, 7*, 1–15.

Liu, X., Zou, H., Slaughter, C., & Wang, X. (1997). DFF, a heterodimeric protein that functions downstream of caspase-3 to trigger DNA fragmentation during apoptosis. *Cell, 189*(2), 175–184. https://doi.org/10.1016/s0092-8674(00)80197-x

Liu, J. R., Ye, Y. L., Lin, T. Y., Wang, Y. W., & Peng, C. C. (2013). Effect of floral sources on the antioxidant, antimicrobial, and anti-inflammatory activities of honeys in Taiwan. *Food Chemistry, 139*, 938–943. https://doi.org/10.1016/j.foodchem.2013.02.015

Marijanović, I., Bevanda Glibo, D., Lasić, T., Kraljević, M., Buhovac, T., Cerić, T., & Sokolović, E. (2021). The use of complementary and alternative medicine among cancer patients undergoing conventional treatment: A single-centre experience. *Psychiatria Danubina, 33*(Suppl 13), 304–307.

Maron, D. M., & Ames, B. N. (1983). Revised methods for the Salmonella mutagenicity test. *Mutation Research/Environmental Mutagenesis and Related Subjects, 113*(3–4), 173–215. https://doi.org/10.1016/0165-1161(83)90010-9

Moniruzzaman, M., Khalil, M. I., Sulaiman, S. A., & Gan, S. H. (2013). Physicochemical and antioxidant properties of Malaysian honeys produced by *Apis cerana*, *Apis dorsata* and *Apis mellifera*. *BMC Complementary and Alternative Medicine, 13*, 43.

Mosmann, T. (1983). Rapid colorimetric assay for cellular growth and survival: Application to proliferation and cytotoxicity assays. *Journal of Immunological Methods, 65*, 55–63.

OECD Guideline 471. (2020). *Guideline for testing of chemicals no: 471: Bacterial reverse mutation test* (pp. 1–11). OECD.

Official Methods of Analysis of The Association of Official Analytical Chemist (AOAC 998.12). (2023). *Official methods of analysis* (22nd ed.).

Pichichero, E., Cicconi, R., Mattei, M., Muzi, M. G., & Canini, A. (2010). Acacia honey and chrysin reduce proliferation of melanoma cells through alterations in cell cycle progression. *Journal of Oncology, 37*, 973–981.

Priyadarsini, V. R., Senthil Murugan, R., Maitreyi, S., Ramalingam, K., Karunagaran, D., & Nagini, S. (2010). The flavonoid quercetin induces cell cycle arrest and mitochondria-mediated apoptosis in human cervical cancer (HeLa) cells through p53 induction and NF-kappaB inhibition. *European Journal of Pharmacology, 649*, 84–91.

Samarghandian, S., Afshari, J. T., & Davoodi, S. (2011). Chrysin reduces proliferation and induces apoptosis in the human prostate cancer cell line PC-3. *Clinics, 66*, 1073–1079.

Sergiel, I., Pohl, P., & Biesaga, M. (2014). Characterization of honeys according to their content of phenolic compounds using high performance liquid chromatography/tandem mass spectrometry. *Food Chemistry, 145*, 404–408.

Shapiro, G. I., & Harper, J. W. (1999). Anticancer drug targets: Cell cycle and checkpoint control. *Journal of Clinical Investigation, 104*(12), 1645–1653. https://doi.org/10.1172/JCI9054

Silici, S. (2018). Antioxidant and physicochemical properties of chestnut honeys from Turkey. *Communications Faculty of Sciences University of Ankara Series C Biology, 27*(2), 104–114. https://doi.org/10.1501/commuc_0000000204

Snedecor, G. W., & Cochran, W. G. (1968). *Statistical methods*. Oxford and IBD Publish.

Tabak, A., Deniz, N. T., Avşar, C., Aslan, H., & Yumak, T. (2024). Some physicochemical properties of pure and pollution-free Sinop chestnut honey. *Spectroscopy Letters, 57*(7), 420–427. https://doi.org/10.1080/00387010.2024.2362937

Tsiapara, A. V., Jaakkola, M., Chinou, I., Graikou, K., Tolonen, T., Virtanen, V., & Moutsatsou, P. (2009). Bioactivity of Greek honey extracts on breast cancer (MCF-7), prostate cancer (PC-3) and endometrial cancer (Ishikawa) cells: Profile analysis of extracts. *Food Chemistry, 116*, 702–708. https://doi.org/10.1016/j.foodchem.2009.03.024

Turkish Food Codex Notification No. 2012/58 on honey. (2012). *FAO, FAOLEX Official Journal No. 28366.*

Waheed, M., Hussain, M. B., Javed, A., Mushtaq, Z., Hassan, S., Shariati, M. A., Khan, M. U., Majeed, M., Nigam, M., Mishra, A. P., & Heydari, M. (2019). Honey and cancer: A mechanistic review. *Clinical Nutrition, 38*(6), 2499–2503. https://doi.org/10.1016/j.clnu.2018.12.019

Xie, F., Su, M., Qiu, W., Zhang, M., Guo, Z., Su, B., Liu, J., Li, X., & Zhou, L. (2013). Kaempferol promotes apoptosis in human bladder cancer cells by inducing the tumor suppressor, PTEN. *International Journal of Molecular Sciences, 14*, 21215–21226. https://doi.org/10.3390/ijms141121215

Zhang, J., & Xu, M. (2000). DNA fragmentation in apoptosis. *Cell Research, 10*, 205–211. https://doi.org/10.1038/sj.cr.7290049

Zhang, Q., Zhao, X. H., & Wang, Z. J. (2008). Flavones and flavonols exert cytotoxic effects on a human oesophageal adenocarcinoma cell line (OE33) by causing G2/M arrest and inducing apoptosis. *Food and Chemical Toxicology, 46*(6), 2042–2053. https://doi.org/10.1016/j.fct.2008.01.049

Živanović, S., Pavlović, D., Stojanović, N., & Veljković, M. (2019). Attitudes to and prevalence of bee product usage in pediatric pulmonology patients. *European Journal of Integrative Medicine, 27*, 1–6.

Chapter 12
Flower Honey and Sustainable Development of Beekeeping in Azerbaijan

Elsevar Sabir Asadov, Sayyara Jamshid Ibadullayeva, Rauf Lutfali Sultanov, and Mursel Seyidov

1 Introduction

Honey is a very valuable natural bee product known to people since ancient times. Nectar collected by honeybees from the flowers of plants, after mixing with enzymes synthesized in their bodies, undergoes certain changes and turns into honey. Bees evaporate the water contained in the honey collected in the honeycomb cells and store it as a reserve by coating it with wax (Dafni et al., 1988; Layek et al., 2020).

The main source of honey is the nectar of flowering plants, various sap, and juicy substances secreted by some sap-sucking insects living on the vegetative organs of plants. The variety of plants in different areas creates differences in the physical properties, chemical composition, and quality of honey. From this point of view, there are differences in the composition, properties, and quality indicators of honey produced in different parts of the world (Pavlova et al., 2018; Bogdanov et al., 2002).

At present, honey standards have been established in several countries of the world. Since Azerbaijan was part of the Soviet Union for many years, bee honey, along with other products, was based on generally accepted standards. Even after gaining independence, these standards were taken as a basis. Despite this, Azerbaijani honey has always been highly valued and appreciated, even in the union (Degirmenci et al., 2020).

E. S. Asadov (✉) · M. Seyidov
Nakhchivan State University, Nakhchivan, Azerbaijan

S. J. Ibadullayeva
Institute of Botany, Ministry of Science and Education of the Republic of Azerbaijan, Baku, Azerbaijan

R. L. Sultanov
Department of Biology, Azerbaijan State Pedagogical University, Baku, Azerbaijan
e-mail: ilqar.sultanli@noc.az

M. Zafar et al. (eds.), *Pure Honey: Assurance & Authentication*,
https://doi.org/10.1007/978-3-031-98913-1_12

Favorable natural-geographical conditions in our republic and rich vegetation cover have led to the formation of unique characteristics in the composition, physical-chemical properties, and quality of the honey produced in the bee colonies existing here. Therefore, the produced honeys are mostly named according to regions (for example, Guba honey, Ismayilli honey, Lerik honey, Nakhchivan honey, etc.).

There are a number of indicators for classifying honey. Based on these indicators, the standards of the honey are determined.

2 Types of Honey

Classification of honey can be done according to its source, production mechanism, color, amount of water in honey, and commercial form. According to the source of formation, honeys are divided into flower honey, honeydew honey, poisonous honey, sugar honey, artificial honey, and express honey.

2.1 *Natural Flower Honey*

It is called the thick juice—bees obtain the nectar of various plants, mix it with the enzymes of the digestive glands in their bodies, and empty it into the honeycomb cells and form it there (Siavoshi et al., 2018). The taste, aroma, and colors of the flower from which the nectar is collected can be felt in these honeys.

2.2 *Honeydew Honey (Secretion)*

It is a type of honey obtained by collecting the secretions of trees such as pine, fir, and oak by bees, as well as juice secreted by some insects (aphids, cochineals, greenfly, etc.) living on trees and collected by bees (Wäckers et al., 2008; Douglas, 2009). Depending on the source, it is distinguished as pine honey, fir honey, etc. The most common of these, pine honey, is a type of honey obtained by bees collecting the sweet secretions secreted by giant pine scale (Marchelina hellenica), which live on pine trees. These insects live inside a cocoon they build and suck sap by inserting their proboscis into the conducting veins of the bark of trees (Oğuzoğlu et al., 2021). The juice they suck contains up to 80% sugar and a small amount of protein. When insects suck this juice, they begin to secrete what they have already collected. Bees collect the red, fragrant, and tasty secretions of insects, carry them to the hive, and turn them into honey. Honeydew honey is widely produced in many countries of the world. It can be of plant origin as well. Bees can also collect secretions on stems and leaves of plants when there is no nectar supply. These honeys usually have antibacterial properties.

2.3 Intoxicating or Poisonous Honey

This honey is produced from the nectar collected by bees from plants such as deadnettle (Lamium) and stramonium (Datura). In many cases, people who consume poisonous honey may experience dizziness, nausea, blurred vision, ringing in the ears, sweating, drowsiness, and fainting (Wu et al., 2019).

2.4 Sugar Honey

Some beekeepers constantly feed their bees with sugar solutions in special containers around or inside the hives. As a result of this kind of feeding, the unnatural honey that bees fill in the honeycomb cells is called sugar honey (Liebezeit & Liebezeit, 2013). The amount of sucrose is very high in such honeys, which are liquid and light in color. Sugar honey is difficult to distinguish from natural flower honey. In local conditions, such honey is perceived as fake honey.

2.5 Artificial Honey

It is a form of honey obtained without bees or plants, made using sugar (sucrose), starch, flavorings, colors, and aromas, similar in appearance, taste, and form to natural honey (Jiang et al., 2022). Such artificial honey is not only not beneficial for the human body but can also cause a number of diseases.

2.6 Express Honey

Recently, in some countries of the world, in order to increase the medical value of honey, bees are given sugar syrup with various medicinal substances added (Lenis et al., 2013). Such obtained honey is referred to as "express honey." Such honey obtained on purpose is used for therapeutic and scientific research purposes. Depending on the plant sources from which natural flower honeys are obtained, two types are defined: polyfloral and monofloral honey.

2.7 Polyfloral Honeys

This type of honey is collected from different plants, and none of these plant nectars are very superior (Cozmuta et al., 2011). If there is no nectar-producing specific plant field in the geographical area where the honey is collected, the honey obtained from it will be polyfloral. Most honey obtained in the territory of Azerbaijan is polyfloral. Polyfloral is named after the geographic area where the honey is collected, as the source of honey is not primarily any one plant species. For example, Guba honey, Lankaran honey, Ismayilli honey, Nakhchivan honey, etc.

2.8 Monofloral (Unifloral) Honey

This honey contains more than 51% of the nectar of any plant and more than 45% of pollen (Hulea et al., 2022). Monofloral honey is mostly named after the plant from which it is collected: acacia honey, chestnut honey, linden honey, sunflower honey, etc. Monofloral honeys have the aroma and color characteristic of nectar sources.

2.9 Linden Honey

Bees collect this honey from linden flowers (Qiao et al., 2020). Linden honey is highly valued for its aroma and good taste. Linden honey is transparent or light yellow in color (Fig. 12.1).

2.10 Acacia Honey

Acacia honey is very light in color, transparent, fragrant, and delicious (Fig. 12.2). Bees collect this honey where there is a lot of acacia. Acacia honey crystallizes late and is considered high-quality honey (Muhammad et al., 2016).

Fig. 12.1 Linden honey

Fig. 12.2 Acacia honey

2.11 Sunflower Honey

The color of sunflower honey is dark yellow, and the taste is pleasant. This honey crystallizes quickly (giving a coarse-grained sediment) and gradually changes taste as it crystallizes (Živkov Baloš et al., 2023).

2.12 Cotton Honey

This honey is transparent in color and has a unique taste and aroma. As it crystallizes, it turns white and gives a large granular sediment (Moffett et al., 1975).

2.13 Chestnut Honey

Bees collect this honey from the flowers of common chestnut and horse chestnut (Kropf et al., 2010). The honey collected from the horse chestnut flower is colorless and watery, while the honey collected from the common chestnut flower is dull in color, very liquid, and bitter, so it is used only in the confectionery industry.

According to the form of sale and consumption of honey, there are three forms: honeycomb honey, filtered honey, and crystallized honey.

1. Honeycomb honey is a natural honey collected in honeycombs and sealed with wax, as shown in Fig. 12.3 (Oeder & Schwabe, 2021).
2. Filtered honey is honey extracted from honeycombs by various methods or filtered in special honey-filtering machines (centrifuges). Filtered honey is liquid and does not contain any foreign matter (dead bees, wax crumbs, wood chips, etc.), as shown in Fig. 12.4 (Peng & Marston, 1986).
3. Crystallized honey. Filtered honey begins to crystallize after remaining for some time. Its structure is like a fat-like. There is no difference in quality from filtered honey (Scripca & Amariei, 2018; Rababah et al., 2024).

Fig. 12.3 Honeycomb honey

Fig. 12.4 Honeycomb honey (left) and filtered (right)

Natural honey is a sweet, liquid product with a unique taste and aroma, made by bees from the nectar of flowers, honeydew of plants, and fruits. Honey is a food product with a very complex composition due to its physical and chemical properties. It has been determined that honey contains more than 70 substances. It contains glucose, fructose, enzymes, vitamins, trace elements, hormones, mineral salts, inorganic acids, and other substances that can be easily absorbed by the body.

3 Mechanism of Honey Formation

Nectar is the name given to sugary liquids that bees obtain from nature to make honey. This tasty substance, which is often synthesized in the nectaries of flowers, is sometimes secreted in other parts of the plant or in the body of plant pests. Plants secrete nectar in order to attract bees that organize their cross-pollination (Krishnan et al., 2021; Rzepecka-Stojko et al., 2015). Bees start their honey production by collecting the sweet-tasting fluids found in the flowers of nectar plants or secreted by the insects that live on some plants. These saps are turned into honey by the action of special enzymes in the honey stomach, which is part of the digestive system of bees. Then the water is evaporated, and the final shape is formed. The bee that brings the nectar to the nest starts walking on the hive, among the bees (Seeley,

1989). If it is the period when the nectar yield is low, then the bee feeds the other bees in the nest with this nectar. When the nectar supply is high, the bee passes it on to another bee that has entered the hive. After receiving the nectar, the worker bee, operating inside the hive, goes to a quiet place and performs a series of complex actions to form honey. After the honey is formed in the bee's body, it vomits it into an empty cell. Nectar undergoes both physical and chemical changes while turning into honey. The physical change is mainly due to the decrease in water content and relative thickening of the nectar. The chemical change involves the conversion of sucrose, nectar sugar, into the simple sugars—glucose and fructose—by the action of the bee's digestive enzyme—invertase. The breakdown of sucrose into simple sugars begins as soon as the bee collects nectar from the flower. The complete formation of honey ends in the honeycomb cells.

4 Physical Properties of Honey

Honey is divided into certain groups according to its technological and chemical characteristics, as well as its color, taste, and smell.

4.1 The Color of Honey

Honey starts from transparent to dark red, yellow, brown, golden, reddish, etc. in colors (Fig. 12.5). According to the color, the honey is divided into four groups: transparent, white, light amber, and dark colors. The substances that give honey its color are carotene, chlorophyll, xanthophyll, and a number of plant pigments that produce the yellow and green color of unknown composition. Researchers have found that there is a relationship between the color and composition of honey (Szabó et al., 2016). Thus, dark-colored honey contains more amino acids, sugars, and minerals—iron, copper, and manganese. As the mineral content of honey increases, its color becomes darker. Other researchers report that the color of honey depends more on carotenoids. In addition to honey, known for its color, different-colored honeys are also known due to their origin. For example, in the North Carolina region of America, blue-colored honey is found, the source of which is not exactly known (White Jr & Riethof 1959).

4.2 Taste of Honey

Each honey has its own unique taste. The main taste of honey is sweet. But when honey is put in the mouth, many flavors are felt in it. Even one hive can sometimes have several different flavors of honey (De Brito Sanchez et al., 2007). The taste of

Fig. 12.5 Honey of different colors

honey can vary depending on the complex ethers in the nectar of different flowers. The real sources of taste in honey: complex esters, aldehydes, ketones, alkaloids, and free acids. The most abundant of these substances are alkaloids. Since aromatic substances come with more nectar, it is possible to feel the taste of the plant from which that nectar is collected in the honey.

4.3 The Smell of Honey

Honey has a specific smell depending on the pollen in it. The smell of honey is felt mainly when it is put in the mouth and eaten (Marcazzan et al., 2018). Heated honey loses a large part of its aromatic compounds. At the same time, if honey is placed next to a strong-smelling substance, it can absorb the smell of that substance. Dark-colored honeys tend to be more odorous than light-colored honeys.

4.4 Fluidity of Honey

The fluidity of honey is normally slow. Slow-flowing honey is difficult to remove from the comb cells and clean from containers during filtering (Kolayli et al., 2013). The fluidity of dark-colored honey is weak, and the fluidity of light-colored honey is relatively fast.

4.5 Consistency of Honey

The plant source from which honey is obtained plays a large role in the consistency of honey. Whether honey is clear or thick depends on the amount of air bubbles, the amount of water, and less of colloidal substances. While plants grown in hot zones and light soils produce solid honey, honey produced in temperate and mountainous zones is liquid, tasty, and aromatic (Marcazzan et al., 2018).

4.6 Specific Weight of Honey

The specific weight of honey depends on the amount of water and temperature and is on average 1.41–1.45 g/cm^3 (Al-Rajhi & EL-Sheikha, 2014).

4.7 Hygroscopic Property of Honey

Honey is a hygroscopic substance; it has the ability to absorb moisture from the air under its conditions. The ability to absorb moisture from the air depends on the source of honey, the amount of sugar, and the water in it.

4.8 Light Reflection (Polarization) Property of Honey

The direction and amount of light reflection of honey vary according to the type of honey (Pavlova et al., 2018). Flower honey deflects light to the left, and honeydew to the right. By using this feature, it is possible to clarify whether the honey is honeydew honey or not.

4.9 Crystallization of Honey

Crystallization (sugaring) is a phenomenon in which glucose in honey, to a greater or lesser extent, loses its fluidity due to the formation of granules (Fig. 12.6). Most honeys tend to crystallize. This tendency can change depending on the amount of water, glucose, fructose, and sugar in honey, and temperature. Plant sources from which honey is obtained also play a role in crystallization (Krishnan et al., 2021). Crystallized honey needs to be heated until it becomes liquid.

Although some methods have been suggested to prevent honey from crystallization, the simplest method to use is to store honey at 0 °C for 5 weeks and then at

Fig. 12.6 Crystallized honey

14 °C after filtering. To liquefy crystallized honey, the method that does not reduce its quality is to keep the container with honey in water at a temperature of 40–45 °C. However, temperatures above this lead to a change in the composition and quality of honey. Usually, sunflower, clover, melon, and cotton honeys crystallize very quickly, while acacia, mustard, dead-nettle, and honeydew honeys crystallize late. Sage honey can stay for years without crystallizing. The factors affecting the crystallization of honey are as follows.

4.10 The Ratio of Sugars in Honey

The speed of crystallization is closely related to the glucose-fructose ratio and water content in honey. The amount of fructose in honey is mainly higher than glucose. If the amount of glucose is close to fructose, the honey will crystallize quickly; if the difference is large, it will crystallize later. Since unripe honey has a lot of sucrose and little glucose, the crystallization is slow. There is not much glucose in ripe honey. When the ratio of fructose to glucose is between 1:1.2, crystallization is fast; as the ratio of glucose exceeds 1:1.3, crystallization slows down (Siddiqui, 1970; Ng, 2022).

4.11 Effect of Glucose-Water Ratio

As a regular pattern, honey never becomes crystallized when glucose/water = 1.70. When glucose/water = 2.10, crystallization is accelerated (Doner, 1977).

4.12 Effect of Temperature

Crystallization of honey begins at a temperature of 5–7°, and crystallization is delayed as the temperature drops (Kurta et al., 2021). There are different ways to prevent crystallization. The most commonly used of these is heating. Recently, ultrasonic waves and homogenization methods have also been used. For this purpose, it is possible to prevent crystallization by using chemicals that do not change the composition of honey.

4.13 The Effect of Packaging Tools

In order to delay and prevent crystallization, glass jars, lids, or other containers must be clean and completely dry when filling with honey. When the honey is strained through a finer sieve, the crystallization is greatly delayed or stopped.

5 Composition and Chemical Properties of Honey

Honey has a rather complex composition due to the variety of substances that make up its basis (Table 12.1). The compounds in honey vary according to the variety of plants in the region where it is produced and when the honey was obtained. Despite this, almost 80% of all honey is various sugars (sucrose, glucose, and fructose), and 17–18% is water. The remaining 3% are enzymes, vitamins, nitrogenous compounds, and acids. It is not fully determined whether enzymes, vitamins, and nitrogenous compounds are introduced into the composition of honey by nectar or whether it is a product of the bees themselves (Soares et al., 2017).

Little is known about the sources of a number of substances, such as acids and vitamins. However, it is thought that some of the amino acids and vitamins are related to the pollen in honey.

From the moment the bee collects the nectar in the honey's stomach and returns to the nest, it begins to turn into honey. The sucrase enzyme in the bee's stomach converts the sucrose in the nectar into fructose and glucose. Other carbohydrates, such as maltose, isomaltose, and erlose, are also formed from the breakdown of sucrose. This causes an increase in the amount of sugar in honey. Bee arrives at the

Table 12.1 Chemical composition of nectar and honeydew honey (according to Whitel)

Nectar (flower) honey				Honeydew honey			
Humidity	17.20%	Fructose	38.19%	Humidity	16.30%	Fructose	31.80%
Glucose	31.28%	Sucrose	1.31%	Glucose	26.08%	Sucrose	0.80%
Maltose	7.31%	Polysaccharides	1.50%	Maltose	8.80%	Polysaccharides	4.70%
Other substances	3.10%	PH	3.90	Other substances	10.10%	PH	4.45
Total acidity (mEq/kg)	29.12	Ash	0.169%	Total acidity (mEq/kg)	54.88	Ash	0.73%
Nitrogen	0.041%	Diastase number	20.80	Nitrogen	0.100%	Diastase number	31.90

When the above indicators are analyzed, it can be seen that the total carbohydrate (sugar) content is 79.59% in nectar honey and 72.18% in secretion honey hive and empties the product from its stomach into a comb cell or another bee's mouth. The second bee that receives the honey evaporates the water in the honey and processes it to thicken and mature. Ripe honey has a specific smell and taste and can be stored for a long time without spoiling. Unripe honey spoils quickly, that is, it ferments (Sanz et al., 2005).

The chemical composition of honey changes depending on a number of factors. The most important of these factors is the natural source of nectar. At the same time, the climate and honey-making characteristics of bees also have an effect on the formation of the chemical composition of honey. In terms of its chemical compounds, honey contains fructose, sucrose, dextrin, nitrogen compounds, enzymes, volatile oils, organic acids, pollen grains, minerals, dyes, etc. (White & Doner, 1980).

5.1 *Fermentation of Honey*

Glucose and fructose, which are the main ingredients of honey, are broken down by sugar yeasts. As a result, alcohol and carbon dioxide are formed. Alcohol also decomposes in an oxygenated environment, turning into acetic acid and water. Fermentation causes the honey to spoil and taste unpleasant. Crystallized honey never loses its quality, but during the fermentation process, it becomes of poor quality (White & Doner, 1980). The main reasons that cause fermentation are the amount of water, storage conditions, yeast fungi, and the immaturity of honey. Experiments have proven that if the amount of water in honey is less than 17%, the honey will not ferment for at least one year, and yeast fungi will not grow in the honey. The source of yeast that ferments honey is soil and flowers. Yeast fungi are also abundant in the honey left over from the previous year in the hive. For this reason, this type of honey should not be added to a new product when filtering honey.

5.2 *The Amount of Water in Honey*

If the concentrations of sugars, which are the basis for the characteristic taste and quality of honey, are high enough, fermentation will not occur. Fermentation occurs when the water content is above 18.5%. Acetic acid and carbon dioxide produced by fermentation change the taste and color of honey (Wedmore, 1955). Honey is divided into three groups according to the amount of water: honey with low, medium, and high water content. These characteristics are considered to be the most important quality indicators of the collected honey. The amount of water in honey included in the first group is up to 17%, in group II, 18–19%, and in group III, more than 20%. The maximum amount of water in honey should not exceed 21%. The amount of water in mountain honey is lower than in plain honey (Yanniotis et al., 2006). An excess amount of water indicates that the honey is not ripe or that water has been added to it. This also accelerates the growth of fungi in honey and reduces its quality.

5.3 *Amount of Carbohydrates in Honey*

Sugar is very high in honey; 95–99.9% of its dry matter is various carbohydrates. While 9 types of 15 types of sugars were present in all honeys, six types of sugars were found only in some honeys (Doner, 1977; Borum, 2021).

Sugars in honey—mostly fructose and glucose. These monosaccharides, which give honey its taste, are formed from sucrose in plant nectar as a result of the action of invertase, a bee's digestive enzyme. The taste, energy value, moisturizing properties, and other physical properties of honey mainly depend on these two carbohydrates. The amount of sucrose in honey varies according to the degree of ripeness of the honey and the composition of the nectar. Unripe honeys harvested prematurely contain a lot of sucrose.

5.4 *Enzymes in Honey*

Honey is very rich in enzymes. The main enzymes of honey are amylase (diastase), invertase (sucrose), catalase, and phosphatase. Some of the enzymes are produced from nectar and the secretions of greenfly, the source of honeydew honey, and a large part from the secretions of the salivary glands of bees (Subramanian et al., 2007). The enzyme invertase carries out most of the chemical changes that convert nectar into honey. This enzyme ensures the conversion of sucrose in the nectar into fructose and glucose. An important enzyme, glucose oxidase, acts on glucose to form hydrogen peroxide and gluconolactone. The antimicrobial effect of honey is also related to the hydrogen peroxide that is formed. Honey also contains the enzyme lysozyme, which has an antimicrobial effect. In addition to these, honey contains a special substance that is synthesized in a very small amount in the human body. This substance, called Chocin, regulates fat metabolism in the human liver and prevents fattening of the liver (Swain et al., 2021).

5.5 *Acids in Honey*

The most abundant acid in honey is gluconic (hydroxyacetic) acid, which is produced as a result of the activity of the enzyme glucooxidase. In addition, honey contains acetic, fatty, citric, formic, lactic, apple vinegar, succinic, gluconic, oxalic, caprylic, gallic, grape (wine), and carbonic acids. However, sources of acids other than gluconic acid are still unknown. Acids in honey increase resistance to microorganisms. Bees also add formic acid to the honey to ensure its maturation (Mato et al., 2003). It is known that the amount of acid that keeps the honey's pH low is due to the formic acid that is poured into it by the bees' needles before their cells are sealed. Honey is mostly acidic and has a pH of 3.5–5.5 (Adugna et al., 2020). The presence of high acidity in honey causes it to ferment after a certain period of time, and, as a result, the alcohol turns into acetic acid under the influence of bacteria.

5.6 Proteins in Honey

Nitrogen substances are approximately 0.3% in flower honey and up to 1% in secretion honey. If the amount of nitrogenous compounds in flower honey is high, it indicates that it has been mixed with honeydew honey. The amount of proteins in honey is not only important for feeding the bee family, but it is also one of the indicators of whether the honey is natural or fake (Nazarian et al., 2010). Although honey is not a protein-rich food, it is very rich in amino acids. Approximately 17 amino acids have been identified in honey. While tyrosine and tryptophan are present in dark-colored honey, they are not found in light-colored honey. Honey contains the most proline, lysine, and glutamine amino acids. These are followed in descending order by histidine, arginine, threonine, serine, valine, methionine, leucine, alanine, and phenylalanine.

5.7 Vitamins in Honey

Although it has long been believed that honey contains no or very few vitamins, research conducted in recent years has revealed that it contains various vitamins. While honey does not contain vitamin A, it contains C, E, K, and B group vitamins (B1, B4). Thiamin, riboflavin, ascorbic acid, pyridoxine (vitamin B6), pantothenic acid (vitamin B5), nicotinic acid (vitamin B3), biotin (vitamin B7), or vitamin H and folic acid (vitamin B9) have been found in honey in varying amounts. Watts and Merrill's research found that honey contains a small amount of vitamin B1, 0.4 mg of vitamin B2, and 10 mg of vitamin C. During the filtration of honey, many of these vitamins are fragmented. For this, it is necessary to carry out the filtering process more carefully so that the vitamin content of honey does not decrease (Kitzes et al., 1943; Ciulu et al., 2011; Valverde et al., 2022).

5.8 Mineral Substances in Honey

The mineral substances in honey content vary between 0.02% and 1.0%. Of chemical elements, honey contains more potassium, calcium, phosphorus, small amounts of sodium, chlorine, sulfur, magnesium, silicon, manganese, copper, iron, and zinc. Minerals found in honey, which is a rich source of minerals, include potassium, sodium, calcium oxide, magnesium, iron, copper, manganese, chlorine, phosphorus, sulfur, silicon, and silica. Rare mineral substances include chromium, lithium, nickel, lead, tin, zinc, osmium, beryllium, vanadium, zirconium, silver, barium, gallium, bismuth, gold, germanium, strontium, and others (Bogdanov et al., 2007; Barreiros et al., 2024).

6 Processing and Storage of Honey

The work carried out during the processing of honey has a significant impact on its quality. In this regard, the storage place, humidity, heat, and other conditions should be considered when processing honey.

During the processing of honey, the glucose crystals, ordinary dust, or pollen grains are the starting points in honey crystallization. For this reason, the resulting honey should be filtered very well and left for a certain period of time so that the particles passing through the filter collect above and below the honey. Honey should not be stored in the open, as it ferments in contact with moisture. The best way to remove air bubbles in honey is to let it rest.

6.1 Filtering and Storing Honey

Filtering of honey is carried out with special bee-filtering machines (centrifuges). It is desirable that the temperature of the air during filtration should be around 35 °C. At this time, the honey is quickly and easily separated from the honeycomb cells (Bryant, 2017). The filtered honey is allowed to rest to rinse it and remove impurities. When honey is left to rest, substances with a low density and air bubbles rise to the top. Substances that are denser than honey and in crystalline form sink to the bottom of the honey. However, the actual cleaning is done by passing through a special sieve with narrow meshes.

The factors that will affect the quality of honey in storage are the temperature of the storage place, humidity, storage containers, and storage period. When the temperature of the storage place of honey is below 11 °C, the activity of fungi causing fermentation stops (Radtke & Lichtenberg-Kraag, 2018). In order to avoid unwanted situations, it is necessary to store honey in a cool place and in tightly closed containers so that it does not absorb moisture. Glass jars are the most suitable containers for storing honey. Since honey has an acidic reaction, when stored in plastic containers, it can react with the ingredients there, and its quality will decrease. Packaged honey should be stored at room temperature. To prevent a decrease in the sugar and enzyme content of honey, it is necessary to lower the temperature of the storage place.

7 Determining the Quality of Honey

Natural honey is both the best food and a valuable product used in the treatment of many diseases. Some "businessmen" who want to make a lot of money easily falsify this wonderful, aromatic product by mixing different ingredients. Therefore, it is necessary to determine whether foreign substances are mixed with honey: powdered sugar, cane sugar, potato and corn starch, water, flour, chalk, etc. For this, it is useful to know simple ways to distinguish natural honey from artificial honey. To detect artificial impurities in honey, a sample of honey taken from the bottom of the container is poured into a test tube and then dissolved by adding distilled water. At this time, if the honey is clean, the solution becomes transparent; if it is not clean, a sediment appears. To determine the quality of honey, it is necessary to determine the number of its components. Simple methods used for this purpose are.

7.1 Determination of Starch

A few drops of iodine are added to the honey solution dissolved in distilled water. When honey contains starch, its color turns bluish.

7.2 Determination of Starch Syrup

Cold-processed starch syrup mixture is easy to identify by its appearance. Such honey is runny and does not crystallize. In addition, mix one part honey with three parts distilled water and add 96% alcohol to 1/4 of the solution and shake it. When there is a mixture in honey, a white milky solution is obtained. When such a solution is kept for some time, a sediment of a translucent liquid mass (dextrin) is formed. If the honey does not contain starch syrup, the solution becomes transparent, and only when the alcohol and honey mix, a barely noticeable turbidity is formed, which disappears when the solution is shaken (Naila et al., 2018).

7.3 Determination of Starch Syrup

To determine this, a little silver nitrate solution is added to a 5–10% solution of honey in water. Obtaining a white sediment indicates that the solution contains impurities. To determine the starch syrup by another method, 2.5 g of lead acetate and 22.5 3 cm^3 of methyl (wood) alcohol are added to 5 g of a 20% honey solution prepared in distilled water. The formation of a large yellowish-white sediment indicates the presence of starch syrup in honey (Wang et al., 2015).

7.4 Determination of Chalk

To determine chalk in honey, a few drops of any acid or vinegar solution are added to a sample of honey dissolved in water. When honey contains chalk, the solution boils and releases carbon dioxide.

7.5 Determination of Honeydew Honey

Honeydew honey is determined in two ways. In the first method, it is determined whether there is any honeydew in the composition of honey at all. With this method, the honeydew honey is checked for quality. The second method is the quantitative method, that is, a quantitative check of honeydew honey. With this method, it is found out how much honeydew honey is contained in honey (Sanz et al., 2005). With the first method, that is, to determine the presence of honeydew in the honey, a little honey is poured into a test bottle, and a little water is added to it, and it is shaken vigorously and mixed. If the solution becomes cloudy and a brown, foamy sediment forms, this indicates that the honey contains honeydew. In order to determine the amount of honeydew in honey by the second method, 2.1 g of honey is poured into a cup, 3 ml of distilled water and 15 ml of lime water are added, heated to a boil (but not boiled), and then cooled. The mixture in the cup is mixed well and poured in equal portions into two centrifuge tubes. It is necessary to try not to leave the mixture in the cup. After this, the test tubes are placed in an electric centrifuge and rotated for 3 minutes at a speed of 1.2–1.5 thousand rpm. Then the transparent part from both test tubes is carefully poured into the previous cup. The sediment in one of the test tubes is mixed with the remaining liquid part and poured into the second test tube. Thus, all the sediment is poured into one test tube. Then the solution from the cup is poured back into the test tube, its weight is equalized, and it is again placed in the centrifuge and rotated for 3 min. Then the level of the sediment is measured with the divisions of the test tube. The amount of honeydew (X) is determined as a percentage according to the following formula:

$$X = V \times 100 / 1,5$$

Here, V is the size of the sediment in the centrifuge test tube in millimeters. If a hand-rotating centrifuge is used, then the result obtained according to the formula is reduced by 20% (Bogdanov et al., 2002; Bicudo de Almeida-Muradian et al., 2020; Muradian et al., 2020).

7.6 Determination of Honey Crystallization Period

To determine the duration of honey crystallization, experimental groups consisting of honeycombs of different quality are created, and five honeycomb frames in each group are placed in 5 hives, separated during the period of the main nectar yield. In the first experimental group, newly knitted and unused combs are used; in the second experimental group, white-colored combs knitted last year and used only for honey collection; in the third experimental group, white-brown-colored combs used in honey production and beekeeping; in the fourth experimental group, combs are used in the cultivation of brown honey and ten to ten generations of bees; and in the fifth experimental group, washed with water and dried brown-colored combs are used. To create equal conditions between the experimental groups, the same colony is provided with honeycombs belonging to all five experimental groups. The rate of crystallization of honey is determined by visual observations and under a microscope (Sultanov, 1993; Scripca & Amariei, 2018; Tappi et al., 2019).

8 The Main Indicators of Azerbaijani Honey

The honeybee differs from other creatures in its ability to collect winter reserve food in advance. During their active life, they process reserve food and collect it in the honeycomb cells in the form of honey. Depending on various natural conditions, the study of the placement of food brought to the nest in the honeycomb, the sealing of the honeycomb cells, and the color and quality characteristics of honey at a comparative speed acquires particular importance. Here, it is necessary to take into account the individual characteristics of each bee colony and the genus to which it belongs (Tahirov et al., 2010).

8.1 Sealing of the Honeycomb Cells

During the active period, bees fill the honeycombs with food supplies for the winter period and seal the cells with wax. In beekeeping, this is called sealing of honeycomb cells. There are different opinions about the sealing of honeycomb cells in different forms. In bee families living in the mountainous and foothill regions of the Caucasus, there are also cases of wet sealing of the honeycomb. Honeycomb's eyes are white when sealed in dry form and dull in color when sealed in wet form. Some beekeepers also choose honeycombs based on color characteristics. The study of the identification of the seal, color, and shape of the honeycomb in different natural and geographical conditions of the Republic is of special scientific importance. It is possible to compare bees by this sign. It is interesting that the bee species in local conditions affect the shape of the seal of the honeycomb. Therefore, observations were made in different natural and geographical conditions of the republic, and the collected materials were analyzed. In mountainous, foothill, and subtropical bee-houses, 92–98% of honeycomb cells, and 85–90% in the lowland region, are sealed in wet form and dull color. In the rest of the cases, mostly on the upper and side parts of the honeycomb, there are dry and white sealed honeycomb cells. Honeycomb cells in 96% of Bozdag Caucasian bees and 87% of Yellow Caucasian bees are sealed in a wet form and in a dull color. Thus, in the conditions of Azerbaijan, in bee colonies, sealing of honeycomb cells in a damp state and dull color is taken as a basis.

8.2 Crystallization Feature of Azerbaijani Honey

Crystallization of honey is its transition from a liquid state to a solid state. This process does not degrade the quality of honey (Table 12.2). Honey is a saturated solution formed by the presence of glucose, fructose, dextrins, etc. When honey crystallizes, the grape sugar crystals form a sediment, and the fruit sugar (fructose) collects in solution on top of the sediment, which coats the grape sugar crystals and crystallizes the unripe honey. Therefore, crystallized honey has a sticky form. Crystallization of honey mostly starts from its upper part. Because honey thickens as the water in it evaporates, which causes glucose to crystallize. The specific gravity of crystallized honey is lower than the specific gravity of liquid honey.

Table 12.2 Crystallization of Azerbaijani honey depending on natural and climatic conditions, quality of honeycomb, and storage conditions (by day)

	Quality and age of honeycomb				
Crystallization of honey	Newly knitted	White knitted last year	White-brown color	Washed with brown water	Brown color
In filtered case					
t = 20–25 °C					
Begins	84	63	38	40	21
Ends	122	122	96	88	48
t = 8–14 °C					
Begins	48	25	21	13	12
Ends	89	67	55	32	28
In unfiltered					
t = 8–10 °C					
Mountainous	55	36	23	13	14
Foothills	63	38	25	17	15
Lowland	75	57	44	24	12
Subtropical	61	41	27	19	13

Crystallized honey particles are divided into three groups depending on their size:

(a) Small-grained particles are visible to the naked eye and are smaller than 0.5 mm in size.
(b) The size of large-grained particles exceeds 0.5 mm.
(c) Fat-like particles are visible to the naked eye and look like honey fat.

Small particles of grape sugar form the germ of further crystallization. Subsequently, it grows more and more, and all the honey crystallizes. Germ particles can also form while honey matures. There are many factors that affect honey crystallization. The chemical composition of honey itself, the ratio of particles to each other, botanical origin, storage conditions, and its processing are among the factors that directly affect the honey's crystallization. During winter, depending on the conditions inside the nest, honey crystallization can be completed late or early. The process of crystallization of honey plays a special role in the life of bees as a biological factor. In particular, the crystallization of honey during the wintering period leads to the destruction of the family. During the winter period, bees cannot use crystallized honey. To do this, there must be water inside the nest, or it must be brought from the outside. During winter, bees die of hunger because they cannot leave the nest and bring water.

It would be wrong to perceive honey crystallization as a negative phenomenon. All honey begins to crystallize 30–50 days after it is filtered through the honeycomb.

Old or new honeycombs, their quality and different temperature conditions show themselves at a sharp rate in the crystallization of honey. In the conducted experiments, while the operation of crystallization of honey filtered from this year's and last year's knitted white combs was not started for 2 months, it has been determined that the process of crystallization occurs in the honey filtered from the brown combs processed four to five times and the light brown combs processed three times. The honey in honeycombs with cultured bee growth begins to crystallize after 36 days, that is, 46 days earlier than in honeycombs without cultured bee growth. Filtered honey stored at a low temperature (8–14 °C) crystallizes 1.2–1.5 times faster than honey stored at a high temperature (20–25 °C).

Thus, the honey that is filtered from the honeycombs that are old and have been used in the collection of reserve honey from the beginning is more easily crystallized. Under the same temperature conditions, honey brought from the highlands and foothills becomes crystallized faster than the honey from the lowland region. Late crystallization of honey has both biological and economic significance. Bees hardly use honey that has quickly crystallized as winter food. The packaging and use of crystallized honey in the household also create a number of difficulties. Therefore, the honey put up for sale is packaged after it has been slightly warmed and made runny. When honey is heated, the enzymes contained in it, which ensure the crystallization of honey, are broken down. Such honey does not crystallize for a long time, even when stored at low temperatures. The rate of honey crystallization depends on the natural climatic conditions, the age and quality of the honeycombs,and the temperature of honey storage.

Table 12.3 Standard crystallization period of honey (days) based on natural geographical conditions and filtration time

Filtered (in months and days)	Regions			
	Mountainous	Foothills	Subtropical	Lowland
In May (25.V–30.V)	–	–	For 55–60 days (20VII–30.VII)	For 55–60 days (20.VII–30.VII)
In June (25. VI–30.VI)	–	For 40–50 days (20.VII–30.VII)	For 55–60 days (25.VII–5.VIII)	–
In July (25. VII–30.VII)	In 35–40 days (01. VIII–10.VIII)	In 35–40 days (01.VIII–10.VIII)	In 40–45 days (05.VIII–15.VIII)	–
In August (10.V–20.V)	–	–	–	40–45 days (20. IX–30.IX)

8.3 Ripening of Honey

After collecting the nectar and various juices of the flowers, the honeybee turns it into honey by adding a number of enzymes synthesized in the digestive glands. The prepared honey is preserved in honeycomb cells and used by bees as food (Table 12.3).

Foods containing protein, fat, and carbohydrates are required for the maturation and development of bees and the normal life of the bee family. In addition, the food must contain minerals and vitamins necessary for life. In this regard, honey can be considered a universal source of nutrients. Due to its rich nutritional value as well as medicinal value, people often try to use honey. The production mechanism of honey, which is considered a natural food product, is very complex. After getting acquainted with the mechanism of making honey by bees, every person begins to think that it is an unusual food product.

Nectar is a tasty juice secreted by the flowers of plants and is considered the main source of food for bees. Its composition consists mainly of various sugars. Depending on the species of plants and environmental conditions, the nectar brought to the nest by bees contains 20–70% complex and simple carbohydrates. The nectar of most plants contains sucrose, small amounts of glucose, and fructose. Therefore, nectar is considered a carbohydrate-containing food due to its composition. Nectar is mainly an acidic reaction product. In nature, there are some plants that secrete nectar with neutral and low alkalinity.

8.4 The Conversion of Nectar into Honey

Nectar undergoes both physical and chemical changes while turning into honey. The physical change is mainly due to the decrease in the water content of the nectar and the relative thickening of the honey. The chemical change consists in the conversion of nectar sugar—sucrose—into simple sugars—glucose and fructose—under the action of the bee's digestive enzyme—invertase. The breakdown of sucrose into simple sugars begins as soon as the bee collects nectar from the flower. The complete formation of honey ends in the honeycomb cells. Bees use a part of the nectar collected during their active life as food, and the rest of it is used to prepare reserve honey for the wintering period in the nest. The conversion of nectar into honey consists of four main processes:

- Condensation by evaporation of the water content.
- Breakdown of sugars in the composition.
- Giving a sour reaction.
- Sealing of the honeycomb cells.

8.5 Thickening of Nectar

The process of evaporation of the water in the nectar takes place in honeycomb cells. Nectar brought to the nest is filled into the honeycomb cells, up to 25–30% of their capacity. In this case, the contact surface of the nectar with air is increased, and as a result, suitable conditions for water evaporation are created. In order to accelerate the intense evaporation of the water in the nectar, the bees collect it around the larvae in the warm places of the nest. They evaporate excess moisture by creating an air current inside the nest. To do this, some of the bees sit on the comb frames and continuously flap their wings. 50% of the water in the nectar is evaporated on the first day, and the rest is gradually evaporated over 5–6 days. This process is continued until the amount of water in the nectar reaches 15–18%.

8.6 Breakdown of Sugars

When bees suck the nectar of a flower, it is mixed with the secretions of the pharyngeal gland. These secretions contain enzymes such as invertase, amylase, maltase, and catalase. The enzyme invertase converts sucrose into glucose and fructose, and amylase converts starch into dextrin and maltose. The catalase enzyme neutralizes the generated hydrogen peroxide. The secretion of the pharyngeal gland also contains enzymes that convert glucose into glucuronic acid after it is oxidized.

As the bee brings the collected nectar to the hive, it repeatedly returns each drop to its throat and back to the honey stomach. Each time, enzymes from the bee's saliva are introduced. Bees collect nectar juice from one cell of the honeycomb to the other and from there to the upper cell. During these processes, they put a drop of juice into the honey stomach 8–15 times each time. Thus, invertase and amylase enzymes secreted by bees' salivary glands are mixed into nectar juice. As a result, the complex carbohydrates contained in it break down into simple carbohydrates—glucose and fructose. The process of full maturation of honey continues even after the cells of the honeycomb are sealed. Newly sealed honey contains 6%, and when ripe 1–3% and even less sucrose (Fig. 12.7).

Fig. 12.7 Honey is collected in the honeycomb cells

8.7 Giving a Sour Reaction to Bee Honey

Honey is an acidic product. Although the acidity of honey comes with a small amount of nectar, the rest is formed during its processing. The acidity of nectar juice is quite weak. The bee is also fed with alkaline foods (with sugar syrup).

As a result of the action of oxidizing enzymes secreted by the pharyngeal gland in bees, part of the glucose from the nectar turns into glucuronic acid. Glucuronic acid combines with oxygen and turns into hydrogen peroxide. When the oxygen necessary for this process enters the honeycomb cell, it enters the honey stomach together with the air. Hydrogen peroxide is neutralized in a short time by the action of the catalase enzyme secreted by the mammary gland of the bee.

8.8 Sealing of the Honeycomb Cells

Giving honey an acidic reaction slows down the development of microorganisms and ensures its long-term storage. Once the process of converting nectar into honey is complete, the honey is considered ripe. After that, the bees seal the honeycomb.

It is possible to store honey for a long time after its water content is less than 18%, the amount of simple carbohydrates is 70–80%, and it has an acidic reaction. Due to the enzymes in the composition of honey, it is enriched with substances necessary for life.

9 Obtaining Honey Product

As a result of the strong reproduction of bees, the number of individuals in the family increases excessively. At this time, bees often do not fit into frames. To expand the internal capacity of the hive and create areas where the bees can work comfortably, a second, third, and more layers are added to the hive. Layers can be added to the hives as needed. As the frames are processed and filled with bees, new frames are provided. As each frame is inserted into the hive, the existing frame is moved aside a little, and a new one is inserted; that is, it is given second to last. After the bees fill the honeycomb cells with nectar and reduce the water content below 18%, they cover the honeycomb cells with wax. It is also called sealing. Sealing with wax starts at the top of the frames and continues downwards. When two-third of the cells in the frames are closed with wax, the honey is considered ripe and taken away. Instead of collecting all the frames of ripe honey at once, it is wiser to filter them one by one. This way, there are no bee attacks during honey collection. Honey is obtained either early in the morning or in the evening. When the honey is taken, the beekeeper places an empty hive or honey box behind the hive. From the bee entrance of the beehive, a large amount of smoke is blown into the interior from below. After waiting for 1–2 min, the lid of the hive and the cover curtain are lifted. After giving a lot of smoke from the top side, the frames are examined, and the frames with honey are taken. The obtained frames with honey are placed in a honey box or an empty hive. Before removing frames with honey, the frame with the queen bee is separated, or the queen is moved to frames without honey or with bee growth. The honey frames removed from the hive are first shaken inside the hive, and the bees on them are poured into the nest. However, still a few bees remain on the frame. Therefore, the frame is gently tapped from the upper part with the hand. If there are any bees on the frame, they are carefully swept into the hive with a beekeeper's brush. These are done quickly but quietly. After the frames to be filtered are placed in the honey box, they are covered with a cloth. Honey frames are also transported to a place where bees cannot enter, where the honey is filtered.

10 Variation of Color and Taste of Food Honey in Azerbaijan due to Environmental Factors

The ecological quality of honey varies depending on the species of plants and natural-geographical conditions. Of the 4500 plant species in our republic, 15–17% are honey and pollen plants. About 500 species of nectar-secreting plants have been identified in the flora of Azerbaijan, of which more than 30% are nectar-bearing plants. In our country, there are up to 100 first-class honey plants that produce the best nectar juice (Bogdanov et al., 2007; Kaškonienė & Venskutonis, 2010; Wang et al., 2022). The nutritional value and quality of honey depend greatly on the origin of the nectar. Bees make honey from the nectar of flowering plants in spring, summer, and autumn. The natural forest cover of our republic is the main source for producing honey. Forest cover covers a total of one million hectares and is 12% of the total area. Here, linden, rowanberry, maple, cotoneaster, meadow clover, wild strawberry, ivy, garden gooseberry, sainfoin, rosehip, self-heal, catmint, and other plants dominate. High mountain meadows and steppes are at an altitude of 2000–2500 m above sea level, located between the upper borders of forests and short high mountain meadows. There is white clover, medick, Syrian sainfoin, geranium, elecampane, plume thistles, raspberry, etc. When analyzing Azerbaijani honey by pollen, it was found that in the lowlands, 64% of it is produced from one plant, and in the foothills and mountainous regions, 76.8–91.1% is produced from the nectar of different plants. Studies have shown that 54.3% of honey in spring, 84.6% in summer, and 71.1% in autumn is made from the nectar of many plants, i.e., it is polyfloral. In our republic, at certain times, monofloral honey is also obtained mainly from acacia, linden, cotton, sunflower, clover, and other plants, prepared from the nectar of one plant. Azerbaijani honey, in addition to being complex in composition, should be considered a product made from the nectar of many plants. In the Kura-Araz and plains around the Caspian of our republic, in the semi-deserts of Gobustan-Absheron, and in the Arazboyu plains and semi-deserts, bees make honey from the nectar of flowering honey plants (sarzasan, Russian thistle, brome, common licorice, milk thistle, Caucasian blackberry, small clover, dead-nettle, felty germander, cotton, corn, clover, sainfoin, and garden plants).

10.1 The Color of Azerbaijani Honey

The color of honey varies in spring, summer, and autumn in lowland, subtropical, and mountainous regions of the Greater and Lesser Caucasus of our republic. The color of honey varies depending on the flora and natural-geographical conditions. Bees collect the nectar of one plant and sometimes of different plants to make honey in the nest. Dark yellow, golden honey (83.3%) predominates in the mountainous regions of the Greater and Lesser Caucasus, dark brown (61.6%) in the foothills, and light yellow (53.4%) in the lowlands. In honey, light yellow and slightly whitish (74.3%) or yellow (88.3%) color prevails in spring, and dark brown color (64.7%) in autumn. In the summer, the color of honey in the mountainous, foothill, and lowland zones of the subtropical region was subjected to changes in the same manner as described above, and when compared, there were no sharp differences here. Analyzing Azerbaijani honey according to pollen, it was determined that there are sharp changes in vegetation cover in the foothills and lowlands. The color of the honey collected by the Bozdag Caucasian bees is mainly dark yellow (41.1%) or dark brown (52.9%), and the composition is polyfloral (86%). The color of the honey filtered from the Bozdag Caucasian bees changes according to the origin of the vegetation cover of the mountain and foothill regions. The color of the honey collected by the Yellow Caucasian bee is light yellow (33.4%) and dark yellow (55.6%), and the composition is polyfloral (66.3%). The color of the honey collected by the Bozdag Caucasian bee is darker than the honey collected by the Yellow Caucasian bee, and it is made from the nectar of many plants. It is clear from the individual observations that the Yellow Caucasian bee is more inclined to collect the pollen of various plants during the morning flight than the Bozdag Caucasian bee.

10.2 Smell of Azerbaijani Honey

The smell of Azerbaijani honey is different, fragrant, and pleasant. Honey has a specific smell depending on the pollen in it. The smell of honey is felt mainly when it is put in the mouth and eaten. Heated honey loses a large part of its aromatic compounds. At the same time, if honey is placed next to a strong-smelling substance, it can absorb the smell of that substance. Dark-colored honey has a sharper odor than light-colored honey.

10.3 Taste of Azerbaijani Honey

The taste of Azerbaijani honey is mostly sweet or sweet-astringent. Meadow honey is very sweet. Azerbaijani honey is less sweet in spring and autumn and very sweet in summer. Honey obtained from common tobacco and other plants is slightly bitter. Honeydew honey differs in that it is not completely pleasant. During the inspection, honey's color, smell, taste, consistency, mechanical impurities, water content, etc., are checked. The first type of honey is transparent, with a yellowish tint, or dull brown with a red tint. Meadow (polyfloral) honey is considered high-quality honey. The taste of some types of honey is slightly bitter. However, honey should not have a bitter, sour, or other taste. Honey impurities may be liquid or solid. Freshly filtered honey has semi-liquid, syrup-like impurities. As honey is stored, it crystallizes, and the impurities harden. However, crystallization does not reduce the quality of honey. Standard indicators of the organoleptic properties of Azerbaijani honey have been determined by us (Table 12.4).

11 Ecological Quality Indicators of Nutritional Quality of Azerbaijani Honey

Products made by honeybees are considered high-grade food products due to their quality and composition. These bee products prepared inside the nest can remain for a long time without changing their composition, even at a temperature of 30–35 °C. Under proper storage conditions, honey does not lose its quality for years.

People use ecologically clean, high-quality honey as a food product and for medicinal purposes. The quality of honey largely depends on the origin of the plant and the geographical conditions. Different plants exist in different geographical conditions of our republic, i.e., subtropical, lowland, mountainous, and foothill zones. That is why the properties of honey are highly dependent on the natural-geographical conditions and flora of those areas.

Table 12.4 Organoleptic properties of Azerbaijani honey

S/n	Indicators	Mountainous				Foothills	Lowland
		%	Fluctuation	%	Fluctuation	%	Fluctuation
	I. Color of Honey						
1.	Dark yellow, golden (in summer)	83.3	75.8–88.4				
2	Dark brown (in summer)			61.6	57.4–65.5		
3	Light yellow (in spring)					53.4	45.1–58.1
	By seasons						
(a)	In spring: light yellow, slightly whitish or yellow					74.3	61.3–64
(b)	In autumn: dark brown	64.7	60.8–68.2	64.7	60.8–68.2		
	II. The Origin of Honey According to Flower Pollen						
1	Monofloral honey					64	61.8–67.4
2.	Polyfloral honey	85	76.8–	75	76.8–		
			91.1		81.1		
	In the spring					54.3	51.2–58.8
	In the summer			84.6	81.3–86.4		
	In the autumn	71.1	68.4–74.8				
	III. Taste of Honey						
1.	Flower honey						
(a).	In the spring			Little sweet			
(b)	In the summer	Much sweet	Much sweet				
(c)	In the autumn			Little sweet			
2	Honeydew honey	Well-pleasing	Well-pleasing				
	IV. Smell of Honey						
	Smell of honey	Pleasant free of foreign smell				Pleasant free of foreign smell	Faintly pleasant, free of foreign smell
	V. Honey Fermentation						
	Signs of fermentation and sour appearance of honey	No				No	No
	Signs of yeast-like fungi	No				No	No

(continued)

Table 12.4 (continued)

		Mountainous				Foothills	Lowland
S/n	Indicators	%	Fluctuation	%	Fluctuation	%	Fluctuation
	VI. Smell of Honey						
	Honey obtained from the nectar of a plant's flower	Gives off the scent of that plant				Gives off the scent of that plant	Gives off the scent of that plant
	Honey made from the nectar of flowers of various plants gives off a	Pleasant				Pleasant	Herbal scent

12 Physico-Chemical Properties of Azerbaijani Honey

It can be seen from the standard indicators of honey that the water content of Azerbaijani honey is 16.39–18.87%, and simple carbohydrates—73.5–76.5%. The activity of diastase in Azerbaijani honey is high (Degirmenci et al., 2020). The acidity of honey according to formic acid fluctuates between 0.1598% and 0.1735%; the amount of sugar-free part—6.33–7.51%; the amount of vitamin C—5.61–9.13%; and the amount of sucrose is 1.60–1.90% (5–6% in freshly filtered honey). Honey samples taken from mountainous and foothill regions differ from lowland honeys due to some physicochemical indicators and quality. Easily digestible carbohydrates (glucose and fructose) are absorbed into the blood through the intestines unchanged and satisfy the body's needs. Therefore, by prescribing honey as a food product, especially to children, the doctor created conditions for satisfying the needs of their constantly developing organisms with energy substances. Honey is very rich in vitamins. More than 12 vitamins have been found in its composition. One of the most important aspects of honey is that it has antibacterial, antiviral, and antifungal properties. In this regard, honey has the ability to destroy microbes that have entered the body. It has been established that the effect of honey stops the development of bacterial flora, and some pathogenic microbes, such as dysentery and paratyphoid bacilli and typhoid bacteria, can survive for only 48 hours. Mineral substances in honey have high biological activity. These substances come into contact with vitamins, hormones, and enzymes and affect the increase in activity of the body, especially the awakening of the nervous system, tissue respiration, and other processes. Standard indicators of the physical and chemical properties of Azerbaijani honey have been determined (Table 12.5).

Table 12.5 Standard indicators of physical and chemical properties of Azerbaijan honey (M ± m)

S/n	Indicators	Unit of measurement	Regions			
			Mountainous	Foothills	Subtropical	Lowland
1.	**Dry substance**					
	(a) In flower honey	%	81.91 ± 0.31	81.13 ± 0.26	82.16 ± 0.29	82.01 ± 0.26
	Fluctation		81.30–82.1	80.87–81.39	81.86–82.45	81.75–82.27
	(b) in honeydew honey	%	82.93 ± 0.34	82.77 ± 0.29	17.84 ± 0.18	17.99 ± 0.76
	Fluctation		82.59–83.27	82.48–83.06	17.66–18.02	17.23–18.75
2.	**Water**					
	(a) In flower honey	%	18.09 ± 0.15	18.87 ± 0.18	17.84 ± 0.18	17.99 ± 0.76
	Fluctation		17.44–18.24	18.69–19.05	17.66–18.02	17.23–18.75
	(b) in honeydew honey	%	17.07 ± 0.13	17.23 ± 0.14	16.73 ± 0.15	17.20 ± 0.51
	Fluctation		16.94–17.20	17.9–17.37	16.58–16.88	16.69–17.71
3.	**Simple carbohydrates** (glucose and fructose)	%	76.51 ± 0.65	75.21 ± 0.44	73.53 ± 0.34	77.01 ± 0.71
	Fluctation		75.10–77.20	74.30–76.11	72.80–74.20	76.10–78.20
4.	**Sucrose** (in ripen honey)	%	1.60 ± 0.24	1.84 ± 0.34	1.80 ± 0.24	1.75 ± 0.31
	fluctation		1.35–1.80	1.50–2.30	1.60–2.10	1.45–2.20
	In the new filtered honey		5–6	5–6	5–6	5–6
5.	**Diastase activity**	unit	5–6	5–6	5–6	4–5
	Fluctation		4–7	4–7	4–7	5–6
6.	**Vitamin C**	mg	9.13 ± 0.39	8.66 ± 0.41	5.61 ± 0.36	6.03 ± 0.29
	Fluctation		8.80–9.40	8.20–9.20	4.50–6.10	5.70–6.50
7.	**PH**		3.98 ± 0.10	3.87 ± 0.09	3.96 ± 0.07	3.91 ± 0.06
	Fluctation		3.80–4.00	3.70–3.90	3.90–4.00	3.85–4.00
8.	**Acidity due to formic acid**		0.1613 ± 0.03	0.1735 ± 0.01	0.162 ± 0.02	0.1598 ± 0.02

(continued)

Table 12.5 (continued)

S/n	Indicators	Unit of measurement	Regions			
			Mountainous	Foothills	Subtropical	Lowland
9.	**Sugar-free part**	%	7.51 ± 1.43	6.33 ± 3.43	6.46 ± 3.11	6.43 ± 2.21
10.	**Mechanical mixtures** (most)		0.05–0.1	0.05–0.1	0.05–0.1	0.05–0.1
11.	**Ash**	%	0.6	0.6	0.6	0.6
12.	**Proteins**	%	0.12	0.12	0.1	0.1
13.	**Presence of pollen grains**		Available	Available	Available	Available
14.	**Color**		Homogeneous. No color variety	Homogeneous. No color variety	Homogeneous. No color variety	Homogeneous. No color variety

13 Technical Indicators of Honey

13.1 Marking (Labeling) of Honey

A label is pasted onto the body or lid of the packaging units, or a lithograph is attached reflecting the following relevant regulatory documents:

- Product name.
- Type of product (botanical origin) at the request of the manufacturer.
- Collection year.
- The name of the location of the manufacturer, packer, exporter, importer (legal address indicating the country), and place of origin (at the request of the manufacturer).
- Manufacturer's trademark (if any).
- Net mass.
- Energy value.
- Storage period.
- Storage conditions.
- Packaging date in consumer containers.
- Marking with a regulatory document reflecting the preparation and certification of the product.
- Certification information.

The following information should be indicated in the transport marking:

- Name and address of the sending company.
- Serial number of the batch.
- Product name.
- Botanical origin of honey (at the request of the producer).
- Collection year.
- Packing dates.
- Container weight (gross) and net weight (net).
- True standard mark (Latorre et al., 2013; Escriche et al., 2017; Soares et al., 2017; Mureşan et al., 2022).

When marking the boxes, the quantity of the product unit should be indicated in addition. A packing sheet with a packing number must be placed inside each box. The lid of the box that contains glass or ceramic containers should have a warning label that says, "Be careful, breakable."

13.2 Packaging of Honey

Honey is packaged in consumption and transport containers with a capacity of 0.03–200 L. The shipping containers consist of cans and barrels made of beech, birch, willow, cedar, linden, plane, white poplar, and alder trees (wood density not exceeding 16%). The inner surface of barrels should be paraffined, or a polystyrene bag should be put in (Roman et al., 2013; Juan-Borrás et al., 2015a, b). At the same time, the following can be used to transport honey:

- Cans with a capacity of 25 and 38 L made of sheet stainless steel, aluminum, and aluminized alloy.
- Boxes made of solid material covered with paraffin parchment paper with normative documentation.
- With a normative document, a special hole for honey.
- With a regulatory document, lithographed metal glazed jars coated on the inside with edible varnish, with a capacity of no more than 500 L.
- With a regulatory document, cups or tubs made of aluminum sheet coated with edible varnish with a capacity of 3.0–4.5 L.
- Glass jars and other types of glass containers.
- Corrugated cardboard cups, molded or moisture-resistant pressed, for use in the food industry with permission from the state sanitary and epidemiological supervision authority.
- With regulatory documents, honeycomb frames made of waxed small paper packaging and boxes, parchment, and artificial polymeric materials, cardboard, paper, and combined materials.
- Ceramic tubes with regulatory documentation, enameled inside.

All types of packaging materials intended for use in the food industry must be approved by the state sanitary and epidemiological supervision authority. Consumable and returnable containers must ensure the safety of the products.

Note When packaging natural honey in small consumer containers (up to 0.2 kg), each packaged unit must indicate the name of the product, production date, storage date, and notes of the current standard. Additional information on the energy value should be shown on the attached label.

When packing honey, a deviation of +2% for a net mass of 0.03–1.5 kg and + 1% for a net mass above 1.5 kg is allowed.

Honey should be filled into containers, provided that it does not exceed 95% of the volume of the container. Consumable containers must be hermetically sealed, tightly closed with a metal cap, or screwed tightly. The container is reinforced with polymer materials and thermally welded. With the permission of the state sanitary-epidemiological control body, it is allowed to use the rubber intermediate layer used in the food industry. According to the standards, consumer containers with a capacity of 0.03–1.5 kg should be placed in wooden or cardboard boxes (net mass should not exceed 30 kg). A multi-use container should ensure the storage of the product. When packing, dry and clean material is placed on the bottom of the box, the side walls, and the space between the unit being packed, which ensures that the container does not shift in the box, does not change its place, and does not deform. A packing list is included with the consumer container, packed in a corrugated cardboard box. If the goods are packed in a corrugated cardboard box and a glass container, it is necessary to provide a box, a cage, and a spacer. The lid and mesh cage of the product packaging box are secured with polymer-based tape or fixed with paper-based tape at least 70 mm wide. Boxes with glued lids should be sealed in two places with polymer materials or metal tape.

13.3 Reception of Honey

Honey is supplied in batches. The document on the quality of honey should indicate:

- The name and legal address of the enterprise (organization) (for an individual—surname, first name, patronymic).
- Product name and botanical origin (according to the manufacturer).
- Honey collection year.
- Collection place.
- Serial number of batches.
- Quantity of places in batches.
- Gross and net mass of batches.
- Date of issuance of documents (veterinary certificate, receipt, etc.)
- Analysis results information of honey.
- Packing date (for honey packing enterprises).
- Records of the actual standard.
- Manufacturer's trademark.
- Energy value.
- Storage period.
- Information about certification.
- Information on the veterinary and sanitary health of the beehouse (Roman et al., 2013; Juan-Borrás, 2015a, b).

13.4 Transportation Of Honey

Honey must be transported in compliance with established sanitary rules.

Transportable barrels should not be stacked higher than two to three tiers. Each tier must be separated by a wooden spacer. Boxes and cans should be piled up. The height of the pile should not exceed 1.5 m for cans, 3 m for wooden boxes, and 2 m for cardboard boxes. Boxes, cans, and barrels must be closed and secured when being transported.

Honey can be transported by all relevant modes of transport according to the rules of transportation. When transporting honey by car, containers with honey should be covered with a tarpaulin.

13.5 *Storage of Honey*

Honey is stored in buildings protected from sunlight. It is not allowed to store honey together with toxic, dust-emitting products and substances that emit odors (Eyer et al., 2016). Barrels and cans filled with honey are stacked in two to three tiers with the lids facing up, and wooden spacers are placed between the tiers. The boxes are stacked with intermediate boards up to 2 m high. The storage period of honey in pits and cans with a capacity of 25 kg and above should be up to 8 months from the time of conducting examinations. The storage period of honey packed in hermetically sealed containers made of glass polymer materials should not exceed 1 year (from the date of production), and in non-hermetically sealed containers, no more than 8 months. The storage period of honey in paraffined paper cups should not exceed 6 months from the date of production. Honey stored in the state reserve should be stored in glass containers and stainless steel cans made for honey at a temperature not exceeding 18 °C, and the storage period is 2 years. Storage temperature of honey: 20 °C when the mass share of water is 19.0%; 4–10 °C when the mass share of water is from 19.0% to 21%.

14 Classification of Honey According to Plant Origin

To name and determine the quality of honey, it is important to study the pollen it contains. By determining the pollen in honey, their classification is carried out. Honey is named after the plant that contains the most pollen of the plant. Thus, according to the ratio of pollen, it is determined from which plants the honey is collected.

Honey is divided into four groups according to the amount of pollen in it. The amount of pollen in honey:

- If more than 45%—dominant (superior).
- If between 15% and 45%—high.
- If between 3% and 15%—average.
- If less than 3%—less.

Microscopic analysis of honey is carried out according to the method accepted in the international world for many years (Maurizio & Hodges, 1951; Lutier & Vaissière, 1993).

Preparations are made for work in the following order:

The honey stored in the container is well mixed with a glass spoon first, then 10 g is taken and placed inside the glass flask. Then, 20 g of distilled water is added to it. To dissolve honey in water, the flasks are kept in a water bath at a temperature of about 450 °C for 10–15 min, and after removing them, they are thoroughly stirred with a glass spoon so that the honey dissolves completely in water. If the honey samples to be analyzed are crystallized, then they should be kept in a 30–400 ml water bath until they become completely liquid, and then the operations should be carried out in sequence (Bicudo de Almeida-Muradian, 2020; Jensen et al., 2019).

The dissolved honey is centrifuged in a centrifuge at 3500 rpm for 45 min, and the liquid in the centrifuged flasks is filtered. A small amount (1–2 mm^3) of glycerin-gelatin based on fuchsin, taken on the tip of a needle, is spread onto the bottom of the flask; the resulting material is collected with a needle and distributed over an object glass.

The object glass is heated in a thermostat at a temperature of 30–400°, ensuring the melting of glycerin-gelatin based on fuchsin, and at this time, care is taken not to boil it so that air bubbles do not form and the shape of the pollen does not change. With a needle, the melted glycerine-gelatin and pollen are mixed on the object glass, ensuring a homogeneous distribution of the honey content. Then it is covered with an 18 × 18 mm^2 cover glass. A label is attached to one side of the object glass of the product; the number of the beehouse where the honey was obtained, the name of the region, the date of preparation, and the name of the person who prepared the preparation are shown on it.

To obtain a complete picture when examining the preparation, the preparation is turned over and placed on two glass rods, ensuring that the pollen in the honey approaches the cover glass.

After approximately 12 h, the preparation is ready for examination.

Pollen preparations are examined under a microscope, and an immersion objective (x100) is used to identify the pollen. In the study, the entire area of 18 × 18 mm^2 under the cover glass should be inspected. When determining pollen, comparison preparations prepared from suitable sources and plant pollen are used (D'Albore, 1997; Lutier & Vaissière, 1993). When determining pollen, preparations prepared in two separate object glasses are examined separately. After the plant species recognition activity is over, starting from the upper left corner and considering the variety of plants, it continued until 200 pollen grains were counted in each preparation. A 10x ocular and a 40x lens were used for pollen counting.

Photographs of determined pollen are taken using a x100 eyepiece on a microscope.

As a result, after subtracting the average number and percentages of the counting results of the two preparations, the proportions of pollen found in the honey are determined: dominant (more than 45%), medium (16–44%), little (3–15%), and very little (less than 3%).

Plants that are dominant in honey content and have a large number of pollen are considered first-class melliferous plants (Esterhuizen, 2020; Anklam, 1998; Cuevas-Glory et al., 2007; Adamchuk et al., 2020; Kaškonienė & Venskutonis, 2010). These plants include:

Dominant:

1. Chestnut (*Castanea sativa*).
2. Eucalyptus (*Eucalyptus camaldulensis*).
3. Oak (*Quercus robur*).
4. Willow (*Salix*).

5. Clover (*Trifolium*).
6. Sainfoin (*Onobrychis*).
7. Ivy (*Hedera helix*).
8. Cornelian cherry tree (*Cornus mas*).
9. Raspberry (*Rubus*).
10. Sage (*Salvia*).
11. Milk gowan (*Taraxacum officinale*).
12. Rockrose (*Cistus*).
13. Garden parsnip (*Pastinaca sativa*).
14. Persimmon (*Diospyros lotus*).
15. Erika (*Erica*).
16. Hedysarum (*Hedysarum*).
17. Purple loosestrife (*Lythrum salicaria*).

Many

1. Maple (*Acer*).
2. Wild chamomile (*Chamomilla recutita*).
3. Strawberry madrona (*Arbutus unedo*).
4. Musk bristlethistle (*Carduus nutans*).
5. Caraway (*Carum carvi*).
6. Cornflower(*Centaurea*).
7. Yellow centaury (*Centaurea solstitialis*).
8. St.-John's-bread (*Ceratonia siliqua*).
9. Purple bugloss (*Echium plantagineum*).
10. Rosebay willowherb (*Epilobium angustifolium*).
11. Common sunflower (*Helianthus annuus*).
12. Plum (*Prunus*).
13. Sweet bay (*Laurus nobilis*).
14. Lavender (*Lavandula*).
15. Wild apple (*Malus sylvestris*).
16. Horehound (*Marrubium*).
17. Medic (*Medicago sativa*).
18. Poplar (*Populus*).
19. Blackthorn (*Prunus spinosa*).
20. Firethorn (*Pyracantha*).
21. Pear (*Pyrus*).
22. Linden (*Tilia*).
23. Vetch (*Vicia*).
24. Thyme (*Thymus*).
25. (*Cephalaria*)

15 Main Nectar Plants of Azerbaijan

Azerbaijan is rich in nectar plants. Out of 12,000 plant species in Europe, up to 4500 are available in our republic. Among them, there are many natural plant species that are important for beekeeping. To be successful in beekeeping, it is necessary to have plants rich in nectar and pollen in different regions. It is impossible to practice beekeeping where there are no flowering plants. For this reason, beekeeping should be carried out in places where there are long-flowering, nectar- and pollen-rich plants, or the beehouse should be moved to these places (Degirmenci et al., 2020).

Nectar productivity of plants is significantly influenced by factors such as plant species existing in a certain area, their ability to secrete nectar, state and time of flowering, and environmental factors such as sunlight, air temperature, humidity, and soil composition. Bees prefer nectars with a high sugar content. In other words, the degree of importance of a nectar plant and its attractiveness to bees is highly dependent on the flowering period of the plant, the amount of nectar it secretes during this period, and the amount of sugar in the nectar.

Every country, even its natural geographic zones, has its own honey plants. At the same time, trees and planting areas are created in order to provide a source of nectar for bees. For this purpose, the planting of trees such as chestnut or acacia, or grasses such as clover, sainfoin, and phacelia was more appropriate in terms of creating a source of nectar.

Plants that are an important source of nectar for beekeeping and have high-quality honey are divided into three groups:

- Cultivated herbaceous plants.
- Wild herbaceous plants.
- Tree and Bush.

15.1 Cultivated Herbaceous Plants

This group includes fodder plants and some technical plants included in the leguminous family. Among fodder plants, clover and esparcet are the most important, and among technical plants, cotton and sunflower are at the top. Because they are cultivated plants, planting times can be adjusted, flowering periods vary, and they provide a long-lasting and rich source of nectar for bees. The flowering periods and nectar-producing ability of some of these plants are as follows:

1. Crimson clover (*Trifolium incarnatum*): Many species of plants belonging to the leguminous family grow naturally in our country. Cultivation of cultural forms such as garden peas, clover, and esparcet is carried out in very large areas. The number of honey plants belonging to this family is more than other families. The flowering period of scarlet clover, which is widespread in our republic, differs according to the climatic conditions of the region where it grows. In places with

a temperate climate, it begins to bloom in April and lasts until September. Scarlet clover honey has a very nice taste and smell. The honey is light yellow in color and crystallizes quickly. Bees can get 10 kg of honey from one decare of scarlet clover field.

2. White clover (*Trifolium repens*): White clover, which is widely grown as fodder for livestock, is also an important source of nectar for bees. Flowering from March to September, white clover honey is colorless or light yellow when freshly filtered. After filtering, it is crystallized. White clover honey tastes great. 10 kg of honey can be obtained from one decare of white clover field.
3. Sunflower (*Helianthus*): Sunflower is one of the important field crops widely planted in our republic. When bees collect nectar from sunflowers, they also pollinate them. Thus, they play an important role in the productivity of sunflowers. The pollination period of sunflowers is July. When sunflower honey is taken from the hive, it has a golden-yellow color and a unique taste, and it crystallizes quickly. It looks like wax when crystallized. The nectar yield of one decare of sunflower field is 5 kg.
4. Alfalfa (*Medicago sativa*): Different species of this plant are spread naturally in our republic, and several species are also planted as fodder plants. Alfalfa is a perennial herbaceous plant. It has blue-purple flowers. It is found on stony slopes, meadows, or fields from 200 to 2000 m above sea level. The flowering period starts in May and lasts until September. Alfalfa honey is pale yellow in color when it is freshly filtered, and it crystallizes quickly. Crystallized alfalfa honey is thick and cream-like. This honey has a very nice smell and a unique taste. More than 400 kg of honey is obtained from one hectare. Due to the abundance of nectar and quality honey, alfalfa is considered one of the important honey plants.
5. Esparcet (*Onobrychis*): Esparcet is an important agricultural plant for the use of dry and bogharic lands, where many cultivated plants cannot grow. Esparcet is an indispensable plant for many places with water shortages. Under such conditions, there is no other leguminous fodder crop to be grown in place of esparcet. Esparcet is one of the important fodder plants grown in the conditions of Azerbaijan. Since this plant gives a lot of nectar, it is also considered an important feed source in beekeeping.

15.2 Wild Herbaceous Plants

The nature of Azerbaijan is very rich in wild herbs. Most of them are nectar plants, which are important in beekeeping. The most important nectar-yielding plants are considered to be thyme, sage, dead-nettle, melilot, chicory, esparcet, astragal, garden pea, and others. They grow in mixed pastures, grow on their own, spread seeds, and increase their offspring. Below are the main characteristics, distribution, and nectar production ability of some of these plants.

6. Thyme (*Thymus*): It is a very widespread plant in Azerbaijan. Different species of this plant are considered nectar plants. Thyme, which grows on roadsides, hills, and mountains, is perennial and emits a pleasant fragrance. Thyme contains a volatile oil called thymol. The flowering period lasts from the beginning of May to the end of August. Thyme honey has a light golden color and a very nice taste and smell. Its composition is very rich in enzymes.
7. Sage (*Salvia*): Many species are naturally distributed in our republic; some of them are known as nectar plants. Sage is a perennial with a height range between 30 and 60 cm. The leaves are densely covered with hair, and the flowers are light yellow in color. The flowering period is from the beginning of July to the end of September. The color of sage honey is light yellow. The taste and smell are very nice. Honey is easily distinguished from other honeys due to its characteristic smell. Crystallization of sage honey is quite slow. 6.5 kg of honey is obtained from one decare area where sage has grown. Cultivation of this plant is important for beekeeping.
8. Dead-nettle (*Lamium*): Day-nettle is a nectar-giving plant. However, since nectar glands are located at the base of flowers, not all bee varieties are able to collect nectar from these plants. Therefore, the white day-nettle is considered to be a greater source of pollen. White day-nettle has beautiful-smelling white flowers. The flowering period lasts from the beginning of May to the end of August.
9. Cornflower *(Centaurea cyanus*): Different species of this plant, which is included in the family of Compositae, grow naturally in our republic. Bluebottle is 40–90 cm tall, and the flowers are blue and very beautiful. Perennial plant. The flowering period starts from May to the end of August. Honey is light greenish-yellow in color. The taste is unique and has a weak bitterness. It crystallizes too late. It is an important source of nectar for beekeeping in our republic.
10. Astragal (*Astragalus*): It is a perennial herbaceous plant. A large number of species of astragal grow naturally in Azerbaijan. However, only a few of these plants are nectar plants. Flowers of Astragal with nectar were more attractive than those without nectar. Astragalus is a herbaceous perennial plant. This plant differs in honey quality. The honey of astragal is watercolor. It is also shown in the literature that some species are poisonous to honeybees. Since the peak flowering time of astragal coincides with July, they are considered the melliferous plants of the main honey collection period of the bees. Depending on the type of astragal, the amount of nectar secreted by one flower during the day varies from 0.35 mg to 0.98 mg. The sugar content of nectar ranges from 30 to 80 percent. Depending on the species of plant, the nectar yield that can be taken from one hectare can vary from 36 to 113 kg, and the approximate amount of honey can vary from 18 to 56 kg.
11. Sainfoin (*Onobrychis*): This perennial herb is widespread in the mountains and foothills of Azerbaijan. Sainfoin mostly grows among the bushes in the high mountain meadows. It is more common in Ganja, Agstafa, Gadabey, Kalbajar, Agdam, Shusha, and other regions of the republic. Many species of sainfoin are

valuable as fodder plants as well as vitamin and melliferous plants. The sainfoin plant begins to bloom in the second half of June. The flowering period lasts 25–35 days. The life of a flower is 2–3 days. The flower secretes nectar for 1 day and, under favorable weather conditions, for 2 days. Among the species of sainfoin that honeybees readily settle on, the most common are Syrian sainfoin, Caucasian sainfoin, tall sainfoin, common sainfoin, and radiant sainfoin. As melliferous plants, common sainfoin and tall sainfoin species of the sainfoin family are cultivated. It was determined that the nectar yield of one hectare of sainfoin crops varies between 72 and 273 kg. It was determined that the cultivated species of sainfoin is one of the first-rate melliferous plants in both the mountainous regions and the lowland cotton-growing regions of Azerbaijan. On average, a sainfoin flower can secrete 0.48 mg of nectar, which contains 45 percent sugar, per day. The average amount of nectar secreted by a single flower of wild sainfoin species is the same as that of cultivated sainfoin flowers. However, the nectar yield per hectare is slightly higher. In these proportions, the absolute amount of sugar and the amount that can be taken from one hectare are more. The excess nectar yield of wild sainfoin species depends on the number of flowers per hectare and a relatively long-flowering period.

12. Echium (*Echium*): Many species of this plant are found in Azerbaijan. They are mostly found on roadsides, meadows, bushes, and forests, as well as in cultivated areas. Echium is a 30–50 cm tall perennial herb with hairy leaves. Its flowers are blue in color and quickly attract the attention of bees with their beauty. Flowers are collected in panicled inflorescence or cluster inflorescence. Echium is a plant that gives a lot of nectar. This plant blooms for a long time. So, it can stay in flower from May to September. Echium is an important source of nectar. The honey produced from this plant is of high quality and crystallizes late. Its honey is yellow in color and smells very good. It is possible to get 300–400 kg of honey from one hectare of cultivated echium field.
13. Common bugloss (*Anchusa officinalis*): In many parts of Europe, some species of common bugloss are cultivated as ornamental plants. Common bugloss is a perennial herb, 20–150 cm tall. Its flowers are dark blue, very beautiful. Blooms from April to August. It is often found in the field and the steppes. Common bugloss honey is light yellow in color, slow to crystallize, pleasant smelling, and delicious.
14. *Brunella*: Only two species of this perennial grass grow in Azerbaijan: common garden burnet and sinuous garden burnet. The first of these species is spread to the upper belt, and the second to the middle belt. The distribution of both species is forest edges, meadows, and sometimes littered areas. Common garden burnet is a vitamin-rich, fatty plant. It blooms in Azerbaijan at the beginning of May. Flowering lasts 30–35 days. The life of a flower is 3–4 days. The literature shows that common garden burnet is a flowering plant that gives honey in the last months of summer. Our research shows that both species are among the best melliferous plants under the conditions of Azerbaijan. The nectar yield of a common garden burnet on one hectare is up to 415 kg. Sinuous garden burnet

yields 175 kg of nectar per hectare. Thus, the common garden burnet is a more melliferous plant than the sinuous garden burnet. It is possible to keep one to three normal bee colonies in one hectare of garden burnet.

15.3 Trees and Shrubs

In our republic, there are many trees and shrubs that give nectar and honeydew juice. Among them, acacia, linden, pine, brier, eucalyptus, willow, maple, blackberry, chestnut, hawthorn, rosehip, oak, and various fruit plants are important. A number of trees and shrubs, such as pine, fir, oak, and juniper, do not secrete nectar. The juices secreted by some greenfly living on them, as well as the juicy secretions of the leaves and stems of plants, are a source of honey for bees. The secretion of juice by fliers living on pine trees lasts for a long time. The collection of these secretions by bees and turning them into pine honey is important in terms of honey production.

15. Acacia (*Acacia*): Acacia, which easily adapts to different climatic conditions, is one of the most common tree species in our republic. Acacia has an important place in beekeeping. The areas of planting of this beautiful plant in parks and on roadsides are increasing every day. Acacia is among the plants planted in the protective state forest strips of Azerbaijan. Acacia is widespread in the Ganja, Shusha, Agdash, Shamakhi, Guba, Zagatala, and Sheki regions of the republic and in the Nakhchivan Autonomous Republic. It is of great importance due to the fact that it is a plant with vitamins and honey, as well as its aesthetic appearance. The flowering period of acacia is related to the region and altitude and is from April to June. The taste and smell of acacia honey are very nice, and it crystallizes very quickly. It does not lose its unique brightness and fluidity for a long time. According to the literature, a strong bee family can collect up to 8 kg of honey from one acacia tree grown in favorable climatic conditions. 150 kg of honey is obtained from one decare of acacia.
16. Linden (*Tilia*): In the territory of Azerbaijan, in the foothills of the Greater and Lesser Caucasus mountains, in the territories of the Zagatala and Balakan regions, several species of linden are common. In the conditions of Azerbaijan, Caucasian linden and small-leaved linden are considered to be first-class melliferous plants. In addition to being considered an important medicinal plant, linden is also an important feed source in beekeeping. Linden is a plant that produces both pollen and nectar. Flowering begins in early June and lasts up to 2 weeks. Linden is considered one of the first-class melliferous plants. According to calculations, the yield of linden is equal to 1 ton/ha. With this calculation, 8–10 bee colonies can be kept on each hectare of linden.
17. Oleaster (*Eleagnus*): Several species of this family grow naturally in Azerbaijan. It is tree- and shrub-like. Its flowers are light yellow in color and have a very nice smell. Bees fly over the oleaster flowers from early morning until the end

of the day. Flowering time varies by region. It blooms in April in temperate regions and in May in other places. Its nectar is not that much. It is important for beekeeping only if the cultivated area is large.

18. Chestnut (*Castanea*): Chestnut, which is naturally distributed in the forests at the foothills of the Greater Caucasus Mountains in our republic, is a plant that gives very valuable nectar and pollen. The color of honey obtained from chestnut trees, which bloom in June to July, is dark brown, has a bitter taste, and has a characteristic pungent smell. Chestnut honey, accepted among natural honeys, crystallizes late. Chestnut honey is different from other honeys due to its taste, smell, and color; it is also called bitter honey. Although it is known as low-quality honey in local conditions, in fact, this honey has a very important medicinal value. Chestnut honey is obtained directly from the nectar of the staminate flowers collected in the amentiform inflorescence of the chestnut plant and is used in the treatment of a number of diseases. This honey is widely used in the treatment of cough, asthma, bronchitis, and a number of respiratory diseases. In this regard, chestnut honey is also called a "natural antibiotic." It has a very good effect on diseases such as pharyngitis, asthma, and anemia. It blooms in June in Azerbaijani conditions. Flowering lasts 10–14 days. The nectar yield of chestnut plants per hectare is, on average, up to 55 kg.
19. Willow (*Salix*): It is an important source of nectar and pollen for bees, especially in early spring. It spreads naturally in the swamp areas of our republic. Bees can get 10–15 kg of honey from 1 decare of willow stand. Willow honey is yellow in color and crystallizes late.
20. Hawthorn (*Crataegus*): Hawthorn can be found in both shrubby and tree forms. Several species of hawthorn are common in Azerbaijan. These species are found in the forests of the Lesser and Greater Caucasus of our republic. Hawthorn is one of the best honey plants. One hawthorn flower can secrete an average of 0.56 mg of nectar per day. Nectar contains 77.4% sugar. 63.3% of this sugar is sucrose, and 0.76% is glucose and fructose. One hectare of hawthorn plants yields 58.2 kg of nectar, and the absolute sugar content is 54.04 kg. The amount of honey obtained from one hectare of hawthorn reaches 34.9 kg.
21. Citrus plants: Among citrus plants, oranges and lemons are important nectar sources for bees. In our republic, citrus plants are typical for the Lankaran-Astara region and grow naturally. The honey of these plants has a unique smell and taste and is very rich in vitamin C.

State Standard of the Republic of Azerbaijan

Natural honey	**AZS 849-2015**
Specifications	
	Application time

1. **Application Area**

This standard applies to the natural honey product processed by honeybees using nectar, fruit juices, honeydew, and pollen. Safety and quality requirements of natural honey are defined in subsections 4.1.6 and 4.1.7. This standard can be used for certification purposes. The requirements of this standard are mandatory.

2. **Normative References**

The following normative references are used in this standard:

GOST 1341-97 Vegetable parchment. Technical conditions
GOST 5037-97 Metal flasks for milk and dairy products. Technical conditions
GOST 5717.1-2014 Glass containers for canned food products. General technical conditions
GOST 5717.2-2003 Glass jars for canned food. Main parameters and dimensions
GOST 8777-80 Wooden barrels for filling and dry tare. Technical conditions
GOST 9142-90 Corrugated cardboard boxes. General technical conditions
GOST 9569-2006 Waxed paper. Technical conditions
GOST 12303-80 Packs of cardboard, paper, and combined materials. General technical conditions
GOST 13358-84 Wooden boxes for canned goods. Technical conditions
GOST 13512-91 Corrugated cardboard boxes for confectionery products. Technical conditions
GOST 13516-86 Corrugated cardboard boxes for canned goods, preserves, and liquid food products; technical conditions
GOST 13950-91 Welded and rolled steel barrels with corrugations on the body. Technical conditions
GOST 14192-96 Marking of cargo
GOST 16214-86 Polyvinyl chloride electrical insulating tape with an adhesive layer. Technical conditions
GOST 18251-87 Paper-based tape. Technical conditions
GOST 19792-2001 Natural honey. Technical conditions
GOST 25629-2014 Beekeeping: Terms and Definitions
GOST 26930-86 Raw materials and food products. Method for the determination of arsenic
GOST 26932-86 Raw materials and food products. Method for the determination of lead
GOST 26933-86 Raw materials and food products. Method for the determination of cadmium
GOST 26935-86 Canned food products. Method for the determination of tin
GOST 30178-30178-9696 Raw materials and food products. Atomic absorption method for the determination of toxic elements
GOST 31766-2012 Monofloral honeys. Technical conditions
GOST R53126-2008 Refractometric method for determination of water
GOST R53877-2010 Med. Method for the determination of hydrogen index and free acidity
GOST R53883-2010 Med. Method for the Determination of Sugars

GOST R54386-2011 Med. Methods for determining the activity of sucrose, diastase number, and insoluble matter
GOST R54644-2011 Natural honey. Technical conditions
CODEX STAN 1-1985 (Rev. 2-1999) Codex General Standard for the Labeling of Prepackaged foods
CODEX STAN 12-1981 (Rev. 2-2001) Revised Codex standard for honey "Hygienic requirements for the safety and nutritional value of food products," approved by Order No. 25 of the Ministry of Health of the Republic of Azerbaijan dated 30.04.2010. Sanitary-epidemiological rules and regulations

3. **Terms and Definitions**

    3.1. In this standard, the following terms and definitions are used in accordance with GOST 25629:

    3.1.1. **Natural honey:** A product produced by honeybees from nectar and honeydew.
    3.1.2 **Flower honey:** Honey produced by honeybees from flower nectar.
    3.1.3 **Honeydew honey:** Aphids and cochineals feed on the sap of the leaves and stems of plants and excrete the excess sugary liquid produced in their bodies onto the plant's stem. Honey is made by bees using that sugary liquid.
    3.1.4 **Mixed honey:** A natural mixture of flower honey and honeydew honey.
    3.1.5 **Monofloral honey:** Honey made from the nectar of the flower of a plant. For example, linden honey, acacia honey, clover honey, citrus fruit honey, etc.
    3.1.6 **Polyfloral honey:** Honey made from the nectar of several different plant flowers.
    3.1.7 **Honey filtered by centrifugation:** Honey filtered from combs by centrifugation.
    3.1.8 **Honey obtained by the compression method:** Honey obtained by the method of compression with the condition of a small amount of heating or no heating of the combs.
    3.1.9 **Mechanical impurities:** Dead bees and their body parts, wax particles, larvae, insects, pollen, wood, straw, metal, glass, and other foreign particles.
    3.1.10 **Signs of fermentation:** the presence of foam on the surface or in the content of honey, the expulsion of gas, a sour appearance, signs of yeast or mold, and the presence of a specific smell and taste.
    3.1.11 **Batch of honey:** Any quantity of honey of the same botanical origin, packaged and documented by a quality certificate.
    3.1.12**Total acidity:** Acidity determined by the amount of sodium hydroxide (cubic cm) used to titrate 100 g of honey.

3.1.13 **White acacia pollen grains:** White acacia pollen grains are round-triangular at the poles, flattened at the equator, and three-furrowed and porous. Pores are round or oval in shape. The pollen is yellow in color.
3.1.14 **Storage period of honey:** The period during which honey can retain all its properties by observing the specified storage conditions. Honey stored before the expiration of the shelf life should be usable, provided that its consumption characteristics are not lower than the requirements of the current standard.
3.1.15 **Consumer tare:** Tare that does not have a free transport unit and delivers the product to the consumer (jar, cup, box, etc.).
3.1.16 **Transport tare:** Canteen with a free transport unit (canteen, tank, barrel, etc.).
3.1.17 **Selecting and sorting:** A collection of product units taken from batches for inspection.
3.1.18 **Volume of selected**: The quantity of shipping or consumer containers with the product that make up the selection.
3.1.19 **Spot samples:** A non-quantitative sample taken at the same time from a specified point of production (samples from a barrel, canteen, tank, etc.).
3.1.20 **Combined sample**: Combined sample, consisting of a series of incremental samples, collected and thoroughly mixed in a single container to produce an average sample.

4. **Technical Requirements**

   4.1. **Characteristics**

4.1.1 According to its botanical origin, natural honey is divided into floral (monofloral, polyfloral), honeydew, and mixed honey.
4.1.2 According to the methods of obtaining it, honey is divided into filtered, centrifuged, and compressed honey.
4.1.3 In honeycomb honey, at least 2/3 of the surface of the comb must be sealed. The combs should be similar to each other, white or yellow in color.
4.1.4 Organoleptic and physicochemical indicators of natural honey must meet the requirements.
4.1.5 The mass fraction of tin in honey is determined when the product is packaged in tinned iron containers with a storage period of no more than 6 months and when corrosion is detected in the containers.
4.1.6 The amount of toxic elements, oxymethylfurfural, pesticides, and radionuclides should not exceed the limits established by the "Hygienic requirements, sanitary-epidemiological rules and regulations for the safety and nutritional value of food products" approved by the Order No. 25 of the Ministry of Health of the Republic of Azerbaijan dated 30.04.2010 and should comply with the norms.
4.1.7 Acceptance rules and storage conditions for the quality of natural honey are determined by the necessary regulatory requirements.

4.2. **Marking**

4.2.1. A label is affixed to the area where the lid of the packaging unit's body connects, or a lithography is added in accordance with regulatory documents that reflects the following:

- Product name.
- Type of product (botanical origin) at the request of the supplier.
- Collection year.
- The name of the location of the manufacturer, packer, exporter, importer (legal address indicating the country), and geographical origin (at the request of the manufacturer).
- Manufacturer's trademark (if any).
- Netto and gross weight.
- Net mass.
- Energy value (calorieness).
- Storage period.
- Storage conditions.
- Packaging date in consumer containers.
- The mark of the regulatory document relevant to the production and certification of the product.
- Information about certification.

4.2.2. The design of the label can be determined by the manufacturer.

4.2.2.1 When honey is packaged in glass jars with a capacity of 0.3–3.0 kg, the rectangular part of the corresponding label is attached to the jar, and the adjacent circular part is attached to the lid.
4.2.2.2 When honey is packed in tares with a capacity of more than 3.0 kg, the square part of the corresponding label is attached to the part of the container close to the lid—the body, and the adjacent circular part is attached to the lid. The label should be affixed in such a way that it is not possible to open the cover of the tare without damaging the label.

4.2.3. When making labels, the material should be selected so that the inscription on it is clear and easy to read, and it retains its purpose during loading, transportation, storage, and other processes (it should not be torn, the writing should not be damaged, etc.).
4.2.4. Marking of transport tare is carried out according to GOST 14192, indicating the following information:

- Name and address of the sending company.
- Serial number of the batch.
- Product name.
- Botanical origin of honey (at the request of the producer).

- Collection year.
- Packing dates.
- Total mass and net weight.
- Mark of this standard.

4.2.5. Marking should be in the state language of the republic. In addition, the quantity of the product unit should be indicated on the marking of the boxes. A packing sheet with the packer's number must be placed inside each box. On the top lid of the box where the glass or ceramic tare is placed, the warning words *"Be careful. Breakable"* should be written.

4.2.6. Honey production is carried out in accordance with the rules of the Codex Alimentarius "Marking of packaged food products (Codex Stan 1—1985, Rev. 2—1999)."

4.2.6.1 If the product is exclusively produced in a limited geographical area, the name of the landscape (forest, field, clover, etc.) or the name of the region may be indicated in its name. In this case, the name of the origin is indicated as a combination of the name of the product (for example, Lesser Caucasus forest, mountain field, etc.).

4.2.6.2 If the product is collected from a plant of the same species in a limited area and has appropriate organoleptic, physicochemical, and morphological characteristics, it can be called by its plant origin. In such a case, the name of the plant is shown as a combination of the name of the product (for example, linden honey, honeydew honey, or white acacia honey).

4.3. **Packaging**

4.3.1. The process of filtering honey from honeycomb cells for packaging must be organized in places that meet the following requirements:

- The process of filtering honey from cells should be carried out in an isolated closed space where bees, other fliers, and insects cannot enter.
- The place where the honey is filtered from the combs must be bright and meet the requirements of sanitary-hygienic norms and rules.
- Comb frames full of honey taken from bee families should be transported in a special isolated closed container.
- After the filtering process is completed, the comb frames should be returned to the nest of their bee families.

4.3.2. Filtered honey for the purpose of packaging must be filtered so that there are no comb fragments, dead bee bodies, and other impurities.

4.3.3. According to the color, honey is white, yellow, and dark. Filtered honey should be packed without mixing so that the color of the honey in the tare is uniform.

4.3.4. The process of settling filtered honey in containers for packaging is carried out for 3 days at 18–20 °C, 6 days at 15 °C, and 15 days at 10°.

4.3.5. Before packaging, filtered honey is stored in a closed container in a dry, ventilated room at a temperature of 5–10 °C.
4.3.6. Honey is packaged in containers that meet special requirements to be sent to suppliers.
4.3.7. Containers intended for packaging must ensure the maintenance of product quality.
4.3.8. The material of the containers intended for packaging should be dry and heated under the sun's rays before use.
4.3.9. When reusing used containers, they should be washed with clean hot water, rinsed, and dried completely.
4.3.10. When packing in barrels, the joints of their bottom and side boards should be fixed with hot wax.
4.3.11. When packing honey into barrels, a round hole with a diameter of 5–6 cm should be opened at the top and used to empty the honey into the container. The level of honey inside the barrel should be 2–3 cm below the top cover. Because in the process of expansion of honey in hot weather, the barrel may collapse.
4.3.12. During the realization of the barrels, the hole in the lid is firmly and reliably closed by means of wood. The excess part of the plug is cut off, covered with an iron sheet, and fixed with small nails.
4.3.13. In packaging, metal containers, milk cans, and metal barrels can be used after processing. Barrels made of non-oxidizing metal materials can be used directly. It is dangerous to use oxidizing metal containers for packaging honey.
4.3.14. Depending on local conditions, standard glass containers of different sizes should be used during the packaging of honey at the consumer's tare.
4.3.15. Honey can be sold in a crystallized state or in a liquefied form after packaging, at the consumer's discretion. To turn the honey into a liquid, the crystallized honey should be heated a little (in 30–40 °C 30 min) and poured into containers.
4.3.16. Unripe honey and honey that has absorbed moisture in an open container are not packaged for commercial purposes, as they ferment and lose their quality. This type of honey is heated in a water bath at a temperature of 60 °C for 30 min, and its quality is restored in a small amount. They can be used in the bee feeding process and for other purposes in industry.
4.3.17. Honey is packaged in consumer and transport wares with a capacity of 0.3–3 kg in retail form:

- Metal barrels and wooden barrels with an internal paraffin or polystyrene bag according to GOST 8777.
- Canteens made of stainless steel sheet, aluminum, and aluminized stainless alloy according to GOST 5037.

- Boxes made of solid wood according to the regulatory document, covered with corrugated parchment paper from the inside.
- Special tank for honey according to the regulatory document.
- Lithographed metal cans, coated on the inside with edible varnish according to GOST 3950.
- Cups or tubes made of aluminum foil coated with edible varnish, with a capacity of 3.0–4.5 liters according to the regulatory document.
- Dark-colored glass jars and other types of glass tares, according to GOST 5717.1 and GOST 5717.2.
- Cups made of molded or moisture-proof pressed corrugated cardboard used in the food industry according to the rules established by the Ministry of Health.
- On regulatory documents, honeycomb frame sets are made of waxed small paper packages and boxes according to GOST 9569, parchment according to GOST 1341, and artificial polymer materials, cardboard, paper, and combined materials according to GOST 12303.
- Internally enameled ceramic tubes according to the normative document.

- All types of packaging materials for use in the food industry must comply with the norms and rules established by the Ministry of Health of the Republic of Azerbaijan. Consumable tare and returning tare must ensure product storage.

- **Note**: When natural honey is packaged in small consumer containers (up to 0.2 kg), the name of the product, production date, storage period, and the mark of this standard must be displayed on each packaging unit. Additional information on energy value (calories) should be shown on a special label.

4.3.18. When packaging honey, a deviation of ±2% is allowed for a net mass of 0.3–1.5 kg, and ± 1% for a net mass of more than 1.5 kg.

4.3.19. It is filled with honey, provided that the tare volume does not exceed 95%.

4.3.20. The consumer tare must be hermetically, tightly closed, or tightly fastened or screwed with a metal cap. The tare made of polymer material is closed with thermal welding. The rubber interlayer used in the food industry must comply with the norms and rules approved by the Ministry of Health of the Republic of Azerbaijan.

4.3.21. According to GOST 13358, GOST 13512, and GOST 13516, consumer tares with a capacity of 0.3–1.5 kg must be placed in wooden or cardboard boxes with a net mass of no more than 30 kg. Reusable tares must ensure the preservation of the product. During packing, dry, clean, and homogeneous material must be placed on the bottom of the box, on the side wall, and in the space between the packing units, which ensures that the tare does not move and is not deformed.

4.3.22. A packing sheet according to GOST 9142 is added to the consumer tare packed in corrugated cardboard boxes. If a product packaged in a glass container is packaged in a corrugated cardboard box, the box must be provided with a cage and spacer. The lid and the grid of the product packaging box are secured with polymer-based tape in accordance with GOST 16214 or fixed with paper-based tape at least 70 mm wide in accordance with GOST 18251. Boxes with glued covers should be sealed in two places with a tape made of polymer material or metal.

5. **Reception Rules**

5.1. Honey is supplied in batches.
5.2. The quality document must indicate:

- The name and legal address of the enterprise (organization) (for an individual—surname, first name, patronymic).
- Product name and botanical origin (according to the manufacturer).
- Year of supply of honey.
- Origin.
- Serial number of the batches.
- Quantity of places in batches.
- Gross and net mass in batches.
- Date of issuance of documents (veterinary certificate and compliance certificate, receipt, etc.)
- Information on the analysis results of the honey.
- Packing date (for honey packing enterprises).
- The mark of this standard.
- Manufacturer's trademark (if any).
- Energy value (calorieness).
- Storage period.
- Storage conditions.
- Information about certification.
- Information about the veterinary-sanitary health of the beehouse.

6. **Test Methods**

6.1. Sample selection according to GOST 19792 (GOST P 52451).
6.2. Pollen analysis method according to GOST 19792.
6.3. Determination of the mass fraction of water according to GOST 19792 (GOST P 53126).
6.4. Determination of the mass fraction of reducing sugars and sucrose according to GOST 19792 (GOST P 53883).
6.5. Determination of diastase number according to GOST 19792.
6.6. Qualitative reaction to oxymethylfurfural according to GOST 19792.
6.7. Determination of oxymethylfurfural according to GOST 19792.
6.8. Determination of mechanical mixtures according to GOST 19792 (GOST P 54386).
6.9. Determination of total acidity according to GOST 19792 (GOST P 53877).

6.10. Determination of toxic elements according to GOST 26930, GOST 26932, and GOST 26933.
6.11. Pesticides and radionuclides are determined according to the sanitary norms and rules approved by the Ministry of Health of the Republic of Azerbaijan.
6.12. Determination of toxic elements by atomic absorption method according to GOST 30178.

7. **Transportation and Storage**

7.1. **Transportation**

7.1.1 Honey is transported according to accepted sanitary-hygienic norms and rules.
7.1.2 Transportable barrels should not be stacked higher than 2–3 tiers. Each tier should be separated by a wooden spacer, and the boxes and canteens should be stacked in a pile. The height of the pile should not exceed 1.5 m for canteens, 3 m for wooden boxes, and 2 m for cardboard boxes.
7.1.3 Boxes, canteens, and barrels must be tightly closed and secured when being transported.
7.1.4 Honey can be transported by all relevant modes of transport according to the rules of transportation.
7.1.5 When transporting by car, containers full of honey should be covered with a tarpaulin to protect them from the sun and other influences.

7.2. **Storage of honey**

7.2.1 Honey is stored in places protected from direct sunlight. It is not allowed to keep it together with toxic, dust-emitting products and products that emit odors that are not characteristic of honey.
7.2.2 Honey-filled barrels and canteens are stacked in two or three tiers, with the lids facing up. Wooden boards are placed between the floors and tiers. The boxes are stacked with intermediate boards up to 2 m high.
7.2.3 The storage period of honey in tanks and canteens with a capacity of 25 kg or more should be a maximum of 8 months from the time of the examination.
7.2.4 The storage period of honey packed in hermetically sealed glass wares or polymer tares, should not exceed one year from the date of production and the day of the examination, and in non-hermetically sealed jars, it should not exceed 8 months.
7.2.5 The storage period of honey in paraffined paper cups should be no more than 6 months from the date of production.
7.2.6 For the state reserve, honey should be stored in glass jars, special tanks for honey, and stainless steel canteens at a temperature not exceeding 18 °C for 2 years.

7.2.7 The storage temperature of honey should be 20 °C if the mass fraction of water in its composition is 18.01% and 4–10 °C if the mass fraction of water—18.01–20.0%.
7.2.8 The places where honey is stored should be arranged in a place where there are no conditions for the collection of rain and snow water. Those places should be fully provided with drinking water, hot water, and sewage, and their territory should be greened to the maximum extent.
7.2.9 It is not permissible to store and transport honey together with toxic, specific-smelling goods (Sultanov and Asadov 2017).

Interstate Standards of Honey
(Interstate Council for Standardization, Metrology and Certification (Minsk)
Natural Honey
Technical conditions

1. **Scope of application**
   This standard applies to natural honey—honey is a product obtained as a result of the processing of flower nectar by honeybees. This standard may be used for certification purposes.
2. **Basic concepts**

Flower honey—honey produced by honeybees from the nectar of flowers.
Mixed honey—a mixture of honey produced by honeybees from the nectar of flowers and honeydew honey.
Honey filtered by centrifugation—honey filtered from combs by centrifugation.
Honey obtained by the method of compression—honey obtained by the method of heating the combs in a small amount or the compressing method without heating.
Mechanical mixtures—a mixture of dead bees and their body parts, wax particles, larvae, pollen, metal, glass, and other foreign particles.
Signs of fermentation—the presence of foam on the surface or in the volume of honey, the release of gas, and the presence of a specific smell and taste.
Honey batch—any quantity of honey of the same botanical origin, packaged and documented with one quality certificate.
Total acidity is the acidity determined by the amount of sodium hydroxide (cubic cm) used to titrate 100 grams of honey.
Pollen grains of white acacia—Pollen grains of white acacia are round-triangular from the poles, flattened at the equator, furrowed at the tip, and porous. Pores are round or oval in shape. Pollens are yellow.
Pollen grains of cotton—Pollen grains of cotton are two-celled, round, spiny, porous, and 90–120 μm in size.
The storage period of honey—the period during which all the qualities of honey are kept by observing the storage rules. At the end of the storage

period, the honey is suitable for use if the consumption qualities are in accordance with the current standards.
Containers for consumption—Containers delivered to the consumer without a free transport unit (cans, plastic, paper, glass containers, containers, etc.).
Containers for transport—containers with a free transport unit (flask, barrel, etc.).
Selection—A collection of product units selected for inspection from a batch.
Selection quantity—the volume of product units selected for inspection.
Point sample—a sample of a product that is not measured in quantity (a flask or a barrel) taken from a specific point.
Combined sample—A sample of honey taken from several points and mixed and placed in a container to calculate an average.

3. **Technical requirements**

3.1. Characteristics

3.1.1 According to the botanical origin, natural honey is divided into floral (monofloral, polyfloral), honeydew, and mixed honey.
3.1.2 According to the method of honey obtaining, it is divided into comb honey, filtered honey, and pressed honey.
3.1.3 At least 2/3 of the comb surface must be sealed, and combs must be of the same color, yellow or white.
3.1.4 According to its organoleptic and physicochemical indicators, natural honey must comply with the requirements specified.
3.1.5 The amount of tin in honey is determined no earlier than 6 months after the product is packaged in tin-alloyed iron containers and when corrosion is detected in the container.
3.1.6 The residual amount of pesticides (dichlorodiphenyltrichloroethane and hexachlorocyclohexane) in 1 kg of honey should not exceed 0.005 mg. Residual amounts of other pesticides are not allowed.
3.1.7 The presence of toxins in honey (lead, cadmium, oxymethylfurfural) should be within the limits of hygienic norms.

4. **Marking**

4.1. A regulatory document containing the following information is affixed to the body or cover of the packaging unit:

- Product name.
- Product type (botanical origin).
- Harvest year of honey.
- Legal address of the manufacturer, packer, importer, or exporter.
- Manufacturer's trademark.
- Weight (net).
- Energy value.

- Storage period.
- The date the product was packed into a consumer container.
- Information about the certificate.

4.2. Transport marking should have the following indicators:

- The name of the sending entity.
- Serial number of the batch.
- Product name.
- Botanical origin of the product.
- Harvest year of honey.
- Weight (net, gross).
- Indicators of the standard.

When marking the boxes, their number is additionally indicated, and a packing list is placed on each box. The word "beware" should be written on the boxes where glass and ceramic containers are placed.

5. **Packaging**

5.1. Honey is packaged in barrels made of beech, maple, willow, sidr, linden, or plane trees with a volume of 0.03 liters to 200 liters (the inner surface of the barrels must be waxed or covered with polystyrene film), in 25- and 38-liter stainless steel flasks, in wooden boxes covered with paraffin parchment paper inside, in glass containers, inside lacquered iron jars intended for food products, and in solid or corrugated containers made of pressed cardboard. All types of containers must be approved by the relevant authorities for use in the food industry. When packing honey, its weight is allowed to be 0.03–1.5 kg ± 2% of the net weight and 1.5 kg ± 1% of the gross weight. Containers are not filled to more than 95% of their total volume. Consumer containers must be hermetically sealed. Containers made of polymeric materials must be closed using hot melt adhesive. When the cover is closed, the approved spacer can be used. The bottom and sides of the box should be treated with appropriate material so that the containers in the boxes are not deformed. Consumer containers are placed in boxes made of corrugated cardboard.

6. **Reception**

6.1. Honey is taken in batches. The following should be noted in the honey quality document:

- Name of the enterprise, legal address, name, surname, and patronymic for individuals.
- Product name and botanical origin.
- Harvest year of honey.
- Harvest place of honey.
- Serial number of the batch.
- Number of places in the batches.

- Net and gross weight of the batch.
- Date of issuance of documents (veterinary certificate, receipt).
- Honey analysis documents.
- Packing date (for packing establishments).
- Indicators of the standard.
- Manufacturer's trademark.
- Energy value of the product.
- Storage period.
- Storage conditions.
- Information about the certificate.
- Veterinary-sanitary document on the health of the beehouse.

7. **Test methods**

7.1. Sampling

7.1.1 Non-crystallized honey packed in containers with a volume of 25 or more liters is mixed. The sample is taken through an aluminum tube with a diameter of 10–12 mm. For this, the tube is placed vertically in the container with honey, and when it is removed, the top part of the honey is allowed to flow, and the remaining honey in the tube is placed in a previously prepared clean and dry container. A sample of the crystallized honey is collected with a 50 cm long conical sampler with a longitudinal cross-section. For this, the sampler is placed at an angle, and after removal, samples are taken from the upper, middle, and lower parts of the sampler with a clean tool. From containers with a volume of 0.03–1 L, samples for the combined sample are taken with a special tool. A sample of the comb is taken every fifth frame in the following order. In the upper part of the frame, a 5×5 cm-sized comb is cut and filtered through a sieve with 0.5 mm pores. If the honey has crystallized, it is heated.

7.1.2 The combined sample consists of samples taken from different parts of the honey mass; its weight should not be less than 1.5 kg. The sample is divided into two parts, one part is sent to the laboratory for analysis, and the other part is used if a repeat inspection is necessary.

7.1.3 A label with the following information is attached to the container containing the sample:

- Place and date of sampling.
- Weight of honey and batches.
- Month and year of packaging.
- Last name and first name of the person who took the sample.
- Sampling method (heated or unheated).

7.1.4 A sample must be taken from each unit of natural honey packed in containers of 25 kg or more. In order to check the quality of natural honey packed in small containers, a container should be taken from each batch brought.

7.2. For sampling, product units are taken from different parts of the batch.
7.3. The sample is taken from undamaged containers.
7.4. If the results of the inspection are not satisfactory, a re-inspection is scheduled.
7.5. The results of the re-inspection apply to the entire batch.
7.6. The smell, taste, and signs of fermentation of honey are checked organoleptically from selected containers.

8. **Transportation and storage**

8.1. Transportation

8.1.1 Honey is transported according to accepted sanitary rules.
8.1.2 Barrels should not be stacked more than two to three layers during transportation. A wooden spacer should be placed between the tiers. Stacked barrels must not exceed 1.5 m in height, boxes—3 m, and cardboard boxes—2 m.
8.1.3 Honey containers must be tightly closed during transportation.
8.1.4 Honey can be transported by all available means of transport, subject to compliance with the rules.

8.2. Storage

8.2.1 Honey is stored in places protected from sunlight.
8.2.2 Honey cannot be stored together with toxic, specific-smelling goods.
8.2.3 Honey barrels and flasks can be stacked in 2–3 tiers. At this time, special wooden spacers should be placed on the floor and intertiers.
8.2.4 Honey with a volume of 25 kg or more cannot be stored for more than 8 months starting from the day of the examination.
8.2.5 Hermetically packaged honey in glass and polymer containers cannot be stored for more than 1 year from the date of production, and non-hermetically packaged honey for more than 8 months.
8.2.6 Honey packed in paraffin glasses cannot be stored for more than 6 months from the date of production.
8.2.7 Honey for state reserves can be stored in glass containers, flasks, and special volumes for up to 2 years in places with a temperature no higher than 18°.
8.2.8 Honey with a water content of up to 19% should be stored at a temperature of less than 20, 19% at a temperature of less than 21, and 21% at a temperature of 4–10° (Bogdanov et al., 2002, 1999; Wei et al., 2012; Hermanns et al., 2020; Bal-Prylypko et al., 2022; Podgórska, 2019; Raweh et al. 2023).

References

Adamchuk, L., Sukhenko, V., Akulonok, O., Bilotserkivets, T., Vyshniak, V., Lisohurska, D., et al. (2020). Methods for determining the botanical origin of honey. *Slovak Journal of Food Sciences, 14.*

Adugna, E., Hymete, A., Birhanu, G., & Ashenef, A. (2020). Determination of some heavy metals in honey from different regions of Ethiopia. *Cogent Food & Agriculture, 6*(1), 1764182.

Al-Rajhi, M. A., & EL-Sheikha, A. M. A. (2014). Development and evaluation of a tangential honey-extractor. *Misr Journal of Agricultural Engineering, 31*(4), 1501–1522.

Anklam, E. (1998). A review of the analytical methods to determine the geographical and botanical origin of honey. *Food Chemistry, 63*(4), 549–562.

Bal-Prylypko, L., Nikolaenko, M., Zheplinska, M., Vasyliv, V., & Mushtruk, M. (2022). Comparative data on the content of harmful impurities in honey on the example of Ukrainian standards and foreign documents. *Journal of Hygienic Engineering & Design, 41.*

Barreiros, J., Cepeda, A., Franco, C., Nebot, C., & Vázquez, B. (2024). Analysis of minerals in honey and their nutritional implications. *Journal of Food Composition and Analysis, 106733.*

Bicudo de Almeida-Muradian, L., Monika Barth, O., Dietemann, V., Eyer, M., Freitas, A. D. S. D., Martel, A. C., et al. (2020). Standard methods for Apis mellifera honey research. *Journal of Apicultural Research, 59*(3), 1–62.

Bogdanov, S., Lüllmann, C., Martin, P., Von Der Ohe, W., Russmann, H., Vorwohl, G., et al. (1999). Honey quality and international regulatory standards: review by the International Honey Commission. *Bee World, 80*(2), 61–69.

Bogdanov, S., Martin, P., & Lullmann, C. (2002). *Harmonised methods of the international honey commission* (Vol. 5, Issue 1). Swiss Bee Research Centre, FAM.

Bogdanov, S., Haldimann, M., Luginbühl, W., & Gallmann, P. (2007). Minerals in honey: Environmental, geographical and botanical aspects. *Journal of Apicultural Research, 46*(4), 269–275.

Borum, A. E. (2021). Microbiota and its importance in honey bees. *Bee Studies, 13*(1), 23–30.

Bryant, V. (2017). Filtering honey almost every filter removes some pollen. *Bee Culture The Magazine of American Beekeeping, 145*, 75–76.

Ciulu, M., Solinas, S., Floris, I., Panzanelli, A., Pilo, M. I., Piu, P. C., et al. (2011). RP-HPLC determination of water-soluble vitamins in honey. *Talanta, 83*(3), 924–929.

Cozmuta, A. M., Cozmuta, L. M., Varga, C., Marian, M., & Peter, A. (2011). Effect of thermal processing on quality of polyfloral honey. *Romanian Journal of Food Science, 1*(1), 45–52.

Cuevas-Glory, L. F., Pino, J. A., Santiago, L. S., & Sauri-Duch, E. (2007). A review of volatile analytical methods for determining the botanical origin of honey. *Food Chemistry, 103*(3), 1032–1043.

D'Albore, G. R. (1997). Methods for the characterization of the botanical and geographical origin of some bee products and for their quality control. In *Bee products: Properties, applications, and apitherapy* (pp. 253–262). Springer.

Dafni, H., Lensky, Y., & Fahn, A. (1988). Flower and nectar characteristics of nine species of Labiatae and their influence on honeybee visits. *Journal of Apicultural Research, 27*(2), 103–114.

De Brito Sanchez, G., Ortigao-Farias, J. R., Gauthier, M., Liu, F., & Giurfa, M. (2007). Taste perception in honeybees: just a taste of honey? *Arthropod-Plant Interactions, 1*, 69–76.

Degirmenci, A., Can, Z., Boyraci, G. M., Yildiz, O., Asadov, E., & Kolayli, S. (2020). Honeys from some different regions of Azerbaijan: bioactive characteristics based on phenolic profile and antioxidant activity. *Journal of Apicultural Research, 59*(4), 390–397.

Doner, L. W. (1977). The sugars of honey – A review. *Journal of the Science of Food and Agriculture, 28*(5), 443–456.

Douglas, A. E. (2009). Honeydew. In *Encyclopedia of insects* (pp. 461–463). Academic Press.

Escriche, I., Sobrino-Gregorio, L., Conchado, A., & Juan-Borrás, M. (2017). Volatile profile in the accurate labelling of monofloral honey. The case of lavender and thyme honey. *Food Chemistry, 226*, 61–68.

Esterhuizen, N. (2020). *The botanical origin, physicochemical properties and antibacterial activity of selected West Coast honeys* (Doctoral dissertation, Stellenbosch: Stellenbosch University).

Eyer, M., Neumann, P., & Dietemann, V. (2016). A look into the cell: Honey storage in honey bees, Apis mellifera. *PLoS One, 11*(8), e0161059.

Hermanns, R., Mateescu, C., Thrasyvoulou, A., Tananaki, C., Wagener, F. A., & Cremers, N. A. (2020). Defining the standards for medical grade honey. *Journal of Apicultural Research, 59*(2), 125–135.

Hulea, A., Obiştioiu, D., Cocan, I., Alexa, E., Negrea, M., Neacşu, A. G., et al. (2022). Diversity of monofloral honey based on the antimicrobial and antioxidant potential. *Antibiotics, 11*(5), 595.

Jensen, A. B., Evans, J., Jonas-Levi, A., Benjamin, O., Martinez, I., Dahle, B., et al. (2019). Standard methods for Apis mellifera brood as human food. *Journal of Apicultural Research, 58*(2), 1–28.

Jiang, X., Lin, A., Li, S., Shi, Y., Zhou, F., Felix Gomez, G. G., et al. (2022). Effects of artificial honey and epigallocatechin-3-gallate on streptococcus pyogenes. *BMC Microbiology, 22*(1), 207.

Juan-Borrás, M., Domenech, E., Conchado, A., & Escriche, I. (2015a). Physicochemical quality parameters at the reception of the honey packaging process: Influence of type of honey, year of harvest, and beekeeper. *Journal of Chemistry, 2015*(1), 929658.

Juan-Borrás, M., Periche, A., Domenech, E., & Escriche, I. (2015b). Routine quality control in honey packaging companies as a key to guarantee consumer safety. The case of the presence of sulfonamides analyzed with LC-MS-MS. *Food Control, 50*, 243–249.

Kaškonienė, V., & Venskutonis, P. R. (2010). Floral markers in honey of various botanical and geographic origins: A review. *Comprehensive Reviews in Food Science and Food Safety, 9*(6), 620–634.

Kitzes, G., Schuette, H. A., & Elvehjem, C. A. (1943). The B vitamins in honey. *The Journal of Nutrition, 26*(3), 241–250.

Kolayli, S., Yildiz, O., Sahin, H., & Aliyazicioglu, R. (2013). Biochemistry and physicochemical properties of honey. In *Honey in traditional and modern medicine* (pp. 21–35). CRC Press/ Taylor & Francis Group.

Krishnan, R., Mohammed, T., Kumar, G. S., & Arunima, S. (2021). Honey crystallization: Mechanism, evaluation and application. *The Pharma Innovation Journal, 10*(5S), 222–231.

Kropf, U., Korošec, M., Bertoncelj, J., Ogrinc, N., Nečemer, M., Kump, P., & Golob, T. (2010). Determination of the geographical origin of Slovenian black locust, lime and chestnut honey. *Food Chemistry, 121*(3), 839–846.

Kurta, S., Khatsevich, O., Solonytska, I., Basiurkina, N., Boyko, N., & Horzov, L. (2021). Biopolymer food compositions based on caramelized honey, wax and propolis for prevention of periodontosis. *Scientific Works, 85*(2).

Latorre, C. H., Crecente, R. P., Martín, S. G., & García, J. B. (2013). A fast chemometric procedure based on NIR data for authentication of honey with protected geographical indication. *Food Chemistry, 141*(4), 3559–3565.

Layek, U., Mondal, R., & Karmakar, P. (2020). Honey sample collection methods influence pollen composition in determining true nectar-foraging bee plants. *Acta Botanica Brasilica, 34*, 478–486.

Lenis, L. A., Ruales, C., & Benitez, R. (2013). Determination of glycosides, in express honey of Stevia (Stevia Rebaudiana) by liquid chromatography. In *Biotecnología en el Sector Agropecuario y Agroindustrial*.

Liebezeit, G., & Liebezeit, E. (2013). Non-pollen particulates in honey and sugar. *Food Additives & Contaminants: Part A, 30*(12), 2136–2140.

Lutier, P. M., & Vaissière, B. E. (1993). An improved method for pollen analysis of honey. *Review of Palaeobotany and Palynology, 78*(1–2), 129–144.

Marcazzan, G. L., Mucignat-Caretta, C., Marina Marchese, C., & Piana, M. L. (2018). A review of methods for honey sensory analysis. *Journal of Apicultural Research, 57*(1), 75–87.

Mato, I., Huidobro, J. F., Simal-Lozano, J., & Sancho, M. T. (2003). Significance of nonaromatic organic acids in honey. *Journal of Food Protection, 66*(12), 2371–2376.

Maurizio, A., & Hodges, F. E. D. (1951). Pollen analysis of honey. *Bee World, 32*(1), 1–5.

Moffett, J. O., Stith, L. S., Burkhart, C. C., & Shipman, C. W. (1975). Honey bee visits to cotton flowers. *Environmental Entomology, 4*(2), 203–206.

Muhammad, A., Odunola, O. A., Ibrahim, M. A., Sallau, A. B., Erukainure, O. L., Aimola, I. A., & Malami, I. (2016). Potential biological activity of acacia honey. *Frontiers in Bioscience (Elite Edition), 8*, 351–357.

Muradian, L. B. D. A., Barth, O. M., Dietermann, V., Eyer, M., Freitas, A. D. S. D., Martel, A. C., ... & Sattler, J. A. G. (2020). *Standard methods for Apis mellifera honey research.*

Mureşan, C. I., Cornea-Cipcigan, M., Suharoschi, R., Erler, S., & Mărgăoan, R. (2022). Honey botanical origin and honey-specific protein pattern: Characterization of some European honeys. *LWT, 154*, 112883.

Naila, A., Flint, S. H., Sulaiman, A. Z., Ajit, A., & Weeds, Z. (2018). Classical and novel approaches to the analysis of honey and detection of adulterants. *Food Control, 90*, 152–165.

Nazarian, H., Taghavizad, R., & Majd, A. (2010). Origin of honey proteins and method for its quality control. *Pakistan Journal of Botany, 42*(5), 3221–3228.

Ng, W. J. (2022). *Bioactivities and botanical origin differentiation of malaysian stingless bee honey produced by heterotrigona itama and geniotrigona thoracica* (Doctoral dissertation, UTAR).

Oeder, R., & Schwabe, D. (2021). The upward tilt of honeycomb cells increases the carrying capacity of the comb and is not to prevent the outflow of honey. *Apidologie, 52*, 174–185.

Oğuzoğlu, Ş., Avcı, M., & İpekdal, K. (2021). Predators of the giant pine scale, Marchalina hellenica (Gennadius 1883; Hemiptera: Marchalinidae), out of its natural range in Turkey. *Open Life Sciences, 16*(1), 682–694.

Pavlova, T., Stamatovska, V., Kalevska, T., Dimov, I., Assistant, G., & Nakov, G. (2018). *Quality characteristics of honey: A review.* In Proceedings of University of Ruse (p. 57).

Peng, Y. S., & Marston, J. M. (1986). Filtering mechanism of the honey bee proventriculus. *Physiological Entomology, 11*(4), 433–439.

Podgórska, K. S. (2019). Polonya mevzuatında bal kalitesi standartları. *Bilimsel ve Teknik Araştırma Biyomedikal Dergisi, 19*(5), 14669–14671.

Qiao, J., Chen, L., Kong, L., Dong, J., Zhou, Z., & Zhang, H. (2020). Characteristic components and authenticity evaluation of rape, acacia, and linden honey. *Journal of Agricultural and Food Chemistry, 68*(36), 9776–9788.

Rababah, T., AL-U'datt, M., Naqresh, A., Gammoh, S., Almajwal, A., Saleh, M., et al. (2024). Effect of temperature and time on the physicochemical and sensory properties of crystallized honey. *ACS Omega, 9*(18), 20243–20252.

Radtke, J., & Lichtenberg-Kraag, B. (2018). Long-term changes in naturally produced honey depending on processing and temperature. *Journal of Apicultural Research, 57*(5), 615–626.

Raweh, H. S., Badjah-Hadj-Ahmed, A. Y., Iqbal, J., & Alqarni, A. S. (2023). Physicochemical composition of local and imported honeys associated with quality standards. *Food, 12*(11), 2181.

Roman, A., Popiela-Pleban, E., Kozak, M., & Roman, K. (2013). Factors influencing consumer behavior relating to the purchase of honey part 2. Product quality and packaging. *Journal of Apicultural Science, 57*(2), 175–185.

Rzepecka-Stojko, A., Stojko, J., Kurek-Górecka, A., Górecki, M., Kabała-Dzik, A., Kubina, R., et al. (2015). Polyphenols from bee pollen: Structure, absorption, metabolism and biological activity. *Molecules, 20*(12), 21732–21749.

Sanz, M. L., Gonzalez, M., De Lorenzo, C., Sanz, J., & Martınez-Castro, I. (2005). A contribution to the differentiation between nectar honey and honeydew honey. *Food Chemistry, 91*(2), 313–317.

Scripca, L., & Amariei, S. (2018). Research on honey crystalization. *Revista de Chimie, 69*(10), 2953–2957.

Seeley, T. D. (1989). Social foraging in honey bees: how nectar foragers assess their colony's nutritional status. *Behavioral Ecology and Sociobiology, 24*, 181–199.
Siavoshi, F., Sahraee, M., Ebrahimi, H., Sarrafnejad, A., & Saniee, P. (2018). Natural fruits, flowers, honey, and honeybees harbor Helicobacter pylori-positive yeasts. *Helicobacter, 23*(2), e12471.
Siddiqui, I. R. (1970). The sugars of honey. In *Advances in carbohydrate chemistry and biochemistry* (Vol. 25, pp. 285–309). Academic Press.
Soares, S., Amaral, J. S., Oliveira, M. B. P., & Mafra, I. (2017). A comprehensive review on the main honey authentication issues: Production and origin. *Comprehensive Reviews in Food Science and Food Safety, 16*(5), 1072–1100.
Subramanian, R., Umesh Hebbar, H., & Rastogi, N. K. (2007). Processing of honey: A review. *International Journal of Food Properties, 10*(1), 127–143.
Sultanov, R. L. (1993). Azərbaycan balının kristallaşması haqqında. Az. ETETİ və TİTİ informasiya vərəqi Bakı 1993 N56.
Sultanov, R. L., & Asadov, E. S. (2017). *Azerbaycan balının dovlet standartlari*. https://www.researchgate.net/publication/377931058_Azerbaycan_balinin_dovlet_standartlari
Swain, S. K., Ail, S. K. S., Jena, S. K., Bairwa, M. K., & Sahoo, S. N. (2021). Preference of breeding substratum, embryonic development and seed production of honey gourami, Trichogaster chuna (Hamilton, 1822)-An indigenous ornamental fish in demand. *Aquaculture, 542*, 736874.
Szabó, R. T., Mézes, M., Szalai, T., Zajácz, E., & Kovács-Weber, M. (2016). Colour identification of honey and methodical development of its instrumental measuring. *Columella: Journal of Agricultural and Environmental Sciences, 3*(1), 29–36.
Tahırov, A., Huseynov, H., & Esedov, E. (2010). Nahcivan özerk cumhuriyetinde bal arısı (Apis mellifera L.) kolonilerinin gelişim süresinin hızlandırılması yollarının araştırılması. *Kafkas Üniversitesi Veteriner Fakültesi Dergisi, 16*(5).
Tappi, S., Laghi, L., Dettori, A., Piana, L., Ragni, L., & Rocculi, P. (2019). Investigation of water state during induced crystallization of honey. *Food Chemistry, 294*, 260–266.
Valverde, S., Ares, A. M., Elmore, J. S., & Bernal, J. (2022). Recent trends in the analysis of honey constituents. *Food Chemistry, 387*, 132920.
Wäckers, F. L., Van Rijn, P. C., & Heimpel, G. E. (2008). Honeydew as a food source for natural enemies: making the best of a bad meal? *Biological Control, 45*(2), 176–184.
Wang, S., Guo, Q., Wang, L., Lin, L., Shi, H., Cao, H., & Cao, B. (2015). Detection of honey adulteration with starch syrup by high performance liquid chromatography. *Food Chemistry, 172*, 669–674.
Wang, X., Chen, Y., Hu, Y., Zhou, J., Chen, L., & Lu, X. (2022). Systematic review of the characteristic markers in honey of various botanical, geographic, and entomological origins. *ACS Food Science & Technology, 2*(2), 206–220.
Wedmore, E. B. (1955). The accurate determination of the water content of honeys: Part I. Introduction and results. *Bee World, 36*(11), 197–206.
Wei, G. X., Huang, J. K., & Jun, Y. (2012). Honey safety standards and its impacts on China's honey export. *Journal of Integrative Agriculture, 11*(4), 684–693.
White, J. W., & Doner, L. W. (1980). Honey composition and properties. *Beekeeping in the United States Agriculture Handbook, 335*, 82–91.
White, J. W., Jr., & Riethof, M. L. (1959). The composition of honey. III. Detection of acetylandromedol in toxic honeys. *Archives of Biochemistry and Biophysics, 79*, 165–167.
Wu, H., Luo, H., Huang, F., Zhou, X., Huang, X., & Chen, J. (2019). The study of the constituents and source of toxicants in poisonous honey. *Analytical Biochemistry, 569*, 10–15.
Yanniotis, S., Skaltsi, S., & Karaburnioti, S. (2006). Effect of moisture content on the viscosity of honey at different temperatures. *Journal of Food Engineering, 72*(4), 372–377.
Živkov Baloš, M., Popov, N., Jakšić, S., Mihaljev, Ž., Pelić, M., Ratajac, R., & Ljubojević Pelić, D. (2023). Sunflower honey – Evaluation of quality and stability during storage. *Food, 12*(13), 2585.

Chapter 13
Next-Generation Sequencing (NGS) and Electronic Nose (eNose) Technologies for Honey Safety and Authentication

Aneta A. Ptaszyńska

1 Application of Next-Generation Sequencing (NGS) in Honey Research

Next-generation sequencing (NGS) represents a revolutionary advance in DNA and RNA sequencing technology that has transformed molecular biology and genetic research. Next-generation sequencing refers to high-throughput DNA sequencing technologies capable of sequencing millions of DNA fragments simultaneously. As demonstrated in the study by Laha et al. (2017), NGS enables massive parallel sequencing of genetic material, allowing researchers to analyze entire genomes or targeted regions with unprecedented speed and accuracy.

The NGS process typically involves several key steps. According to Smith et al. (2021), these include:

1. Library preparation: DNA or RNA samples are fragmented into smaller pieces and prepared using specialized adapters. Their research showed that this step is crucial for successful sequencing, especially when working with complex samples such as honey, which contains genetic material from multiple sources.
2. Sequencing: The prepared fragments are sequenced simultaneously using different chemical and optical detection methods. As described by Utzeri et al. (2018), modern NGS platforms can generate millions of sequence reads in a single run, providing extensive coverage of the genetic material present in the sample.
3. Data analysis: The resulting sequence data undergoes sophisticated bioinformatic analysis to assemble and interpret the genetic information. Özkök et al. (2023) emphasized that this step requires powerful computational tools and carefully designed analysis pipelines.

A. A. Ptaszyńska (✉)
Institute of Biological Sciences, Maria Curie-Skłodowska University, Lublin, Poland
e-mail: aneta.ptaszynska@mail.umcs.pl

M. Zafar et al. (eds.), *Pure Honey: Assurance & Authentication*,
https://doi.org/10.1007/978-3-031-98913-1_13

NGS offers several significant advantages over traditional Sanger sequencing methods. Chavan et al. (2022) highlighted that NGS can process millions of sequences simultaneously, dramatically reducing the time and cost per base sequenced; detect rare genetic variants due to its high sensitivity; identify unknown sequences through comprehensive analysis; and provide quantitative information on the abundance of genetic material. The applications of NGS is diverse and growing. In food authentication alone, Smith et al. (2021) demonstrated its use in species identification in complex mixtures, food fraud detection, quality control and safety assessment, and food origin verification.

2 Honey Authentication Using NGS

Next-generation sequencing (NGS) has emerged as a revolutionary tool in honey authentication, providing unprecedented insight into the composition and origin of honey. This technology allows comprehensive analysis of the genetic material present in honey, providing a more detailed understanding than traditional methods.

NGS technology applied to honey analysis typically focuses on several key genetic markers. According to Laha et al. (2017), the most effective approach uses multiple genetic loci, including rbcL, matK, and ITS2. Their research demonstrated that this multi-locus approach provides superior discriminatory power compared to single-marker methods. Smith et al. (2021) made significant advances in understanding honey composition using NGS, revealing that honey contains different types of RNA, including microRNAs and tRNA fragments. Their work showed that these molecular components originate from multiple sources, including plants, invertebrates, and prokaryotes, providing a complex molecular signature for authentication purposes.

Recent methodological improvements have increased the effectiveness of NGS in honey analysis. Utzeri et al. (2018) developed a semiconductor-based sequencing approach specifically for the detection of botanical composition in different types of honey. Their method successfully identified the dominant botanical contribution in both monofloral and polyfloral honeys, demonstrating the versatility of the technology. A major breakthrough came from Chavan et al. (2022), who successfully developed methods to isolate and sequence pollen-free DNA from filtered honey. This development addressed a major limitation in honey authentication, particularly for processed honey products where traditional pollen analysis proves ineffective.

NGS is also an invaluable tool for detecting adulteration of honey and verifying its purity. The technology can identify pollen DNA and other organic components to authenticate the botanical and geographical origin of honey, which is critical for label accuracy. Given the prevalence of adulteration in the honey market—often involving the addition of cheaper sugars and syrups—NGS provides a robust method for authenticating and tracing the origin of honey, protecting the interests of consumers and supporting genuine producers.

3 The Challenge of Honey Authentication

The global honey industry faces significant challenges with adulteration and mislabeling, which have become widespread concerns affecting market integrity and consumer confidence. A recent study found that up to 27% of honey samples tested showed evidence of dilution with sugar syrup substitutes (Zhou et al., 2018). This widespread fraud has led to the development of increasingly sophisticated authentication methods to protect both consumers and legitimate producers.

4 Traditional Authentication Methods and Their Limitations

Melissopalynology, the microscopic analysis of pollen in honey, has traditionally served as the gold standard for determining the botanical and geographical origin of honey. However, as demonstrated by Özkök et al. (2023), this method has several significant limitations. Their research showed that while microscopic analysis can identify pollen from different plant families, it requires extensive expertise and training and becomes particularly challenging when dealing with closely related plant species.

Furthermore, the method proves ineffective when analyzing filtered honey or in cases where pollen from more desirable plants has been artificially added to mask the true origin. Utzeri et al. (2018) highlight that this limitation has become particularly problematic as honey processing techniques have become more sophisticated, making it easier to manipulate pollen content.

5 Emergence of Molecular Authentication Approaches

Recent advances in molecular biology have revolutionized honey authentication methods. DNA-based approaches, particularly next-generation sequencing (NGS), have emerged as powerful complementary tools. Laha et al. (2017) conducted a landmark study comparing traditional microscopic analysis with NGS-based methods and found that molecular approaches could detect significantly more plant species than traditional methods.

A study by Chavan et al. (2022) made a major breakthrough by successfully isolating pollen-free DNA from filtered honey using three types of adsorbents: anti-dsDNA antibodies coupled to magnetic microspheres, anion-exchange adsorbents, and ceramic hydroxyapatite. This development opened up new possibilities for authenticating processed honey products that could not be effectively analyzed using traditional methods. NGS offers several advantages over traditional authentication methods (Kim et al., 2015; Kim et al., 2019; Lanh et al., 2024; Li et al., 2013, Paula 2021; Rustemoglu 2023). Özkök et al. (2023) demonstrated that NGS can detect 48% more plant families than traditional melissopalynological analysis. Their research highlighted the superior sensitivity of NGS and its ability to identify species that might be missed by microscopic examination. The success of NGS in honey authentication relies heavily on sophisticated bioinformatics pipelines. Laha et al. (2017) highlighted the importance of robust data analysis protocols, including quality filtering of raw reads and careful species annotation. Their work demonstrated that proper bioinformatics analysis is critical for accurate interpretation of sequencing results.

Despite its advantages, NGS faces certain challenges in honey authentication. Smith et al. (2021) found that a significant proportion of sequenced genetic material remains unidentified due to incomplete reference databases, particularly for regional flora. This highlights the continuing need to expand genetic databases of honey-relevant plant species. The implementation of NGS in routine honey authentication requires standardized protocols. Özkök et al. (2023) proposed a novel bioinformatics approach for the analysis of honey samples, emphasizing the need for reproducible and reliable methods that can be adopted by different laboratories.

Smith et al. (2021) made an important discovery regarding the molecular composition of honey by identifying different types of small RNA molecules, including microRNAs and tRNA fragments. Their research showed that these molecules come not only from plants, but also from invertebrates and prokaryotes, providing a complex molecular signature that could potentially distinguish different types of honey with high accuracy. The field continues to evolve, with several promising areas of development.

6 Technological Developments Have Increased the Potential of Molecular Authentication Methods

DNA marker selection: Research by Utzeri et al. (2018) showed that the combination of multiple DNA markers, particularly the rbcL and trnH-psbA regions, provided better discriminatory power for identifying the botanical source of honey. Their work demonstrates how different genetic markers can complement each other to provide more accurate results. Bioinformatics advances: Laha et al. (2017) developed improved bioinformatics pipelines that increased the accuracy and reliability of NGS-based authentication methods. Their work established protocols for processing and analyzing the complex data generated by molecular methods.

A major challenge identified by several researchers is the need for standardization. Özkök et al. (2023) highlighted how different protocols and methods can lead to inconsistent results between laboratories. They emphasized the need for standardized procedures that combine traditional and molecular methods into robust authentication frameworks. NGS-based methods were able to detect 48% more plant families than traditional melissopalynological analysis. However, each method has its strengths and limitations, suggesting that an integrated approach may provide the most reliable authentication results (Özkök et al. 2023).

The economic impact of improved authentication methods is significant. Zhou et al. (2018) noted that honey fraud affects not only consumer confidence but also market prices and fair competition. The development of more reliable authentication methods could help protect premium honey products and support sustainable beekeeping practices. Honey authentication methods reflect the industry's ongoing efforts to combat fraud and ensure product integrity. Combining traditional expertise with cutting-edge molecular techniques offers promising solutions to this complex challenge, although continued research and development will be crucial to address emerging authentication challenges.

7 Implementation of Electronic Nose (eNose) Technology in Honey Analysis

Electronic nose (eNose) technology represents a significant advance in the field of sensory analysis and quality control. This biomimetic system attempts to replicate the human olfactory system using an array of chemical sensors sensitive to volatile organic compounds (VOCs) (Borgonovo et al., 2021). The basic components of an eNose system typically include a sensor array, a signal processing unit, and pattern recognition software. Metal oxide semiconductor (MOS) sensors are commonly used because of their broad sensitivity to various chemical compounds and relatively low cost. These sensors undergo conductivity changes when exposed to volatile compounds, generating unique response patterns that can be analyzed using various statistical methods (González María et al., 2023). eNose technology has applications in a wide range of industries, from food quality assessment to medical diagnostics. In food authentication studies, researchers have demonstrated particular success in distinguishing products based on origin and composition. One of the key advantages of eNose technology is its ability to provide rapid, nondestructive analysis without the need for complex sample preparation. However, the technology requires careful calibration and standardized operating conditions to ensure reproducible results. The sensitivity of the sensors to environmental factors such as temperature and humidity requires controlled test conditions (Borgonovo et al., 2021; Correa et al., 2017).

Current research efforts are focused on improving sensor stability, improving pattern recognition algorithms, and developing more cost-effective systems. The integration of artificial intelligence and machine learning techniques has shown promising results in improving the accuracy of eNose systems, with some studies reporting classification accuracies of over 87% for certain applications (González María et al., 2023).

8 Honey Authentication Using eNose

Designed to detect and discriminate complex odors through an array of sensors and a sophisticated pattern recognition system, the electronic nose mimics the human olfactory system. In the context of honey research, the eNose analyzes aroma profiles, which are key indicators of honey quality, flavor nuances, and freshness. This technology skillfully classifies different types of honey based on their floral origin by identifying specific volatile organic compounds (VOCs) characteristic of each type. This capability is particularly useful for verifying the botanical origin of honey, which is a key determinant of market value. For example, the unique aromatic compounds of manuka honey, renowned for its health benefits, can be clearly identified, distinguishing it from other types of honey. Research has shown that electronic noses, combined with appropriate statistical analysis methods, can effectively classify honey samples based on their geographical and botanical origin (Borgonovo et al., 2021; Benedetti et al., 2004a, b; Cuevas-Glory et al., 2007; Di Rosa et al., 2018; Gan et al., 2016; Hong et al., 2011; Huang et al., 2015; Johnson et al., 2018; Kenjeric et al., 2009; Subari et al., 2012). Preliminary studies showed that electronic noses equipped with metal oxide semiconductor (MOS) and metal oxide semiconductor field effect transistor (MOSFET) sensors could generate distinctive patterns of volatile compounds present in honey samples. Through principal component analysis (PCA) and artificial neural network (ANN) processing, these patterns allowed accurate classification of honey varieties (Benedetti et al., 2004a, b). Čačić Kenjerić et al. (2009a, b) achieved significant results in differentiating honey samples from different geographical locations using eNose analysis combined with principal component analysis (PCA). Research by Romano et al. (2016) compared two different eNose approaches—a metal oxide semiconductor (MOS) sensor array and proton transfer reaction time-of-flight mass spectrometry (PTR-ToF-MS). Both methods achieved around 85% accuracy in classifying honey samples by geographical origin. The MOS-based eNose provided a low cost, portable solution, while PTR-ToF-MS allowed detailed analysis of volatile compounds. This is consistent with other research demonstrating the capabilities of the eNose for honey authentication. Tian et al. (2018) found that eNose systems, combined with multivariate statistical analysis, could effectively characterize honey samples from different botanical origins based on their aroma profiles. Their work showed strong correlations between eNose sensor responses and traditional sensory evaluation results.

The electronic nose provides complementary detection capabilities, achieving 94.45% detection accuracy and 80.83% prediction accuracy in distinguishing authentic from adulterated samples. Although slightly less sensitive than the electronic tongue, which achieved 99.22% accuracy, particularly at lower adulteration levels between 10% and 20%, the eNose remains a valuable analytical tool when used in conjunction with other methods (Dymerski et al., 2014, Bodor et al., 2020). Both studies show that traditional sensory analysis by human tasters, while important for verification, has lower sensitivity compared to electronic methods. Tasters were able to detect significant differences in four parameters for acacia honey and eight parameters for lime honey but struggled to detect adulteration at lower concentrations (Bodor et al., 2020). The research reveals several physicochemical indicators that help confirm adulteration, such as lower total soluble dry matter content, lower pH values, reduced antioxidant capacity, and significant changes in color parameters, with adulterated samples typically appearing lighter and less yellow. These measurements provide supporting evidence for the verification of electronic sensor as eNose and eTongue results (Bodor et al., 2020; Bougrini et al. 2016; Juan-Borrás et al., 2017; do Nascimento et al., 2022; Escriche et al.,2012; Kenjeric et al., 2007; Kiselev et al., 2018; Linder & Pöppl 2003; Liu et al., 2013; Liu et al., 2015; Lu et al., 2022; Maamor et al. 2014; Mahmoud et al., 2024; María et al., 2023; Masoomi et al., 2024; Panseri et al., 2023; Wang & Liu 2019; Wei & Wang 2014; Zuluaga-Domínguez et al. 2018).

Recent research has explored hybrid sensing approaches that combine electronic nose data with other analytical methods such as Fourier transform infrared (FTIR) spectroscopy. Studies have shown that data fusion techniques, particularly low-level fusion (LLF) and intermediate-level fusion (ILF), can achieve higher classification accuracies than single modality approaches. For example, the LLF of electronic nose and FTIR data achieved classification accuracies of up to 92.2% for pure and adulterated honey samples (Subari et al., 2012). The technology offers several advantages for industrial applications, including rapid analysis times, nondestructive testing, and the ability to perform routine quality control checks without specialized laboratory equipment or personnel. This makes electronic nose systems particularly valuable for honey producers and regulators working to verify product authenticity and detect potential adulteration.

Studies by Čačić et al. (2009a, b) used an electronic nose with metal oxide semiconductor sensors combined with artificial neural network analysis. This approach successfully classified different unifloral honey varieties with over 90% accuracy based on their volatile compound profiles. The researchers noted that the eNose technology offered rapid screening capabilities compared to traditional analytical methods. More recent work by Rivai et al. (2022) developed an eNose system using semiconductor gas sensors and convolutional neural networks for pattern recognition. Their system achieved 100% accuracy in detecting honey adulteration at different concentration levels. The researchers used an array of MQ series gas sensors to detect volatile organic compounds characteristic of pure and adulterated honey samples. Other researchers achieved 87.5% accuracy in classifying honey varieties using support vector machine (SVM) analysis of eNose data combined with physicochemical parameters (González María et al., 2023). The technology seems to be particularly effective in discriminating honey samples with higher phenolic content, such as chestnut honey.

According to Gonçalves et al. (2023), eNose systems based on electrodes covered with ionogel films can effectively discriminate between authentic and adulterated honey samples by analyzing their volatile organic compounds. In the study, both authentic and adulterated honey samples were investigated using sensory evaluation, gas chromatography-mass spectrometry (GC-MS), and eNose analysis.

Principal component analysis of the eNose data showed a clear separation between authentic and adulterated honey clusters with no overlap. Machine learning algorithms applied to the sensor response data achieved high accuracy, with the Kstar algorithm reaching 100% accuracy in classifying honey samples (Gonçalves et al., 2023).

9 The Use of eNose in Assessing the Freshness of Honey

In addition, the eNose can be used to detect spoilage or fermentation in honey, which can result from improper storage or premature harvesting when honey still has a high moisture content. By monitoring changes in VOC profiles, the eNose helps to assess the quality and safety of honey, ensuring that only products that meet high-quality standards are marketed (Casalinuovo et al. 2006).

The nondestructive nature and rapid analysis capabilities of the eNose technology make it particularly attractive for routine quality control applications. However, researchers note that proper calibration and standardization protocols are essential for reliable results. Current limitations include the need for controlled environmental conditions and potential sensor drift over time.

In summary, both NGS and eNose technologies are key to the advancement of honey research and production. Not only do they enhance product integrity and consumer confidence, but they also contribute to the sustainable management of bee populations, which is essential for maintaining ecological balance and supporting global food security. Current research continues to focus on optimizing sensor arrays, improving pattern recognition algorithms, and developing more sophisticated data fusion approaches to improve classification accuracy. The technology shows promise as a practical tool for honey authentication, although further validation studies may be required before widespread industrial adoption.

References

Benedetti, S., Mannino, S., Sabatini, A. G., & Marcazzan, G. L. (2004a). Electronic nose and neural network use for the classification of honey. *Apidologie, 35*(4), 397–402.

Benedetti, S., Mannino, S., Sabatini, A. G., et al. (2004b). Electronic nose and neural network use for the classification of honey. *Apidologie, 35*(4), 397–402. https://doi.org/10.1051/apido:2004025

Bodor, Z., Kovacs, Z., Rashed, M. S., et al. (2020). Sensory and physicochemical evaluation of acacia and linden honey adulterated with sugar syrup. *Sensors, 20*(17), 4845. https://doi.org/10.3390/s20174845

Borgonovo, F., Marzorati, S., Piana, L., Rizzi, R., Panseri, S., Verotta, L., & Guarino, M. (2021). Electronic nose for the classification of honeys of different floral origins. *Chemical Engineering Transactions, 85*, 205–210.

Bougrini, M., Tahri, K., Saidi, T., et al. (2016). Classification of honey according to geographical and botanical origins and detection of its adulteration using voltammetric electronic tongue. *Food Analytical Methods, 9*(8), 2161–2173. https://doi.org/10.1007/s12161-015-0393-2
Čačić Kenjerić, F., Mannino, S., Bennedetti, S., Primorac, L., & Čačić Kenjerić, D. (2009a). Honey botanical origin determination by electronic nose. *Journal of Apicultural Research, 48*(2), 99–103.
Čačić Kenjerić, F., Primorac, L., Kenjerić, D., Benedetti, S., & Mandić, M. L. (2009b). Application of electronic nose in honey geographical origin characterisation. *Journal of Central European Agriculture, 10*(1), 19–26.
Casalinuovo, I. A., Di Pierro, D., Coletta, M., & Di Francesco, P. (2006). Application of electronic noses for disease diagnosis and food spoilage detection. *Sensors, 6*(11), 1428–1439. https://doi.org/10.3390/s6111428
Chavan, et al. (2022). *Isolation and barcoding of trace pollen-free DNA for authentication of honey*. https://doi.org/10.1021/acs.jafc.2c04309.
Correa, A. R., Cuenca, M. M., Zuluaga, C. M., et al. (2017). Validation of honey-bee smelling profile by using a commercial electronic nose. *Ingenieria E Investigacion, 37*(3), 45–51. https://doi.org/10.15446/ing.investig.v37n3.59656
Cuevas-Glory, L. F., Pino, J. A., Santiago, L. S., et al. (2007). A review of volatile analytical methods for determining the botanical origin of honey. *Food Chemistry, 103*(3), 1032–1043. https://doi.org/10.1016/j.foodchem.2006.07.068
Di Rosa, A. R., Leone, F., Scattareggia, C., et al. (2018). Botanical origin identification of Sicilian honeys based on artificial senses and multi-sensor data fusion. *European Food Research and Technology, 244*(1), 117–125. https://doi.org/10.1007/s00217-017-2945-8
do Nascimento, A. S., Silva, F. D., Machado, C. S., et al. (2022). Use of the electronic tongue as a tool for the characterization of Melipona scutellaris Latreille honey. *Journal of Apicultural Research, 61*(1), 79–90. https://doi.org/10.1080/00218839.2020.1842579
Dymerski, T., Gebicki, J., Wardencki, W., et al. (2014). Application of an electronic nose instrument to fast classification of Polish honey types. *Sensors, 14*(6), 10709–10724. https://doi.org/10.3390/s140610709
Escriche, I., Kadar, M., Domenech, E., et al. (2012). A potentiometric electronic tongue for the discrimination of honey according to the botanical origin. Comparison with traditional methodologies: Physicochemical parameters and volatile profile. *Journal of Food Engineering, 109*(3), 449–456. https://doi.org/10.1016/j.jfoodeng.2011.10.036
Etzold, E., & Lichtenberg-Kraag, B. (2008). Determination of the botanical origin of honey by Fourier-transformed infrared spectroscopy: An approach for routine analysis. *European Food Research and Technology, 227*(2), 579–586. https://doi.org/10.1007/s00217-007-0759-9
Gan, Z., Yang, Y., Li, J., et al. (2016). Using sensor and spectral analysis to classify botanical origin and determine adulteration of raw honey. *Journal of Food Engineering, 178*, 151–158. https://doi.org/10.1016/j.jfoodeng.2016.01.016
Gonçalves, W. B., Teixeira, W. S. R., Cervantes, E. P., et al. (2023). Application of an electronic nose as a new technology for rapid detection of adulteration in honey. *Applied Sciences-Basel, 13*(8), 4881. https://doi.org/10.3390/app13084881
González María, E., Luna, A. M., Celdrán, A. C., Muñoz, G. M., Oates, M. J., & Ruiz-Canales, A. (2023). Classification of Monofloral honeys by measuring a low-cost electronic nose prototype based on resistive metal oxide sensors. *Agronomy, 13*, 2183.
Hong, E. J., Park, S. J., Lee, H. J., et al. (2011). Analysis of various honeys from different sources using electronic nose. *Korean Journal for Food Science of Animal Resources, 31*(2), 273–279. https://doi.org/10.5851/kosfa.2011.31.2.273
Huang, L., Liu, H., Zhang, B., et al. (2015). Application of electronic nose with multivariate analysis and sensor selection for botanical origin identification and quality determination of honey. *Food and Bioprocess Technology, 8*(2), 359–370. https://doi.org/10.1007/s11947-014-1407-6

Johnson, G., Patel, D., Alluhayb, A., et al. (2018). *Preliminary studies of honey queen bee conditions using cyranose 320 nose technology*. In ACMSE'18: Proceedings of the ACMSE 2018 conference. https://doi.org/10.1145/3190645.3190696.

Juan-Borrás, M., Soto, J., Gil-Sánchez, L., et al. (2017). Antioxidant activity and physico-chemical parameters for the differentiation of honey using a potentiometric electronic tongue. *Journal of the Science of Food and Agriculture, 97*(7), 2215–2222. https://doi.org/10.1002/jsfa.8031

Kenjeric, D., Mandic, M. L., Primorac, L., et al. (2007). Flavonoid profile of Robinia honeys produced in Croatia. Food Chemistry, 102(3), 683–690. https://doi.org/10.1016/j.foodchem.2006.05.055

Kenjeric, F. C., Mannino, S., Bennedetti, S., et al. (2009). Honey botanical origin determination by electronic nose. *Journal of Apicultural Research, 48*(2), 99–103. https://doi.org/10.3896/IBRA.1.48.2.03

Kim, H. K., Kim, I., Lee, M. L., et al. (2015). Microsatellite markers developed by next-generation sequencing differentiate inbred lines of Apis mellifera. *Journal of Asia-Pacific Entomology, 18*(4), 801–805. https://doi.org/10.1016/j.aspen.2015.10.003

Kim, J. S., Kim, M. J., Kim, H. K., et al. (2019). Development of single nucleotide polymorphism markers specific to Apis mellifera (hymenoptera: Apidae) line displaying high hygienic behavior against Varroa destructor, an ectoparasitic mite. *Journal of Asia-Pacific Entomology, 22*(4), 1031–1039. https://doi.org/10.1016/j.aspen.2019.08.005

Kiselev, I., Sysoev, V., Kaikov, I., et al. (2018). On the temporal stability of Analyte recognition with an E-nose based on a metal oxide sensor Array in practical applications. *Sensors, 18*(2), 550. https://doi.org/10.3390/s18020550

Laha, et al. (2017). *Meta-barcoding in combination with palynological inference is a potent diagnostic marker for honey floral composition*. https://doi.org/10.1186/s13568-017-0429-7.

Lanh, P. T., Duong, B. T. T., Thu, H. T., et al. (2024). Comprehensive analysis of the microbiome in Apis cerana honey highlights honey as a potential source for the isolation of beneficial bacterial strains. *PeerJ, 12*, e17157. https://doi.org/10.7717/peerj.17157

Li, Y., Li-Byarlay, H., Burns, P., et al. (2013). TrueSight: A new algorithm for splice junction detection using RNA-seq. *Nucleic Acids Research, 41*(4), e51. https://doi.org/10.1093/nar/gks1311

Linder, R., & Pöppl, S. J. (2003). A new neural network approach classifies olfactory signals with high accuracy. *Food Quality and Preference, 14*(5–6), 435–440. https://doi.org/10.1016/S0950-3293(03)00016-8

Liu, N., Shi, B., Zhao, L., et al. (2013). Optimizing an electronic nose for analysis honey from different nectar sources. *Sensor Letters, 11*(6–7), 1145–1148. https://doi.org/10.1166/sl.2013.2872

Liu, N., Shi, B., Zhao, L., et al. (2015). Research on pattern recognition method for honey nectar detection by electronic nose. In *Computer and computing technologies in agriculture VIII* (pp. 393–403). https://doi.org/10.1007/978-3-319-19620-6_44

Lu, L., Hu, Z., Hu, X., et al. (2022). Electronic tongue and electronic nose for food quality and safety. *Food Research International, 162*, 112214. https://doi.org/10.1016/j.foodres.2022.112214

Maamor, H. N., Rashid, F. N. A., Zakaria, N. Z. I. et al. (2014). *Bio-inspired taste assessment of pure and adulterated honey using multi-sensing technique*. In 2014 2nd International Conference on Electronic Design (ICED) (pp. 270–274).

Mahmoud, M. A. A., Kiliç-Büyükkurt, Ö., Fotouh, M. M. A., et al. (2024). Aroma active compounds of honey: Analysis with GC-MS, GC-O, and molecular sensory techniques. *Journal of Food Composition and Analysis, 134*, 106545. https://doi.org/10.1016/j.jfca.2024.106545

María, E. G., Luna, A. M., Celdrán, A. C., et al. (2023). Classification of Monofloral honeys by measuring a low-cost electronic nose prototype based on resistive metal oxide sensors. *Agronomy-Basel, 13*(8), 2183. https://doi.org/10.3390/agronomy13082183

Masoomi, S., Sharifi, H., & Hemmateenejad, B. (2024). A paper-based optical tongue for characterization of iranian honey: Identification of geographical/botanical origins and adulteration detection. *Food Control, 155*, 110052. https://doi.org/10.1016/j.foodcont.2023.110052

Özkök, A., Bilgic, H. A., Kosukcu, C., et al. (2023). Comparing the melissopalynological and next generation sequencing (NGS) methods for determining botanical origin of honey. *Food Control, 148*, 109630. https://doi.org/10.1016/j.foodcont.2023.109630

Panseri, S., Borgonovo, F., Guarino, M., et al. (2023). Monitoring volatile organic compounds and aroma profile of Robinia pseudoacacia L. honey at different storage temperatures during shelf life. *Food, 12*(16), 3105. https://doi.org/10.3390/foods12163105

Paula, D. P. (2021). Next-generation sequencing and its impacts on entomological research in ecology and evolution. *Neotropical Entomology, 50*(5), 679–696. https://doi.org/10.1007/s13744-021-00895-x

Rivai, M., Misbah, M., Kurniawan, F., Purwanto, D., Aulia, S., & Tasripan, T. (2022). *Electronic nose using convolutional neural network to determine adulterated honeys*. In International conference on computer engineering, network, and intelligent multimedia (pp. 55–59).

Romano, A., Cuenca, M., Makhoul, S., Biasioli, F., Martinello, L., Fugatti, A., & Scampicchio, M. (2016). Comparison of e-noses: The case study of honey. *Italian Journal of Food Science, 28*(2), 326–337.

Rustemoglu, M. (2023). Next generation sequencing of bee gut microbiota in urban and rural environments. *Diversity-Basel, 15*(9), 1016. https://doi.org/10.3390/d15091016

Smith, et al. (2021). *Cataloguing the small RNA content of honey using next generation sequencing*. https://doi.org/10.1016/j.fochms.2021.100014.

Subari, N., Saleh, J. M., Shakaff, A. Y. M., et al. (2012). A hybrid sensing approach for pure and adulterated honey classification. *Sensors, 12*(10), 14022–14040. https://doi.org/10.3390/s121014022

Tian, H., Shen, Y., Yu, H., et al. (2018). Aroma features of honey measured by sensory evaluation, gas chromatography-mass spectrometry, and electronic nose. *International Journal of Food Properties, 21*(1), 1755–1768. https://doi.org/10.1080/10942912.2016.1213744

Utzeri, et al. (2018). *Application of next generation semiconductor based sequencing to detect the botanical composition of monofloral, polyfloral and honeydew honey*. https://doi.org/10.1016/j.foodcont.2017.11.033.

Wang, W., & Liu, Y. (2019). Electronic tongue for food sensory evaluation. In *Evaluation Technologies for Food Quality* (pp. 23–36). https://doi.org/10.1016/B978-0-12-814217-2.00003-2

Wei, Z., & Wang, J. (2014). Tracing floral and geographical origins of honeys by potentiometric and voltammetric electronic tongue. *Computers and Electronics in Agriculture, 108*, 112–122. https://doi.org/10.1016/j.compag.2014.07.014

Zakaria, A., Shakaff, A. Y. M., Masnan, M. J., et al. (2011). A biomimetic sensor for the classification of honeys of different floral origin and the detection of adulteration. *Sensors, 11*(8), 7799–7822. https://doi.org/10.3390/s110807799

Zhou, X., Taylor, M. P., Salouros, H., et al. (2018). Authenticity and geographic origin of global honeys determined using carbon isotope ratios and trace elements. Scientific Reports, 8, 14639.https://doi.org/10.1038/s41598-018-32764-w

Zuluaga-Domínguez, C. M., Nieto-Veloza, A., & Quicazán-de-Cuenca, M. (2018). Classification of Colombian honeys by electronic nose and physical-chemical parameters, using neural networks and genetic algorithms. *Journal of Apicultural Research, 57*(1), 145–152. https://doi.org/10.1080/00218839.2017.1339521

Chapter 14
Global Perspectives and Challenges: Global Perspectives on Honey Standard, Market Challenges, and Opportunities in the Honey Industry

Iram Liaqat, Nazish Mazhar Ali, Saiqa Andleeb, Sajida Naseem, Sikander Ali, and Muhammad Nauman Aftab

1 Introduction

1.1 Honey

Honey is the only sweet substance that can be stored and used according to its natural occurrence (Edo et al., 2023; Kumar et al., 2024a, b). There is no need for any refining or processing before enjoying honey. Honey is essentially a sugar solution (either regular or supersaturated) that is produced by bees and some other insects (Bhattarai, 2021). Bees collect nectar from flowers and convert it into honey by adding enzymes. Then they store it as food for future use (Dalio, 2021; Paray et al. 2021; Kavimugilan et al., 2024).

The European Union Council Directive 2001/110/EC, and the *Codex Alimentarius*, define Honey is a sweet product obtained by bees from the nectar of plants, the secretions of living plants or the pus of insects that feed on plants, collected by honeysuckle bees, with its specific characteristics, its modification, its storage, drying and preserve it. It ripens in the honeycomb (Otero & Bernolo, 2020;

I. Liaqat (✉) · N. M. Ali
Microbiology Lab, Department of Zoology, Government College University, Lahore, Pakistan
e-mail: dr.iramliaqat@gcu.edu.pk

S. Andleeb
Department of Zoology, University of Azad Jammu and Kashmir, Muzzafarabad, Pakistan

S. Naseem
Department of Zoology, University of Education, Division of Science and Technology, Lahore, Pakistan

S. Ali · M. N. Aftab
Ikram-ul –Haq Institute of Industrial Biotechnology, Government College University, Lahore, Pakistan

M. Zafar et al. (eds.), *Pure Honey: Assurance & Authentication*,
https://doi.org/10.1007/978-3-031-98913-1_14

Zerrouk & Bahloul, 2023; Brar & Nanda, 2024). Honey is praised for its medicinal properties, and it is produced by bees in their hives through the collecting, processing, and storing of nectar. Honey is a sweet substance that bees gather from fresh water sources near their habitats, transform, and enhance (Anumba et al., 2020; Saklani & Kumar, 2021; Mustar & Ibrahim, 2022).

Bees produce a natural, sweet, and flavorful liquid known as honey by sucking nectar from flowers (Sharma et al., 2023a, b; Dar et al., 2024). Honey is one of the prevailing natural sweet liquid. From a chemical point of view, it can be described as a food that contains sugar and water, minerals, vitamins, amino acids, organic acids, flavonoids, and other phenolic compounds, as well as pleasant substances. Composition of honey particularly varies, and it depends on the floral and geographical origin of the honey (Ayoub et al., 2023; Bobiş et al., 2024; Mahesar et al., 2024).

Honey is primarily produced through the bees (social insects) whose life cycle is a constant cycle, although some other insects also contribute to honey production (Johnson, 2023; Walton et al., 2024). Honeybees are mostly divided into different groups, including all honeybees (*Apis* species), stingless bees (*Melipona* and *Trigona* species) found in South America, and various species of bees, especially *Melophorus inflatus*, which are found in Australia (Ayers & Rehan, 2021; Leclercq et al., 2023). In addition, there are other social wasps and bumblebees (bumble bees) present, but they only produce a small amount of honey (Figueroa et al., 2023; Babin et al., 2024).

1.2 Classification of Honey

Honey has the unique property of its color depending on the type of flowers it comes from. Naturally, a blend of honey is available based on consistent foundations, such as various specific flowers that are harvested from different regions, including alfalfa, clover, and mixed flowers (Crosson et al., 2023; Zhang et al., 2023a, b, c). The color of honey is generally estimated in the honey industry through a honey color grading scale known as Pfund Honey Color Grader. Honey is primarily divided into two categories (Fig. 14.1):

(i) Nectar honey
(ii) Honeydew honey (Antunes et al., 2024; Raweh et al., 2023; Kibar, 2024)

1.2.1 Nectar Honey

Honey is naturally produced from the nectar of flowers gathered by bees, who then transform it by mixing it with specific enzymes, drying it, storing it, and leaving it in honeycombs. This process is described as a sweet compound in the directive of

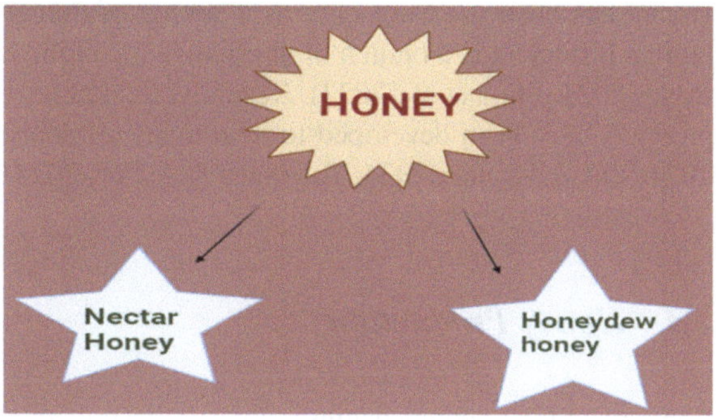

Fig. 14.1 Different types of honey

the European Commission Council, prepared through the *Apis mellifera* (Dar et al., 2024; Gernt et al., 2024; Kessi et al., 2024).

1.2.2 Honeydew Honey

The European Commission has stated that honey is a source of nutrition, obtained from the secretion of plants or the excretions of insects that suck on plants. These insects cut leaves or other parts of plants and feed on plant sap, excreting excess amounts in the form of droplets, which later accumulate as honey through the bees' process (Zaheen et al., 2020; Oliver, 2021).

The difference between nectar honey and honeydew honey (Fig. 14.1) can be determined through chemical analysis, although using a pollen analysis alone may not yield a clear distinction (Recklies et al., 2021; Machado et al., 2022).

Honey contains high electrical conductivity, pH, ash, acidity, and deep color. Additionally, it has high diastase and trypticase activity, along with low monosaccharide concentration. Furthermore, honey contains yeast and spore cells, but their presence is unrelated to its origin (Debela, 2020; Ondogo, 2023; Alu'datt et al., 2024).

1.3 Importance of Honey

Honey has always been an important food for Homo sapiens, as well as for bees. In fact, the relationship between bees and humans began in ancient times (Hart-Davies, 2024; Tian et al., 2024). To obtain sweet honey, humans were willing to risk their lives. The first written records of honey date back to Sumerian tablets dating back to 2100–2000 BC, which mention its medicinal and cosmetic uses (Afroz et al., 2023; Hurley, 2024). Honey has both nutritional and medicinal uses in most ancient cultures. According to the Bible, King Solomon said, Eat honey, my son, for it is good (Kumar, 2024; Manoj et al., 2024). Honey is a food, a medicine, and a remedy which has been used in human history for a long time. Since sugar production began only after 1800, honey was the only widely available sweetener on a large scale.

Honey has been used not only as a source of nutrition but also as a medicine in human history. It is a branch of alternative medicine called apitherapy (de Andrade Royo, 2023, Pluta et al., 2023). Recently, treatments based on honey and other bee products have been developed to treat many ailments (Boukraâ, 2023; Sahri et al., 2024; Rák & Csutak, 2024; Yohannes & Asfaw, 2024).

1.4 Honey Production

The production of honey begins with bees that collect nectar from flowers (Fig. 14.2), which is then transformed into honey through the use of enzymes. However, only nectar is used to make honey (Cengiz & Erdogan, 2023; Bareke & Addi, 2024). Mostly, nectar is the mixture of sugar and water. Quantity of sugar varies, but generally it is between 25% and 70%. Honey is produced by bees from the nectar of flowers and is considered a reward from plants (Zhang et al., 2023a, b, c; Villagómez et al., 2024). Nectar is collected by bees from flowers and its processing is done in their bodies and ultimately delivers it to the hive through their digestive system (Ma et al., 2023; Wei et al., 2023). Bees transported back the pollens to the hive and nectar is transferred to the stomach (Halvacı et al., 2023a, b; Londei & Marzi, 2024). Honey is stored in wax cells at the tip of the comb, and excess water is converted into steam vapors until the honey is approximately 83% sugar and 17% water. This process takes several days (Shahbaz et al., 2024). Honeybees cover the cells with a waxy layer which is removed later when bees have to consume honey. Bees use their wings to fan the hive and convert nectar into honey through evaporation (Malovrh et al., 2024; Rodríguez-Vásquez et al., 2024). Bees produce an enzyme (invertase) that converts excess water into vapors, breaking down every sucrose molecule into glucose and fructose, and sucrose is converted into smaller sugars. Therefore, beeswax concentrate of honey for the proliferation of yeast and other bacteria. To extract honey, the top is removed using a sharp knife and then warmed in hot water.

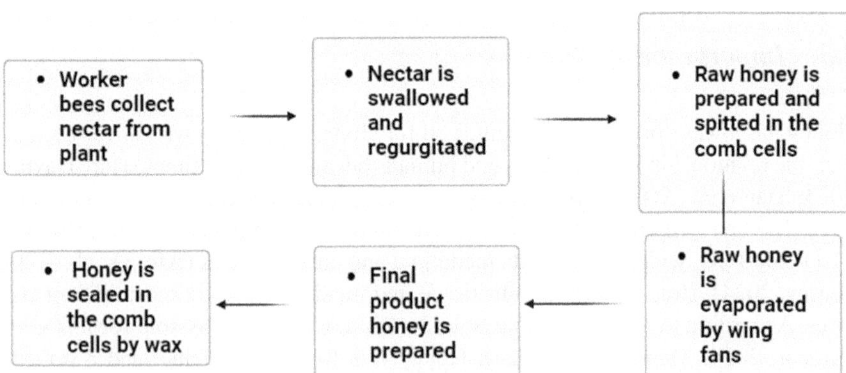

Fig. 14.2 Honey production by bee]

Afterward, the honeycomb is placed in a centrifuge to remove the honey. Sometimes, pressure is also used to extract honey, which can potentially damage the comb and reduce the amount of wax in the honey (Singh, 2022; Sharma & Saxena, 2024; Thaliath, 2024).

1.5 Microbes in Honey

Knowledge of the production of honey and the process of collecting honey is necessary to understand the various methods, including how bacteria and other germs can enter honey (Majtan et al., 2021; Tsadila et al., 2023). The bacteria present in honey are those that can withstand the concentrated sugar, acidity, and other antimicrobial properties of honey. The digestive pathways of honeybees, dirt, dust, air, and flowers are possible sources of microbial contamination (Catania & Vallone, 2020; Sujanto et al., 2021; Tragnone et al., 2023). Primarily produced bacteria and yeast found in the stomachs of bees are from raw material (nectar) or external sources. The larvae can initially be free of germs, but they are exposed to nectar and pollen by worker bees, which is why they are vaccinated through nectar, pollen, and the plants of the workers (Wirta et al., 2023; Luca et al., 2024; Scarfone et al., 2024).

Bacillus, *Micrococcus*, and *Saccharomyces* varieties can be easily distinguished from honey combs and adult bees. Several microbial species have been isolated from the larval feces of bees. *Bacillus* spp. are the most common mold, actinomycetes, gram-negative rods (possibly anaerobic bacteria), and yeast have also been isolated separately, while *streptomyces* spp. have been detected from larval feces of honeybee (Tsadila et al., 2023). Comparison of this microbial load is done with the microflora of adult bees' intestines, which often includes a diverse mix of gram-variable pleomorphic bacteria, *Bacillus* species, *Enterobacteria*, Penicillium species, Aspergillus species, and sometimes predominance of *Torulopsis* species (Bora et al., 2023; Nowak & Nowak, 2023).

The actual source of bacteria in the intestines of bees could be pollen. It has been suggested that compared to flowers and soil, hives are more significant source for microbes. The most prevalent bacteria found on the external surface, crop, and intestines of honeybees were those that produce aerobic spores (Haag et al., 2023; Ebeling et al., 2023; Rodríguez et al., 2023; Aguado-López et al., 2024).

In the intestines of honeybees, 1% yeast and 27% gram-positive bacteria, including *Bacillus*, *Bacteroides*, *Streptococcus*, and *Clostridium* species, are found. Additionally, 70% gram-negative or gram-variable bacteria, including *Acromobacter*, *Citrobacter*, *Enterobacter*, *Erwinia*, *Escherichia coli*, *Flavobacterium*, *Klebsiella*, *Proteus*, and *Pseudomonas*, are present (Faustino & Pinheiro, 2021; Alam et al., 2023; Bakour et al., 2024). The fundamental sources of honey production are flowers. Other sources of microbial impurity in honey include human handling, equipment, containers, air, dust, etc. Yeast in honey can be introduced through contaminated equipment in beekeepers' homes. Possible routes of transmission in extracted honey include household contamination or during honey processing, involving food handlers (due to skin infections, sneezing, or contamination from wounds). The bacteria, yeast, and molds found in honey have been identified (Szulc, 2020; Díaz et al., 2023; Ayoub et al., 2023; Painkra, 2023; Sharma et al., 2023a, b).

Most bacteria and other germs cannot grow or increase in honey, meaning they are inactive, and the reason for this is the antibacterial activity of honey. Different bacteria were inoculated in honey at 200 °C (Khataybeh et al., 2023; Salaria et al., 2024; Tiusanen et al., 2024). It is evident from the results that within 8–24 days there has been a decrease in the effectiveness of bacteria. These are only the bacteria that can survive in low temperatures in honey. Even after 4 months, the number of spores remained the same. *Bacillus cereus*, *Clostridium botulinium*, and *Clostridium perfringes* spores were inoculated in honey and stored at 25 °C. The population of *Clostridium botulinum* in honey at 4 °C did not change in 1 year. However, no spores found after storage of 5 days at 65 °C (Fig. 14.3) (Scepankova et al., 2022; Garrido et al., 2024).

This analysis shows that honey diluted with water helps to inhibit the growth of nonpathogenic bacteria and kill dangerous bacteria (Fig. 14.3). Honey medicine containing less than 50% honey can support bacteria for a long time, but not for more than 40 days (Arredondo et al., 2023; Daisley et al., 2023; Alves et al., 2024). Therefore, it can be concluded that honey has a low potential to be a vector of typhoid, hepatitis, and diarrhea (Chekecha, 2023; Shivani & Sathiavelu, 2023; Hasan et al., 2024; Luca et al., 2024).

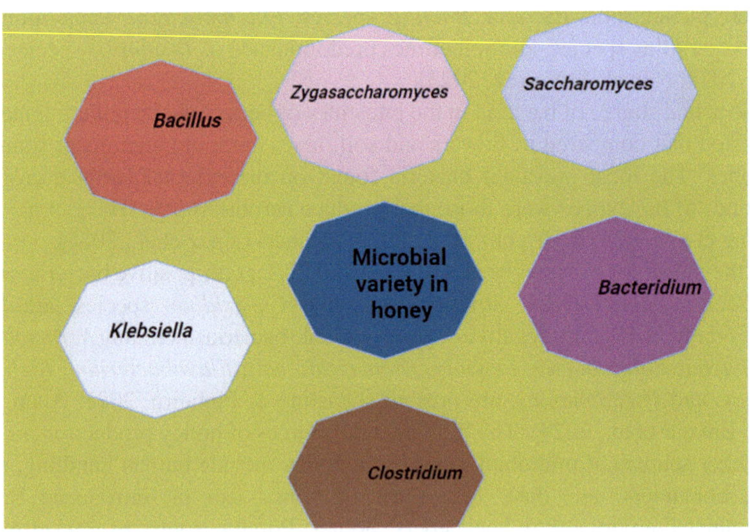

Fig. 14.3 Microbial variety in honey

1.6 Composition and Quality of Honey

During this process, the yield and quality of honey depend on many environmental factors, such as the air and humidity in the hive, the botanical history, the condition of the nectar, and the processing of the honey during extraction and clearance. The honey composition varies among different bee species (Tafi et al., 2024). More than 180 substances are present in honey (Table 14.1) and are important part of traditional medicine (Catania & Vallone, 2020; Chirsanova et al., 2021; Landaverde et al., 2023).

1.6.1 Carbohydrates

Carbohydrates are important components, which make up approximately 95% of the dry weight of honey. Honey does not contain just one type of sugar, but rather it consists of three types of sugars (Raweh et al., 2023; Akanda et al., 2024). It is primarily composed of fruit sugar (fructose), which makes up the highest percentage (41%), followed by grape sugar (glucose) at approximately 34%, and common sugar (sucrose) which falls between 1% and 2% (Khoo et al., 2023; Nhlapho, 2023).

A special type of enzyme found in flowers, which is dependent on the ratio of sucrose in nectar and is also somewhat dependent on invertase, breaks down sucrose consistently in grapes and fruits. This enzyme is present in flowers, allowing bees to collect nectar, but it is also found within the bee's own body (Bogo et al., 2021; Minami et al., 2021; Malovrh et al., 2024).

1.6.2 Amino Acids and Proteins

Honey contains essential components from pollen and nectar, including proteins. Proteins in honey can be in a complex structure or in the form of simple compounds, such as amino acids (Végh & Csóka, 2024). Amino acids and protein content in honey are relatively low, up to a maximum of 0.7% (Hossain et al., 2023). Almost all essential amino acids for the body are present in honey. Proline is a fundamental amino acid used to measure the maturity of honey. For regular honey, the proline content should be more than 200 mg per kilogram. If it is less than 180 mg per kilogram, it indicates adulteration with added sugar (Bryś & Strachecka, 2024).

Table 14.1 Composition of honey (Cappellari et al., 2024)

Components	Concentration (%)
Water	17.1
Sugar (monosaccharides, disaccharides, trisaccharides)	82.4
Proteins	0.25
Amino acids	0.5

1.6.3 Aroma Compounds and Phenolics

Honey comes in various flavors and colors, dependent on its botanical origin. Responsible for the fragrance of honey are compounds found in the nectar. Through separate extraction, it has been discovered that most of the compounds responsible for fragrance are likely produced by plants, but some are contributed by bees through their secretions (Svyatnenko et al., 2023; Kumar et al., 2024a, b). Extensive research has been conducted in recent decades on the compounds responsible for fragrance, and various types of honey have been identified with more than 500 different volatile compounds (Mahmoud et al., 2024). In fact, most of the compounds that create fragrance in honey are dependent on its botanical origin. The taste of honey is an important criterion in the food industry and also a factor in consumer preference. So far, approximately 600 compounds have been identified in different types of honey (Nakib et al., 2024).

Phenolic acids and polyphenols are secondary metabolites obtained from plants. These compounds have been used as chemotaxonomic markers in plant systems. They have also been suggested as possible indicators for determining the botanical origin of honey. It is reported that dark-colored honey contains higher levels of phenolic acids, whereas light-colored honey has lower levels of flavonoids (Bhatla & Lal, 2023; Tananaki et al., 2024).

1.6.4 HydroxyMethylFurfuraldehyde (HMF)

HMF is a six-membered heterocyclic compound containing carbonyl and alcohol (hydroxymethyl) functional groups. Hydroxymethylfurfural is a solid, yellow substance with a low melting point, but surprisingly water soluble. HMF is produced by the breakdown of fructose, which is formed slowly and naturally in the honey storage and rapidly when the honey is heated. The HMF content in honey is used to measure calories: honey is unhealthy if the HMF content is high. Some countries have limited the maximum moisture content of honey (sometimes 40 mg per kilogram), and honey with a higher moisture content cannot be used. (Kashbor, 2023; Živkov Baloš et al. 2023; Zembrzuska et al. 2024).

1.6.5 Minerals and Trace Elements

Honey contains various minerals (Table 14.2). In addition to the essential element potassium, honey also contains many other elements. Research has shown that the trace element composition of honey is primarily dependent on the botanical origin of the honey. It contains approximately 368% minerals (Mărgăoan et al., 2021; Tahboub et al., 2022; Valverde et al., 2022; dos Santos Silva et al., 2024). The majority of minerals found in honey include potassium, chlorine, sulfur, calcium, sodium, phosphorus, magnesium, silicon, iron, manganese, and copper. On average,

Table 14.2 Different minerals and trace elements in honey

Element	Unit	Average amount in 100 g honey
Calcium	Mg	4–30
Copper	Mg	0.01–0.1
Iron	Mg	1–3.4
Sodium	Mg	0.6–40
Chlorine	Mg	2–20
Barium	Mg	0.01–0.08
Cadmium	Mg	0–0.001
Cobalt	Mg	0.1–0.35
Iodide	Mg	1–100
Aluminum	Mg	0.01–2.4
Fluoride	Mg	0.4–1.34
Boron	Mg	0.05–0.3
Zinc	Mg	0.2–0.5

dark honey varieties tend to be richer in minerals compared to lighter varieties (Bodor et al., 2024; Mahesar et al., 2024).

1.6.6 Acidity and pH

Honey also contains traces of acid. Before, it was believed that bees inject honeybee venom into honey through their stingers and preserve it in this way. It was believed that honey contains formic acid, one of the important components of bee venom. Some people even advised against using honey because of this reason. It has been observed through research that they are completely different acids, which are naturally present in honey, mostly apple and lemon acids. Honey is a substance that does not change due to the addition of small amounts of pH acid and base. Its properties are due to the presence of phosphates, carbonates, and other mineral salts (Çelik & Aşgun, 2020; Khalil et al., 2021; Gajski et al., 2024).

1.6.7 Water

Honey contains the first amount of water, which determines its ability to resist spoilage from staying fresh and fermentation. The water content in raw honey can be less than 14%, and the less water it contains, the higher its value. High-quality honey should be processed with less than 20% water content; it is internationally recognized. A low water content is desired because if honey has more than 20% water, it can start fermenting and compromise its freshness. Unpasteurized honey is considered raw because it contains wild yeast. However, due to the higher concentration of natural sugars in honey, it is less likely to ferment when compared to other foods with low water content (Hameed et al., 2024; Rabbani et al., 2024).

1.7 *Different Physiological Effects and Uses of Honey*

Honey is widely and frequently used worldwide for a variety of uses, including food, ritual and magic, human and animal medicine (Fig. 14.4). It is a valuable nutrient used in hundreds of dishes, especially food products, due to its sweetness, color, taste, caramelization, viscosity, and emulsification (Kumar et al., 2024a, b; Laskaris, 2024; Patti et al., 2024).

1.7.1 Food Source

1.7.1.1 Bee Forage

Bees produce honey to make regionally specific food. When there are no flowers or the weather is bad, honeybees rely on their honey reserves to survive (El-Seedi et al., 2020; Palma-Morales et al., 2023; Akanda et al., 2024). For example, during the cold seasons from October to March in temperate countries, very few plants are flowering, and honeybee colonies need honey stores to stay alive during this scarcity of flowers (Aqueel et al., 2023; Gebremedhn et al., 2024). In tropical countries, honeybees need to rely on their honey reserves during times when there are no flowers, during droughts, or when the weather is unfavorable (Wakgari & Yigezu, 2021; Mishra et al., 2023).

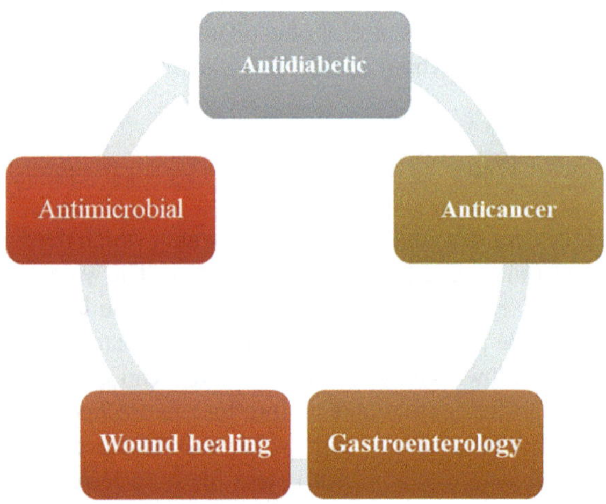

Fig. 14.4 Beneficial effects of honey

1.7.1.2 Food for Humans

Honey is a beneficial source of carbohydrates and generally contains a rich variety of essential nutrients (minerals, proteins, vitamins, etc.) that contribute to dietary diversity in human nutrition (Wali et al., 2020; Cucu et al., 2021).

1.7.2 Medicinal Tonic

1.7.2.1 Antimicrobial, Antiviral, and Antiparasitic Activity

Honey has proven to be effective against microbial activity. It inhibits the growth of microorganisms and prevents the development of fungi (Table 14.3). Honey also has a broad spectrum of activity against various types of bacteria (Ullah et al., 2023; Naskar et al., 2024). Its alcohol content demonstrates an inhibitory effect on both aerobic and anaerobic bacteria, including gram-positive and gram-negative ones. Additionally, honey has strong antibacterial effects against harmful and nonpathogenic bacteria (yeast and mold) to the extent that it has even shown resistance against many antibiotics. The antimicrobial effects of honey can vary depending on the concentration used (Asoutis Didaras et al., 2024; Di Bella et al., 2024).

1.7.2.2 The Antioxidant Property of Honey

The oxidative stress applies to a variance between the free radicals and antioxidant protective activities in a specific organism. Honey has been used for both medicinal and household purposes for a long time, but recently its antioxidant properties have gained public attention. With the increasing demand for antioxidant supplies in food, honey is becoming popular as a source of antioxidants. As a result of oxidative

Table 14.3 Bacterial species sensitive to honey (Deng et al., 2022; Chhun et al., 2024)

Bacterial species	Diseases
Escherichia coli	Enteric infection
Klebsiella pneumonia	Pneumonia
Proteus species	Nosocomial infections, respiratory tract infections
Salmonella species	Salmonellosis
Shigella species	Shigellosis
Pseudomonas aeruginosa	Pneumonia, bloodstream infections
Staphylococcus aureus	Skin infections, osteomyelitis
Streptococcus	Strep throat, pneumonia
Salmonella typhi	Typhoid fever
Bacillus anthracis	Anthrax
Mycobacterium	Tuberculosis
Vibrio cholerae	Cholera

stress, cellular damage and genetic alterations occur, and honey contains important antioxidants such as phenols (Muscolo et al., 2024; Uniyal & Kumar, 2024).

1.7.2.3 Wound-Healing Property of Honey

Honey is one of the natural antidotes that has been used for wound treatment. It actively promotes speedy wound healing through its regenerative properties and antiseptic effects, without causing any significant scarring (Damto et al., 2024; Maina et al., 2024). Honey has been proven safe for use. Its external application on the dressing, as well as on the mucous membranes of the body, does not show any signs of allergic reactions or adverse effects (Yilmaz & Aygin, 2020; Almasaudi, 2021; Angioi et al., 2021).

1.7.2.4 Other Uses

Honey is one of the natural remedies that has been used for wound healing. It promotes rapid wound healing through its regenerative properties and antiseptic effect. It has been proven safe for use without causing any adverse reactions or negative effects, even when applied externally with dressings on mucous membranes or sensitive skin (Yupanqui et al., 2022; Khairan et al., 2024; Lata et al. 2024).

1.7.3 Nutritional and Health Effects

1.7.3.1 Oral Health

Is honey harmful to teeth or not? There is much debate on this. Some reports have shown that honey has very little cariogenic effect compared to sucrose. Due to its antibacterial activity, honey inhibits the growth of bacteria that produce cavities, which can have a carioprotective effect. It has been demonstrated that manuka honey, a very powerful antibacterial honey, has a positive effect on the remineralization of tooth enamel and against gingivitis and can be used in candy preparation instead of refined sugar (Hayei et al., 2024; Heilmann & Watt, 2024; Flores & López, 2024).

1.7.3.2 Gastroenterology

According to the sacred book of Muslims, Hadith, in the eighth century AD, the Prophet Muhammad (peace be upon him) recommended honey as a remedy against constipation. Additionally, the Roman physician Celsius (approximately 25 AD) and different publications from Eastern Europe and Arab countries have also reported the use of honey for the treatment of constipation and gastrointestinal issues (Shlibak & Dalla, 2020; Iqbal, 2022; Boukraâ, 2023). Honey acts as a strong barrier against agents that cause peptic ulcers, gastritis, and Helicobacter pylori. The use of Manuka honey reduces stomach acidity by 56% (Ali et al., 2020; Nweze et al., 2020).

1.7.3.3 Infants

During the past centuries, the use of honey in children's nutrition was commonly recommended, and there are some interesting observations. Children who consumed honey with their meals had better blood formation and greater weight gain compared to those who did not have honey in their diet. Honey was well tolerated by children in comparison to sucrose. The positive effects of honey on digestion are attributed to its inclusion in children's diet (Nikhat & Fazil, 2022; Halvacı et al., 2023a, b; Pehlivan, 2023).

1.8 Adulteration in Honey

Adulteration deteriorates the quality standards, and the safety of products is also questionable. The chemicals used for adulteration reduce the effectiveness of medicines and nutritional value, causing harm to humans. Approximately 128 chemicals have been reported to be used for adulteration. Starch syrup, inverted syrup, starch, or inverted syrup is commonly mixed with honey from bees and sometimes with low-quality honey in the market (Rust, 2020; Thangaraju et al., 2021; Aslam et al., 2024). If a substance is directly added to honey, it is called direct adulteration. When chemical and industrial sugar is fed to honeybees, it is called indirect adulteration (Damto et al., 2024; Ahmed, 2024).

2 Beekeeping

It is said about Albert Einstein that "if bees disappeared from the surface of the earth, humans would not survive for more than four years." This statement highlights the crucial role of bees in environmental conservation and maintaining human life. Bees are an important global resource, providing honey and other products, and also playing a vital role in pollinating crops. However, only a few people understand the importance of bees beyond just producing honey (Brar et al., 2023; Dodd et al., 2025). In fact, bees play a crucial role in indicating and maintaining a balanced ecological system, serving as natural pollinators, and consequently contributing to food supply and preservation for human beings. In addition to honey, bee products such as propolis, bee venom, and royal jelly offer various applications, including natural sweetness and medicinal benefits (Samoraj et al., 2018; Joshi, 2023).

It is believed that the tradition of beekeeping for honey has been practiced for a long time. Beekeepers would directly harvest honey from rural honeybees. Essentially, honey hunters collect honey from wild bees, specifically from the hives of *Apis dorsata,* which reside in tall trees deep within the forest. Honey is called as "Tualang honey" in this way, although bees collect nectar from various plants in the surrounding area. They harvest honey in the middle of the night using traditional tools including a bamboo ladder, a cow bone knife, and a container made from cowhide, and they perform a special ceremony. This activity takes place once a year, usually between January and April (Ibrahim et al., 2021; Jinsong, 2023; Ngussie & Hailu, 2023).

Beekeeping, also known as apiculture, is an interesting and exemplary agricultural activity. It provides additional income and sometimes substantial profits to people in rural areas. Honey is a miraculous food of nature in terms of nutritional composition; honey is close to fruits compared to sugar or other sweeteners. Bees are not only important for honey and other products, but they also play a crucial role as pollinators for agricultural and horticultural crops (de Salcedo, 2023; Iftikhar et al., 2023; Perichon et al., 2024).

Before adopting beekeeping as a profession, it is essential to understand the fundamental aspects of beekeeping, such as the various types of bees, their life cycles, methods of breeding and management, enemies and diseases of bees, and factors affecting honey production and quality (Van Espen et al., 2023; Delena & Kayamo, 2024). The purpose of agricultural development is to assist people in better managing their livelihoods through the sustainable use of available resources. It provides them with greater social and economic power and creates opportunities for them to work according to their abilities, without disrupting the environmental services provided to them (Harianja et al. ,2023; Redclift, 2023).

2.1 Modern Beekeeping

The cultivation of modern honeybees in Malaysia began in the 1970s when a beekeeper imported commercial honeybees (*Apis mellifera*) from Taiwan. In 1981, the Malaysian Bee Research and Development Team (MBRDT), funded by the International Development Research Centre (IDRC), was established to lead research and development efforts in beekeeping. Under the leadership of Dr. Mukhtaruddin, the team aimed to promote organized honeybee farming in Malaysia. Honeybees raised for commercial honey production include domesticated honeybees, just like *A. mellifera, A. cerana*, and a number of stingless bee of species such as *Geniotrigona thoracica*, *Heterotrigona itama*, *Lepidotrigona terminata,* and *Tetragonula leaviceps* (Abrol, 2023; Srinivasan & Haran, 2023).

2.2 History of Beekeeping

Beekeeping and the collection of honey have been practiced in various societies since ancient times. Honey hunting is an ancient practice. The history of beekeeping dates back to the capture of bees and related methods. As agriculture expanded, the practice of keeping bees in hives also became common (Fajari et al., 2023; Asma et al., 2024). However, instead of taking wild bees from their natural habitats, the focus shifted toward cultivating them. Even today, collecting wild honey remains a common practice in many parts of the world. In India, it is estimated that beekeepers annually collect 22,000 tons of wild honey. Honey collectors produce twice as much as beekeepers (Hailu et al., 2023; Mouillard-Lample et al., 2023; Thyaka, 2023).

Various societies in Asia, Europe, and North America have developed their own methods for beekeeping, including genetic research on honeybees, hive design, management operations, organized pollination, and investment in diverse fields such as honey processing and marketing (Van Espen et al., 2023; Sless et al., 2024). *Apis mellifera* is a particularly fortunate species of bee found in honey. Scientists and researchers have studied it for over 150 years. Continuous selection and breeding have improved the genetics of honeybees. The design of the hive, which provides a better space for honeybee movement (essential for honeybee reproduction and movement), has helped beekeepers and commercial honey production organizations produce more honey sustainably (Johnson, 2023; Owen et al., 2023).

2.3 Meliponiculture

Honeybees, along with their colorless honey, are social insects. Stingless honeybees are famous for their honey and pollen production, but nowadays their role in providing environmental services is also recognized. Beekeeping is known as the practice of raising bees to produce honey (apiculture). It is commonly carried out by traditional communities through local and traditional knowledge. Honey and beeswax (which is not pure beeswax from the glands of stingless honeybees, but a compound of wax/resin from plants) were traditional products. Occasionally, beeswax also serves as an important source of income for beekeepers, along with renting colonies for pollination services (Hashifah & Haryanto, 2023; de Cordova, 2024).

2.4 Value of Beekeeping

Various benefits of raising bees are honey, which are related to health (both in terms of medicinal products and dietary consumption), economy (directly through the sale of honey and other bee products, as well as indirectly through increased production of pollinated crops and beekeeping business activities), employment (in honey production and pollination services), and environment (by ensuring the conservation of wild pollinator populations). Beekeeping can play a role in transforming weaknesses into stability for sustainable livelihoods (Gondar, 2023; Honeycutt, 2023).

Apis mellifera beekeeping has become an integral part of various initiatives and programs, from the northern to the southern regions. These efforts facilitate the transfer of technology and ensure consistent global honey production. However, as part of this endeavor, *Apis mellifera* has also been introduced and promoted in areas outside its natural habitat (Büchler et al., 2024; Kükrer, 2024; Page & Williams, 2023). There are four indigenous species of bees in the Himalayan region, one of which (*Apis cerana*) can be kept in hives. These bees provide special benefits for farmers at higher altitudes where they are naturally found (Raina et al., 2024; Singh et al., 2024).

However, the benefits of the vital diversity of honeybees remained unrecognized for many years, and very little effort was made to conserve native honeybee species. Due to a lack of information and ignorance, some attempts in beekeeping, including the care of honeybees in rural development, occasionally proved detrimental, focusing attention on "northern" blueprints rather than local knowledge or indigenous expertise (Maderson, 2023; Yang et al., 2023). Miranpur is a traditional area for beekeeping *Apis dorsata* species. During the biannual honey harvesting seasons, local farmers used to collect more than five tons of honey each year (Khan, 2020).

The Department of Agriculture started beekeeping with *Apis mellifera* bees in the 1980s, and now more than 7000 colonies produce honey in the region during the autumn and spring seasons. This allows for the production of honey from the flowers of *Zizyphus jujuba, Zizyphus sativus, Adhatoda vasica, Acacia Arabica,* and *Robinia pseudoacacia* (Delaplane, 2023). Beekeepers who raise honeybees in this manner produce more honey than those local farmers who are not beekeepers but rather honey hunters. They have not welcomed this intervention. They believe that organized beekeeping has taken away their honey and is the reason why the population of *Apis dorsata* is decreasing (Landaverde et al., 2023; Mg et al., 2023).

Entomologists and biologists have different perspectives. They explain that the types of native bees migrate from north to south and toward the rear after the blooming of indigenous flowers and seasonal changes (Torezan-Silingardi et al., 2021). The use of changes in agricultural practices and insecticides in the landscape has consistently affected the migration patterns, resulting in a decrease in the population of migratory birds and environmental services for both agriculture and traditional landscapes (Brain & Anderson, 2020; Reif et al., 2024).

However, introducing domestic honeybees to familiarize them with wild honeybees can also reduce the availability of native nectar (Khalfan et al., 2024). Environmentalists believe the introduction of bees has had a negative impact on the species, leading to the growth of *Tropilaelaps clareae,* an insect that now continues to thrive in bee colonies, and poor methods of killing the bees. It seems that in this area, there is a decrease in the number of *Apis dorsata*, at least partially due to improper introductions of foreign bees alongside changes in agricultural practices (Sladonja et al., 2023; Narciso et al., 2024).

Natural environments can be maintained in a healthy state only through the interest and active participation of local people. Beekeeping is a good way to generate income for individuals without harming the environment. Additionally, bees and other pollinators play a crucial role in preserving endangered plants (Pluchinotta et al., 2024). Rearing bees for honey production can provide economic stability to weaker communities. There are several economic benefits to beekeeping, including honey production, pollination services, agriculture, and forests. Honeybee products like propolis, royal jelly, honey, and honeybee venom are high-value, small-volume green products. Besides direct income from honeybee products, beekeeping creates opportunities for agricultural livelihoods in various sectors, including carpentry for beehives, honey trading, renting out bee colonies, and microenterprises based on honeybees (Haroon et al., 2024; Kortsch et al., 2024; Sahu et al., 2024).

In the field of honey production, competition is intensifying, and several stakeholders, especially beekeepers, have started using the concept of value chain to gain access to renowned markets (Hando, 2023). Value chain can be described as the provision of specific inputs, from raw materials to transformation, marketing, and distribution, for a particular product, ultimately leading to the final product (Ingram et al., 2020; Harianja et al., 2023). Honey is an important commodity, and it is being investigated by buyers, traders, and marketers through other actors in the industry. A value chain systematically considers all stages of production. It analyzes the flow of relationships and information within the chain, highlighting strengths and weaknesses, including losses (Tigistu, 2020; Mclean, 2024).

Empowering families and communities to cultivate bees for honey production is an important condition for achieving sustainable and stable livelihoods. It can play a significant role in empowering the poor and also promotes the fair and equitable distribution of benefits within communities (Febrianto et al., 2023; Ilham et al., 2024). Beekeeping fraternities that raise honeybees understand the structure of the honeybee hive, where life is organized with meaning and discipline. Keeping honeybees in a community encourages people to work together for mutual benefits and promotes organized production and trade of their products (Johnson, 2024).

The Headlines of honeybee colony losses have given an impression of large-scale global decline of the bee population. As a result, the care of honeybees is now at risk, and the world is facing a major threat of hunger. However, stories are generally based on limited research reports from one or a few countries, in which observations are relatively scarce (Stubbs, 2020). Despite this, during the winter season, there have been severe losses to colonies in some parts of Europe and North America (Hristov et al., 2020). Bees are socially and economically important because they play a crucial role in pollination of crops and produce various types of goods that are essential for many industries and businesses (Tang et al., 2023; Rajagopalan et al., 2024).

Honeybee products, which hold social and economic importance, include honey, beeswax, propolis, pollen, royal jelly, and honeybee venom (Gupta et al., 2024). Living honeybees are also bought and sold in the form of queens or on a large scale. In some cases, honeybees are used as a source of protein in human and animal nutrition. Honey is used in some cultural traditions as medicine, for crop pollination, and as an important ingredient (Niklasson et al., 2024). Not only are honeybees important for honey production, but they are also used in over 300 other industrial processes. The production includes cosmetics, medicines, candles, electronic components, polish, and special industrial lubricants. Therefore, the possibility of the decline of honeybee populations or ongoing global extinction could have significant social and economic impacts (Kieliszek et al., 2023). Beekeeping has the great ability to decrease the poverty level by raising bees and preventing deforestation. Climate change negatively affects the productivity of bees. During tree cutting, changes in water pressure due to irrigation, limited availability of pollen and nectar, disruptions in bee movement, physical harm to colonies, and controlling beekeeping activities in colonies are all important measures (Yilmaz Kaya et al., 2023; Mahankuda & Tiwari, 2024).

In response to the changes in weather, beekeepers are relocating their beehives to abundant areas in pollen, providing food and water for the bees, altering the types of roofs, changing the location of ant hills, placing the beehives under the shade of trees, increasing the use of wooden roofs, and modifying cutting methods and timing to minimize the impact (Jašarević, 2024). Beekeepers must face various challenges and intervene to strengthen the capabilities of beekeepers in order to align local knowledge and practices with water and weather services, so that we can work in harmony (Burke & Corrigan, 2024).

3 Honey Industry: Global Challenges and Opportunities

Due to increasing awareness about the benefits of honey, a significant growth in the global honey market size is predicted in the coming years. The world that is conscious about health is currently focusing on alternatives, and honey is a hopeful natural product that is widely used worldwide. In addition, honey is believed to have anti-inflammatory, antioxidant, antibacterial, and antimicrobial properties. Honey is a rich source of nutritional components, which makes it an important ingredient in various applications such as energy drinks, alcoholic beverages like mead, bakery products, and bars (Benjamin, 2024; Gustafson et al., 2024; Rowe, 2024).

In 2018, the global honey market was worth 84 billion US dollars, and it is expected to reach approximately 103 billion US dollars by 2025. Globally honey market is categorized into different regions such as North America, Asia Pacific, Europe, and the Rest of the globe. The prediction has been made to maintain dominance in the field of honey production and beekeeping in the Asia-Pacific region due to an expected increase in honey production and the volume of trade. It is expected to reach approximately 11,628 tons by 2023 in terms of volume in the Asia-Pacific region.

There are some potential players in the honey market globally:

(i) Capilano Honey (Australia)
(ii) Dabur (India)

(iii) Comvita Ltd. (New Zealand)
(iv) Bee Maid Honey (Canada)
(v) Billy Bee Honey Products (Canada)
(vi) Lamex Food Group (USA)
(vii) Beeyond the Hive (USA) (Jahangir et al., 2020; Hirpara, 2023; Puranik et al., 2023)

In terms of trade, only 312% of global honey production entered international trade this year. While in 2013, Africa produced only 102% (169, 30,600 tons) of total natural honey, of which 155% entered international trade. In 2013, the estimated total honey production from member countries of COMESA was 81,454 tons, and the total beeswax production was 9811 tons. Only 274% of honey enters international markets from commercial markets, which is influenced by the dietary standards of partially developed countries (Kraisoraphong, 2023; Landaverde et al., 2023).

Honey production from beekeeping is one of the most important resources that can be utilized for economic development and livelihood in the Eastern and Southern African (COMESA) countries, informed by suitable agricultural and environmental conditions, as well as intensity possibilities (Ngunjiri et al., 2023).

However, for many countries, the production and trade of honeybee products pose significant challenges in terms of introducing modern honeybee production, limited technical expertise, laboratory facilities, operational capacity, inadequate planning and implementation in both private and government sectors, inadequate implementation of market support systems, and inadequate compliance with standards. In addition, deficiencies in tax services sectors, ineffective cross-border trade facilitation and the limited organizational capacity of operators on a small-scale pose significant challenges (Mutua & Felix, 2023; Yang et al., 2023).

The lack of national policies and strategic interventions by development agencies, as well as the absence of coordination among member countries, are significant reasons for the underdevelopment of the subsectors under honey. It also appears that industry-related issues affecting the industry are not addressed through various programs (Zhang et al., 2023a, b, c; Seyoum, 2024). Effective investment in agriculture and commerce is essential for controlling current developmental challenges and achieving desired results. It is necessary to establish trade relations with China for the supply of honeybee products, maintaining revenue, improving commercial efficiency, and enhancing the quality of goods. Establishing an effective and harmonized regulatory framework, as well as implementing standardized administrative measures, is essential through the collaboration of member states and all relevant stakeholders (Bazargani & Deemyad, 2024).

Beekeeping is an important agricultural activity in Southern and Eastern African countries (Amanuel & Alemayehu, 2024). There is a need for small-scale industry in it, and it can play an important role in household income and food security. At the regional level, COMSA has established a policy framework that creates commercial opportunities for this sector throughout the region, resulting in economic, social, and environmental benefits. Despite this, there is a lack of information about the subsector, which poses a challenge in accurately assessing the industry's true capabilities and in understanding the impact of government and private sector interventions (Maneli, 2024).

Due to the support of various programs in the subsectors, the industry of honey is relatively better in developed countries. The technique of beekeeping, along with technology, input financing, and market support systems, is included in the implementation of quality standards. For example, the European Union (EU) has established standards and conditions under which beekeepers are provided assistance in raising bees (Rahmad et al., 2024; Tambi, 2024).

Honey production, quality, income, and exports vary in different countries. In China, honey production is a significant source of income. Chinese honey production accounts for more than 20% of the world's total honey production. Various Chinese journals, including the Herbal Classics of Shen Nong and Materia Medica, document the medicinal properties of honey. Economically and ecologically, it produces important profits to society. Honey production has significant social impacts and vital biodiversity. Economically and environmentally, it generates important benefits for society (Harianja et al., 2023; Yue et al. ,2024).

The estimated global honey production was 172 million metric tons, with China being the largest contributor, followed by Turkey and Canada. The global market is highly competitive because it depends on environmental conditions, and even minor changes in climate can adversely affect production. The global demand for honey depends on the quality, packaging, and location of production. Honey is used in personal care and cosmetic industries as face wash, moisturizer, soap, lip balms, and other products (Trak & Chauhan, 2022; Puranik et al., 2023).

4 Presence of International, Regional, and National Organizations

There are many organizations that help and provide services to the honey sector in various ways. Some of the key stakeholders in beekeeping that can influence the future of bees include the African Union Bureau of Animal Resources for Africa (AU-IBAR), the International Center for Insect Physiology and Ecology (ICIPE), the Food and Agriculture Organization (FAO), the World Health Organization (OIE), the International Animal Research Institute (ILRI), regional trade associations (such as COMESA, IGAD, SADC, EAC), and the East and Central Africa Permaculture Research Association. There are also partnerships with agricultural research and development institutions in South Africa, regional farmers or private organizations, and national organizations such as the Institute of Animal Care, national research organizations, extension services, community organizations and schools (Wakgari & Yigezu, 2021; Landaverde et al., 2023).

4.1 *Presence of Supporting Institutions in COMESA*

The presence of Alliance for Commodity Trade (ACTESA) and African Trade Insurance (ATI) in East and South Africa is a significant credit agency for Africa. ACTESA provides insurance against political risks and trade credit risks with the aim of reducing business costs and risks in Africa. The African, Caribbean, and Pacific Group of States (ACP) supports bilateral trade flows between Africa and the world. It facilitates domestic and foreign investment, promoting grassroots commerce. The COMSEA Business Council (CBC) is a business membership organization that provides three core services: business assistance and connections, policy advocacy, and membership development. Its purpose is to eliminate barriers to business and competition in the region, influence policy agendas from promoting private sector participation in the perspective of local integration agendas. Private sectors and regional organizations can utilize these organizations to enhance cultural value and promote the honey industry in the region (Ngomirakiza, 2022).

5 Challenges Facing the Honey Industry in the Region

The honey industry in the region faces some key challenges and obstacles that need to be addressed if we want to fully utilize beekeeping abilities.

5.1 *Lack of an Enabling Regulatory and Policy Framework*

The honey industry faces a fundamental challenge at both regional and national levels in terms of advancing and prospering. To provide essential guidelines and directives for the growth and development of the industry, there is a need for a robust policy framework, practical wisdom, and regulatory system. The organization and implementation of industrial development have been poorly defined in terms of the roles of private and government sectors. Due to this difference, besides other factors, lack of attention, distribution of efforts, and reduced support have resulted in the deterioration of industrial efficiency.

5.2 *Get Financial Aid*

Access to financial services is a big challenge for the growth of the honey industry. Inadequate access to finance has been a major obstacle in establishing production and processing packaging plants for honey. Switching from traditional methods and tools to the use of modern equipment like box tops and others is a big challenge for small farmers, which can double their productivity.

5.3 Lack of Standards and Quality Control

The business does not have good management and control procedures, resulting in incomplete production, processing, packaging, and labeling. The lack of proper facilities for laboratory testing and quality control leads to a deficiency in standards and their management. It affects local and international markets and also reduces honey production and sales in local markets and wholesale markets.

Despite the lack of assistance in proper processing, packaging, and labeling, local honey production still does not achieve good market prices. This is because it is natural and free from human influence, making it a good product that appeals to a specific market. Unnecessary non-tariff barriers related to standards are impediments to public health and food safety requirements in international trade. While it is important to develop and maintain high standards for international business needs, it is equally important to establish regional standards based on current realities.

5.4 Limitation in Business Management

Mostly, beekeepers raise bees for honey production as a part-time economic activity, which involves minimal labor. Taking care of beekeeping as a business and increasing honey production is a big challenge for those who raise bees. Generally speaking, income from beekeeping is not reinvested in improving production and quality. On the other hand, processors, especially big marketing agents, need a specific amount and standard of guaranteed supply. Therefore, beekeepers cannot participate in competitive markets where they can obtain premium prices for their products.

Lack of cooperation among beekeepers leads to a relationship between small, medium, and large producers alongside systems. Limited information is available at international, regional, and national levels to support standards for beekeepers and traders to set up bee farms. The overall impact of the challenges mentioned above appears to be less productivity, in complete crops, limited market access, low income, and underutilization of bee honey for wealth generation.

5.5 *Constraints for Beekeeping in the Study Areas*

Rearing honeybees for honey production in research areas was a challenge (Kumar et al., 2020; Thakur et al., 2023; Tefera & Melaku, 2024). It required addressing issues like scarcity of bee forage, lack of insecticides and herbicides, and water shortage (Fig. 14.5).

5.5.1 Climate and Bee Productivity

5.5.1.1 Rainfall

Honey production is a sensitive process related to weather conditions. Excessive rain washes away nectar, an important part of honey, disrupting the bees' activity and ultimately leading to stored honey being used as food. Moderate rainfall, on the other hand, promotes higher yields by reducing plant stress and facilitating the availability of nectar and pollen. Directly proportional to rainfall, honey production also depends on the availability of nectar. In dry years, bee colonies suffer more compared to rainy seasons.

5.5.1.2 Temperature

The increase in temperature negatively affected the survival of honeybee colonies. Due to higher temperatures, water scarcity occurred, prompting worker bees to search for water. The fluctuating temperatures also impact the foraging behavior, especially during nectar and pollen collection. In extreme heat, honeybees tend to flee and relocate to more suitable environments. Moderate temperatures support

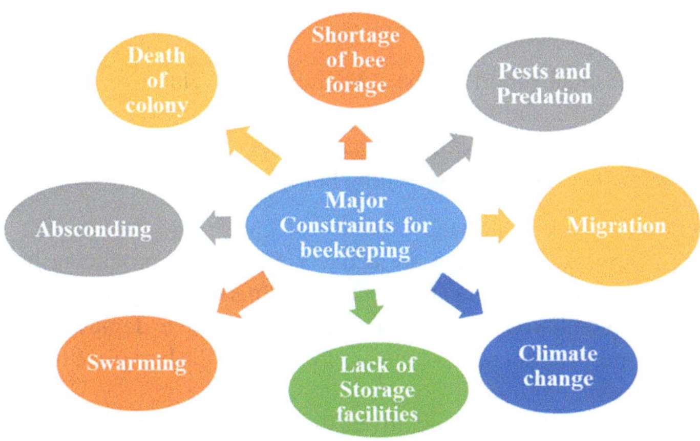

Fig. 14.5 Different major constraints for beekeepers

higher honey production and consistent temperament. To prevent disruption of nocturnal activities, temperatures below 10 °C are avoided.

5.5.1.3 Effect of Wind on Beekeeping

Fast and strong winds negatively affect both bees and colonies. Although wind is important for keeping the hive cool, they also disrupt the bees' communication and hinder their interactions. Honeybees use waggle dances and round dances to communicate about food, water, and pollen locations and distances. They also release pheromones during critical situations. Fast winds interfere with this communication process. They also cause physical damage to the hive and put the queen at risk, especially in swarms.

5.6 Adaptation to Climate Change and Its Challenges

Climate change is affecting beekeeping and is one of the reasons for the decline in honey. Beekeepers have plans to mitigate these effects, including relocating hives toward pollen sources, ensuring proper nutrition and water supply for bees, using different cover trees, increasing bee numbers, planting trees, painting bee cover trees in your colors, using larger cover trees, and adjusting the harvesting process and timing. Limited access to financial resources, loans, and markets, recurring droughts, weakened foundational structures, and restricted access to information about weather and climate have created significant obstacles for beekeepers in adapting to climate change and its impacts.

6 Ethiopian Honey Value Chain and Competition

According to the final report of the AMDE Honey Value Chain project held in Ethiopia in July 2012, the following four labels are considered central to understanding honey actors: (1) beekeepers, (2) sellers, (3) buyers, and (4) important players in the market for buying or selling honey.

Level 1: Producers (beekeepers)

Honeybee farmers in Ethiopia are actively engaged in producing high-quality honey, which is in high demand in the local market. They are benefiting relatively more (compared to supply) from the lucrative domestic demand for Ethiopian honey. Beekeepers are always looking for the best price for their honey.

Level 2: direct buyers of honey

Honey producers/traders, cooperatives, Taj House, and agricultural businesses/processors directly buy honey from beekeepers (for example, Biza Mar in

SNNPR buys honey from beekeepers). At this stage, there are many price cooperatives in the honey sector that compete with each other in terms of purchasing quantity, quality, and price. Some players, such as Biza Mar, are trying to integrate the honey industry by establishing their own beekeeping operations to meet their own needs. In addition, some partners with competitive prices at this level have worked to ensure the continuity of honey supply by creating business models and reducing the risks caused by price changes in household businesses. (Scarbrough, 2024).

Level 3: Agricultural business and honey wholesaler selling honey in local and export market in Addis Ababa (Mercato)
Multiple partners are also involved in China's honey value chain. Agri-business companies in Addis Ababa (such as Mercato) compete with these rural businesses in terms of quantity (reliable and timely supply), quality, and pricing for honey markets (Wakgari and Yigezu, 2021).

Level 4: Retail stores (supermarkets, retail stores) and retail shops (markets/processors) Additionally, some agricultural businesses/processors supplying honey to the domestic market also compete with level 3 wholesalers.

Honey value chain actors are as follows:

Producers
Producers are important actors who take actions from purchasing raw materials to marketing, adding more value to the chain. In the honey industry, important value-added activities include extracting, filtering, packaging, and transportation. In the honey-producing areas, producers sell their products in local markets and neighboring markets.

Collectors
These are the farmers and traders who collect honey from farms, gather it at the level obtained from producers, and then sell it to wholesale and retail markets in the district.

Wholesalers
They buy large quantities of honey from producers, and other traders, and then resell it.

Retailers
They play an important role in the honey industry. They connect wholesalers with urban customers by sourcing products according to the seller's needs and purchasing power (Wakgari & Yigezu, 2021).

Processors
These urban and rural areas are both food vendors who buy raw honey from wholesalers and locally provide consumers with honey in the form of a famous wine house called Tej.

Cooperatives

These are the prominent actors who directly participate in the marketing of honey and fully support the producers of honey. These actors systematically provide filtered honey from farming households and then arrange for its sale to consumers (Bariyah, 2020). They also provide education about its usage and supply them with standardized honey (Scarbrough, 2024).

Consumers

Consumers are the final actors in the chain who purchase products for their own needs (Bariyah, 2020; Schouten, 2020; Vercelli et al., 2021; Wakgari & Yigezu, 2021).

7 Major Opportunities for Honey Marketing in Ethiopia

Honey production and sale can be linked to water conservation through water management projects and unemployed youth can be trained to participate in honey production and sale, creating wonders of honey houses and fabric production, and increasing the number of tourists and guests celebrating the GENA festival.

There are opportunities for beekeeping in the field of study:

(i) Availability of local honeybees with different agricultural environments and suitable conditions.
(ii) Currently, the government is promoting self-made projects that organize cooperatives of local beekeepers who use locally available materials and raise honeybees at the grassroots level. These projects aim to connect them with local markets and produce affordable modern hives by linking them with local authorities.
(iii) Farmers in the area demonstrate readiness to improve beekeeping methods.
(iv) Although there is a shortage of food for bees in the dry season, there are several types of flowering plants in the rainy season that are used as food for bees (Amanuel & Alemayehu, 2024; Yue et al., 2024).

References

Abrol, D. P. (2023). Beekeeping for sustainable economic development of India: Challenges and opportunities. *Journal of the Indian Institute of Science, 103*(4), 997–1017.

Afroz, R., Tanvir, E. M., & Hossain, M. M. (2023). Physical properties of honey. In *Honey: Composition and health benefits* (pp. 12–31).

Aguado-López, D., Urbieta Magro, A., Higes, M., Rodríguez, J. M., & Martín-Hernández, R. (2024). Influence of age of infection on the gut microbiota in worker honey bees (Apis mellifera iberiensis) experimentally infected with Nosema ceranae. *Microorganisms, 12*(4), 635.

Ahmed, E. (2024). Detection of honey adulteration using machine learning. *PLOS Digital Health, 3*(6), e0000536.

Akanda, M. K. M., Mehjabin, S., & Parvez, G. M. (2024). Honey for nutrition and health benefits: An overview. In *Honey in food science and physiology* (pp. 33–56).

Alam, F., Maroof, K., Choon Ming, N., Khalil, M. I., & Gan, S. H. (2023). Other possible contaminants, toxic compounds, and microbial growth. In *Honey: Composition and health benefits* (pp. 167–185).

Ali, A., Sajood, S., Taban, Q., Mumtaz, P. T., Rather, M. A., Paray, B. A., & Ganie, S. A. (2020). Honey as component of diet: Importance and scope. In *Therapeutic applications of honey and its phytochemicals: Vol. 1* (pp. 215–236).

Almasaudi, S. (2021). The antibacterial activities of honey. *Saudi Journal of Biological Sciences, 28*(4), 2188–2196.

Alu'datt, M. H., Rababah, T., Alhamad, M. N., Gammoh, S., Dardakeh, M., Abujelban, D., Alrosan, M., Kubow, S., & Al Khateeb, W. (2024). Physiochemical properties, bioactive compounds and aroma profile of honey. In *Advanced techniques of honey analysis* (pp. 93–127). Academic Press.

Alves, V. F., Chaul, L. T., Bueno, G. C., Reinecke, I., Silva, T. C., Brito, P. V., & De Martinis, E. C. (2024). Associated bacterial microbiota of honey and related products from stingless bees as novel sources of bioactive compounds for biotechnological applications. *Current Opinion in Food Science, 55*, 101122.

Amanuel, E., & Alemayehu, Y. (2024). Beekeeping practices and challenges at Lemo district, Hadiya zone, Southern Ethiopia. *Journal of Apicultural Research, 63*(3), 462–474.

Angioi, R., Morrin, A., & White, B. (2021). The rediscovery of honey for skin repair: Recent advances in mechanisms for honey-mediated wound healing and scaffolded application techniques. *Applied Sciences, 11*(11), 5192.

Antunes, A. C. N., Gomes, V. V., Seraglio, S. K. T., Schulz, M., Silva, B., da Luz, C. F. P., de Moraes, A. L., Müller, M. R. R. P., Gonzaga, L. V., Fett, R., & Costa, A. C. O. (2024). Canudo-de-pito (Escallonia sp.) honey: A comprehensive analysis of quality, composition, and pollen identification. *European Food Research and Technology, 250*(4), 1239–1251.

Anumba, I. A., Akunne, C. E., Ononye, B. U., & Chidi, C. A. (2020). Assessment on colonization, absconding and honey yield by African honeybee colonies reared in hives with different colours in Awka, south-eastern Nigeria. *Journal of Apiculture, 35*(4), 199–204.

Aqueel, M. A., Shurjeel, H. K., Muhammad, A. B., Raza, M. Y., Akram, I., & Ahsan, M. H. (2023). Seasonal management of honeybees for their improved honey production. In *Advances in insect pollination technology in sustainable agriculture*.

Arredondo, D., Añón, G., Campá, J., Harriet, J., Castelli, L., Zunino, P., & Antúnez, K. (2023). Supplementation of honey bee production colonies with a native beneficial microbe mixture. *Beneficial Microbes, 1*(aop), 1–16.

Aslam, N., Fatima, R., Altemimi, A. B., Ahmad, T., Khalid, S., Hassan, S. A., & Aadil, R. M. (2024). Overview of industrial food fraud and authentication through chromatography technique and its impact on public health. *Food Chemistry, 460*, 140542.

Asma, S. T., Acaroz, U., Alim, H., & Pandohee, J. (2024). The origin and evolution of beekeeping. In *Honey bees, beekeeping and bee products* (pp. 1–9). CRC Press.

Asoutis Didaras, N., Karaiskou, I., Nikolaidis, M., Siaperopoulou, C., Georgi, I., Tsadila, C., Karatasou, K., Amoutzias, G. D., & Mossialos, D. (2024). Contribution of microbiota to bioactivity exerted by bee bread. *Pharmaceuticals, 17*(6), 761.

Ayers, A. C., & Rehan, S. M. (2021). Supporting bees in cities: How bees are influenced by local and landscape features. *Insects, 12*(2), 128.

Ayoub, W. S., Zahoor, I., Dar, A. H., Farooq, S., Mir, T. A., Ganaie, T. A., Srivastava, S., Pandey, V. K., & Altaf, A. (2023). Exploiting the polyphenolic potential of honey in the prevention of chronic diseases. *Food Chemistry Advances, 3*, 100373.

Babin, A., Schurr, F., Delannoy, S., Fach, P., Huyen Ton, N., Nguyet, M., Bougeard, S., de Miranda, J. R., Rundlöf, M., Wintermantel, D., Albrecht, M., & Attridge, E. (2024). Distribution of infectious and parasitic agents among three sentinel bee species across European agricultural landscapes. *Scientific Reports, 14*(1), 3524.

Bakour, M., El Ghouizi, A., Laaroussi, H., Ousaaid, D., Touzani, S., Lyoussi, B., & Nayik, G. A. (2024). Microbial, thermal, and rheological analysis of honey. In *Advanced techniques of honey analysis* (pp. 129–152). Academic Press.

Bareke, T., & Addi, A. (2024). Floral nectar secretion dynamics of Pavonia urens (Malvaceae) and honey production potential. *Nusantara Bioscience, 16*(1).

Bariyah, N. (2020). Developing a model of employment creation in border region: Gaharu cultivation and honey bee farming in Bengkayang, West Kalimantan, Indonesia. *Biodiversitas Journal of Biological Diversity, 21*(11).

Bazargani, K., & Deemyad, T. (2024). Automation's impact on agriculture: Opportunities, challenges, and economic effects. *Robotics, 13*(2), 33.

Benjamin, S. (2024). *Fun foods of America: Outrageous delights, celebrated brands, and iconic recipes*. Rowman & Littlefield.

Bhatla, S. C., & Lal, M. A. (2023). Secondary metabolites. In *Plant physiology, development and metabolism* (pp. 765–808). Springer.

Bhattarai, B. (2021). *Effect of honey proportion and pH on the sensory quality of mead* (Doctoral dissertation, Department of Food Technology Central Campus of Technology Institute of Science and Technology Tribhuvan University, Nepal 2019).

Bobiş, O., Asma, S. T., Acaroz, U., Shah, S. R. A., Arslan-Acaroz, D., Kolayli, S., & Nayik, G. A. (2024). Honey. In *Honey bees, beekeeping and bee products* (pp. 69–87). CRC Press.

Bodor, Z., Zaukuu, J. L. Z., Aouadi, B., Kovács, Z., & Benedek, C. (2024). Analytical methods for nutritional composition of honey. In *Sustainable and functional foods from plants* (pp. 323–369). Apple Academic Press.

Bogo, G., Fisogni, A., Rabassa-Juvanteny, J., Bortolotti, L., Nepi, M., Guarnieri, M., Conte, L., & Galloni, M. (2021). Nectar chemistry is not only a plant's affair: Floral visitors affect nectar sugar and amino acid composition. *Oikos, 130*(7), 1180–1192.

Bora, F. D., Babeş, A. C., Călugăr, A., Jitea, M. I., Hoble, A., Filimon, R. V., Bunea, A., Nicolescu, A., & Bunea, C. I. (2023). Unravelling heavy metal dynamics in soil and honey: A case study from Maramureş region, Romania. *Food, 12*(19), 3577.

Boukraâ, L. (2023). *Honey in traditional and modern medicine*. CRC Press.

Brain, R. A., & Anderson, J. C. (2020). Anthropogenic factors affecting wildlife species status outcomes: Why the fixation on pesticides? *Environmental Science and Pollution Research*, 1–24.

Brar, D. S., & Nanda, V. (2024). A comprehensive introduction to honey adulteration. In *Advanced techniques of honey analysis* (pp. 63–91). Academic Press.

Brar, D. S., Nayik, G. A., Aggarwal, A. K., Kaur, S., Nanda, V., Saxena, S., Gautam, S., Ramniwas, S., & Tolcha, T. D. (2023). Chemical and functional characteristics to detect sugar syrup adulteration in honey from different botanical origins. *International Journal of Food Properties, 26*(1), 1390–1413.

Bryś, M. S., & Strachecka, A. (2024). The key role of amino acids in pollen quality and honey bee physiology – A review. *Molecules, 29*(11), 2605.

Büchler, R., Andonov, S., Bernstein, R., Bienefeld, K., Costa, C., Du, M., Gabel, M., Given, K., Hatjina, F., Harpur, B. A., & Hoppe, A. (2024). Standard methods for rearing and selection of Apis mellifera queens 2.0. *Journal of Apicultural Research*, 1–57.

Burke, J., & Corrigan, S. (2024). Bee well: A positive psychological impact of a pro-environmental intervention on beekeepers' and their families' wellbeing. *Frontiers in Psychology, 15*, 1354408.

Cappellari, A., Malagnini, V., Fontana, P., Zanotelli, L., Tonidandel, L., Angeli, G., Ioriatti, C., & Marini, L. (2024). Impact of landscape composition on honey bee pollen contamination by pesticides: A multi-residue analysis. *Chemosphere, 349*, 140829.

Catania, P., & Vallone, M. (2020). Application of a precision apiculture system to monitor honey daily production. *Sensors, 20*(7), 2012.

Çelik, K., & Aşgun, H. F. (2020). *Apitherapy: Health and healing from the bees*. Tudás Alapítvány.

Cengiz, M. M., & Erdogan, Y. (2023). Nutrient needs and food gathering activities of honeybees. In *Bee and beekeeping* (p. 3).

Chekecha, C. (2023). *Ethnobotany and antibacterial effects of golden berry (physalis peruviana l.) on Salmonella typhi in Mbeya rural district, Tanzania* (Doctoral dissertation, NM-AIST).

Chhun, A., Moriano-Gutierrez, S., Zoppi, F., Cabirol, A., Engel, P., & Schaerli, Y. (2024). An engineered bacterial symbiont allows noninvasive biosensing of the honey bee gut environment. *PLoS Biology, 22*(3), e3002523.

Chirsanova, A., Capcanari, T., & Boistean, A. (2021). *Quality assessment of honey in three different geographical areas from Republic of Moldova.*

Crosson, M., Zaman, N., Rajchel, D., Charbonneau, J., Walker, J. D., Ritchie, D., Kambos, J., Barrette, E., Mellas, S., Bellemore, M. I., & Ress, S. (2023). *Llewellyn's 2024 herbal almanac: A practical guide to growing, cooking & crafting*. Llewellyn Worldwide.

Cucu, A. A., Baci, G. M., Moise, A. R., Dezsi, Ş., Marc, B. D., Stângaciu, Ş., & Dezmirean, D. S. (2021). Towards a better understanding of nutritional and therapeutic effects of honey and their applications in apitherapy. *Applied Sciences, 11*(9), 4190.

Daisley, B. A., Pitek, A. P., Mallory, E., Chernyshova, A. M., Allen-Vercoe, E., Reid, G., & Thompson, G. J. (2023). Disentangling the microbial ecological factors impacting honey bee susceptibility to Paenibacillus larvae infection. *Trends in Microbiology, 31*(5), 521–534.

Dalio, J. S. (2021). Bee flora and biology of honey production. In *Honey* (pp. 23–86). CRC Press.

Damto, T., Zewdu, A., & Birhanu, T. (2024). Impact of different adulterants on honey quality properties and evaluating different analytical approaches for adulteration detection. *Journal of Food Protection, 87*(4), 100241.

Dar, S. A., Farook, U. B., Rasool, K., & Ahad, S. (2024). Honey: Classification, composition, safety, quality issues and health benefits. In *Advanced techniques of honey analysis* (pp. 1–37). Academic Press.

de Andrade Royo, V. (2023). Aroeira honey: New product for apitherapy. *Journal of Apitherapy, 10*(9).

de Cordova, C. M. M. (2024). Chemical diversity of Propolis from Meliponinae: An ancestral treasure to be preserved. *London Journal of Research In Science: Natural and Formal, 24*(4).

de Salcedo, A. M. (2023). *In Defense of processed food*. Reaktion Books.

Debela, H. (2020). *Physicochemical properties and biofunctional activities of two monofloral honeys from southwest ethiopia: The case of jimma zone.*

Delaplane, K. (2023). *Crop pollination by bees, volume 2: Individual crops and their bees*. CABI.

Delena, M. F., & Kayamo, S. E. (2024). Beekeeping opportunities, challenges and technology adoption in Gedeo zone, Southern Ethiopia. *Journal of Apicultural Research, 63*(3), 456–461.

Deng, Y., Yang, S., Zhao, H., Luo, J., Yang, W., & Hou, C. (2022). Antibiotics-induced changes in intestinal bacteria result in the sensitivity of honey bee to virus. *Environmental Pollution, 314*, 120278.

Di Bella, S., Sanson, G., Monticelli, J., Zerbato, V., Principe, L., Giuffrè, M., Pipitone, G., & Luzzati, R. (2024). Clostridioides difficile infection: History, epidemiology, risk factors, prevention, clinical manifestations, treatment, and future options. *Clinical Microbiology Reviews*, e00135–e00123.

Díaz, M. A., O'Connell, D. P., Jordan, S., O'Connor, C., Martin, P., Jones, J. C., & Garvey, J. (2023). Analysis of pesticide levels in honey and Pollen from Irish honey bee colonies using a modified Dutch mini-luke method with gas and liquid chromatography–Tandem mass spectrometry detection. *Journal of Agricultural and Food Chemistry, 71*(34), 12657–12667.

Dodd, S., Kevei, Z., Karimi, Z., Parmar, B., Franklin, D., Koidis, A., & Anastasiadi, M. (2025). Detection of sugar syrup adulteration in UK honey using DNA barcoding. *Food Control, 167*, 110772.

dos Santos Silva, A., Maciel, M. C., de Oliveira, A. A. F., & de Oliveira, T. F. (2024). Evaluation of the content of macro and trace elements and the geographic origin of honey in North Brazil through statistical and machine learning techniques. *Journal of Food Composition and Analysis*, 106050.

Ebeling, J., Reinecke, A., Sibum, N., Fünfhaus, A., Aumeier, P., Otten, C., & Genersch, E. (2023). A comparison of different matrices for the laboratory diagnosis of the epizootic American foulbrood of honey bees. *Veterinary Sciences, 10*(2), 103.

Edo, G. I., Onoharigho, F. O., Akpoghelie, P. O., Akpoghelie, E. O., Agbo, J. J., Agoh, E., & Lawal, R. A. (2023). Natural honey (raw honey): Insights on quality, composition, economic and health effects: A comprehensive review. *Food Science and Engineering*, 265–293.

El-Seedi, H. R., Khalifa, S. A., Abd El-Wahed, A., Gao, R., Guo, Z., Tahir, H. E., Zhao, C., Du, M., Farag, M. A., Musharraf, S. G., & Abbas, G. (2020). Honeybee products: An updated review of neurological actions. *Trends in Food Science & Technology, 101*, 17–27.

Fajari, M. F., Budhi, G., & Umezaki, M. (2023). A mini-ethnography of honey gathering: The practice and its contribution to livelihood systems in rural areas. *ETNOSIA: Jurnal Etnografi Indonesia, 8*(2), 212–229.

Faustino, C., & Pinheiro, L. (2021). Analytical rheology of honey: A state-of-the-art review. *Food, 10*(8), 1709.

Febrianto, N., Akhiroh, P., Masyithoh, D., Helmi, M., & Hartono, B. (2023). Factors affecting consumers' loyalty and purchase decisions on honey products: An emerging market perspective. *Open Agriculture, 8*(1), 20220235.

Figueroa, L., Sadd, B., Tripodi, A., Strange, J., Colla, S., Adams, L., Duennes, M., Evans, E., Lehmann, D., Moylett, H., & Richardson, L. (2023). Endosymbionts that threaten commercially raised and wild bumble bees (Bombus spp.). *Journal of Pollination Ecology, 33*, 14–36.

Flores, D. A. F., & López, E. N. C. (2024). Relationship between Oral health and nutrition during pregnancy, nursing, and the first two years of life. In *The role of nutrition in integral health and quality of life* (pp. 409–454). Apple Academic Press.

Gajski, G., Leonova, E., & Sjakste, N. (2024). Bee venom: Composition and anticancer properties. *Toxins, 16*(3), 117.

Garrido, P. M., Porrini, M. P., Alberoni, D., Baffoni, L., Scott, D., Mifsud, D., Eguaras, M. J., & Di Gioia, D. (2024). Beneficial bacteria and plant extracts promote honey bee health and reduce nosema ceranae infection. *Probiotics and Antimicrobial Proteins, 16*(1), 259–274.

Gebremedhn, H., Gebrewahid, Y., Haile, G. G., Hadgu, G., Atsbha, T., Hailu, T. G., & Bezabih, G. (2024). Projecting the impact of climate change on honey bee plant habitat distribution in northern Ethiopia. *Scientific Reports, 14*(1), 15866.

Gernt, P., Dittes, J., Vervuert, I., & Emmerich, I. U. (2024). Nutritional needs of honeybees and legislation on apiculture by-products in animal nutrition. *Animals, 14*(15), 2208.

Gondar, E. (2023). *Challenges, opportunities and prospects of honeybee production in Ethiopia.*

Gupta, G., Rawat, D., & Aggarwal, R. (2024). The role of bee products in cosmetic and skincare industry: Current trends and future prospects. *Uttar Pradesh Journal of Zoology, 45*(16), 360–371.

Gustafson, C. R., Champetier, A., Tuyizere, O., & Gitungwa, H. (2024). The impact of honey fraud information on the valuation of honey origin: Evidence from an incentivized economic experiment. *Food Control, 155*, 110070.

Haag, K. L., Caesar, L., da Silveira Regueira-Neto, M., de Sousa, D. R., Montenegro Marcelino, V., de Queiroz Balbino, V., & Torres Carvalho, A. (2023). Temporal changes in gut microbiota composition and pollen diet associated with colony weakness of a stingless bee. *Microbial Ecology, 85*(4), 1514–1526.

Hailu, T. G., Chagunda, M., & Rosenkranz, P. (2023). Sustainable development outlooks to subsistent apiculture in a transition: The case of Ethiopia. *Journal of Apicultural Research, 62*(4), 730–740.

Halvacı, E., Kozak, T., Gül, M., Kars, H., & Şen, F. (2023a). Bee anatomy: A comprehensive overview of bee morphology and physiology. *Journal of Scientific Reports-B, 008*, 1–19.

Halvacı, E., Kozak, T., Gül, M., Kars, H., Bayat, R., & Şen, F. (2023b). Journey to the sweet world of beekeeping: Historical development, honey harvesting and overview of bee products. *Journal of Scientific Reports-B, 9*, 39–58.

Hameed, O. M., Shaker, O. M., Ben Slima, A., & Makni, M. (2024). Biochemical profiling and physicochemical and biological valorization of Iraqi honey: A comprehensive analysis. *Molecules, 29*(3), 671.
Hando, F. H. (2023). Market value chain of honey production in northern Ethiopia. *African Journal of Agricultural Research, 19*(11), 1019–1031.
Harianja, A. H., Adalina, Y., Pasaribu, G., Winarni, I., Maharani, R., Fernandes, A., Saragih, G. S., Fauzi, R., Tampubolon, A. P., Njurumana, G. N., & Sukito, A. (2023). Potential of beekeeping to support the livelihood, economy, society, and environment of Indonesia. *Forests, 14*(2), 321.
Haroon, S. A., Hassan, B., Attia, M., & Rady, M. (2024). Nematicidal effectiveness of propolis against the root-knot nematode Meloidogyne spp. *Fayoum Journal of Agricultural Research and Development, 38*(2), 300–310.
Hart-Davies, C. (2024). *The herbal year: Folklore, history and remedies*. Yale University Press.
Hasan, M., Hasan, M. R., & Rashid, T. U. (2024). Bioremediation of microplastics by microorganisms: Trends, challenges, and perspectives. In *Microbial bioremediation and multiomics technologies for sustainable development: Recent trends* (Vol. 13, p. 49).
Hashifah, F. N., & Haryanto, T., (2023, July). *Check for updates potential natural enemies of stingless bee in Sleman Hortikultura Plantation, Yogyakarta*. In Proceedings of the International conference of tropical studies and its applications (ICTROPS 2022) (Vol. 31, p. 278). Springer.
Hayei, N. A. A., Reduan, M. H., Zulkaflee, N. S., Baharin, N. H. M., Sabarudin, M. A. & Azis, N. M. N. (2024). *The effect of stingless bee honey mouthrinse on dental plaque accumulation: A randomised clinical trial*.
Heilmann, A., & Watt, R. G. (2024). Strategic public health considerations for caries control in populations. In *Dental caries: The disease and its clinical management* (p. 17).
Hirpara, P. (2023). *Honey: A functional food and its application in food products* (Doctoral dissertation, Xidian University).
Honeycutt, C. M. (2023). *Profitable, alternative income generation and improved quality of life among global beekeepers* (Doctoral dissertation, East Tennessee State University).
Hossain, M. M., Barman, D. N., & Rahman, M. A. (2023). Amino acids, proteins, and enzymes. In *Honey: Composition and health benefits* (pp. 50–65).
Hristov, P., Shumkova, R., Palova, N., & Neov, B. (2020). Factors associated with honey bee colony losses: A mini-review. *Veterinary Sciences, 7*(4), 166.
Hurley, K. (2024). *Lichenpedia: A brief compendium* (Vol. 11). Princeton University Press.
Ibrahim, K. H., Abba, M. A. S., Margret, K. C. O., Ogechi, I. V., & Chinenye, A. V. (2021). Apiculture (beekeeping), an easy economic venture irrespective of age, gender, religion, and profession. *International Journal of Applied Agricultural Sciences, 7*(4), 169.
Iftikhar, A., Nausheen, R., Mukhtar, I., Iqbal, R. K., Raza, A., Yasin, A., & Anwar, H. (2023). The regenerative potential of honey: A comprehensive literature review. *Journal of Apicultural Research, 62*(1), 97–112.
Ilham, Q.P., Purnomo, H., Rohadi, D., Puspitaloka, D., Herawati, T., Brady, M. A., Dermawan, A., & Okarda, B. (2024, March). Value chain analysis for community livelihoods in peatlands: A case study in Southern Riau. In *IOP conference series: Earth and environmental science* (Vol. 1315, No. 1, p. 012009). IOP Publishing.
Ingram, V., Hansen, M. E., & Bosselmann, A. S. (2020). To label or not? Governing the costs and benefits of geographic indication of an African forest honey value chain. *Frontiers in Forests and Global Change, 3*, 102.
Iqbal, M. (2022). *Complementary and alternative medicinal approaches for enhancing immunity*. CRC Press.
Jahangir, M. A., Muheem, A., Anand, C., & Imam, S. S. (2020). Traditional and modern applications of honey: An insight. In *Therapeutic applications of honey and its phytochemicals: Vol. 1* (pp. 151–169). Springer.
Jašarević, L. (2024). *Beekeeping in the end times*. Indiana University Press.
Jinsong, L. (2023). Agricultural and mineral processing. In *Chinese handicrafts* (pp. 115–226). Singapore.

Johnson, B. R. (2023). *Honey bee biology*. Princeton University Press.
Johnson, M. C. (2024). *Illuminating our true nature: Yogic practices for personal and collective healing*. Shambhala Publications.
Joshi, A. (2023). *Artificial intelligence and human evolution: Contextualizing AI in human history*. Apress.
Kashbor, M. (2023). *Doped zeolite-based catalysts for selective glucose isomerisation and fructose dehydration* (Doctoral dissertation, University of Sheffield).
Kavimugilan, S., Kalyanasundaram, A., & Ambethgar, V. (2024). Managing bee's health during dearth periods: A review of pollen and nectar substitutes for tropical climates. *Uttar Pradesh Journal of Zoology, 45*(16), 472–489.
Kessi, O., Mekious, S., Aouadi, A., Houdeib, J., & Megatli, S. (2024). Global assessment of Algerian honeys quality by palynological, physicochemical analyses, trace elements and potentially toxic elements screenin. *Acta Agriculturae Slovenica, 120*(2), 1–14.
Khairan, K., Mudatsir, M., Diah, M., Rizal, S., Putra, M. I. A., Jannah, S. M., & Chairani, I. (2024, June). Therapeutic activities of honey in wound care: a narrative review. In *IOP conference series: Earth and environmental science* (Vol. 1356, No. 1, p. 012025). IOP Publishing.
Khalfan Saeed Alwali Alkindi, F., El-Keblawy, A., Lamghari Ridouane, F., & Bano Mirza, S. (2024). Factors influencing the quality of royal jelly and its components: A review. *Cogent Food & Agriculture, 10*(1), 2348253.
Khalil, A., Elesawy, B. H., Ali, T. M., & Ahmed, O. M. (2021). Bee venom: From venom to drug. *Molecules, 26*(16), 4941.
Khan, K. (2020). *Beekeeping in Pakistan (history, potential, and current status).*
Khataybeh, B., Jaradat, Z., & Ababneh, Q. (2023). Anti-bacterial, anti-biofilm and anti-quorum sensing activities of honey: A review. *Journal of Ethnopharmacology*, 116830.
Khoo, H. E., Chen, B. J., Li, J., Li, X., Cheng, S. H., & Azrina, A. (2023). Emerging natural and high-phenolic sweet substances: A review. *International Food Research Journal, 30*(2).
Kibar, M. (2024). A cheaply non-destructive technique to estimate honey quality: Thermal imaging and machine learning. *Uludağ Arıcılık Dergisi, 24*(1), 79–92.
Kieliszek, M., Piwowarek, K., Kot, A. M., Wojtczuk, M., Roszko, M., Bryła, M., & Trajkovska Petkoska, A. (2023). Recent advances and opportunities related to the use of bee products in food processing. *Food Science & Nutrition, 11*(8), 4372–4397.
Kortsch, S., Timberlake, T. P., Cirtwill, A. R., Sapkota, S., Rokoya, M., Devkota, K., Roslin, T., Memmott, J., & Saville, N. (2024). Decline in honeybees and its consequences for beekeepers and crop pollination in Western Nepal. *Insects, 15*(4), 281.
Kraisoraphong, K. (2023). The Destiny of China's honeybees in the Anthropocene. In *Human-animal interactions in Anthropocene Asia* (pp. 184–206). Routledge.
Kükrer, M. (2024). *From diversity to conservation: insights for a national genetic monitoring program for honey bees in the face of climate change*.
Kumar, R. (2024). Historical and religious perspective of honey. In *Honey* (pp. 1–16). CRC Press.
Kumar, Y., Peshin, R., Nain, M. S., Rather, B. A., Namgyal, D., Kanwar, M. S., & Yangchan, J. (2020). Constraints perceived by the beekeepers of Jammu Province in adoption of scientific beekeeping practices. *Indian Journal of Extension Education, 56*(4), 49–53.
Kumar, R., Kumar, S., & Kanwar, S. S. (2024a). Honey: Introduction, history, composition, and its uses. In *Biomedical perspectives of herbal honey* (pp. 1–17). Springer.
Kumar, R., Kumari, I., & Hajam, Y. A. (2024b). Honey beyond medicine. In *Honey in food science and physiology* (pp. 229–253). Springer.
Landaverde, R., Rodriguez, M. T., & Parrella, J. A. (2023). Honey production and climate change: Beekeepers' perceptions, farm adaptation strategies, and information needs. *Insects, 14*(6), 493.
Laskaris, J. (2024). Galen's pharmacology in context. In *The Oxford handbook of Galen* (p. 186).
Lata, P., Hajam, Y. A., & Kumar, R. (2024). Honey 8 dressing as and wound healing agent. In *Honey: Nutraceutical and therapeutic significance* (p. 127).

Leclercq, N., Marshall, L., Caruso, G., Schiel, K., Weekers, T., Carvalheiro, L. G., Dathe, H. H., Kuhlmann, M., Michez, D., Potts, S. G., & Rasmont, P. (2023). European bee diversity: Taxonomic and phylogenetic patterns. *Journal of Biogeography, 50*(7), 1244–1256.
Londei, T., & Marzi, G. (2024). Honey bees collecting pollen from the body surface of foraging bumble bees: A recurring behaviour. *Apidologie, 55*(1), 4.
Luca, L., Pauliuc, D., & Oroian, M. (2024). Honey microbiota, methods for determining the microbiological composition and the antimicrobial effect of honey–A review. *Food Chemistry: X*, 101524.
Ma, B., Chen, Q., Lin, F., Wang, G., & Ren, B. (2023). Lepidoptera proboscis pollens are mainly derived from nectar. *bioRxiv*, pp. 2023–03.
Machado, A. M., Tomás, A., Russo-Almeida, P., Duarte, A., Antunes, M., Vilas-Boas, M., Miguel, M. G., & Figueiredo, A. C. (2022). Quality assessment of Portuguese monofloral honeys. Physicochemical parameters as tools in botanical source differentiation. *Food Research International, 157*, 111362.
Maderson, S. (2023). Co-producing agricultural policy with beekeepers: Obstacles and opportunities. *Land Use Policy, 128*, 106603.
Mahankuda, B., & Tiwari, R. (2024). Impact of climate change on honeybees and crop production. In *Adapting to climate change in agriculture-theories and practices: Approaches for adapting to climate change in agriculture in India* (pp. 211–224). Springer.
Mahesar, S. A., Sidhu, A. R., Naz, S., & Kandhro, A. A. (2024). Phenolic and mineral contents in honey and their associated health benefits. In *Honey in food science and physiology* (pp. 155–179). Springer.
Mahmoud, M. A., Kilic-Buyukkurt, O., Fotouh, M. M. A., & Selli, S. (2024). Aroma active compounds of honey: Analysis with GC-MS, GC-O, and molecular sensory techniques. *Journal of Food Composition and Analysis*, 106545.
Maina, M. W., Mburu, J., Maina, F. W., & Chimoita, E. L. (2024). Determinants of consumers' perceptions of honey quality attributes which influence purchasing decisions in Nyandarua County, Kenya. *Bee World*, 1–6.
Majtan, J., Bucekova, M., Kafantaris, I., Szweda, P., Hammer, K., & Mossialos, D. (2021). Honey antibacterial activity: A neglected aspect of honey quality assurance as functional food. *Trends in Food Science & Technology, 118*, 870–886.
Malovrh, K., Ravnjak, B., Bavcon, J., & Križman, M. (2024). Nectar production and three main sugars in nectar of salvia pratensis and salvia glutinosa in correlation with abiotic factors. *Agriculture, 14*(5), 668.
Maneli, S. (2024). *The African continental free trade area and developmental regionalism: leveraging transport corridors to actualise the AfCFTA*.
Manoj, K., Phukan, M. M., Kumar, R., Jamir, L., Roy, V. K., & Pankaj, P. P. (2024). Therapeutic applications of honey: Traditional and contemporary approaches in human health management. In *Honey in food science and physiology* (pp. 1–31). Springer.
Mărgăoan, R., Topal, E., Balkanska, R., Yücel, B., Oravecz, T., Cornea-Cipcigan, M., & Vodnar, D. C. (2021). Monofloral honeys as a potential source of natural antioxidants, minerals and medicine. *Antioxidants, 10*(7), 1023.
Mclean, L. (2024). *More than milk and honey: Alternative farmer's practices, motivations, and aspirations in Aotearoa New Zealand* (Doctoral dissertation, ResearchSpace@ Auckland).
Mg, D., Khan, M. S., Kumar, G., & Udikeri, A. (2023). Pollinator interaction with selected 'weeds' flora, Asteraceae, in the context of land use. *Oriental Insects, 57*(3), 935–950.
Minami, A., Kang, X., & Carter, C. J. (2021). A cell wall invertase controls nectar volume and sugar composition. *The Plant Journal, 107*(4), 1016–1028.
Mishra, M., Bhunia, P., Sen, R., Bhunia, R., Gulia, J., Mondal, T., Zidane, Z., Saha, T., Chacko, A., Raj, R., & Kaushal, L. (2023). The impact of weather change on honey bee populations and disease. *Journal of Advanced Zoology, 44*.

Mouillard-Lample, L., Gonella, G., Decourtye, A., Henry, M., & Barnaud, C. (2023). Competition between wild and honey bees: Floral resources as a common good providing multiple ecosystem services. *Ecosystem Services, 62*, 101538.

Muscolo, A., Mariateresa, O., Giulio, T., & Mariateresa, R. (2024). Oxidative stress: The role of antioxidant phytochemicals in the prevention and treatment of diseases. *International Journal of Molecular Sciences, 25*(6), 3264.

Mustar, S., & Ibrahim, N. (2022). A sweeter pill to swallow: A review of honey bees and honey as a source of probiotic and prebiotic products. *Food, 11*(14), 2102.

Mutua, D., & Felix, K. (2023). Effects of coordinated border management techniques on trade facilitation in Kenya. *African Tax and Customs Review, 6*(2), 25–25.

Nakib, R., Ghorab, A., Harbane, S., Saker, Y., Ouelhadj, A., Rodríguez-Flores, M. S., Seijo, M. C., & Escuredo, O. (2024). Sensory attributes and chemical composition: The case of three monofloral honey types from Algeria. *Food, 13*(15), 2421.

Narciso, L., Topini, M., Ferraiuolo, S., Ianiro, G., & Marianelli, C. (2024). Effects of natural treatments on the varroa mite infestation levels and overall health of honey bee (Apis mellifera) colonies. *PLoS One, 19*(5), e0302846.

Naskar, A., Chatterjee, K., Roy, K., Majie, A., Nair, A. B., Shinu, P., Morsy, M. A., Jain, N., Pandey, M., & Gorain, B. (2024). Mechanistic roles of different varieties of honey on wound healing: Recent update. *Journal of Pharmacology and Pharmacotherapeutics, 15*(1), 5–18.

Ngomirakiza, E. (2022). Introduction of Burundi commercial justice and potentials of investment with the integration in east African community. *The Journal of Law, the Postgraduate Course of Kansai University, 2022*(100), 151–199.

Ngunjiri, M. W., Birech, J., & Muleka, J. (2023). *Assessing the contributions of the women's Movements & Women's self-mobilization, to women's economic empowerment in Kenya between 1963 and 2010*. UoN, AWSC, Women's Economic Empowerment Hub.

Ngussie, H. G., & Hailu, T. G. (2023). Tigray war put honey bees in peril. *Bee World, 100*(2), 51–55.

Nhlapho, S. (2023). *Investigation of sugars in honey for authentication and detection of adulterants using selected analytical techniques* (Doctoral dissertation, University of Johannesburg).

Nikhat, S., & Fazil, M. (2022). History, phytochemistry, experimental pharmacology and clinical uses of honey: A comprehensive review with special reference to Unani medicine. *Journal of Ethnopharmacology, 282*, 114614.

Niklasson, M., Svensson, E., Leidenberger, S., Norrström, N., & Crawford, E. (2024). Free-living colonies of native honey bees (Apis mellifera mellifera) in 19th and early 20th century Sweden. *Journal of Insect Conservation, 28*(3), 389–400.

Nowak, A., & Nowak, I. (2023). Review of harmful chemical pollutants of environmental origin in honey and bee products. *Critical Reviews in Food Science and Nutrition, 63*(21), 5094–5116.

Nweze, A. J., Olovo, C. V., Nweze, E. I., John, O. O., & Paul, C. (2020). Therapeutic properties of honey. In *Honey analysis – New advances and challenges* (Vol. 332, pp. 1–21). IntechOpen.

Oliver, R. (2021). Honey bee nutrition. In *Honey bee medicine for the veterinary practitioner* (pp. 93–123).

Ondogo, L. K. (2023). *Determination of the physicochemical properties of honey from suba region, homa bay county-kenya* (Doctoral dissertation, Kenyatta University).

Otero, M. C. B., & Bernolo, L. (2020). Honey as functional food and prospects in natural honey production. In *Functional foods and nutraceuticals: Bioactive components, formulations and innovations* (pp. 197–210).

Owen, R., Scheerlinck, J. P., & Stevenson, M. (2023). *Honey bee pests and diseases: A complete guide to prevention and management*. Exisle Publishing.

Page, M. L., & Williams, N. M. (2023). Honey bee introductions displace native bees and decrease pollination of a native wildflower. *Ecology, 104*(2), e3939.

Painkra, G. P. (2023). *Commercial beekeeping*. BFC Publications.

Palma-Morales, M., Huertas, J. R., & Rodríguez-Pérez, C. (2023). A comprehensive review of the effect of honey on human health. *Nutrients, 15*(13), 3056.

Paray, B. A., Kumari, I., Hajam, Y. A., Sharma, B., Kumar, R., Albeshr, M. F., Farah, M. A., & Khan, J. M. (2021). Honeybee nutrition and pollen substitutes: A review. *Saudi Journal of Biological Sciences, 28*(1), 1167–1176.

Patti, M., Musarella, C. M., Postiglione, S. M., Papalia, F., Falcone, M. C., Mammone, G., Zappalà, M. A., Laface, V. L. A., & Spampinato, G. (2024). Ethnobotanical studies on the Tyrrhenian side of the Aspromonte massif (Calabria, Southern Italy). *Plant Biosystems-An International Journal Dealing with All Aspects of Plant Biology, 158*(3), 545–562.

Pehlivan, T. (2023). Comparison study on honey in the Islamic nutrition culture; the status of prophet Muhammad's (Pbuh) and Avicenna's applications according to current scientific studies. *Journal of Tourism & Gastronomy Studies, 11*(1), 279–297.

Perichon, S., Adamchuk, L., Biber, L., Bozic, J., Chlebo, R., Filipi, J., Leidenberger, S., Mavrofridis, G., Ozgor, E., Pocol, C., & Porporato, M. (2024). Perception of threats to bee colonies and the future of local beekeeping by beekeepers in various European countries. *Bulletin de la Société Géographique de Liège.*

Pluchinotta, I., Zhou, K., Moore, G., Salvia, G., Belesova, K., Mohajeri, N., Hale, J., Davies, M., & Zimmermann, N. (2024). Co-producing knowledge on the use of urban natural space: Participatory system dynamics modelling to understand a complex urban system. *Journal of Environmental Management, 353*, 120110.

Pluta, R., Miziak, B., & Czuczwar, S. J. (2023). Apitherapy in post-ischemic brain neurodegeneration of alzheimer's disease proteinopathy: Focus on honey and its flavonoids and phenolic acids. *Molecules, 28*(15), 5624.

Puranik, S. I., Akbar, A. A., & Ghagane, S. C. (2023). Economic benefits of honey and honey products. In *Honey: Composition and health benefits* (pp. 330–339).

Rabbani, S., Anvar, S. A. A., Allahyaribeik, S., Jannat, B., & Ahari, H. (2024). Effect of ultrasound technique to improve quality of Iranian industrial honey by controlling crystallization process. *Food Science & Nutrition, 12*(4), 2932–2946.

Rahmad, B., Damiri, N., Hanafiah, Z., & Adriani, D. (2024). Feasibility analysis and contribution of beekeeping to the welfare of beekeepers in Gunung Megang, Muara Enim Regency, South Sumatra Province, Indonesia. *Dinasti International Journal of Education Management and Social Science, 5*(4), 531–542.

Raina, R. H., Jangid, T., Choudhary, P., & Sharma, I. (2024). Distributional pattern and food plants of Western Himalayan bumble bees from Nanda Devi biosphere reserve. *Indian Journal Of Entomology*, 824–829.

Rajagopalan, K., DeGrandi-Hoffman, G., Pruett, M., Jones, V. P., Corby-Harris, V., Pireaud, J., Curry, R., Hopkins, B., & Northfield, T. D. (2024). Warmer autumns and winters could reduce honey bee overwintering survival with potential risks for pollination services. *Scientific Reports, 14*(1), 5410.

Rák, T., & Csutak, A. (2024). Complementary practices in pharmacy and their relation to glaucoma – Classification, definitions, and limitations. *Scientia Pharmaceutica, 92*(1), 16.

Raweh, H. S., Badjah-Hadj-Ahmed, A. Y., Iqbal, J., & Alqarni, A. S. (2023). Physicochemical composition of local and imported honeys associated with quality standards. *Food, 12*(11), 2181.

Recklies, K., Peukert, C., Kölling-Speer, I., & Speer, K. (2021). Differentiation of honeydew honeys from blossom honeys and according to their botanical origin by electrical conductivity and phenolic and sugar spectra. *Journal of Agricultural and Food Chemistry, 69*(4), 1329–1347.

Redclift, M. (2023). The role of agricultural technology in sustainable development. In *Technological change and the rural environment* (pp. 81–103). Routledge.

Reif, J., Gamero, A., Hološková, A., Aunins, A., Chodkiewicz, T., Hristov, I., Kurlavičius, P., Leivits, M., Szép, T., & Voříšek, P. (2024). Accelerated farmland bird population declines in European countries after their recent EU accession. *Science of the Total Environment, 946*, 174281.

Rodríguez, M. A., Fernández, L. A., Daisley, B. A., Reynaldi, F. J., Allen-Vercoe, E., & Thompson, G. J. (2023). Probiotics and in-hive fermentation as a source of beneficial microbes to support the gut microbial health of honey bees. *Journal of Insect Science, 23*(6), 19.

Rodríguez-Vásquez, S., Bejar, A. A. G., Romo-Chacón, A., García-Hernández, J., González-Ríos, H., & Orozco-Avitia, J. A. (2024). Hive temperature regulation by honeybees in response to extreme conditions. *Agrociencia*.

Rowe, C. J. (2024). *Organic residue analysis and archaeology: A critical evaluation of ora methods* (Master's thesis, The University of Wisconsin-Milwaukee).

Rust, A. E. (2020). *Honey authentication: Effect of irradiation and ageing on near-infrared (NIR) spectroscopy classification models* (Doctoral dissertation, Stellenbosch: Stellenbosch University).

Sahri, N., Alaoui, A., Elherradi, E., Mahmoud, M. F., Louaradi, D., & Sobeh, M. (2024). Euphorbia honey: A comprehensive compile of its traditional use, quality parameters, authenticity, adulteration, and therapeutic merits. *Phytochemistry Reviews, 23*(3), 721–748.

Sahu, D., Dansena, D., Sharma, K., & Gupta, P. (2024). Strategies and outcomes in bee conservation: Evaluating impact on pollination ecosystems. *Journal of Advances in Biology & Biotechnology, 27*(2), 57–66.

Saklani, S., & Kumar, N. (2021). Quality honey production, processing, and various mechanisms for detection of adulteration. In *Honey* (pp. 87–118). CRC Press.

Salaria, N., Kumari, I., & Kumar, R. (2024). Micro-organisms and foodborne pathogens in honey along with their impact on human health. In *Honey* (pp. 57–75). CRC Press.

Samoraj, M. J. Ł., et al. (2018). Why is the world sleeping? Ancient, ancestral spirit of Gaia suggesting holistic healing solutions for our planet based on Sir David Attenborough's (2018) observation:"Helping Cool Earth to halt tropical deforestation makes a real difference. Perhaps the biggest difference we will make in our whole lives.".

Scarbrough, J. L. (2024). *Analysis of the current state of the beekeeping industry in Alabama* (Master's thesis, Auburn University).

Scarfone, A., Cammerata, A., Romano, E., Vinciguerra, V., Marabottini, R., Gallucci, F., Paris, E., Carnevale, M., Vincenti, B., Palma, A., & Bergonzoli, S. (2024). Effects of ground transport on the presence of heavy metals in selected honeybee products. *Environmental Science and Pollution Research*, 1–12.

Scepankova, H., Pinto, C. A., Estevinho, L. M., & Saraiva, J. A. (2022). High-pressure-based strategies for the inactivation of Bacillus subtilis endospores in honey. *Molecules, 27*(18), 5918.

Schouten, C. N. (2020). Factors influencing beekeepers income, productivity and welfare in developing countries: A scoping review. *Journal of Apicultural Research, 60*(2), 204–219.

Seyoum, B. (2024). International trade policy for growth and prosperity. In *State fragility, business, and economic performance: An Ethiopian perspective* (pp. 275–314). Springer.

Shahbaz, M., Naeem, H., Hamza, A., Murtaza, S., Nayik, G. A., Wachkoo, A. A., Nanda, V., & Singh, R. (2024). Beeswax. In *Honey bees, beekeeping and bee products* (pp. 170–188). CRC Press.

Sharma, B., & Saxena, P. (2024). Honey granulation and fermentation Vis-à-Vis nutritional and commercial value. In *Honey* (pp. 30–56). CRC Press.

Sharma, A., Pant, K., Brar, D. S., Thakur, A., & Nanda, V. (2023a). A review on Api-products: Current scenario of potential contaminants and their food safety concerns. *Food Control, 145*, 109499.

Sharma, R., Thakur, M., Rana, K., Devi, D., & Bajiya, M. R. (2023b). Honey, its quality and composition and their responsible factors. *International Journal of Bio-resource and Stress Management, 14*(1), 178–189.

Shivani, T. M., & Sathiavelu, M. (2023). A comprehensive review on functionality of probiotics in edible packaging. *Packaging Technology and Science, 36*(1), 15–30.

Shlibak, A. A. A., & Dalla, L. O. F. B. (2020). The sustainable research Long while between bee pollen and honey bee diversity in Libya: Literature review. *International Journal of Social Sciences and Management Research, 7*(1), 2545–5303.

Singh, L. B. (2022). Cosmetic and medicinal use of bee's honey: A review. In *A monthly peer reviewed magazine for agriculture and allied sciences* (Vol. 79).

Singh, A. P., De, K., Uniyal, V. P., & Sathyakumar, S. (2024). Unveiling of climate change-driven decline of suitable habitat for Himalayan bumblebees. *Scientific Reports, 14*(1), 4983.

Sladonja, B., Tlak Gajger, I., Uzelac, M., Poljuha, D., Garau, C., Landeka, N., Barták, M., & Bacaro, G. (2023). The impact of beehive proximity, human activity and agricultural intensity on diptera diversity in a mediterranean mosaic of agroecosystems, with a focus on pest species. *Animals, 13*(6), 1024.

Sless, T. J., Branstetter, M. G., Mikát, M., Odanaka, K. A., Tobin, K. B., & Rehan, S. M. (2024). Phylogenomics and biogeography of the small carpenter bees (Apidae: Xylocopinae: Ceratina). *Molecular Phylogenetics and Evolution, 198*, 108133.

Srinivasan, M. R., & Haran, M. S. R. (2023). Honey bee diversity. In *Commercial insects* (pp. 146–165). CRC Press.

Stubbs, G. R. (2020). *Honey from the rock poverty alleviation from a missio Dei perspective presented as a missiological case study of Integral Mission through the African Honey Bee project in the uMfolozi area of South Africa* (Doctoral dissertation, North-West University (South Africa)).

Sujanto, I. S. R., Ramly, N. S., Abd Ghani, A., Huat, J. T. Y., Alias, N., & Ngah, N. (2021). The composition and functional properties of stingless bee honey: A review. *Malaysian Journal of Applied Sciences, 6*(1), 111–127.

Svyatnenko, R. S., Marynin, A. I., Litvynchuk, S. I., & Pasichnyi, V. M. (2023). Study on the quality of honey from different botanical sources and one re-gional origin. *Scientific Messenger of LNU of Veterinary Medicine and Biotechnologies. Series: Food Technologies, 25*(99), 56–60.

Szulc, J. (2020). Biological agents. In *Respiratory protection against hazardous biological agents* (pp. 9–58). CRC Press.

Tafi, E., Sagona, S., Meucci, V., Bortolotti, L., Galloni, M., Bogo, G., Gatta, D., Casini, L., Barberis, M., Nepi, M., & Felicioli, A. (2024). Effect of amino acid enriched diets on hemolymph amino acid composition in honey bees. *Archives of Insect Biochemistry and Physiology, 115*(1), e22085.

Tahboub, Y. R., Al-Ghzawi, A. A. M. A., Al-Zayafdneh, S. S., & AlGhotani, M. S. (2022). Levels of trace elements and rare earth elements in honey from Jordan. *Environmental Science and Pollution Research*, 1–12.

Tambi, M. D. (2024). Honey extraction for agribusiness and family income in Bamenda Town, Cameroon. *Journal of Management and Science, 14*(2), 66–77.

Tananaki, C., Rodopoulou, M. A., Dimou, M., Kanelis, D., & Liolios, V. (2024). The total phenolic content and antioxidant activity of nine monofloral honey types. *Applied Sciences, 14*(10), 4329.

Tang, J., Ji, C., Shi, W., Su, S., Xue, Y., Xu, J., Chen, X., Zhao, Y., & Chen, C. (2023). Survey results of honey bee colony losses in winter in China (2009–2021). *Insects, 14*(6), 554.

Tefera, W., & Melaku, M. (2024). Bee flora identification, constraints and opportunities of beekeeping in north Wollo, Amhara, Ethiopia. *Cogent Food & Agriculture, 10*(1), 2292370.

Thakur, M., Sharma, D., Khajuria, M., Kapoor, B., & Yadav, V. (2023). *Prospects and constraints in traditional beekeeping in Jammu and Kashmir*.

Thaliath, A. (2024). *A baker's journey: Science in the art of baking*. Partridge Publishing.

Thangaraju, S., Modupalli, N., & Natarajan, V. (2021). Food adulteration and its impacts on our health/balanced nutrition. In *Food chemistry: The role of additives, preservatives and adulteration* (pp. 189–216).

Thyaka, T. (2023). *Improved mixed-method farming and food security in kenya: A case of kenya cereals enhanced programme in mwingi west sub-county, Kitui county* (Doctoral dissertation, The catholic University of Eastern Africa).

Tian, L., Bilamjian, S., Cuthbertson, D., Anumol, T., De Leoz, L., Julien, C., Giovenazzo, P., Chahal, S., & Bayen, S. (2024). Impact of processing steps (filtration, creaming and pasteurization) on the botanical classification of honey using LC-QTOF-MS. *Food Research International, 194*, 114841.

Tigistu, S. (2020). Review on actors and their value adding activities in honey value chain; challenges and opportunities in Ethiopia. *The International Journal of Business Management and Technology, 4*(5), 128–140.

Tiusanen, M., Becker-Scarpitta, A., & Wirta, H. (2024). Distinct communities and differing dispersal routes in bacteria and fungi of honey bees, honey, and flowers. *Microbial Ecology, 87*(1), 100.

Torezan-Silingardi, H. M., Silberbauer-Gottsberger, I., & Gottsberger, G. (2021). Pollination ecology: Natural history, perspectives and future directions. In *Plant-animal interactions: Source of biodiversity* (pp. 119–174). Springer.

Tragnone, B. M., Arzoumanidis, I., D'Eusanio, M., & Petti, L. (2023). Contribution of the product social impact life cycle assessment (PSILCA) database in assessing the risks and opportunities of a jar of honey production. *The International Journal of Life Cycle Assessment, 28*(8), 1054–1071.

Trak, N. H. D. T. H., & Chauhan, M. F. H. D. D. (2022). *Skin care secrets from herbal world.* Notion Press.

Tsadila, C., Amoroso, C., & Mossialos, D. (2023). Microbial diversity in bee species and bee products: Pseudomonads contribution to bee well-being and the biological activity exerted by honey bee products: A narrative review. *Diversity, 15*(10), 1088.

Ullah, A., Aldakheel, F. M., Anjum, S. I., Raza, G., Khan, S. A., & Gajger, I. T. (2023). Pharmacological properties and therapeutic potential of honey bee venom. *Saudi Pharmaceutical Journal, 31*(1), 96–109.

Uniyal, A., & Kumar, A. (2024). Minor flowers of European and American countries. In *Edible flowers* (pp. 385–430). Academic Press.

Valverde, S., Ares, A. M., Elmore, J. S., & Bernal, J. (2022). Recent trends in the analysis of honey constituents. *Food Chemistry, 387*, 132920.

Van Espen, M., Williams, J. H., Alves, F., Hung, Y., de Graaf, D. C., & Verbeke, W. (2023). Beekeeping in Europe facing climate change: A mixed methods study on perceived impacts and the need to adapt according to stakeholders and beekeepers. *Science of the Total Environment, 888*, 164255.

Végh, R., & Csóka, M. (2024). Amino acids, peptides, and proteins of pollen. In *Pollen chemistry & biotechnology* (pp. 17–49). Springer.

Vercelli, M., Novelli, S., Ferrazzi, P., Lentini, G., & Ferracini, C. (2021). A qualitative analysis of beekeepers' perceptions and farm management adaptations to the impact of climate change on honey bees. *Insects, 12*(3), 228.

Villagómez, G. N., Spaethe, J., & Leonhardt, S. D. (2024). The stingless bee Trigona fulviventris prefers sweet and salty over savory nectar. *Apidologie, 55*(4), 39.

Wakgari, M., & Yigezu, G. (2021). Honeybee keeping constraints and future prospects. *Cogent Food & Agriculture, 7*(1), 1872192.

Wali, A. F., Pillai, J. R., Razmpoor, M., Jabnoun, S., Akbar, I., Rasool, S., Arafah, A., Khan, A., Akhter, R., & Majid, S. (2020). Honey and its Phyto-constituents: From chemistry to medicine. In *Therapeutic applications of honey and its phytochemicals: Vol. 1* (pp. 31–52).

Walton, A., Herman, J. J., & Rueppell, O. (2024). Social life results in social stress protection: A novel concept to explain individual life-history patterns in social insects. *Biological Reviews.*

Wei, J., Rico-Guevara, A., Nicolson, S. W., Brau, F., Damman, P., Gorb, S. N., Wu, Z., & Wu, J. (2023). Honey bees switch mechanisms to drink deep nectar efficiently. *Proceedings of the National Academy of Sciences, 120*(30), e2305436120.

Wirta, H., Jones, M., Peña-Aguilera, P., Chacón-Duque, C., Vesterinen, E., Ovaskainen, O., Abrego, N., & Roslin, T. (2023). The role of seasonality in shaping the interactions of honeybees with other taxa. *Ecology and Evolution, 13*(10), e10580.

Yang, Y., Wu, Y., Long, H., Ma, X., Shariati, K., Webb, J., Guo, L., Pan, Y., Ma, M., Deng, C., & Cao, P. (2023). Global honeybee health decline factors and potential conservation techniques. *Food Security, 15*(4), 855–875.

Yilmaz, A. C., & Aygin, D. (2020). Honey dressing in wound treatment: A systematic review. *Complementary Therapies in Medicine, 51*, 102388.

Yilmaz Kaya, M., Gültekin, Y. S., & Gültekin, P. (2023). Evaluation of honey bees within the scope of sustainable development goals and ecosystem services. *Duzce University Journal of Science & Technology, 11*(5).

YohanneS, G., & Asfaw, T. (2024). Potentially to medical appliances through nutraceutical studies. *Quantum Journal of Medical and Health Sciences, 3*(2), 23–35.

Yue, L., Wu, C., Zhang, M., & Huang, Q. (2024). Survey on trends, challenges, and opportunities of honey production and its marketing system in Hawa Gelan District Kellam Wollega zone Western Ethiopia. *Journal of Theory and Practice of Management Science, 4*(04), 1–10.

Yupanqui Mieles, J., Vyas, C., Aslan, E., Humphreys, G., Diver, C., & Bartolo, P. (2022). Honey: An advanced antimicrobial and wound healing biomaterial for tissue engineering applications. *Pharmaceutics, 14*(8), 1663.

Zaheen, Z., Yatoo, A. M., Ali, S., Ali, M. N., Majid, S., Rasool, S., Rashid, S. M., Ahmad, S. B., Mir, M. U. R., & Zehra, U. (2020). Honey: Types, composition and antimicrobial mechanisms. In *Therapeutic applications of honey and its phytochemicals: Vol. 1* (pp. 193–214).

Zembrzuska, J., Pakulski, Ł., Karbowska, B., Bartoszewicz, J., Janeba-Bartoszewicz, E., Selech, J., & Kurczewski, P. (2024). Comparison of different methods for determining 5-hydroxymethylfurfural in assessing the quality of honey from greater Poland. *Sustainability, 16*(11), 4568.

Zerrouk, S., & Bahloul, R. (2023). Palynological and physicochemical properties of multifloral honey produced in some regions of Algeria. *Journal of Apicultural Research, 62*(2), 345–354.

Zhang, H. P., Wen, S. J., Wang, H., & Ren, Z. X. (2023a). Floral nectar reabsorption and a sugar concentration gradient in two long-spurred Habenaria species (Orchidaceae). *BMC Plant Biology, 23*(1), 331.

Zhang, S., Wu, B., Chen, R., Liang, J., Khan, N., & Ray, R. L. (2023b). Government intervention on cooperative development in poor areas of rural China: A case study of xm beekeeping cooperative in Sichuan. *Land, 12*(4), 731.

Zhang, X. H., Gu, H. W., Liu, R. J., Qing, X. D., & Nie, J. F. (2023c). A comprehensive review of the current trends and recent advancements on the authenticity of honey. *Food Chemistry: X*, 100850.

Živkov Baloš, M., Popov, N., Jakšić, S., Mihaljev, Ž., Pelić, M., Ratajac, R., & Ljubojević Pelić, D. (2023). Sunflower honey – Evaluation of quality and stability during storage. *Food, 12*(13), 2585.

Chapter 15
Ethical and Sustainable Practices in Algerian Beekeeping: A Focus on Honey Production, Authentication, and Beekeeping Practices

Hami Halima, M. Adaouri, and M. L. Berkani

1 Introduction

In the current economic situation, where increasing production and productivity are the subject of fairly extensive research, the need to develop beekeeping becomes imperative. Like all Muslim countries in general and those of the Maghreb in particular, Algeria is considered a nation that is, traditionally, a major consumer of honey. The need to develop this animal production is not only due to economic reasons but is also justified by the presence in our country of natural factors that condition its success. Among the latter are the relative mildness of the climate, the richness of the melliferous flora, etc.

To this end, it would therefore be logical to develop this sector in Algeria in order to avoid imports of beehive products and thus increase the chances of true economic independence. The increase in beekeeping production would undoubtedly lead to the supply of honey, wax, royal jelly, etc., on the domestic market at relatively low prices. This would make them accessible to a very large number of consumers. In addition, the development of beekeeping would contribute to raising the standard of living of the masses by offering them new and highly remunerative jobs.

The essential purpose of beekeeping, which is often forgotten, is not only to produce honey to satisfy the needs of the population or to be exported but also to take into account the amount of wealth that bees create and provide through the pollination of cultivated entomophilous plants. This useful task, the value of which few Algerian farmers and arboriculturists appreciate, indirectly brings to agriculture and arboriculture a benefit of the order of 10 to 15 times that obtained by direct

H. Halima (✉)
Department of Animal Production, National Higher School of Agronomy, Algiers, Algeria
e-mail: halima.hami@edu.ensa.dz

M. Adaouri · M. L. Berkani
National Institute of Agronomic Research of Algeria (INRAA), El-Harrach, Algiers, Algeria

M. Zafar et al. (eds.), *Pure Honey: Assurance & Authentication*,
https://doi.org/10.1007/978-3-031-98913-1_15

beekeeping production. Thus, the need to improve and modernize the beekeeping sector and to extend it to all the northern regions of the country leads us to study a certain number of parameters that are essential to its development.

In the first part, the study of the current situation of national beekeeping and its possibilities for development is essential in order to try to provide initial solutions to the most urgent problems that arise there. The second part of this document will be devoted to experimental research on three main parameters, which are undoubtedly the most decisive in the development of Algerian beekeeping. There is the housing of these insects called the hive; the animal, which is the bee; and finally, the plant environment, which is the source of beekeeping production.

2 History of Beekeeping

Beekeeping is a very old concern. The presence of the bee among the agricultural and pastoral populations of antiquity is evidenced by painted, woven, or engraved drawings found in Egyptian tombs, in churches, and in convents of various religious denominations. Beck-Bodog (1935), Anonymous (1970), Roussy (1973), and Adam (1978) evoke a cave painting, which depicts the honey harvest in the Mesolithic period (20,000–5000 years before Christ), found in the Arama caves in Spain. These same authors inform us of the presence, in the tomb of Pharaoh Menes, of drawings representing the management and beekeeping work dating back to 5000 B.C.

Also, the practice of beekeeping by primitive man consisted of extracting honey from the hollows of trees and building beehives from clay and straw. Several works, dating from antiquity, evoke the life of bees. Among the latter are those of Plinius, Aristotle, and Virgil, to which must be added certain Homeric and biblical books (Roussy, 1973; Weiss, 1985). In the Qur'an, the bee has been the subject of extensive commentaries on the existence of the animal community (Sura 16, Verses 68 and 69 of Sura No. 16 of Ennahal translated by Bucaille in 1978, in which God says:

> Take up your abode in the mountains and in the trees, and in what (men) build (for you), eat of every fruit, and humbly follow the ways of your lord. A liquor of different colors comes out of the interior of his body, in which there is a remedy for men

As far as the biology of the insect is concerned, knowledge only begins to be deepened in the middle of the seventeenth century, with SWAMMERDAM establishing the sex of the queen. At that time, Reaumur, cited by Cady and Arnold (1997), laid the foundations of bee biology; CHIRACH discovered the possibility of transforming common larvae into queens when the mother of the colony is lost. Anthonescu (1973), Crane (1976), and Adam (1980) mention that the blind naturalist FRANCOIS HUBERT established already at the beginning of the eighteenth century that the mating of the queen with the males takes place in the air during the nuptial flight.

In the nineteenth century, enormous progress was made in the improvement of beekeeping equipment, especially with regard to the different models of hives: Langstroth, DADANT, Layens, etc. (Boussouf, 1981; Bencsik, 1994).

3 Beekeeping in Algeria

Beekeeping in Algeria is a very old practice, its origin is lost in the mists of time. Muslims, especially those in the Maghreb, were considered to be big consumers of honey. Many of their baked goods and cooked meals contained honey (Skender, 1972). According to Benhamza (1979), Algerian, beekeeping has gone through several important stages.

4 Algerian Beekeeping During Colonization

It was generally of the traditional type, and there was at the same time the modern type of conduct held by the colonists without any transfer of know-how to the indigenous populations. Skender (1972), cites statistical data from 1891, when there were 27,885 beekeepers, including 26,861 Algerians along with 231,329 traditional hives. The thousand French beekeepers exploited about 10,000 frame hives.

Before the war of national liberation, the French authorities estimated that there were 150,000 traditional hives in Algeria (Berthouse & Rabia, 1973). Benhamza (1979) cites the figure 300,000 traditional hives and 20,000 modern hives. In 1954, the war of liberation contributed to the destruction of a large part of the livestock, whose situation was critical at independence (Belhoues, 1977).

5 Algerian Beekeeping After Independence

After independence, the state addressed the beekeeping problem while solving it through the elaboration of various development programs. He focused his efforts on the multiplication of the herd, on the importation of foreign bees, and also on the construction of a so-called Algerian hive (Benhamza, 1979; Fronty, 1980). Within the framework of the special development programs of the wilayas, significant credits have been allocated to allow the modernization of beekeeping. During this period, beekeeping cooperatives were also established that integrated the three sectors of agriculture (the Agrarian Revolution Sector, the self-managed sector, and the private sector).

As far as the production capacity of the cooperatives is concerned, it appears that out of the 17 cooperatives, only nine are functional, while the others are out of business for various reasons, such as financial problems, sabotage and fires, and losses of livestock. With a production capacity of 90,000 empty hives and 60,000 swarms per year, it is difficult to cover the demand induced by the National Fund for Rural Development in Algeria (FNRDA) (one million hives registered) as well as within the framework of the ANSEJ program.

The dissolution of almost all the cooperatives spread throughout the country paved the way for the private sector, which gradually took over by providing part of the national needs for beekeeping herds (Chenane, 2003). Despite this, Algeria has great skills to develop the beekeeping sector, which lie in the abundant and varied melliferous potentialities, especially in the north, where the climate is favorable, without forgetting the presence of a breed of bee with an interesting genetic potential. Despite all these resources and the state's willingness to develop and improve beekeeping through the PNDA and the FNRDA, the latter remains below the desired level. Several factors could be the cause of this stagnation, such as the lack of rigor in the management by all the actors involved upstream and downstream of this speculation.

6 Current Situation of Algerian Beekeeping

According to estimates by the Institute for the Development of Small Livestock Farms, honey production has increased significantly; it increased from 90 tons in 1973 to 1130 tons in 1980, to 1500 tons in 1990, to 1845 tons in 2001, to 2875 tons in 2004, and in 2018, honey production in Algeria reached 75,000 quintals, marking an increase of 80% compared to 2000.

7 Factors Limiting Beekeeping Production

One of the main limiting factors lies in the nature of the hive. The latter of the common types, made of cork or esparto grass or box-shaped wood, directly affects honey production (1–3 kg per hive) and is not suitable for intensive farming because of its low capacity. Also, the bees are obliged to restore the wax combs after each harvest (during the harvest and to extract the honey, the breeder can possibly destroy the combs). Partiotl (1976), and Prost (1977), report that for 1 kg of wax, bees consume 8–10 kg of honey.

8 The Tellian Race or *Apis mellifera intermissa*

Its area of establishment extends from Morocco to Tunisia. It is bounded to the north and east by the Mediterranean Sea, to the west by the Atlantic Ocean, and to the south by the Sahara. This bee inhabits the region between the Atlas and the

Mediterranean, known as the Tell. A bee specialist, BALDENSPERGER, was the first to call this bee "The Tellian." It is as black as charcoal, but there is sometimes a slight lightening of the abdominal tergites and the scutellum. Its coat is very short, which brings out its black coloration. Adam (1980) describes the Tellian bee as a very aggressive insect that builds many queen cells with a strong tendency to swarm. On top of that, the polarization is abnormally developed. This bee also has advantages, namely longevity, great resistance to flight, its remarkable ability to collect pollen, and a high production of honey, which can reach up to 100 kg per colony provided that modern beekeeping methods are applied (Fresnaye, 1981).

9 The Saharan Breed or *Apis mellifera sahariensis*

It is the golden bee of the Sahara, discovered by BALDENSPERGER at the beginning of the twentieth century. Its area of establishment is the south of Morocco and the southwest of Algeria, especially at the level of the oases. This bee is small in size and yellow in color, and the tongue and hair are short. As far as behavior is concerned, this bee has the ability to be forward-looking during difficult times; it becomes accustomed to reducing its brood and spares its provisions during long scarcity.

9.1 The Hive

It would have been illusory, to say the least, to want to test many models of hives, some of which are only used when very particular conditions in Northern Algeria require it (very restrictive climate, bees resistant to any enriching, beekeeping still traditional in some regions, etc.). After trials, very quickly abandoned, of a so-called Algerian hive when the country's beekeeping had been completely decimated during the war of liberation, the LANGSTROTH hive, common in use in the Mediterranean regions, has been recommended for 30 years without tests other than the experience of colonial or Provençal beekeeping. Until the early eighties, the DADANT hive was never the subject of an exhaustive study of its advantages and disadvantages. However, there were some local initiatives on the part of private beekeepers, who gave it preference and were anxious to produce quality honeys on the national market (the case of beekeepers in Constantine and Medea).

In 1977, Bornes recommended, for the Tlemcen project, the establishment of a beekeeping cooperative in western Algeria which would use only the DADANT model hives. The latter are built from a plan and a prototype from the standard DADANT. Since then, this project has never seen the light of day. This hive seems desirable for various reasons: its ease of management and transhumance.

- Its large body avoids the risk of swarming.
- Its honey is qualitatively and quantitatively superior to that of LANGSTROTH.

In Algeria, it should be noted that work in this direction was carried out in 1980 and 1985 in various ecosystems in eastern and central Algeria and reached conclusions in favor of both models. In order to be able to confirm or refute the results previously obtained, more in-depth studies were carried out in 2001 and 2002 as part of a research project involving the university and beekeeping professionals.

9.2 Description of the LANGSTROTH Hive

It is a hive that consists of a reversible tray forming a flight hole along the entire length. There upon the plateau are posed the two bodies from the same dimensions, each containing ten frames suspended by their shoulders on smooth strips. The spacing of the frames is ensured by a bulge on the vertical upright, and it calls this system "HOFFMANN spacing." In this type of hive, there is a flat roof serving as protection against the weather and certain predators.

10 The Relief

Algeria, part of the Maghreb, extends between Morocco and Tunisia over 1000 km of coastline and penetrates toward the interior of Africa up to 2000 km from the sea. It is bordered to the south by Niger and Mali, to the southwest by Mauritania and Western Sahara, and to the west by Morocco.

The area of Algeria is 2,195,098 km^2 and is divided into two distinct geographical units: Northern Algeria consists of a strip of land along the coastline about 200 km wide. Southern Algeria, called the Sahara, is almost totally desert. The second largest country in Africa, Algeria, is made up of four main sets of reliefs, of very unequal extent, which follow one another from north to south: To the north, along the Mediterranean coast, stretches the narrow plain of the Algerian Tell. This discontinuous plain of variable width (80–190 km) and the adjoining valleys are home to the vast majority of the country's agricultural land. Limiting the coastal plain to the south, there is a vast set of higher reliefs: two important mountain ranges oriented east-west (the Tellian Atlas to the north and the Saharan Atlas and the Aurès massif to the south) frame a region of high desert plateaus, whose basin profile explains the presence of numerous salt lakes (chotts) collecting surface water (the lowest point: Chott Melghir at 40 m). To the south of the Saharan Atlas stretches the immense Sahara Desert, which represents 80% of Algeria's surface.

11 Climate

The Tell region in the north is characterized by a Mediterranean climate, with hot, dry summers and mild, rainy winters. It is the wettest area in Algeria, with annual rainfall varying between 400 and 1000 mm of water (Fig. 15.1). The average summer and winter temperatures are 25 °C and 11 °C, respectively. In the summer, sirocco, a dry and very hot wind (locally called chile), blows from the Sahara in a northerly direction. The further south you go, the drier the climate becomes: the annual rainfall in the highlands and the Saharan Atlas does not exceed 200–400 mm. The pre-steppe and steppe areas are windy and very arid regions, where the temperature variations are often considerable: these temperature variations, extremely high during the day and very low at night, can be explained by the total absence of humidity capable of attenuating the contrasts. The annual rainfall is less than 130 mm in the whole of the Algerian steppe.

12 The Melliferous Flora

The melliferous flora of Northern Algeria is distributed according to the following relief:

- **The flora of the coastal plains:**
  - Citrus fruits: Orange, lemon, and mandarin trees
  - Arboriculture: Medlar, peach, plum, almond, and cherry
  - Shrubs: Tree heather

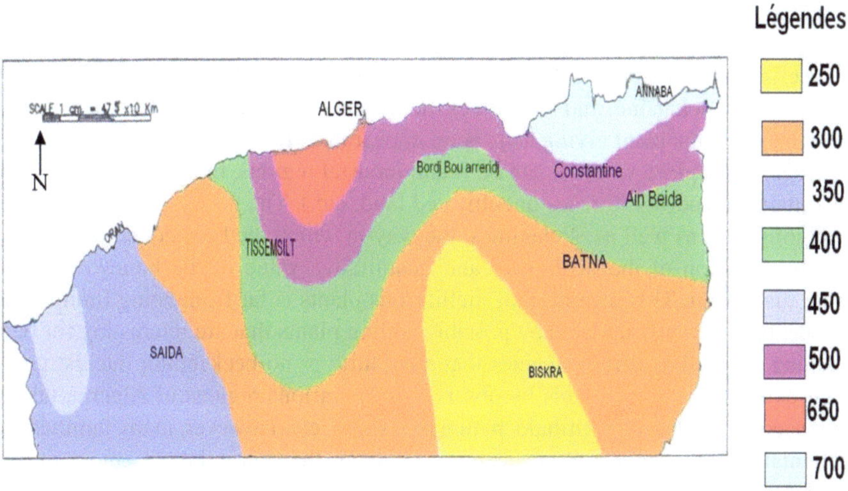

Fig. 15.1 Rainfall map of northern Algeria

- Forest trees: Eucalyptus and carob trees
- Spontaneous plants: Lavender, rosemary, oxalis, borage, mint, viper's viper, cabbage, and mustard
- Market gardening: Melons, watermelons, zucchini, garlic, onions, peas, and broad beans
- Fodder crops: Alfalfa, bersim, and field peas

- **The flora of the mountain ranges:**
  - Forest trees: Eucalyptus, carob, Aleppo pine, maritime pine, and chestnut.
  - Shrubs: Tree heather and strawberry tree
  - Arboriculture: Olive and cherry trees
- **The flora of the high interior plains and the steppe:**
  - Cereal plants: Corn, wheat, and barley
  - Spontaneous plants: Thistles, globulus, white mugwort, harmel, cabbage, cistus, and lobulus
  - Forest trees: Aleppo pine, eucalyptus, and mulberry
  - Arboriculture: Apricot, peach, plum, pear, and almond
- **The flora of the interior mountain ranges:**
  - Forest trees: Aleppo pine, eucalyptus, and zen oak
  - Shrubs: Tree heather

13 Inventory and Description of Algerian Honey Plants

This description is based on a sampling of the melliferous plants that are found in abundance in Algeria and that are essential to the native bee. This work was carried out through an annual inventory over a period of 200 years, which affected almost all the northern regions of Algeria. For each species of plant, the Latin botanical name, the French name, and sometimes the Arabic name are indicated. Then, a brief description of the plant giving its general appearance, its flowering dates, as well as the region and place where it can be found (coastal regions, high plateaus, steppes, mountains, meadows, forests, uncultivated land, etc.). The honey value of the plant is highlighted, as well as, sometimes, the way in which the bees collect nectar and the appreciation of the qualitative and quantitative value of the honey produced from this nectar. This inventory of melliferous plants is far from being limiting, so it must be constantly updated by possibly adding plants that are interesting for beekeeping, as well as certain species that have little or no beekeeping interest given their extent or value; and this by observation in various regions of Algeria, taking into account the latitude, altitude, nature of the soil, etc. However, many families are undeniably interesting for beekeeping, such as the Aurantiaceae, most of the Labiatae, the Borraginae, the Compositae, the Umbelliferae, and the Cruciferae.

In order to be able to determine exactly the different main honey flows which dominate, a palynological study on a large number of Algerian honeys, collected in almost all regions of the country, has proved indispensable. This study will confirm or refute our hypothesis of the presence or absence of a wide variety of Algerian honeys. Melliferous plants produce substances that are collected by foraging insects to be transformed into honey. Many plants are melliferous, but only a part can be foraged by honeybees because of their morphology (body size, length of the trunk, etc.). Beekeeping classifies a plant as melliferous when it can be exploited by the honeybee.

The following table lists some of the melliferous plants in Algeria, indicating the flowering period of these plants, as well as the various resources to which bees have access (nectar, pollen, propolis, and honeydew). It should be noted that each plant provides neither the same quantity nor the same quality of these substances from one species to another, and that even for the same species, this can vary according to the region, the health status of the plant, climate variations, etc. The purpose of this table is therefore only to show whether or not the plant produces one or the other of these raw materials that can be exploited by honeybees (*Apis Mellifera*) (Table 15.1).

Table 15.1 Summary of the results of the palynological analysis of honeys from Algeria

Honey no.	Origin		Extraction	Initial name	Appellation after analysis
1	Algiers	Bouzaréah	Mechanics	Eucalyptus	Eucalyptus
2		Birtouta	Manual	Orange tree	All flowers
3		Cheraga	Mechanics	All flowers	Eucalyptus
4	Boumerdès	Bordj-Ménail	Manual	All flowers	All flowers
5		Dellys	Mechanics	All flowers	Sainfoin
6		Khemis El-Khechna	Manual	All flowers	All flowers
7	Blida	Julia	Mechanics	All flowers	All flowers
8		Mafteh	Mechanics	All flowers	All flowers
9		Boufarik	Manual	Eucalyptus	Eucalyptus
10		Baba-Ali	Manual	All flowers	Eucalyptus
11	Blah	–	Mechanics	All flowers	Sainfoin
12		Djabsia	Mechanics	All flowers	Sainfoin
13		Kadiria	Manual	All flowers	Sainfoin
14		Tikdjda	Mechanics	All flowers	All flowers
15	Bejaia	Toudja	Manual	All flowers	All flowers
16		Sidi-Aich	Mechanics	All flowers	All flowers
17		El-Kseur	Manual	All flowers	All flowers
18		Amizour	Manual	All flowers	All flowers

(continued)

Table 15.1 (continued)

Honey no.	Origin		Extraction	Initial name	Appellation after analysis
19	Tizi-ouzou	Azeffoune	Mechanics	All flowers	Eucalyptus
20		Boghni	Mechanics	Pear tree	Sainfoin
21		Beni-Yenni	Mechanics	All flowers	All flowers
22		Maâtka	Mechanics	Vetch	Sainfoin
23		Mizrana	Mechanics	All flowers	Sainfoin
24		Tighzirt	Manual	Mint Poliot	All flowers
25	Jijel	Julia	Manual	All flowers	All flowers
26	Setif	–	Mechanics	All flowers	All flowers
27	Bordj bourreridj	–	Manual	All flowers	Sainfoin
28	Skikda	–	Manual	All flowers	All flowers
29	Constantine	–	Manual	Eucalyptus	Eucalyptus
30	Annaba	–	Mechanics	All flowers	All flowers
31	Taref	Ain-Assel	Manual	Eucalyptus	Eucalyptus
32	Tébessa	–	Manual	All flowers	All flowers
33	Guelma	–	Manual	Fruit tree	All flowers
34	Batna	Arris	Manual	Fruit tree	All flowers
35	Khenchela	Chachar	Manual	All flowers	All flowers
36	Tipaza	Cherchell	Manual	Fruit tree	All flowers
37	Medea	–	Manual	All flowers	All flowers
38		Tablat	Mechanics	All flowers	All flowers
39	Ain-Defla	–	Manual	All flowers	All flowers

It should also be noted that the flowering of a population of individuals belonging to a species can be spread out considerably over time. Between the beginning of flowering and the end of flowering, several months can pass. Since the dates mentioned here are particularly interested in the periods during which a significant proportion—or even the greatest number—of individuals are in flower. The following flowering calendar includes the species presented above and is supplemented by other beekeeping species of undeniable interest in our region.

14 Algerian Honeys

The analytical table of the 39 honeys with the following information: presumed botanical origin, date and conditions of harvesting, place, also includes the qualitative results of the pollen analysis, in the form of a pollen spectrum which has been established from at least 300 pollen grains.

The taxa identified are grouped together:

"Pollen dominating": (more than 45% of pollen counted)
"Accompanying pollen": (16–45%)
"Significant isolated pollen": (3–15%)
"Rare isolated pollen": (less than 3%)

Louveaux et al. (1970), report that honey comes mainly from a certain nectar source when the corresponding pollen is in the "dominant" stage in the honey sediment (such honey is said to be "unifloral"or "raw honey"). If the honey in question comes from multiple bees and without a clear dominance of any particular plant, it is considered polyfloral, or commonly referred to as vital for beekeeping management.

References

Adam, F. (1978). *My beekeeping method* (pp. 45–47). Le Courrier Du Livre.

Adam, F. (1980). *In search of the best bee breeds* (198 p.). Courrier Du Livre.

Anonymous. (1970). *International encyclopedia of science and technology* (pp. 364–366). Presses De La Cité.

Anthonescu, M. (1973). Beekeeping in the world. *Rev Apicultura, 12*, 202–230.

Berthouse, A., & Rabia, A. (1973). *Transhumance: Introduction to beekeeping* (pp. 44–45). CPR, MARF.

Bornes, G. (1977). The hive in the future. *Rev Franç Api, 401*, 481–484.

Bucaille, M. (1978). *The bee in the animal world: "The Bible, the Koran and science"* (95p). Seghers.

Beck Bodog, F. (1935). *The Bible of Bee Venom Therapy, D. Appleton-Century Co., USA,* pp. 10–14.

Boussouf M. 1981.Comparaison de deux types de ruches en conditionsd'hivernage dans *l'Est Algérien. Mémoire. univer. Constantine*, pp 35–50.

Bencsik, J. (1994). Pragmatique apicole. Rev. *franç. api*, (544), pp. 421–428.

Benhamza, (1979). Perspectives *de développement de l'Apiculture en Algérie: la prophylaxie dans le développement de l'Apiculture dans l'Est algérien. Mémoire ing.université de Constantine*, pp. 4–8.

Belhoues, L. (1977). *Etude des possibilités de transhumance de ruches en vue d'une meilleure utilisation de la base mellifère en Kabylie*. Mémoire ing, Inst. nati agro., El Harrach, pp. 8–30.

Cady, J., & Arnold, G. (1997). *Apis and Osiris – The bees* (32p). INRA.

Crane, E. (1976). *L'apiculture dans le monde.Rev.Gazette .Apicole, Paris* (255), pp. 127–130.

Fresnaye, J. (1981). *Biométrie de l'abeille* (2nd ed., 56 p.). Echauffour, orne, office d'inf. Doc. Apic.

Fronty, A. (1980). *L'Apiculture d'aujourd'hui*, (pp. 45–56). Dargaud.

Guerriat. (1996). *Being efficient in beekeeping* (416 p). Melin.

Louveaux, J. Maurizioa, (1970). *Les méthodes de la palynologie. Apidologie* volume 1 n.2 pp. 221–238.

Partiotl. (1976). LANGSTROTH, the father of American beekeeping. *Rev Gazette Apicole*, 79–81.

Prost, J. (1977). *Apiculture* (6th ed., 579p). Bailliere J. B.

Skender, K. (1972). Situation actuelle de l'apiculture algérienne et ses possibilités de développement". *Mémoire ing, Inst. nati agro,El Harrach*, pp. 12–5427.

Roussy, L. (1973). Four generations of beekeepers. *Rev Gazette Apicole Ann api Mondial*, 26.

Weiss, K. (1985). *The beekeeper of the weekend* (pp. 12–18). European Beekeeping Editions.

Index

M. Zafar et al. (eds.), *Pure Honey: Assurance & Authentication*,
https://doi.org/10.1007/978-3-031-98913-1